AF358672

Advanced Biomaterials and Coatings

Advanced Biomaterials and Coatings

Guest Editors

Richard Drevet
Hicham Benhayoune

Basel • Beijing • Wuhan • Barcelona • Belgrade • Novi Sad • Cluj • Manchester

Guest Editors

Richard Drevet
Department of Plasma Physics
and Technology
Masaryk University
Brno
Czech Republic

Hicham Benhayoune
Institut de Thermique,
Mécanique et Matériaux
(ITheMM)
Université de Reims
Champagne-Ardenne (URCA)
Reims
France

Editorial Office
MDPI AG
Grosspeteranlage 5
4052 Basel, Switzerland

This is a reprint of the Special Issue, published open access by the journal *Coatings* (ISSN 2079-6412), freely accessible at: https://www.mdpi.com/journal/coatings/special_issues/bio_biomaterials_coatings.

For citation purposes, cite each article independently as indicated on the article page online and as indicated below:

Lastname, A.A.; Lastname, B.B. Article Title. *Journal Name* **Year**, *Volume Number*, Page Range.

ISBN 978-3-7258-4049-6 (Hbk)
ISBN 978-3-7258-4050-2 (PDF)
https://doi.org/10.3390/books978-3-7258-4050-2

Contents

Richard Drevet and Hicham Benhayoune
Advanced Biomaterials and Coatings
Reprinted from: *Coatings* 2022, 12, 965, https://doi.org/10.3390/coatings12070965 1

Richard Drevet and Hicham Benhayoune
New Insights into Biomaterials and Coatings
Reprinted from: *Coatings* 2025, 15, 332, https://doi.org/10.3390/coatings15030332 4

Carmen Steluta Ciobanu, Daniela Predoi, Simona Liliana Iconaru, Krzysztof Rokosz, Steinar Raaen, Catalin Constantin Negrila, et al.
Chrome Doped Hydroxyapatite Enriched with Amoxicillin Layers for Biomedical Applications
Reprinted from: *Coatings* 2025, 15, 233, https://doi.org/10.3390/coatings15020233 7

Abdulaziz Alhotan, Saleh Alhijji, Sahar Ahmed Abdalbary, Rania E. Bayoumi, Jukka P. Matinlinna, Tamer M. Hamdy and Rasha M. Abdelraouf
An Advanced Surface Treatment Technique for Coating Three-Dimensional-Printed Polyamide 12 by Hydroxyapatite
Reprinted from: *Coatings* 2024, 14, 1181, https://doi.org/10.3390/coatings14091181 34

Sohee Kang, So Jung Park, Sukyoung Kim and Inn-Kyu Kang
Chemical Bonding of Nanorod Hydroxyapatite to the Surface of Calciumfluoroaluminosilicate Particles for Improving the Histocompatibility of Glass Ionomer Cement
Reprinted from: *Coatings* 2024, 14, 893, https://doi.org/10.3390/coatings14070893 50

Hafedh Dhiflaoui, Sarra Ben Salem, Mohamed Salah, Youssef Dabaki, Slah Chayoukhi, Bilel Gassoumi, et al.
Influence of TiO_2 on the Microstructure, Mechanical Properties and Corrosion Resistance of Hydroxyapatite HaP + TiO_2 Nanocomposites Deposited Using Spray Pyrolysis
Reprinted from: *Coatings* 2023, 13, 1283, https://doi.org/10.3390/coatings13071283 65

Daniil Golubchikov, Tatiana V. Safronova, Elizaveta Nemygina, Tatiana B. Shatalova, Irina N. Tikhomirova, Ilya V. Roslyakov, et al.
Powder Synthesized from Aqueous Solution of Calcium Nitrate and Mixed-Anionic Solution of Orthophosphate and Silicate Anions for Bioceramics Production
Reprinted from: *Coatings* 2023, 13, 374, https://doi.org/10.3390/coatings13020374 81

Letizia Ferroni, Chiara Gardin, Federica Rigoni, Eleonora Balliana, Federica Zanotti, Marco Scatto, et al.
The Impact of Graphene Oxide on Polycaprolactone PCL Surfaces: Antimicrobial Activity and Osteogenic Differentiation of Mesenchymal Stem Cell
Reprinted from: *Coatings* 2022, 12, 799, https://doi.org/10.3390/coatings12060799 96

Harish K Handral, C. Ashajyothi, Gopu Sriram, Chandrakanth R. Kelmani, Nileshkumar Dubey and Tong Cao
Cytotoxicity and Genotoxicity of Metal Oxide Nanoparticles in Human Pluripotent Stem Cell-Derived Fibroblasts
Reprinted from: *Coatings* 2021, 11, 107, https://doi.org/10.3390/coatings11010107 110

Richard Drevet, Joël Fauré and Hicham Benhayoune
Electrophoretic Deposition of Bioactive Glass Coatings for Bone Implant Applications: A Review
Reprinted from: *Coatings* 2024, 14, 1084, https://doi.org/10.3390/coatings14091084 124

Robert B. Heimann
Plasma-Sprayed Osseoconductive Hydroxylapatite Coatings for Endoprosthetic Hip Implants:
Phase Composition, Microstructure, Properties, and Biomedical Functions
Reprinted from: *Coatings* **2024**, *14*, 787, https://doi.org/10.3390/coatings14070787 . . . . **149**

Sergey V. Dorozhkin
Calcium Orthophosphate (CaPO$_4$)-Based Bioceramics: Preparation, Properties, and Applications
Reprinted from: *Coatings* **2022**, *12*, 1380, https://doi.org/10.3390/coatings12101380 . . . . **177**

Editorial

Advanced Biomaterials and Coatings

Richard Drevet [1,*] and Hicham Benhayoune [2]

[1] Department of Physical Electronics, Masaryk University, Kotlářská 2, CZ-61137 Brno, Czech Republic
[2] Institut de Thermique, Mécanique et Matériaux (ITheMM), EA 7548, Université de Reims Champagne-Ardenne, Moulin de la Housse, BP 1039, 51687 Reims CEDEX 2, France; hicham.benhayoune@univ-reims.fr
* Correspondence: drevet@mail.muni.cz or richarddrevet@yahoo.fr

Citation: Drevet, R.; Benhayoune, H. Advanced Biomaterials and Coatings. *Coatings* **2022**, *12*, 965. https://doi.org/10.3390/coatings12070965

Received: 25 June 2022
Accepted: 5 July 2022
Published: 7 July 2022

Publisher's Note: MDPI stays neutral with regard to jurisdictional claims in published maps and institutional affiliations.

Everywhere on Earth, people are living longer and longer. The World Health Organization predicts that by 2050, the world's population of 60 years or older will double to reach 2.1 billion, and the population of 80 years or older will triple to reach 426 million [1]. Due to the aging of the population, the clinical demand for biomaterials used for bone tissue repair increases every year [2,3]. Academic and industrial research is constantly developing innovative biomaterials and coatings to improve the properties and extend the lifetime of bone implants. The main metallic materials used in orthopedic or dental surgeries are made of titanium alloys [4–6], stainless steel [7,8], and CoCr alloys [9,10]. The mechanical properties of these alloys are suitable for bone tissue replacement, and their biocompatibility with the body environment is good. The International Union of Pure and Applied Chemistry (IUPAC) defines biocompatibility as the *ability of a material to be in contact with a biological system without producing an adverse effect* [11]. These alloys are biocompatible, but their biological interaction with bone tissue is very low. The surface bioactivity of these implants must be improved to avoid any bone anchorage failure that would induce mandatory revision surgery. The osseointegration of bone implants can be enhanced by a surface coating of bioactive material, such as calcium phosphate or bioglass. These materials initiate the formation of a strong bond to bone tissue [12–14]. The bioactive coating is a scaffold to bone growth that provides a rapid biological response and improves the adhesion of the implant to bone [15,16]. Several methods can be used to produce a bioactive coating on the surface of implants. This Special Issue aims to describe the latest developments in deposition methods used to synthesize advanced biomaterials and coatings for bone implant applications. The main deposition processes are plasma spraying, magnetron sputtering, pulsed laser deposition, electrospray deposition, electrophoretic deposition, and electrodeposition.

Plasma spraying (PS) is the main industrial process due to its ability to produce large quantities of coatings with good reproducibility. Plasma spraying involves the injection of calcium phosphate or bioglass powder into a hot plasma jet whose temperature is thousands of degrees [17,18]. Inside the plasma, the grains of powder are in a molten or semi-molten state. They are accelerated toward the surface of the bone implant, where they cool down and solidify instantly to form a coating. There are some drawbacks with the high temperatures involved in plasma spraying that produce uncontrolled phase changes, chemical decompositions, and structural modifications. The physicochemical and biological properties of the coating are different from those of the initial powder [19].

Magnetron sputtering (MS) of a calcium phosphate or bioglass target is another process to produce bioactive coatings on bone implants. Magnetron sputtering is a physical vapor deposition (PVD) process. In a vacuum chamber at room temperature, high-energy ions from a plasma collide with the atoms of the target with enough energy to eject and transport them toward the surface of the substrate to form a coating [20]. Magnetron sputtering produces dense, uniform, and adherent coatings. However, the different atoms of a multicomponent target have different sputtering behaviors. The chemical composition of the deposited coating usually differs from that of the target. The experimental parameters of the process can be used to modify the stoichiometry, morphology, and structure of the bioactive coating, corresponding to different physicochemical and biological properties [21].

Pulsed laser deposition (PLD) uses a high-power laser that hits a calcium phosphate or bioglass target inside a vacuum chamber [22]. The laser–matter interaction induces the ablation of the target and generates a plasma plume containing ejected atoms, ions, and electrons. The plasma plume is directed toward the surface of the substrate where the material is collected and condensed to form a coating. The process rapidly produces uniform and adherent thin coatings [23]. As described for magnetron sputtering, the stoichiometry of the coating may differ from that of the target, resulting in variations in the physicochemical and biological properties of the coating.

Electrospray deposition (ESD) uses an aerosol that contains calcium phosphate particles, bioglass particles, or precursors of these materials. The aerosol is produced by injecting a solution through a nozzle connected to high voltage [24]. Charged droplets are produced at the tip of the nozzle, and they are directed toward a grounded and heated substrate. The droplets lose their surface charge, and the solvent is evaporated to produce the bioactive coating. The process produces uniform coatings with different morphologies as a function of the experimental parameters [25].

Electrophoretic deposition (EPD) requires two conductive electrodes connected to a generator and immersed in a stable colloidal suspension of calcium phosphate or bioglass powder. In contact with the solution, the colloidal particles carry a positive or negative electrostatic charge on their surface [26]. They move through the liquid under the influence of the electric field between the two electrodes. They agglomerate on one electrode surface to form a bioactive coating. Post-deposition thermal annealing in a furnace is necessary to evaporate the solvent and improve the cohesive and adhesive properties of the coating [27].

Electrodeposition (ELD) is a low-temperature process that requires two metallic electrodes connected to a generator and immersed in an aqueous solution containing calcium ions and phosphate ions. In academic research, a reference electrode is also usually connected in a three-electrode setup that provides electrochemical measurements. In the electrolytic cell, the anode is the positive electrode, and the cathode is the negative electrode. Electrical energy from the generator is used to trigger a series of chemical reactions at the electrode–electrolyte interfaces. At the cathode, where the bone implant is connected, the main electrochemical reaction is the reduction of water, the solvent of the electrolyte solution. This reaction results in a local pH variation in the vicinity of the cathode that induces surface precipitation of the calcium phosphate coating [28]. As a function of the experimental parameters, various chemical compositions, phases, and morphologies are obtained. Direct current was typically used first, but pulsed current electrodeposition became more interesting in recent years [29]. Since the process takes place in an aqueous medium at room temperature, ionic additives (Na^+, Ag^+, F, Co^{2+}, Cu^{2+}, Mg^{2+}, Sr^{2+}, Zn^{2+}, etc.) or organic components (polymers, proteins, drugs, etc.) can be added to enhance the biological and mechanical properties of the calcium phosphate coating [30]. Post-deposition annealing is also required to evaporate the solvent and improve the cohesive and adhesive properties of the electrodeposited coating [31].

Author Contributions: Conceptualization, R.D. and H.B.; validation, R.D. and H.B.; resources, R.D. and H.B.; writing—original draft preparation, R.D. and H.B.; writing—review and editing, R.D. and H.B. All authors have read and agreed to the published version of the manuscript.

Funding: This research received no external funding.

Institutional Review Board Statement: Not applicable.

Informed Consent Statement: Not applicable.

Data Availability Statement: Not applicable.

Conflicts of Interest: The authors declare no conflict of interest.

References

1. Available online: https://www.who.int/news-room/fact-sheets/detail/ageing-and-health (accessed on 23 June 2022).
2. Gheno, R.; Cepparo, J.M.; Rosca, C.E.; Cotton, A. Musculoskeletal Disorders in the Elderly. *J. Clin. Imaging Sci.* **2012**, *2*, 39. [CrossRef] [PubMed]
3. Li, G.; Thabane, L.; Papaioannou, A.; Ioannidis, G.; Levine, M.A.H.; Adachi, J.D. An overview of osteoporosis and frailty in the elderly. *BMC Musculoskelet. Disord.* **2017**, *18*, 46. [CrossRef] [PubMed]
4. Geetha, M.; Singh, A.K.; Asokamani, R.; Gogia, A.K. Ti based biomaterials, the ultimate choice for orthopaedic implants—A review. *Prog. Mater. Sci.* **2009**, *54*, 397–425. [CrossRef]
5. Ijaz, M.F.; Laillé, D.; Héraud, L.; Gordin, D.M.; Castany, P.; Gloriant, T. Design of a novel superelastic Ti-23Hf-3Mo-4Sn biomedical alloy combining low modulus, high strength and large recovery strain. *Mater. Lett.* **2016**, *177*, 39–41. [CrossRef]
6. He, G.; Hagiwara, M. Ti alloy design strategy for biomedical applications. *Mater. Sci. Eng. C* **2006**, *26*, 14–19. [CrossRef]
7. Chen, Q.; Thouas, G.A. Metallic implant biomaterials. *Mater. Sci. Eng. R* **2015**, *87*, 1–57. [CrossRef]
8. Prokoshkin, S.; Pustov, Y.; Zhukova, Y.; Kadirov, P.; Dubinskiy, S.; Sheremetyev, V.; Karavaeva, M. Effect of Thermomechanical Treatment on Functional Properties of Biodegradable Fe-30Mn-5Si Shape Memory Alloy. *Metall. Mater. Trans. A* **2021**, *52*, 2024–2032. [CrossRef]
9. Tchana Nkonta, D.V.; Drevet, R.; Fauré, J.; Benhayoune, H. Effect of surface mechanical attrition treatment on the microstructure of cobalt–chromium–molybdenum biomedical alloy. *Microsc. Res. Technol.* **2021**, *84*, 238–245. [CrossRef]
10. AlMangour, B.; Luqman, M.; Grzesiak, D.; Al-Harbi, H.; Ijaz, F. Effect of processing parameters on the microstructure and mechanical properties of Co–Cr–Mo alloy fabricated by selective laser melting. *Mater. Sci. Eng. A* **2020**, *792*, 139456. [CrossRef]
11. Williams, D.F. On the mechanisms of biocompatibility. *Biomaterials* **2008**, *29*, 2941–2953. [CrossRef]
12. Paital, S.R.; Dahotre, N.B. Calcium phosphate coatings for bio-implant applications: Materials, performance factors, and methodologies. *Mater. Sci. Eng. R* **2009**, *66*, 1–70. [CrossRef]
13. Hench, L. The story of Bioglass®. *J. Mater. Sci. Mater. Med.* **2006**, *17*, 967–978. [CrossRef] [PubMed]
14. Sergi, R.; Bellucci, D.; Cannillo, V. A Comprehensive Review of Bioactive Glass Coatings: State of the Art, Challenges and Future Perspectives. *Coatings* **2020**, *10*, 757. [CrossRef]
15. Williams, D.F. On the nature of biomaterials. *Biomaterials* **2009**, *30*, 5897–5909. [CrossRef] [PubMed]
16. Dorozhkin, S.V. Calcium orthophosphate deposits: Preparation, properties and biomedical applications. *Mater. Sci. Eng. C* **2015**, *55*, 272–326. [CrossRef]
17. Heimann, R.B. Thermal spraying of biomaterials. *Surf. Coat. Technol.* **2006**, *201*, 2012–2019. [CrossRef]
18. Cañas, E.; Orts, M.J.; Boccaccini, A.R.; Sánchez, E. Microstructural and in vitro characterization of 45S5 bioactive glass coatings deposited by solution precursor plasma spraying (SPPS). *Surf. Coat. Technol.* **2019**, *371*, 151–160. [CrossRef]
19. Heimann, R.B. Structural Changes of Hydroxylapatite during Plasma Spraying: Raman and NMR Spectroscopy Results. *Coatings* **2021**, *11*, 987. [CrossRef]
20. Surmenev, R.A.; Ivanova, A.A.; Epple, M.; Pichugin, V.F.; Surmeneva, M.A. Physical principles of radio-frequency magnetron sputter deposition of calcium-phosphate-based coating with tailored properties. *Surf. Coat. Technol.* **2021**, *413*, 127098. [CrossRef]
21. Berbecaru, C.; Stan, G.E.; Pina, S.; Tulyaganov, D.U.; Ferreira, J.M.F. The bioactivity mechanism of magnetron sputtered bioglass thin films. *Appl. Surf. Sci.* **2012**, *258*, 9840–9848. [CrossRef]
22. Koch, C.F.; Johnson, S.; Kumar, D.; Jelinek, M.; Chrisey, D.B.; Doraiswamy, A.; Jin, C.; Narayan, R.J.; Mihailescu, I.N. Pulsed laser deposition of hydroxyapatite thin films. *Mater. Sci. Eng. C* **2007**, *27*, 484–494. [CrossRef]
23. D'Alessio, L.; Teghil, R.; Zaccagnino, M.; Zaccardo, I.; Ferro, D.; Marotta, V. Pulsed laser ablation and deposition of bioactive glass as coating material for biomedical applications. *Appl. Surf. Sci.* **1999**, *138–139*, 527–532. [CrossRef]
24. Leeuwenburgh, S.; Wolke, J.; Schoonman, J.; Jansen, J. Electrostatic spray deposition (ESD) of calcium phosphate coatings. *J. Biomed. Mater. Res. A* **2003**, *66*, 330–334. [CrossRef] [PubMed]
25. Müller, V.; Jobbagy, M.; Djurado, E. Coupling sol-gel with electrospray deposition: Towards nanotextured bioactive glass coatings. *J. Eur. Ceram. Soc.* **2021**, *41*, 7288–7300. [CrossRef]
26. Bartmański, M.; Pawłowski, Ł.; Strugała, G.; Mielewczyk-Gryń, A.; Zieliński, A. Properties of nanohydroxyapatite coatings doped with nanocopper, obtained by electrophoretic deposition on Ti13Zr13Nb alloy. *Materials* **2019**, *12*, 3741. [CrossRef]
27. Azzouz, I.; Faure, J.; Khlifi, K.; Cheikh Larbi, A.; Benhayoune, H. Electrophoretic Deposition of 45S5 Bioglass® Coatings on the Ti6Al4V Prosthetic Alloy with Improved Mechanical Properties. *Coatings* **2020**, *10*, 1192. [CrossRef]
28. Furko, M.; Balázsi, C. Calcium phosphate based bioactive ceramic layers on implant materials preparation, properties, and biological performance. *Coatings* **2020**, *10*, 823. [CrossRef]
29. Drevet, R.; Benhayoune, H. Electrodeposition of Calcium Phosphate Coatings on Metallic Substrates for Bone Implant Applications: A Review. *Coatings* **2022**, *12*, 539. [CrossRef]
30. Drevet, R.; Zhukova, Y.; Dubinskiy, S.; Kazakbiev, A.; Naumenko, V.; Abakumov, M.; Fauré, J.; Benhayoune, H.; Prokoshkin, S. Electrodeposition of cobalt-substituted calcium phosphate coatings on Ti22Nb6Zr alloy for bone implant applications. *J. Alloy. Compd.* **2019**, *793*, 576–582. [CrossRef]
31. Drevet, R.; Fauré, J.; Benhayoune, H. Structural and morphological study of electrodeposited calcium phosphate materials submitted to thermal treatment. *Mater. Lett.* **2017**, *209*, 27–31. [CrossRef]

Editorial

New Insights into Biomaterials and Coatings

Richard Drevet [1,*] and Hicham Benhayoune [2]

[1] Department of Plasma Physics and Technology, Masaryk University, Kotlářská 2,
CZ-61137 Brno, Czech Republic

[2] Institut de Thermique, Mécanique et Matériaux (ITheMM), EA 7548, Université de Reims
Champagne-Ardenne (URCA), Bât.6, Moulin de la Housse, BP 1039, 51687 Reims Cedex 2, France;
hicham.benhayoune@univ-reims.fr

* Correspondence: drevet@mail.muni.cz or richarddrevet@yahoo.fr

Received: 3 March 2025

Accepted: 7 March 2025

Published: 13 March 2025

Citation: Drevet, R.; Benhayoune, H. New Insights into Biomaterials and Coatings. *Coatings* **2025**, *15*, 332. https://doi.org/10.3390/coatings15030332

Global life expectancy is constantly rising throughout the world. The World Health Organization (WHO) predicts that the world population aged 80 years or over will triple by 2050 to reach 426 million [1]. The constant aging of the population implies an increasing demand for new solutions in bone tissue engineering [2,3]. Innovative biomaterials and bioactive coatings are constantly in development to produce new bone implants with better properties and an extended lifetime [4–11]. Biocompatibility and bioactivity are two major requirements for bone implants. Biocompatibility is the ability of a material to be in contact with a biological system without producing any adverse effects [12–14]. The bioactivity of bone implants defines their ability to develop a direct, adherent, and strong bonding with bone tissue [15–19]. This Special Issue on "Advanced Biomaterials and Coatings" gathers recent research articles dealing with the latest developments on this research topic [20].

The article by Ciobanu et al. describes an innovative chromium-doped hydroxyapatite coating deposited by the dip coating technique [21]. The low temperature of this deposition process allows the coating to be enriched with an antibiotic such as amoxicillin. The biocompatibility of the obtained coatings was studied in contact with bone cells (MG63 cell line) and the antibacterial activity of the coating was evaluated against *Pseudomonas aeruginosa*.

In another article, Alhotan et al. used 3D printing to produce a synthetic polymer material (polyamide 12) commonly used to reconstruct bony defects. To make its surface bioactive, the 3D-printed polymer was coated with hydroxyapatite using light-cured resin cement [22].

In their work, Kang et al. synthesized a composite material made of a glass ionomer cement named calcium fluoroaluminosilicate (CFAS) filled with nano-hydroxyapatite (nHA). This composite material shows enhanced cytocompatibility and histocompatibility, which make it suitable for dental repairs [23].

Dhiflaoui et al. developed a composite coating made of titanium oxide (TiO_2) and hydroxyapatite (HaP) deposited by spray pyrolysis on titanium alloy (Ti6Al4V). They studied the morphology, chemical composition, phase composition, and mechanical properties of this innovative composite coating. They used electrochemical impedance spectroscopy to show that the coating improves the corrosion resistance of Ti6Al4V in simulated body fluids [24].

In their article, Golubchikov et al. used precipitation to produce a bioceramic powder from an aqueous solution containing calcium, sodium, phosphate, and silicate ions. Different annealing temperatures were studied to crystallize and densify the precipitated bioceramic powder. The biocompatible and bioresorbable phases in the powder make it suitable for the treatment of bone defects [25].

In another article, Ferroni et al. developed a composite material made of polycapro-lactone (PCL) combined with reduced graphene oxide (rGO). This innovative material is biocompatible, supports the adhesion and differentiation of human mesenchymal stem cells (MSCs), and shows good antibacterial properties [26].

Handral et al. studied the cytotoxic and genotoxic effects of nanoparticles of zinc oxide (ZnO), titanium dioxide (TiO_2), and silicon dioxide (SiO_2). These experiments were conducted with pluripotent human embryonic stem cell-derived fibroblasts (hESC-Fib) [27].

The article by Drevet et al. comprehensively reviews the electrophoretic deposition of bioactive glass coatings on metallic bone implants. Bioactive glasses are highly reactive materials in physiological environments. They are commonly used to support bone repair. The mechanisms involved in the electrophoretic deposition process are thoroughly described [28].

The contribution by Robert B. Heimann is a detailed review of the literature dealing with the physicochemical and biomedical properties of plasma-sprayed hydroxylapatite (HAp) coatings. These bioactive coatings are routinely deposited on the surface of metallic endoprosthetic and dental root implants [29].

Sergey V. Dorozhkin reviews preparation methods, properties, and applications of calcium orthophosphate ($CaPO_4$)-based bioceramics used for biomedical applications such as artificial bone grafting, bone augmentation, maxillofacial reconstruction, spinal fusion, periodontal disease repair, and bone filling after tumor surgery. Prospective future applications of calcium orthophosphate are also comprehensively discussed [30].

Author Contributions: Conceptualization, R.D. and H.B.; validation, R.D. and H.B.; resources, R.D. and H.B.; writing—original draft preparation, R.D. and H.B.; writing—review and editing, R.D. and H.B. All authors have read and agreed to the published version of the manuscript.

Funding: This research received no external funding.

References

1. Available online: https://www.who.int/news-room/fact-sheets/detail/ageing-and-health (accessed on 1 March 2025).
2. Gheno, R.; Cepparo, J.M.; Rosca, C.E.; Cotton, A. Musculoskeletal Disorders in the Elderly. *J. Clin. Imaging Sci.* **2012**, *2*, 39. [CrossRef]
3. Dai, K.; Geng, Z.; Zhang, W.; Wei, X.; Wang, J.; Nie, G.; Liu, C. Biomaterial design for regenerating aged bone: Materiobiological advances and paradigmatic shifts. *Nat. Sci. Rev.* **2024**, *11*, nwae076. [CrossRef] [PubMed]
4. Drevet, R.; Zhukova, Y.; Kadirov, P.; Dubinskiy, S.; Kazakbiev, A.; Pustov, Y.; Prokoshkin, S. Tunable Corrosion Behavior of Calcium Phosphate Coated Fe-Mn-Si Alloys for Bone Implant Applications. *Metall. Mater. Trans. A* **2018**, *49*, 6553–6560. [CrossRef]
5. Chen, Q.; Thouas, G.A. Metallic implant biomaterials. *Mater. Sci. Eng. R* **2015**, *87*, 1–57. [CrossRef]
6. Kloster, G.A.; Rivero, G.; Ballarre, J.; Herrera Seitz, M.K.; Ceré, S.M.; Abraham, G.A. Innovative pH-triggered antibacterial nanofibrous coatings for enhanced metallic implant properties. *Front. Mater.* **2024**, *11*, 1484465. [CrossRef]
7. Zhou, K.; Simonassi-Paiva, B.; Fehrenbach, G.; Yan, G.; Portela, A.; Pogue, R.; Cao, Z.; Fournet, M.B.; Devine, D.M. Investigating the Promising P28 Peptide-Loaded Chitosan/Ceramic Bone Scaffolds for Bone Regeneration. *Molecules* **2024**, *29*, 4208. [CrossRef]
8. Lanzino, M.C.; Le, L.Q.R.V.; Wilbig, J.; Rheinheimer, W.; Seidenstuecker, M.; Günster, J.; Killinger, A. Thin GB14 coatings on implants using HVSFS. *Front. Mater.* **2024**, *11*, 1522447. [CrossRef]
9. Alluhaidan, T.; Qaw, M.; Garcia, I.M.; Montoya, C.; Orrego, S.; Melo, M.A. Seeking Endurance: Designing Smart Dental Composites for Tooth Restoration. *Designs* **2024**, *8*, 92. [CrossRef]
10. Balasubramani, G.; Premkumar, J.; Paul Pradeep, J. Optimizing bone-metal implant interfaces: The role of bio-ceramic coatings in improving stability and tissue metabolism. *Front. Mater.* **2025**, *11*, 1514559. [CrossRef]
11. Lallukka, M.; Miola, M.; Verné, E.; Baino, F. Silver-doped glass-ceramic scaffolds with antibacterial and bioactive properties for bone substitution. *Ceram. Int.* **2024**, *50*, 30997–31005. [CrossRef]
12. Williams, D.F. On the mechanisms of biocompatibility. *Biomaterials* **2008**, *29*, 2941–2953. [CrossRef] [PubMed]
13. Williams, D.F. Biocompatibility pathways and mechanisms for bioactive materials: The bioactivity zone. *Bioact. Mater.* **2022**, *10*, 306–322. [CrossRef]

14. Nirwan, V.P.; Filova, E.; Al-Kattan, A.; Kabashin, A.V.; Fahmi, A. Smart Electrospun Hybrid Nanofibers Functionalized with Ligand-Free Titanium Nitride (TiN) Nanoparticles for Tissue Engineering. *Nanomaterials* **2021**, *11*, 519. [CrossRef]
15. Modelski Schatkoski, V.; Larissa do Amaral Montanheiro, T.; Rossi Canuto de Menezes, B.; Monteiro Pereira, R.; Faquine Rodrigues, K.; Guimarães Ribas, R.; Morais da Silva, D.; Patrocínio Thim, G. Current advances concerning the most cited metal ions doped bioceramics and silicate-based bioactive glasses for bone tissue engineering. *Ceram. Int.* **2021**, *47*, 2999–3012. [CrossRef]
16. Drevet, R.; Fauré, J.; Benhayoune, H. Calcium Phosphates and Bioactive Glasses for Bone Implant Applications. *Coatings* **2023**, *13*, 1217. [CrossRef]
17. Shaikh, M.S.; Fareed, M.A.; Zafar, M.S. Bioactive Glass Applications in Different Periodontal Lesions: A Narrative Review. *Coatings* **2023**, *13*, 716. [CrossRef]
18. Lukaviciute, L.; Ganceviciene, R.; Tsuru, K.; Ishikawa, K.; Yang, J.C.; Grigoraviciute, I.; Kareiva, A. Cationic substitution effects in phosphate-based bioceramics—A way towards superior bioproperties. *Ceram. Int.* **2024**, *50*, 34479–34509. [CrossRef]
19. Tripathi, S.; Raheem, A.; Dash, M.; Kumar, P.; Elsebahy, A.; Singh, H.; Manivasagam, G.; Sekhar Nanda, H. Surface engineering of orthopedic implants for better clinical adoption. *J. Mater. Chem. B* **2024**, *12*, 11302. [CrossRef]
20. Drevet, R.; Benhayoune, H. Advanced Biomaterials and Coatings. *Coatings* **2022**, *12*, 965. [CrossRef]
21. Ciobanu, C.S.; Predoi, D.; Iconaru, S.L.; Rokosz, K.; Raaen, S.; Negrila, C.C.; Ghegoiu, L.; Bleotu, C.; Predoi, M.V. Chrome Doped Hydroxyapatite Enriched with Amoxicillin Layers for Biomedical Applications. *Coatings* **2025**, *15*, 233. [CrossRef]
22. Alhotan, A.; Alhijji, S.; Abdalbary, S.A.; Bayoumi, R.E.; Matinlinna, J.P.; Hamdy, T.M.; Abdelraouf, R.M. An Advanced Surface Treatment Technique for Coating Three-Dimensional-Printed Polyamide 12 by Hydroxyapatite. *Coatings* **2024**, *14*, 1181. [CrossRef]
23. Kang, S.; Park, S.J.; Kim, S.; Kang, I.-K. Chemical Bonding of Nanorod Hydroxyapatite to the Surface of Calciumfluoroaluminosilicate Particles for Improving the Histocompatibility of Glass Ionomer Cement. *Coatings* **2024**, *14*, 893. [CrossRef]
24. Dhiflaoui, H.; Ben Salem, S.; Salah, M.; Dabaki, Y.; Chayoukhi, S.; Gassoumi, B.; Hajjaji, A.; Ben Cheikh Larbi, A.; Amlouk, M.; Benhayoune, H. Influence of TiO_2 on the Microstructure, Mechanical Properties and Corrosion Resistance of Hydroxyapatite HaP + TiO_2 Nanocomposites Deposited Using Spray Pyrolysis. *Coatings* **2023**, *13*, 1283. [CrossRef]
25. Golubchikov, D.; Safronova, T.V.; Nemygina, E.; Shatalova, T.B.; Tikhomirova, I.N.; Roslyakov, I.V.; Khayrutdinova, D.; Platonov, V.; Boytsova, O.; Kaimonov, M.; et al. Powder Synthesized from Aqueous Solution of Calcium Nitrate and Mixed-Anionic Solution of Orthophosphate and Silicate Anions for Bioceramics Production. *Coatings* **2023**, *13*, 374. [CrossRef]
26. Ferroni, L.; Gardin, C.; Rigoni, F.; Balliana, E.; Zanotti, F.; Scatto, M.; Riello, P.; Zavan, B. The Impact of Graphene Oxide on Polycaprolactone PCL Surfaces: Antimicrobial Activity and Osteogenic Differentiation of Mesenchymal Stem Cell. *Coatings* **2022**, *12*, 799. [CrossRef]
27. Handral, H.K.; Ashajyothi, C.; Sriram, G.; Kelmani, C.R.; Dubey, N.; Cao, T. Cytotoxicity and Genotoxicity of Metal Oxide Nanoparticles in Human Pluripotent Stem Cell-Derived Fibroblasts. *Coatings* **2021**, *11*, 107. [CrossRef]
28. Drevet, R.; Fauré, J.; Benhayoune, H. Electrophoretic Deposition of Bioactive Glass Coatings for Bone Implant Applications: A Review. *Coatings* **2024**, *14*, 1084. [CrossRef]
29. Heimann, R.B. Plasma-Sprayed Osseoconductive Hydroxylapatite Coatings for Endoprosthetic Hip Implants: Phase Composition, Microstructure, Properties, and Biomedical Functions. *Coatings* **2024**, *14*, 787. [CrossRef]
30. Dorozhkin, S.V. Calcium Orthophosphate ($CaPO_4$)-Based Bioceramics: Preparation, Properties, and Applications. *Coatings* **2022**, *12*, 1380. [CrossRef]

Article

Chrome Doped Hydroxyapatite Enriched with Amoxicillin Layers for Biomedical Applications

Carmen Steluta Ciobanu [1], Daniela Predoi [1,*], Simona Liliana Iconaru [1], Krzysztof Rokosz [2], Steinar Raaen [3], Catalin Constantin Negrila [1], Liliana Ghegoiu [1,4], Coralia Bleotu [5,6,7] and Mihai Valentin Predoi [4,*]

[1] National Institute of Materials Physics, Atomistilor Street, No. 405A, MG 07, 077125 Magurele, Romania; ciobanucs@gmail.com (C.S.C.); simonaiconaru@gmail.com (S.L.I.); catalin.negrila@infim.ro (C.C.N.); ghegoiuliliana@gmail.com (L.G.)

[2] Faculty of Electronics and Computer Science, Koszalin University of Technology, Śniadeckich 2, PL 75-453 Koszalin, Poland; rokosz@tu.koszalin.pl

[3] Department of Physics, Norwegian University of Science and Technology (NTNU), Realfagbygget E3-124 Høgskoleringen 5, NO 7491 Trondheim, Norway; steinar.raaen@ntnu.no

[4] Department of Mechanics, University Politehnica of Bucharest, BN 002, 313 Splaiul Independentei, Sector 6, 060042 Bucharest, Romania

[5] Department of Cellular and Molecular Pathology, Stefan S. Nicolau Institute of Virology, Romanian Academy, 030304 Bucharest, Romania; cbleotu@yahoo.com

[6] Research Institute of the University of Bucharest (ICUB), University of Bucharest, 060023 Bucharest, Romania

[7] The Academy of Romanian Scientist, 050711 Bucharest, Romania

* Correspondence: dpredoi@gmail.com (D.P.); predoi@gmail.com (M.V.P.)

Academic Editors: Hicham Benhayoune and Richard Drevet

Received: 30 December 2024
Revised: 12 February 2025
Accepted: 13 February 2025
Published: 15 February 2025

Citation: Ciobanu, C.S.; Predoi, D.; Iconaru, S.L.; Rokosz, K.; Raaen, S.; Negrila, C.C.; Ghegoiu, L.; Bleotu, C.; Predoi, M.V. Chrome Doped Hydroxyapatite Enriched with Amoxicillin Layers for Biomedical Applications. *Coatings* **2025**, *15*, 233. https://doi.org/10.3390/coatings15020233

Abstract: In the last decade, it has been observed that the field of biomaterials has gained the attention of the researchers. This study presents the physicochemical and biological properties of coatings based on chromium-doped hydroxyapatite (CrHAp) and chromium-doped hydroxyapatite enriched with amoxicillin (CrHApAx). The coatings were obtained for the first time using the dip coating technique, beginning from dense suspensions of CrHAp and CrHApAx. The obtained layers were then analyzed by various methods in order to have a comprehensive overview of their physicochemical properties. Stability studies performed using ultrasound measurements showed that the CrHAp suspension has very good stability (S = $6.86 \cdot 10^{-6}$ s^{-1}) compared to double-distilled water. The CrHApAx suspension (S = 0.00025 s^{-1}) shows good but weaker stability compared to that of the CrHAp suspension. Following XRD studies, a single hydroxyapatite-specific phase was observed in the CrHAp sample, while in the case of the CrHApAx sample, an amoxicillin-specific peak was also observed. The AFM results showed that the CrHAp coatings had a surface topography of a homogenous and uniform layer, with no significant cracks and fissures, while the CrHApAx coatings exhibited a surface morphology of homogenous layers formed of particles conglomerates. The biocompatibility of CrHAp and CrHApAx coatings was assessed using the MG63 cell line. The cytotoxicity of the coatings was evaluated by measuring cell viability with the aid of an MTT assay after 24, 48, and 72 h of incubation with the CrHAp and CrHApAx coatings. The results demonstrated that both CrHAp and CrHApAx coatings exhibited good biocompatibility for all the tested time intervals. The in vitro antibacterial activity of the coatings was also assessed against *Pseudomonas aeruginosa* 27853 ATCC (*P. aeruginosa*) bacterial cells. The potential of *P. aeruginosa* bacterial cells to adhere and develop on the surfaces of CrHAp and CrHApAx coatings was also investigated using AFM analysis. The findings of the biological assays suggest that CrHAp and CrHApAx coatings could be considered as promising candidates for biomedical applications, including the development of novel antimicrobial materials.

Keywords: chrome; hydroxyapatite; layers; structural features; antimicrobial activity; ultrasound measurements

1. Introduction

The field of biomaterials has garnered significant attention in recent years, emerging as a priority area with expanding applications in the medical and pharmaceutical sectors [1]. Within the realms of nanotechnology and biotechnology, biomaterials represent a critical class of materials. Nanotechnology has found successful applications in medicine, addressing malignant diseases, regenerating bone structures, and treating dental conditions like periodontal diseases [2].

Calcium apatite in its mineral form, hydroxyapatite (HAp), is notable for its excellent biocompatibility, stemming from its chemical composition, which closely resembles that of biological tissues, particularly connective tissues [3]. Research has highlighted HAp as an effective nanocarrier for delivering antibiotics to hard tissues [4]. Studies have further demonstrated the use of HAp nanoparticles combined with polyvinyl alcohol (PVA) and amoxicillin in treating periodontitis [5]. The structural and chemical composition of hydroxyapatite (HAp) closely resembles that of the mineral components found in mammalian bones and teeth [6–9]. The resemblance in structure and composition between HAp and the mineral components of bones and teeth is what makes HAp highly biocompatible, which means that it can integrate well with bone tissue without causing adverse reactions [6,8]. This property is why HAp is extensively used in medical and dental applications, such as bone grafts and coatings for implants, to promote bone growth and repair. In the last decade, HAp has been utilized as a coating for metallic implants to enhance their biocompatibility, addressing concerns regarding the compatibility and toxicity of bone substitute materials in orthopedic and dental surgeries [6,10]. Due to its unique structure, HAp allows the substitution of calcium ions from its structure with various ions such as zinc [11], copper [12], silver [13], etc. According to previously reported studies, conducted by Uysal, I. et al. [14], the addition of foreign metal ions into the HAp structure leads to the improvement of both antimicrobial and mechanical properties of HAp [14].

Synthetic HAp exhibits excellent properties, including a high ion exchange rate with metals, strong affinity for pathogenic microorganisms, and notable biocompatibility, with approximately 70%–80% of implants being made from biocompatible metals [15–17]. Antibiotics are a class of drugs that target the root causes of infections by either destroying bacterial cell walls or inhibiting their synthesis. Amoxicillin, a broad-spectrum, semi-synthetic beta-lactam antibiotic, is widely known for its efficacy. Incorporating amoxicillin into doped HAp not only provides antimicrobial properties without cellular toxicity [14] but also addresses the limitations of traditional antibiotic administration for bone structures, which can be affected by various factors [18].

Chromium (Cr) exists in different oxidation states, with trivalent Cr (III) and hexavalent Cr (VI) being the most common [6–8,19,20]. While Cr (VI) is known to be toxic and carcinogenic, studies have indicated that Cr (III) can promote biological responses and influence the crystallization behavior of HAp bioceramics [7,8,19,21]. On the other hand, Cr^{3+} is an important trace element found naturally in plants and animals, playing a fundamental role in sugar and fat metabolism [19]. It enhances insulin function and provides stability to collagen [7,8,19,22]. Furthermore, research shows that Cr (III) dopants can improve photocatalytic antibacterial activity under visible light [23]. Previous in vitro studies have demonstrated that Cr (VI) does not adversely affect red blood cells at concentrations up to 1 g/L [24]. Its relatively low toxicity stems from the poor solubility of Cr (III) complexes at physiological pH and their limited ability to penetrate cell membranes, which prevents them from causing toxic effects in cells [7,8,19,22,25,26].

In the previous study reported by [27], it was shown that magnesium-doped hydroxyapatite (MgHAp) and its amoxicillin-enriched variant (MgHApOx) were successfully synthesized for the first time using an adapted co-precipitation method, with confirmed incorporation of both Mg^{2+} ions and amoxicillin into the HAp lattice. Furthermore, both materials demonstrated excellent biocompatibility (>95% cell viability for MgHAp and >88% for MgHApOx) and effective antimicrobial properties against *S. aureus*, *E. coli*, and *C. albicans*, making them promising candidates for biomedical applications [27]. In addition, Predoi, D., and coworkers [28] recently reported that chromium-doped hydroxyapatite ($x_{Cr} = 0.2$) nanocomposite coatings possess strong antifungal properties, making them promising candidates for developing antifungal medical devices and implants [28].

This study presents for the first time the physicochemical and biological properties of coatings made by the dip coating method starting from CrHAp and CrHApAx suspensions. The suspensions used for the film formation were obtained using an adapted sol-gel method. In addition, the stability of the suspensions was analyzed using ultrasound measurements. XPS analysis revealed the presence of the constituent elements of Cr-doped hydroxyapatite. Furthermore, the presence of amoxicillin in the chromium-doped and amoxicillin-enriched hydroxyapatite sample was highlighted. Data regarding the layers' surface morphology were collected by SEM and AFM. Furthermore, the vibrational properties of the CrHAp and CrhapAx layers were studied with the aid of FTIR measurements. Information about the biocompatibility of CrHAp and CrHApAx coatings was obtained with the aid of the MG63 cell line (MTT assay). The in vitro antibacterial activity of the coatings was also evaluated against *Pseudomonas aeruginosa* 27,853 ATCC (*P. aeruginosa*) bacterial cells.

2. Materials and Methods

2.1. Materials

Chrome-doped hydroxyapatite (CrHAp, $x_{Cr} = 0.05$; [Ca + Cr]/P = 1.67) and chrome-doped hydroxyapatite enriched with amoxicillin (CrHApAx, $x_{Cr} = 0.05$; [Ca + Cr]/P = 1.67; Cr/(Ca + Cr) = 0.0526; Ca/P = 1.58) were synthesized using an adapted sol-gel method. The following reagents were used: chrome nitrate (Cr^{3+}, $Cr(NO_3)_3 \cdot 9H_2O$, Alfa Aesar, Karlsruhe, Germany; 99.99% purity), calcium nitrate ($Ca(NO_3)_2 \cdot 4H_2O$, Sigma-Aldrich, St. Louis, MA, USA), ammonium hydrogen phosphate ($(NH_4)_2HPO_4$, Alfa Aesar, Karlsruhe, Germany; 99.99% purity), amoxicillin ($C_{16}H_{19}N_3O_5S$, 95.0%–102.0%, Sigma Aldrich, St. Louis, MO, USA), and ethanol. The CrHAp and CrHApAx layers were deposited onto Si substrates using the dip coating technique.

2.2. Development of Chrome-Doped Hydroxyapatite Enriched with Amoxicillin Layers

Both layers were prepared following the detailed procedure outlined in our previous work [28]. Briefly, ammonium hydrogen phosphate was mixed with ethanol and stirred for 2 h at 40 °C (in order to obtain an 0.5 mol/L solution). Separately, calcium nitrate and chromium nitrate (Cr^{3+}) were also dissolved in ethanol and stirred under the same conditions (in order to obtain a 1.67 mol/L solution). The first solution was then gradually added to the second solution under continuous stirring. The resulting mixture was stirred for 12 h at 80 °C while maintaining a pH of 10 throughout the process. Then, the resulting mixture underwent a washing process, being rinsed five times with deionized water and ethanol. Following this, the gel was dispersed in ethanol and subjected to continuous stirring for 12 h. For the development of CrHApAx gel, the same procedure as the one described above was followed. The only change consisted in adding the amoxicillin (0.2 g, 0.01 M) to the solution that contains calcium nitrate and chromium nitrate. The final gels were used to deposit CrHAp and CrHApAx thin films onto Si substrates via the dip coating technique.

The deposition technique was previously described in detail [29]. According to the procedure, three layers of CrHAp and CrHApAx were deposited on Si substrate. The Si disks were cleaned ultrasonically, and rinsed with acetone and distilled water, before being used for the deposition of CrHAp and CrHApAx coatings. Then, the Si substrate underwent a dip coating process in the CrHAp and CrHApAx gels, respectively. This procedure was repeated three times, with each immersion lasting several minutes. Each layer was treated at 70 °C for 4 h. At the end, the CrHAp and CrHApAx layers were treated at 70 °C for 72 h.

2.3. Physicochemical Characterisation

In this study, the stability of the CrHAp and CrHApAx suspensions was estimated using non-destructive ultrasound (US) measurements. As a reference, bidistilled water was used in this study. Both the protocol and instrument used in the US measurements have been described in our previous studies [30]. In order to have a good homogeneity, before starting the US measurements, the suspensions of CrHAp and CrHApAx were stirred continuously for 15 min at 800 rpm. The stirring of the 100 mL of suspension of CrHAp and CrHApAx was performed at room temperature. Next, for the US measurements, the suspension was poured into a transparent cubic container. The transparent cubic container was specially equipped with two coaxial ultrasonic transducers. The two coaxial ultrasonic transducers were spaced 16 mm apart. The axis of the transducers was at mid-height of the container box. After stirring the suspensions at room temperature for 15 min, the acquisition of the 1000 ultrasonic signals began. The signals were recorded every 5 s on the digital oscilloscope. Each recorded signal is an average of 32 signals on the oscilloscope, reducing the experimental noise.

Information about the structure of the CrHAp and CrHApAx samples was obtained using X-ray diffraction (XRD) studies. A Bruker D8 Advance diffractometer (Bruker, Karlsruhe, Germany) with CuKα (λ = 1.5418 Å) radiation was used to study the samples. The XRD diffractograms were collected in the 2θ range 10–60°, using a step size of 0.02. In this study, the Joint Committee on Powder Diffraction Standards (JCPDS) file for pure hexagonal hydroxyapatite (PDF No. 09-0432) was used as a reference. The diffraction peaks corresponding to planes (002) of CrHAp and CrHApAx samples were used in order to calculate the average crystallite size and d-spacing of the samples. The average crystallite size was calculated from the broadening in the XRD pattern using the Scherrer formula [31–34]:

$$D_{hkl} = (K\lambda)/(\beta\cos\theta)$$

where K is the Scherrer constant, λ is the wavelength of the monochromatic X-ray beam (1.54 A), β is the full width at half maximum (FWHM), and θ is the diffraction angle of (002).

To calculate the d-spacing of the samples, the Bragg equation [35] was used:

$$d = \lambda/2\sin\theta$$

X-ray photoelectron spectroscopy (XPS) studies were performed using a SPECS spectrometer with a PHOIBOS 150 analyzer. A Specs XR-50M RX source operated on a non-monochromatic Mg anode (Ex = 1253.6 eV) at 300 W was used for these measurements. Charge compensation was performed with a Specs FG15/40 flood gun. Acquisition was performed with a pass energy of 20 eV for the individual spectrum and 50 eV for the extended spectrum. CasaXPS 2.3.14 software (using Shirley background type) was used for data analysis [36]. XPS tables were also mentioned [37,38]. All binding energy (BE) values presented in this research were corrected for the charge at C1s at 284.8 eV.

The surface topography of the CrHAp and CrHApAx layers was analyzed using atomic force microscopy (AFM). For this purpose, an NT-MDT NTEGRA Probe NanoLaboratory system (NT-MDT, Moscow, Russia) was used. The AFM measurements were conducted in non-contact mode. For this study, the AFM was equipped with a silicon NT-MDT NSG01 cantilever (35 nm gold-coated tetrahedral tip). The 2D AFM micrographs were recorded on a surface area of 10×10 μm^2. The AFM images and their 3D representations were evaluated using Gwyddion 2.55 software [39]. More than that, information about the surface roughness was obtained by determining the root mean square roughness (R_{RMS}).

Information about the surface morphology of the CrHAp and CrHApAx coatings deposited on Si substrate were obtained by performing scanning electron microscopy (SEM) studies with the aid of HITACHI S4500 equipment. Data about the chemical composition of CrHAp and CrHApAx coatings were also assessed from the energy dispersive X-ray spectroscopy (EDS) studies.

Information about the adhesion of CrHAp and CrHApAx coatings was obtained by performing a tape-pull test. For this purpose, a 3M Performance Flatback Tape 2525 was used. The peel adhesion was 7.5 N/cm.

Fourier-transform infrared (FTIR) spectroscopy was used to examine the structural bonding vibrations of functional groups in CrHAp and CrHApAx. The analysis was performed with a Perkin Elmer spectrometer equipped with a Universal Diamond/KRS-5 accessory (Waltham, MA, USA). FTIR spectra were collected between 450 and 4000 cm^{-1} with a resolution of 4 cm^{-1}. The second derivative (450–700 cm^{-1} and 800–1200 cm^{-1}) and deconvoluted spectra (800–1200 cm^{-1}) of CrHAp and CrHApAx were also obtained following the procedure previously described [40].

2.4. In Vitro Biological Evaluation

Colorimetric test assay 3-(4,5-dimethylthiazol-2-yl)-2,5-diphenyltetrazolium bromide (MTT) assay

The cytotoxicity of the CrHAp and CrHApAx coatings was evaluated using osteosarcoma MG63 (ATCC CRL-1427) cells, following the methodology described in detail by Iconaru et al. [41]. For this purpose, the cells were cultured in DMEM supplemented with fetal bovine serum at 37 °C in a 5% CO_2 atmosphere. The MG63 cells were seeded at a density of 1×10^5 cells/well and incubated with the coatings for 24, 48, and 72 h. The cell viability was assessed using the MTT reduction assay, by measuring the absorbance at 595 nm. The cell viability was calculated relative to a control set at 100%. After incubation, the coatings were rinsed with sterile saline, fixed with cold methanol, and prepared for visualization. The MG63 cells were observed using a $10\times$ objective on an inverted trinocular metallographic microscope, model OX.2153-PLM (Euromex, Arnhem, Netherlands). ImageJ software (Image J 1.51j8) was used for images analysis [42].

In vitro antimicrobial assay

The antibacterial properties of CrHAp and CrHApAx coatings were evaluated against the *Pseudomonas aeruginosa* 27853 ATCC strain using in vitro assays following the protocol in [43]. The coatings were incubated with bacterial suspensions (5×10^6 CFU/mL) at 37 °C for 24, 48, and 72 h. The bacterial survival was quantified for each interval, and the CFU/mL values were graphically represented as log CFU/mL over time. A free bacterial suspension was used as a positive control (C+). The experiments were performed in triplicate, and the results expressed as mean $\pm$ SD. Furthermore, atomic force microscopy (AFM) was used for the qualitative analysis of bacterial adherence and proliferation on

CrHAp and CrHApAx surfaces. After incubation, the coatings were washed with sterile saline, fixed with cold methanol, and prepared for visualization.

Statistical analysis

The data from the biological assays were represented graphically as mean $\pm$ SD, and the statistical analysis was performed using Microsoft Excel. The data were analyzed using ordinary one-way ANOVA, and the significance level was set at $p < 0.05$.

3. Results

In this study, we report for the first time the physicochemical and biological properties of CrHAp and CrHApAx coatings obtained by dip coating method starting from dense aqueous suspensions of CrHAp and CrHApAx. The study of the stability of these dense aqueous suspensions is very important. The stability of concentrated suspensions plays a major role in obtaining coatings that could be successfully used in different bone implants. Thus, in this research, the stability of dense aqueous suspensions of CrHap and CrHApAx was studied by ultrasound measurements.

Figure 1 shows the superposition of the 1000 signals that were recorded for the two analyzed samples of CrHAp and CrHApAx. All these signals, covering 5000 s of process evolution, are represented as water flow from right to left. Figure 1a shows that in the case of the CrHAp sample, the sedimentation process is uniform. The small variations in the total amplitude of the signal each last a few tens of seconds. In the case of the CrHApAx sample, it is observed that the sedimentation process is irregular. Large variations in the amplitude of the signals, which last minutes, are observed (Figure 1b).

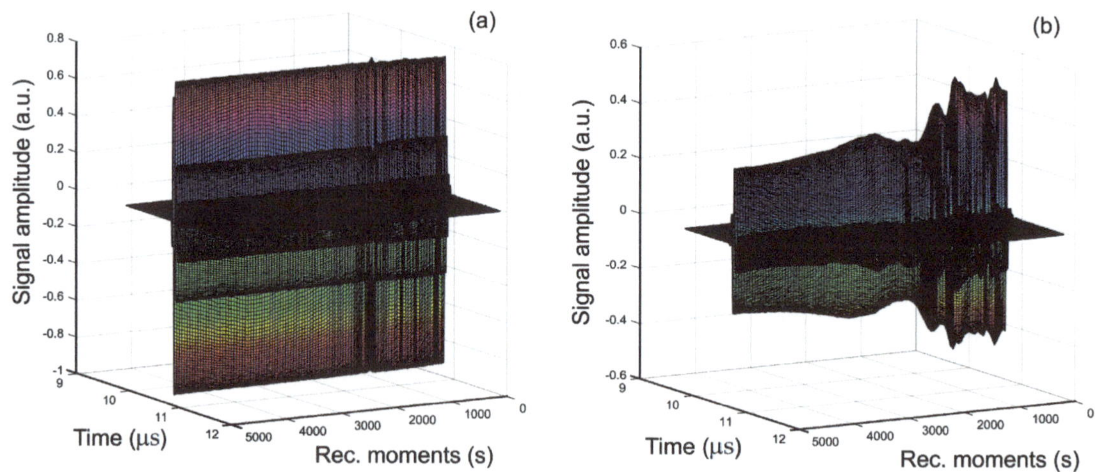

Figure 1. Time evolution of the recorded signals, from left to right over 5000 s, of CrHAp (**a**) and CrHApAx (**b**) samples.

In Figure 2a, a slow and continuous increase of the signal amplitude is observed for the CrHAp sample. A small amplitude decrease recorded at t = 1300 s was attributed to the formation of a particle cluster. In the case of the CrHApAx sample, a rapid variation during 1000 s is observed, followed by a period of slower but ample variation of amplitude with a general decreasing trend (Figure 2b). This evolution of the CrHApAx suspension is associated with the formation of nanoparticle clusters, followed by a rapid temporal variation of the suspension properties. After 2500 s, the evolution is a continuous progressive reduction in amplitude.

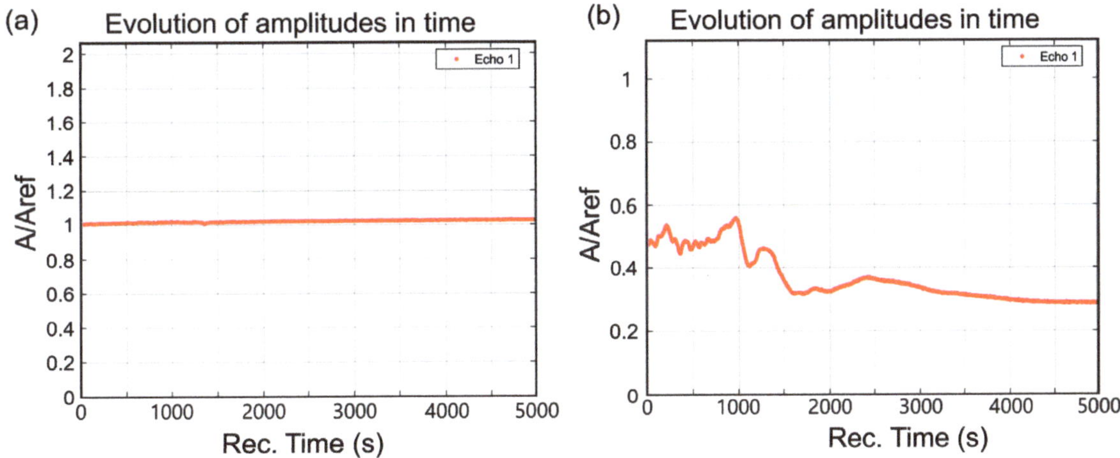

Figure 2. Recorded signals' amplitudes during the experiment for CrHAp (**a**) and CrHApAx (**b**).

The evolution of the frequency spectra for each of the 1000 signals for the two analyzed samples (CrHAp and CrHApAx) is presented in Figure 3. For comparison, the spectrum of the reference liquid (double-distilled water, in dotted blue line) is also represented in Figure 3. It can be noted that in the case of the CrHAp sample, the spectra are very closely packed, indicating a constant composition of the suspension (Figure 3a). The peaks are located at 25.5 MHz. This value is slightly below the 26.2 MHz peak corresponding to the reference liquid. The lack of overlap is due to the attenuation of the ultrasonic signal. In the case of the CrHApAx sample (Figure 3b), it can be noted that the spectra are relatively evenly spread during the monitoring process. The amplitudes are considerably lower than in the reference liquid, indicating a stronger attenuation of the ultrasonic signals.

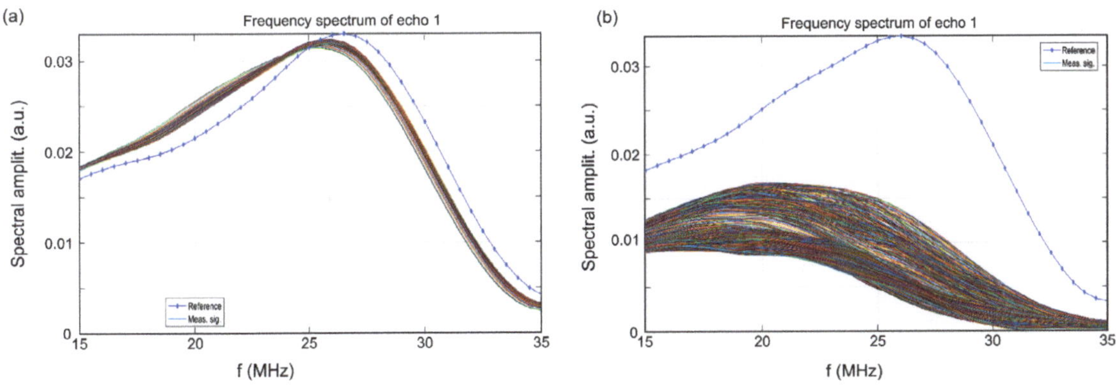

Figure 3. Spectral amplitudes of all recorded signals for CrHAp (**a**) and CrHApAx (**b**) samples.

The temporal evolution of signals' frequency spectra, which is related to the properties of the CrHAp and CrHApAx suspension in front of the transducers, bring more insight to the attenuation process. The time-averaged attenuation plot for both samples is shown in Figure 4. Compared against the standard attenuation in the reference liquid (red dotted line) the attenuation is larger for the CrHAp sample in the higher frequency ranges, reaching 74 nepper/m at 35 MHz (Figure 4a). Moreover, Figure 4a revealed that in the frequency range 15–25.5 MHz, the attenuation is lower than the attenuation in the reference liquid and is even negative for frequencies below 22.5 MHz. The attenuation is obtained by signal comparison with that of the reference liquid. Since the signal propagates with higher amplitude between the transducers compared to the signal in the reference liquid in the same experimental conditions, the determined attenuation is negative. This result indicates the presence of nanoparticles with higher elastic compressibility factor, like in metals for example. The time-averaged attenuation plot for the CrHApAx sample is shown on Figure 4b. Compared against the standard attenuation in the reference liquid (red dotted line), the attenuation is larger for the CrHApAx sample, and this is verified for all analyzed frequencies. There is a specific peak attenuation at 32 MHz, indicating a resonance of the nanoparticles in suspension.

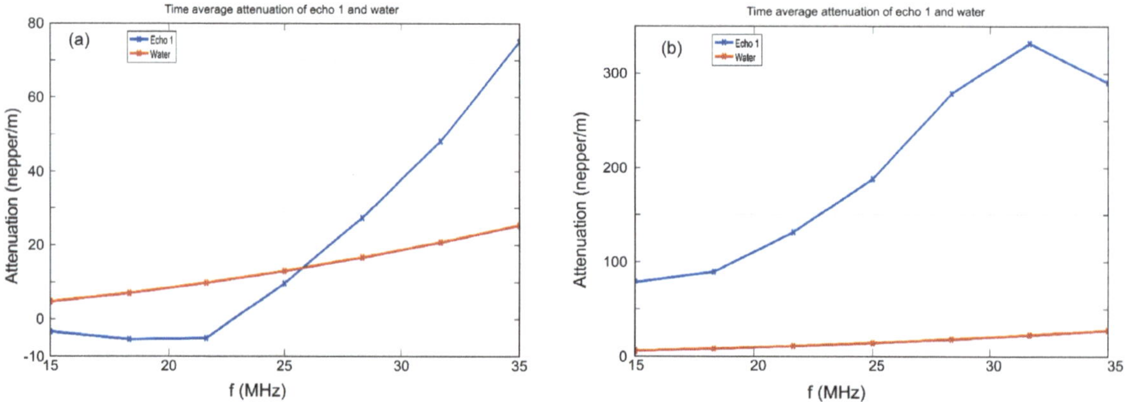

Figure 4. Time-averaged attenuation for the investigated frequency range for CrHAp (**a**) and CrHApAx (**b**) samples.

Another characteristic of the CrHAp and CrHApAx suspensions is their spectral stability, which represents the amplitude of the frequency component of each spectrum as a function of time (Figure 5). In the case of the CrHApAx sample (Figure 5b), it can be observed that during 1500 s, the irregular sedimentation produces a series of ripples for each selected frequency. A slower evolution is recorded up to t = 3500 s, being almost monotonic for lower frequencies (15–18 MHz), but with marked peaks for higher frequencies. The higher the frequency, the later the peak appears in the spectrum, indicating a progressive change in the suspension concentration. After 3500 s, the suspension exhibits a monotonic slowly decreasing amplitude, attributed to the higher attenuation of the smallest nanoparticles remaining in the suspension (Figure 5b).

The CrHAp sample is very stable, proved by the stability parameter: $S = \frac{\overline{dA}}{A dt} = 6.86 \cdot 10^{-6} \text{ s}^{-1}$, in which A is the signal amplitude, with a bar above indicating time averaging. The CrHApAx sample becomes relatively stable after 3500 s, and its stability parameter is: $S = \frac{\overline{dA}}{A dt} = 0.00025 \text{ s}^{-1}$, in which A is the signal amplitude, with a bar above indicating time averaging.

The XRD analysis was used in order to calculate the average crystallite size of CrHAp and CrHApAx samples. The diffraction patterns of CrHAp and CrHApAx with $x_{Cr} = 0.05$ samples are presented in Figure 6. The diffraction pattern of CrHAp was similar to the reference hexagonal pattern (JCPDS 09-0432). In the CrHAp synthesized sample, the formation of a single HAp phase was observed. The absence of secondary phases may be due to the fact that the solubility limit of chromium in HAp has not yet been reached.

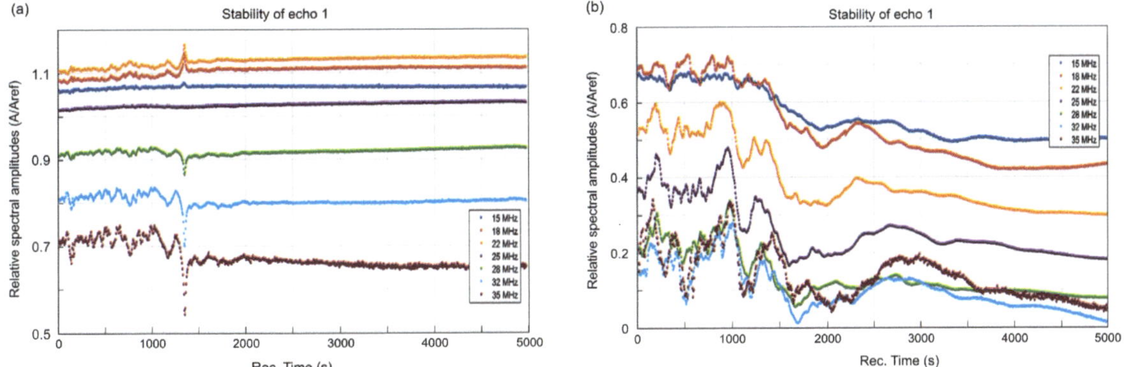

Figure 5. Relative spectral amplitudes vs. time for CrHAp (**a**) and CrHApAx (**b**) samples.

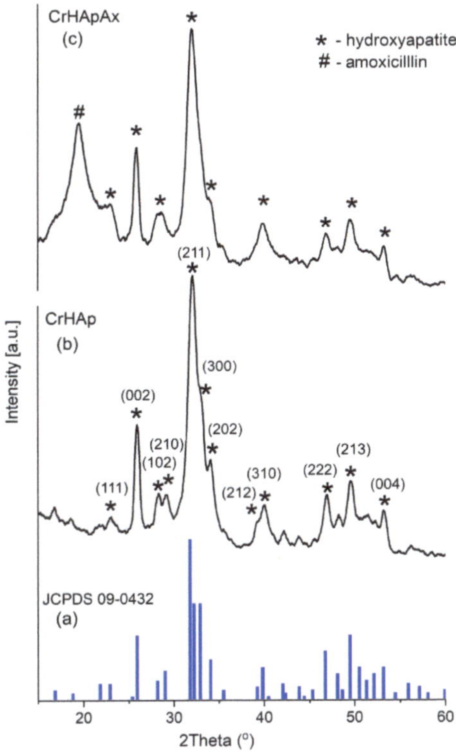

Figure 6. XRD patterns of CrHAp (**b**) and CrHApAx (**c**) samples. XRD patterns of pure hexagonal hydroxyapatite JCPDS 09-0432 (**a**).

On the other hand, in accordance with Bragg's law [44], the observed peaks were slightly shifted to smaller angles for the CrHAp sample. Furthermore, with amoxicillin-doped CrHAp, a slight broadening of the peaks and a decrease in their intensity was observed. The presence of the peak at $2\theta = 19.589$ specific to amoxicillin was observed in the CrHApAx sample. This broadening of the peaks observed in the CrHApAx sample could be caused by both crystallization imperfections and the incorporation of amoxicillin during the synthesis process.

The hydroxyapatite phase as the principal crystalline phase was also identified in the CrHApAx sample. The peaks corresponding to the crystal planes designated by the Miller indices (002), (211), (112), (300), (310), (222), (213), and (004), associated with hexagonal hydroxyapatite (space group of P 63/m), were identified in both samples. The calculated average crystallite size was 13.72 nm for the CrHAp sample and 12.66 nm for the CrHApAx sample. Since the average crystallite size values are <40 nm for both samples, we can speak of the formation of nanoparticles. The calculated d-spacing value was 3.4419 nm for the CrHAp sample and 3.4406 for the CrHApAx sample. These values are consistent with the d-spacing of 0.3441 nm, which corresponds to the reflection of the (002) crystal planes in the hexagonal HAp model [45]. The behavior revealed in the XRD study is in accordance with A. Person et al. [46].

The elemental composition and chemical modifications of chromium-doped and amoxicillin-enriched hydroxyapatite were determined using XPS analysis. As can be seen in Figure 7a,b, peaks associated with carbon (C), oxygen (O), calcium (Ca), phosphorus (P), and chromium (Cr) were observed in the two analyzed samples (CrHAp and CrHApAx). The presence of nitrogen (N) and sulfur (S) was identified only in the CrHApAx sample (Figure 7b), which certifies the presence of amoxicillin. Carbon accidental contamination was used as a charge reference for the XPS spectra. Consequently, the C-C component observed at a binding energy of 284.8 eV was utilized in this study to align the core-level binding energies (EBs).

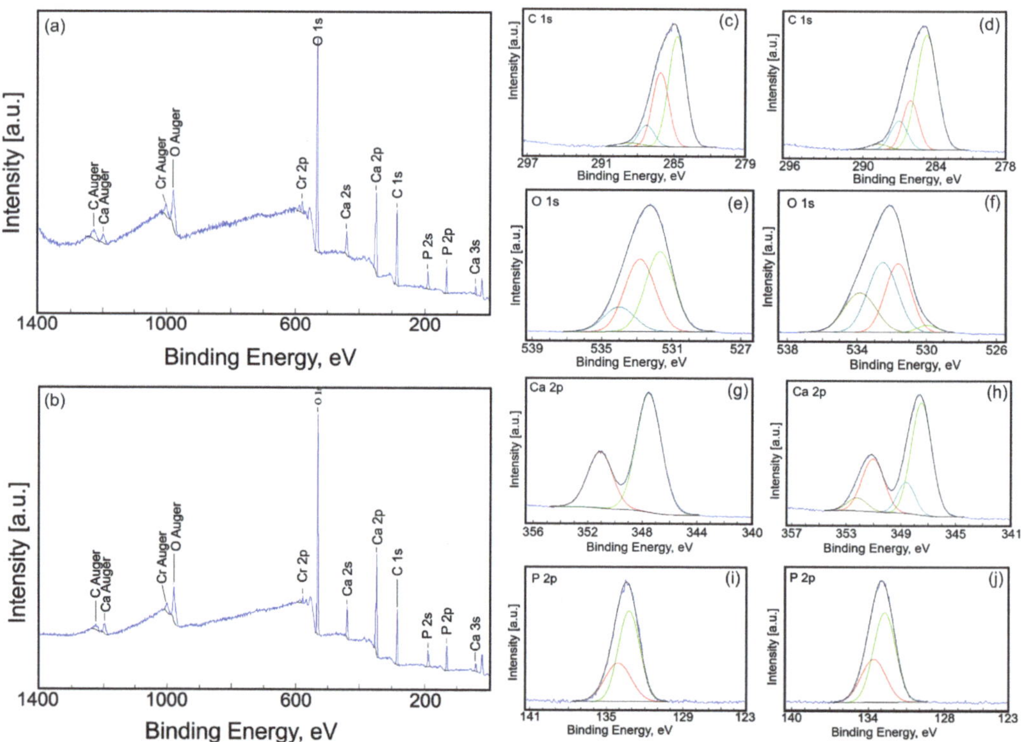

Figure 7. Full XPS spectra of CrHAp (**a**) and CrHApAx (**b**) samples. High-resolution XPS spectra of C 1s for CrHAp (**c**) and CrHApAx (**d**) samples. High-resolution XPS spectra of O 1s for CrHAp (**e**) and CrHApAx (**f**) samples. High-resolution XPS spectra of Ca 2p for CrHAp (**g**) and CrHApAx (**h**) samples. High-resolution XPS spectra of P 2p for CrHAp (**i**) and CrHApAx (**j**) samples.

The surface atomic composition of the studied thin films is shown in Table 1.

Table 1. Surface atomic composition (atomic %).

Sample \ Element	C	O	Ca	P	Cr	N	S
CrHAp	24.7	41.02	16.64	12.84	4.8	-	-
CrHApAx	27.98	46.4	11.15	9.37	4.4	0.2	0.5

The high-resolution XPS spectra of C 1s, O 1s, Ca 2p, P 2p, Cr 2p, N 1s, and S 2p for the two analyzed samples (CrHAp and CrHApAx) are presented in Figure 7c–j. The high-resolution XPS spectra of C for the CrHAp and CrHApAx samples are shown in Figure 7 c–d. Binding energy (BE) was calibrated with C–C peaks at 284.8 eV. The high-resolution XPS spectra of C1s for CrHAp (Figure 7c) and CrHApAx (Figure 7d) samples were deconvoluted into four components. The component at BE of 284.80 eV was assigned to C-C single bonds associated with contaminating C. The second component observed at BE of 286.13 eV (CrHAp) and 286.13 eV (CrHApAx) represents C-O single bonds. The third component at BE of 287.28 eV (CrHAp) and 287.08 eV (CrHApAx) comprises C=O and O-C-O double bonds. The fourth component identified at BE of 288.59 eV (CrHAp) 288.73 eV (CrHApAx) is assigned to –COOR contaminants.

The high-resolution XPS spectra of O1s for CrHAp and CrHApAx samples are shown in Figure 7e,f. The high-resolution spectrum of O 1s for the CrHAp sample shows three components (Figure 7e), while the high-resolution spectrum of O 1s for the CrHApAx sample shows four components (Figure 7f). The peak observed at BE of 531.69 eV (CrHAp) and 531.66 eV (CrHApAx) indicates the presence of oxygen in hydroxyapatite but also includes oxygen in double bonds C=O. The second component observed at BE of 532.83 eV (CrHAp) and 532.55 eV (CrHApAx) could be attributed to the O–H bonds. The third component identified at 534.02 eV (CrHAp) and 533.84 eV (CrHApAx) indicates traces of water. The fourth component observed in the high-resolution XPS spectrum of O 1s for the CrHApAx sample identified at BE of 530.00 eV most likely indicates a metal oxide.

The high-resolution XPS spectra of Ca 2p for CrHAp and CrHApAx samples after deconvolution presented the specific doublet 2p3/2 and 2p1/2 spaced at approximately 3.5–3.6 eV and with an area ratio close to 2:1. The binding energy of the doublet observed at 347.51 eV and 351.07 for the CrHAp sample can be rigorously assigned to HAp (Figure 7g). In addition, the binding energy of the doublet observed at 347.53 eV and 351.10 eV for the CrHApAx sample can be rigorously assigned to HAp (Figure 7h). The two maxima observed at BE 348.6 eV (Ca2p3/2) and 352 eV (Ca2p1/2) are spaced 3.6 eV (with an area ratio close to 2:1), which can be attributed to hydroxyapatite.

The high-resolution XPS spectra of P 2p for the CrHAp (Figure 7i) and CrHApAx (Figure 7j) samples after deconvolution showed two components. The specific 2p3/2 and 2p1/2 doublet spaced at approximately 0.85 eV and with an area ratio close to 2:1 was identified for both samples. The two components observed for the CrHAp sample (Figure 7i) were located at BE of 133.27 eV (P2p3/2) and 134.12 eV (P2p1/2). For the CrHApAx sample (Figure 7j), the two components of the high-resolution XPS spectrum of P 2p were observed at BE of 132.78 (P2p3/2) and 133.63 eV (P2p1/2). The binding energy of the two components associated with the P2p peak of the two samples can be assigned to hydroxyapatite.

The high-resolution XPS spectrum of Cr2p for the CrHAp and CrHApAx samples was fitted with the specific doublet 2p3/2 and 2p1/2 spaced at approximately 9.6 eV and with an area ratio close to 2:1 (Figure 8).

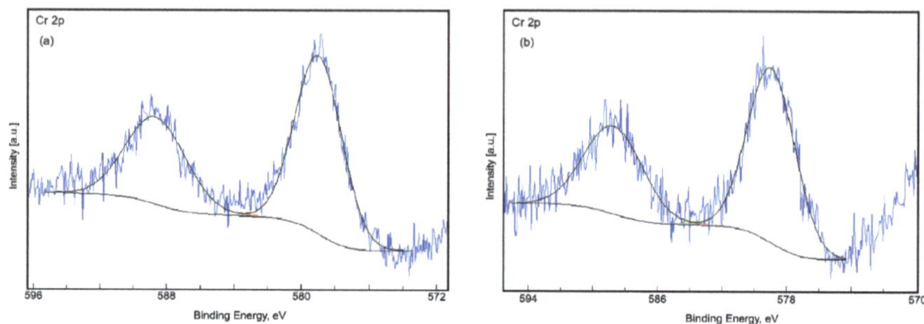

Figure 8. High-resolution XPS spectra of Cr 2p for CrHAp (**a**) and CrHApAx (**b**) samples.

The high-resolution spectra of N1s and S 2p for the CrHApAx sample are shown in Figure 9. As can be seen in Figure 9a, nitrogen is very weak. On the other hand, the high-resolution XPS spectrum of S 2p after deconvolution showed two components located at BE of 168.79 and 169.99 eV. S2p overlaps a loss of P2p. Its binding energy is specific to sulfates.

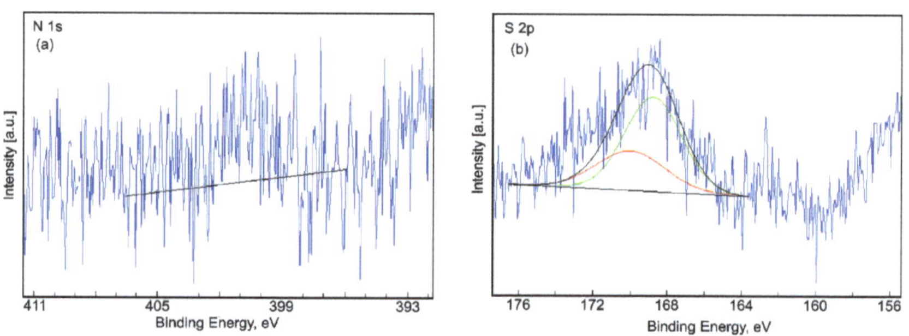

Figure 9. High-resolution XPS spectra of N1s (**a**) and S 2p (**b**) CrHApAx samples.

In Figure 10a,c, the SEM micrographs that were obtained for the CrHAp and CrHApAx samples are presented.

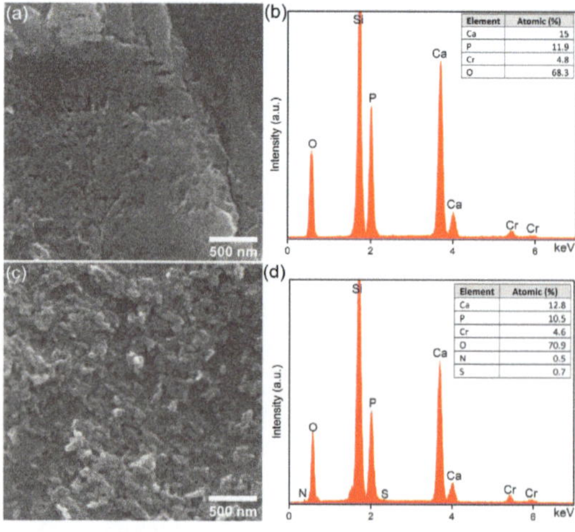

Figure 10. SEM micrograph of CrHAp (**a**) and CrHApAx (**c**). EDS spectra of CrHAp (**b**) and CrHApAx (**d**).

Both SEM micrographs reveal that the coating surfaces (CrHAp and CrHApAx) form a uniform and continuous layer, with no visible cracks or fissures. Furthermore, in the case of CrHAp coatings, it could be noticed that the surface is smoother compared with that of CrHApAx coatings. On the other hand, the SEM images obtained for the CrHApAx coatings underline the presence of a more uneven surface. These results are in agreement with the stability results obtained through ultrasound measurements. It can be seen that from very stable samples (CrHAp), a smooth surface is obtained. In the case of samples with relative stability, it is observed that more uneven layers are obtained. The unevenness of the layer may be due to agglomerations of particles (CrHApAx).

Figure 10b,d reveals the EDS spectra that belongs to the CrHAp and CrHApAx samples. The EDS spectra provide information about the coating's chemical composition and purity. Thus, in the EDS spectra of CrHAp coatings, only the presence of the line of the main chemical elements that are found in their chemical composition is highlighted (Ca, O, P, and Cr).

The presence of the N and S in the EDS spectra characteristic for the CrHApAx underlines the presence of the Ax in the layer. In addition, in Figure 10d, the line that belongs to Ca, P, Cr, and O from the CrHAp composition can be seen. The Si line appears due to the substrate on which the layers are deposited. The purity of the CrHAp and CrHApAx layers is proven by the absence of the additional lines in both EDS spectra.

Based on the results of the EDS semiquantitative analysis, the value of $(Ca + Cr)/P$ for the CrHAp sample was equal with 1.664. The value of $(Ca + Cr)/P$ determined for the CrHApAx was 1.657.

The difference observed in the chemical composition determined by XPS and EDS is due to the different methods of analysis of the two techniques. EDS is an analytical technique used for elemental analysis or chemical characterization of a sample. EDS effectively provides the "bulk" concentration of the elements present in a sample. On the other hand, XPS provides the chemical composition of the near-surface area of the sample, i.e., the elements present in the first few nm (approximately 10 nm) of the sample surface. By comparison, EDS provides information for a depth of a few μm. In the case of EDS, we can talk about the analysis of the surface and a volume below this surface. On the other hand, there is a high probability that the difference between the analyzed areas will be different for the two techniques. Another element that could contribute to these differences is the perfect calibration of the measuring instruments, which leads to different sensitivities to the same elements. The results obtained by the two techniques provide information showing the uniform distribution of chromium ions in the hydroxyapatite structure.

The results of the adherence test conducted on CrHAp and CrHApAx coatings reveal their good adherence on the Si substrate, if we take into consideration the fact that the scotch tape peeled off nearly clean, with only a negligible amount of material remaining on it. Furthermore, the results of this study showed that the best adhesion to the Si substrate was obtained for CrHAp coating (these coatings were obtained from the stable solutions).

It is well known that hydroxyapatite is composed mainly of phosphate and hydroxyl groups that are IR active and can be observed in the HAp infrared spectrum. Thus, the FTIR general spectra obtained for CrHAp and CrHApAx are revealed in Figure 11. In the inset of Figure 11b, the FTIR general spectra obtained for Ax are presented. Thus, in the inset, the peaks that are characteristic for amoxicillin (Ax) structure can be observed [27]. The peak observed around 476 cm^{-1} is attributed to ν_2 vibration of phosphate groups. Meanwhile, the doublet observed around 565 and 602 cm^{-1} is attributed to the ν_4 bending mode of the phosphate group [40,47,48]. A weak band around 964 cm^{-1}, corresponding to the ν_1 symmetric stretching of PO_4^{3-}, is observed in the FTIR general spectra of CrHAp, confirming the presence of the hydroxyapatite in the analyzed sample. The intense band centered at 1032 cm^{-1} together with the broad band observed at around 1105 cm^{-1} are associated with ν_3 vibrations of the phosphate group in HAp [40,47,48]. Additionally, the shoulder observed at around 850 cm^{-1} may result from the carbonate group (ν_2 vibrations) present in the CrHAp [40,47,48]. Usually, the broad band that appears between 3200 and 3600 cm^{-1} belongs to the hydroxyl groups vibration [40,47,48]. More than that, in the CrHApAx spectra, it could be observed that the presence of the Ax in the samples induces a slight shift of the vibrational band's positions. Another effect of the addition of the Ax is represented by the broadening of the specific FTIR maxima accompanied by a slight decrease in the intensity of the CrHApAx maxima compared to the CrHAp maxima. Among the peaks associated with the phosphate and hydroxyl groups, the presence of the peaks characteristic of the vibration of Ax could be noticed (see the inset in Figure 11b), a fact that indicates the interaction of Ax with CrHAp (Figure 11a). Furthermore, the interaction between CrHAp and Ax is also proven by the peak shift, the appearance of the new peaks, and the peak broadening observed in the case of CrHApAx (Figure 11a) compared with CrHAp (Figure 11a) and Ax (inset Figure 11b) samples [27]. In the FTIR spectra of CrHApAx, the peaks that appears around 1772 cm^{-1} correspond to the νC=O (β-lactamic ring) from amoxicillin (Ax) structure [27]. According to our previous study, the presence of the Ax in the CrHApAx sample is also underlined by the presence of the weak maxima in the 1650–1800 cm^{-1} spectral domain. These maxima arise due to the vibration of carbonyl (-C=O) functional groups from Ax [27].

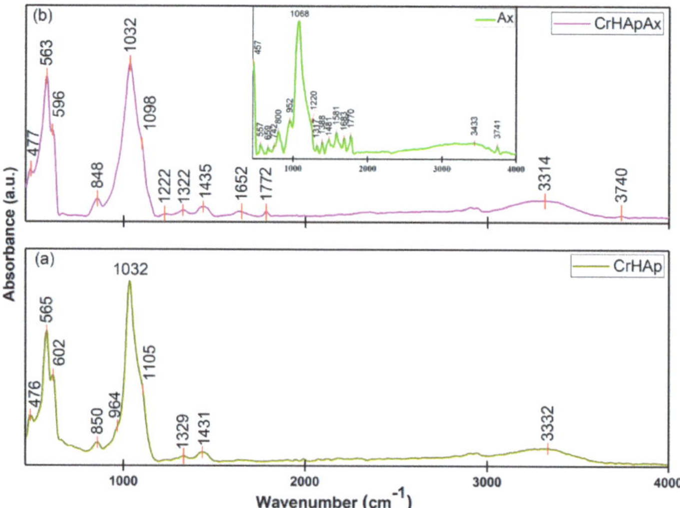

Figure 11. FTIR general spectra obtained for CrHAp (**a**) and CrHApAx (**b**). In the inset, the FTIR general spectra of Ax are presented.

Valuable data regarding the subtle spectral changes resulting from the addition of Ax to the CrHAp were assessed by performing second derivative analysis of the FTIR spectra in the regions of 450–700 cm^{-1} (where the vibration specific to ν_2 and ν_4 of the phosphate groups appears) and 900–1200 cm^{-1} (this region is characteristic to ν_1 and ν_3 vibration of the phosphate groups). The results of second derivative analysis are presented in Figure 12a,b (for the CrHAp) and in Figure 12c,d (for the CrHApAx).

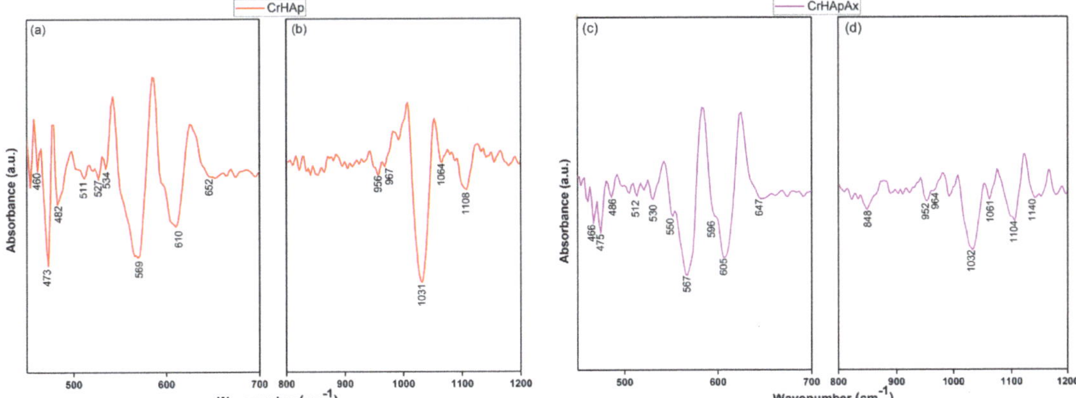

Figure 12. CrHAp-FTIR second derivative curve obtained for 450–700 cm^{-1} (**a**) and 800–1200 cm^{-1} (**b**) spectral domains. CrHApAx-FTIR second derivative curve obtained for 450–700 cm^{-1} (**c**) and 800–1200 cm^{-1} (**d**) spectral domains.

Firstly, both second derivative spectra underline the presence of the ν_1 vibration of the phosphate groups at around 964 cm^{-1} [40,47,48]. The presence of this maxima indicates the presence in both analyzed samples. On the other hand, the peaks that appear between 460 cm^{-1} and 485 cm^{-1} are characteristic of the ν_2 bending mode of the phosphate groups [40,47,48]. Between 500 cm^{-1} and 610 cm^{-1} are observed the maxima that are specific to the ν_4 bending mode of the phosphate groups [47,48]. The intense bands observed in both second derivative spectra between 1000 and 1200 cm^{-1} could be atributed to the ν_3 asymmetric stretching vibration of phosphate groups from HAp [48]. Moreover, in Figure 12, no additional vibrational bands are observed that would suggest the presence of impurities in the analyzed samples.

In order to better highlight the presence of vibration bands that are overlapped in the general FTIR spectra of the two samples, the spectra were deconvoluted and analyzed in a spectral domain in which both maxima associated with carbonate and phosphate groups are found. Thus, in Figure 13, the deconvoluted FTIR spectra obtained in the 800–1200 cm^{-1} spectral region are revealed, which are characteristic of the ν_1 and ν_3 vibration of the phosphate functional groups from the HAp structure and of the ν_2 vibration of carbonate groups. Thus, to achieve a satisfactory fit for the CrHAp sample, ten components were used. Moreover, nine components were used to obtain a good fit for the CrHApAx sample. In the case of the CrHApAx, the deconvoluted FTIR spectra also underlined the decrease in the intensity of the band centred at 1032 cm^{-1} (ν_3 vibration of the phosphate) together with the presence of a weak band at 964 cm^{-1} that could be attributed to the ν_1 vibration of the phosphate from HAp.

Therefore, FTIR studies highlight a synergy between Cr doping and antibiotic enrichment of HAp, also confirming the presence of HAp and antibiotic in the analyzed samples. Our results are in agreement with the studies previously reported by Manoj, M. et al. [47], Antonakos, A., et al. [48], and Cimpeanu C., et al. [27].

Atomic force microscopy (AFM) was used to obtain information about the surface topography of the CrHAp and CrHApAx coatings. The AFM findings, including both 2D and 3D representations, are depicted in Figure 14a–d.

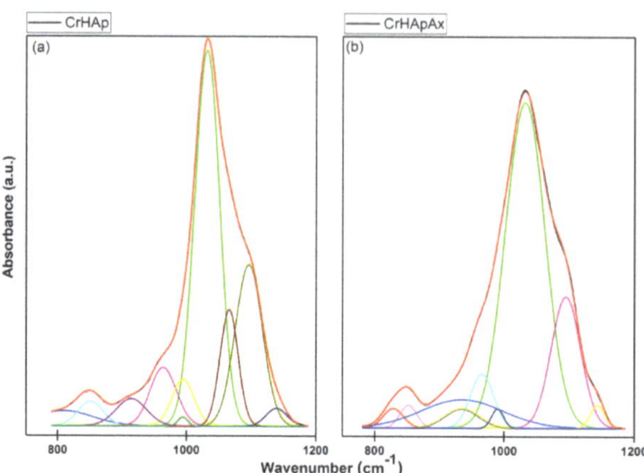

Figure 13. Deconvoluted FTIR spectra of the CrHAp and CrHApAx obtained in the 800–1200 cm^{-1} spectral region.

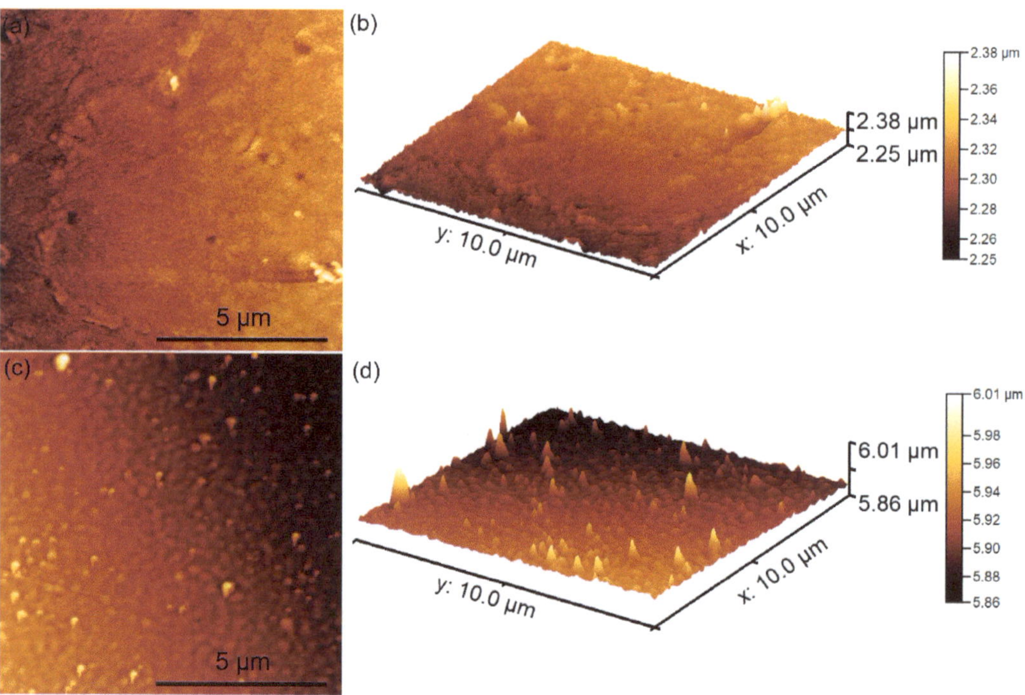

Figure 14. 2D topography of the CrHAp (**a**) and CrHApAx (**c**) coatings' surface recorded on an area of 10×10 μm^2 and their 3D representations (**b**,**d**).

The 2D AFM micrographs of CrHAp coatings illustrated in Figure 14a reveal the presence of a uniform and continuous deposited coating characterized by well-distributed nanoaggregates across the surface. The 2D surface topography, as well as its 3D representation (Figure 14a,b), also indicated the absence of significant cracks or fissures on the surface of the CrHAp coatings. Furthermore, the roughness parameter (R_{RMS}) determined from the AFM micrographs for the CrHAp coatings was measured to be 33.98 nm. The 2D AFM micrograph of CrHApAx and its 3D representation are depicted in Figure 14c,d. The AFM analysis confirms that the coating exhibits a uniform and continuous surface morphology and highlights that there is no evidence of significant cracks or fissures on the surface of the CrHApAx coating. The results of the AFM analysis also suggested that the coating exhibits a homogeneous distribution of nanoaggregates across its surface. The roughness parameter (R_{RMS}) for the CrHApAx surface was also determined from the AFM analysis. The value obtained for the roughness parameter RRMS was found to be 41.82 nm.

The cytotoxicity of CrHAp and CrHApAx coatings was evaluated using the MTT colorimetric assay and MG63 cell line. The osteosarcoma MG63 cells are widely chosen in studies regarding a material's biocompatibility due to their close resemblance to human osteoblasts. Osteoblasts are the cells responsible for bone-forming and are also involved in the synthesis of bone matrix and mineralization regulating. Because they are derived from human osteosarcoma, MG63 cells retain the characteristic osteoblastic properties, such as collagen type I production, alkaline phosphatase (ALP), and osteocalcin, which are known to be important markers for bone formation. These properties make MG63 cells highly relevant in evaluating how biomaterials interact with bone tissue. Furthermore, the MG63 cells exhibit organized behavior and also predictable responses to external stimuli, which allows researchers to accurately evaluate cell adhesion, proliferation, and differentiation on biomaterial surfaces. Their high sensitivity to even slight surface modifications, chemical composition, and mechanical properties helps gather important information regarding cytotoxicity and biocompatibility, which is crucial for evaluating new biomaterials. Moreover, compared to primary osteoblasts, which have often proven difficult to isolate and maintain in vitro, the MG63 cells are easier to cultivate and can also proliferate rapidly, thus making them a cost-effective and efficient option for long-term studies. Additionally, because they are a human-derived cell line, MG63 cells provide results that are more relevant to human biology than animal models, improving the translational potential of in vitro findings for clinical applications. Their extensive use in biocompatibility research has contributed greatly to facilitating comparisons and validation of new biomaterials. These combined advantages make MG63 cells indispensable for developing innovative materials aimed at promoting bone regeneration and integration.

The cell viability of MG63 cells incubated with CrHAp and CrHApAx coatings was measured after 24, 48, and 72 h and expressed as mean $\pm$ standard deviation (SD) relative to a control (100% viability). The results are presented in Figure 15. The results of the MTT assay revealed that both CrHAp and CrHApAx coatings demonstrated good biocompatibility towards MG63 cells. The cell viability determined for CrHAp coatings exceeded 90% for all tested intervals. The results showed a notable increase in cell viability to 92% and 96% after 48 and 72 h, respectively, indicating that they could be considered a good surface for the proliferation of MG63 cells. The results of the MTT studies for the CrHApAx coatings also showed good biocompatibility, having a cell viability above 88% across all the tested intervals, meeting ISO 10993–1:2018 standards. As defined by the ISO 10993–1: 2018 standard [49–51], the biocompatibility of a material refers to its ability to have an appropriate host response for a specific application. A cell viability rate above 88% is regarded as biocompatible, indicating a strong compatibility with living cells, tissues, or organisms. In this context, the cell viability represents the percentage of cells that remain alive and functional after exposure to the tested material. A value above 88% suggests the material does not exhibit significant cytotoxic effects, such as damaging cells or inducing apoptosis. High viability rates above 88% also imply that the material does not provoke adverse biological responses, including inflammation, toxicity, or immune rejection. These factors are critical in determining whether a material can be safely integrated and perform effectively within a biological system. More than that, a slight increase in cell viability was observed over time, reaching 93% after 72 h of incubation, which indicates a sustained compatibility with MG63 cells. The results of the MTT assays highlighted that the presence of Ax (amoxicillin) in CrHAp did not significantly impact cell viability, suggesting its potential for combining antibacterial and osteo-regenerative properties in biomedical applications.

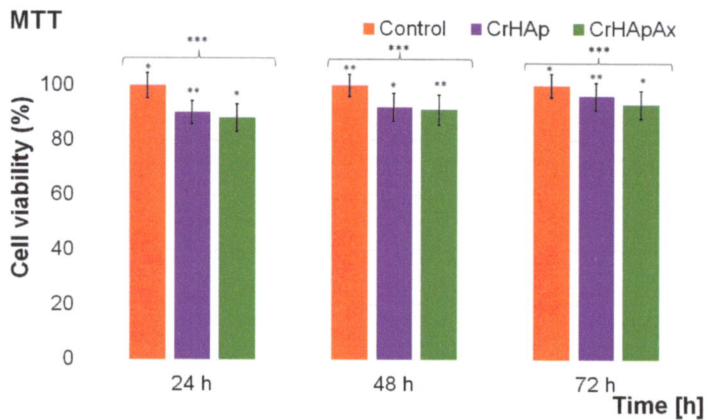

Figure 15. Graphical representation of the cell viability MG63 cells exposed to CrHAp and CrHApAx coatings for 24, 48, and 72 h. The results are depicted as the mean ± standard deviation (SD) and quantified as percentages of the control (100% viability). Statistical analysis was performed by one-way ANOVA. The *p*-values indicated are the following: * $p \leq 0.05$, ** $p \leq 0.005$, *** $p \leq 0.001$.

Additional information about the biological properties of CrHAp and CrHApAx coatings was obtained with the aid of metallographic microscopy. Metallographic microscopy (MM) was used to evaluate the MG63 cells' adherence and development on the surface of CrHAp and CrHApAx coatings. The results of the metallographic microscopy observation are depicted in Figure 16. The MM images indicated that the surfaces of both CrHAp and CrHApAx coatings facilitated the MG63 cell adhesion and development. Moreover, no morphological abnormalities were observed in the MG63 cells that adhered to the coating surfaces. These results are in good agreement with reported literature data [6,28,52–56] and demonstrate that the viability of MG63 cells increases with the increase of incubation time (from 24 h to 72 h) on the CrHAp and CrHApAx coatings. This enhancement in cell viability over time highlights the biocompatibility and bioactivity of the coatings, along with their capacity to support cell adhesion, proliferation, and differentiation. These findings align with existing research on hydroxyapatite-based materials, which are known for their bioactive properties and compatibility with bone tissue [57–59]. The results also emphasize the potential of these coatings for use in bone regeneration and repair applications, as its non-toxic nature and ability to sustain high cell viability make it a suitable candidate for integration into biomedical devices or implants. Furthermore, the consistent biocompatibility of CrHAp across all tested intervals reinforces its role as a material that interacts positively with living cells, without inducing cytotoxic effects or inflammatory responses. This makes CrHAp a promising material for applications requiring long-term compatibility with biological systems, particularly in the field of osteo-regenerative medicine. Chromium-substituted hydroxyapatite (CrHAp) has garnered interest in the field of biomaterials due to its potential applications in bone tissue engineering and regenerative medicine. However, the biocompatibility and cytotoxicity of CrHAp, particularly on human osteosarcoma cell lines like MG63 cells, remain critical concerns that must be thoroughly evaluated. While information about the specific mechanisms influencing cell adhesion and proliferation is still scarce in the literature, the results of our study reflect an overall positive interaction between the cells and CrHAp and CrHApAx coatings. The toxicity of CrHAp on MG63 cells is a significant concern for its use in biomedical applications. While low concentrations of Cr^{3+} may have beneficial effects, high levels of chromium, especially in the hexavalent state, pose substantial risks. Further studies are needed to fine-tune the composition of CrHAp and develop strategies to mitigate its cytotoxicity while leveraging its potential benefits for bone tissue engineering [56,60].

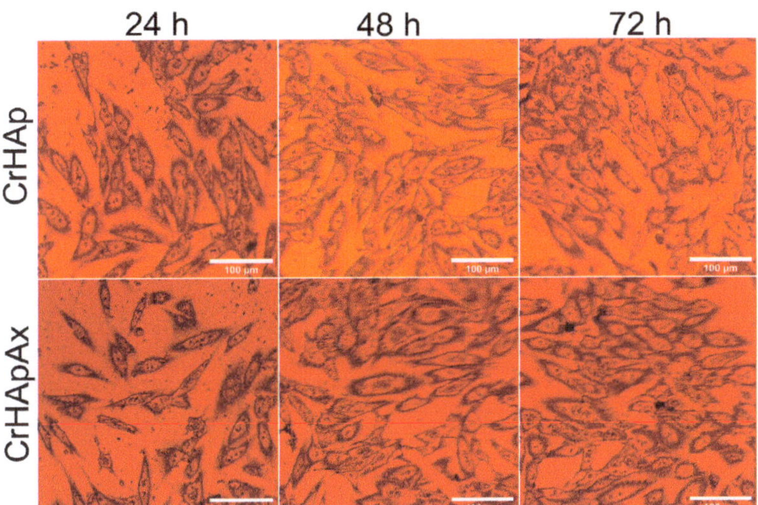

Figure 16. The morphology of MG63 cells grown on the CrHAp coatings (**a**) and CrHApAx coatings (**b**) visualized by metallographic microscopy at different time intervals.

To obtain a more comprehensive understanding of the biological properties of CrHAp and CrHApAx coatings, their antibacterial activity was studied against *Pseudomonas aeruginosa*, a gram-negative bacterial strain that is commonly associated with infections in the blood, lungs, and other body parts that are difficult to treat. The antibacterial activity of CrHAp, CrHApAx, and Ax coatings against *Pseudomonas aeruginosa* was evaluated in vitro at three different time intervals. The results of the antibacterial assay are presented graphically as mean $\pm$ SD in Figure 17.

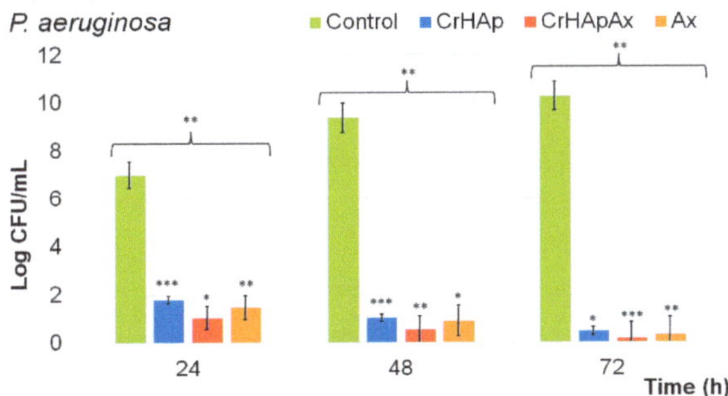

Figure 17. Graphical representation of the log colony forming units (CFUs)/mL of the CrHAp, CrHApAx, and Ax coatings incubated with *Pseudomonas aeruginosa* 27853 ATCC for 24, 48, and 72 h. The statistical analysis was performed using ordinary one-way ANOVA. The *p*-values indicated are * $p \leq 0.001$, ** $p \leq 0.005$, and *** $p \leq 0.0001$.

The results revealed a significant decrease in colony-forming units (CFUs) of *P. aeruginosa* after 24, 48, and 72 h of exposure to the CrHAp and CrHApAx coatings. The quantitative data demonstrated that CrHApAx coatings were particularly effective, showing a more significant decrease in CFUs compared to both the control and CrHAp and Ax coatings alone. Recently, *Pseudomonas aeruginosa* has been extensively studied, as this opportunistic pathogen is a leading cause of healthcare-associated infections and exhibits remarkable resistance to antibiotics. The results of the quantitative antibacterial assays demonstrate that the enhanced antibacterial effects of CrHApAx are attributed to the presence of both chromium ions and amoxicillin. This could be attributed to the fact that the CrHApAx coatings exhibited the best inhibitory effects against *P. aeruginosa* compared to both CrHAp and Ax coatings. Amoxicillin is well known as a broad-spectrum β-lactam antibiotic that has the ability to target the synthesis of the bacterial cell wall, which is an important process for bacterial growth and survival. This way, it helps inhibit the formation of a stable cell wall structure, compromising its integrity and leading to cell lysis by creating an osmotic imbalance. While amoxicillin is highly effective against numerous gram-positive bacteria, its activity against *Pseudomonas aeruginosa* is limited due to the fact that this bacterium has a unique outer membrane structure, which has the ability to reduce antibiotic penetration. Moreover, *P. aeruginosa* produces efflux pumps and possesses an intrinsic resistance mechanism that further hinder the antibiotic's access to its target sites [61–63]. One of the most significant challenges in using amoxicillin against *P. aeruginosa* is considered to be the production of β-lactamases enzymes that have the role of hydrolyzing the β-lactam ring of amoxicillin and rendering it inactive. These enzymes are abundant in *P. aeruginosa*, making this bacterium highly resistant to conventional antibiotics. However, when delivered through surface coatings or in combination with other β-lactamase inhibitors, its efficacy could be improved. The disruption of bacterial growth in such cases highlights the potential of amoxicillin in specialized applications against *P. aeruginosa*, despite the bacterium's formidable defense mechanisms [61–63].

The results of the in vitro antibacterial activity revealed the ability of CrHAp, Ax, and CrHApAx coatings to reduce *P. aeruginosa* CFU counts, highlighting their significant potential as antibacterial agents for infection control in healthcare. In addition, CrHAp provide an intrinsic antibacterial activity attributed to the presence of chromium ions, while Ax exhibits direct antibacterial effects due to the antibiotic's ability to interfere with the bacterial cell wall synthesis. The comparative antibacterial studies showed that CrHApAx coatings demonstrate the greatest CFU reduction, followed by Ax and CrHAp individually.

Therefore, the enhanced antibacterial effects of CrHApAx coatings resulted from a synergistic effect of both CrHAp and Ax, combining the sustained release of amoxicillin with the bioactive properties of CrHAp. Additionally, the results emphasized that CrHApAx can be successfully used for medical devices, implants, and hospital surfaces to inhibit bacterial colonization and biofilm formation more effectively than single-component coatings.

Furthermore, the adhesion and development of *P. aeruginosa* cells on the surfaces of CrHAp and CrHApAx coatings were studied using AFM. These studies aimed to highlight the role of chromium ions and amoxicillin in inhibiting the bacterial growth and adhesion of *P. aeruginosa* on the surfaces of CrHAp and CrHApAx coatings. The AFM topographies of the coatings were recorded after incubation with *P. aeruginosa* bacterial suspensions at different intervals (24, 48, and 72 h) under ambient conditions and room temperature. The 2D surface topographies were captured in non-contact mode over an area of 10×10 μm^2. The 2D AFM topographies of the CrHAp and CrHApAx coatings, as well as 3D representations of the coatings after three different incubation periods with *P. aeruginosa*, are presented in Figures 18 and 19.

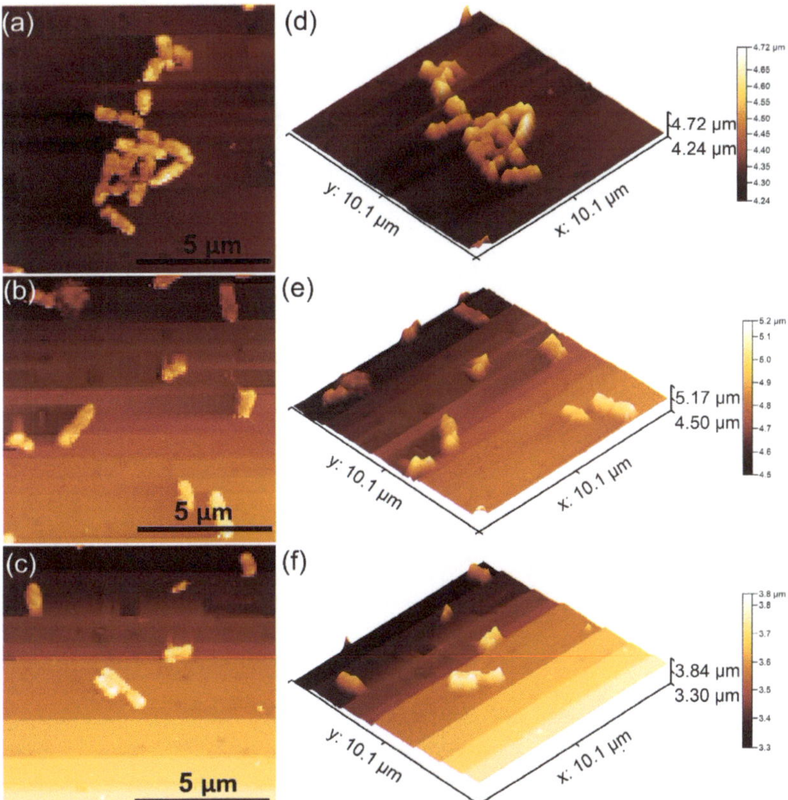

Figure 18. Two-dimensional AFM topography of *Pseudomonas aeruginosa* 27853 ATCC cells attached to the surface of the CrHAp coatings after a 24 (**a**), 48 (**b**), and 72 h (**c**) incubation period and their 3D representation (**d–f**).

The AFM topography data demonstrated that both CrHAp and CrHApAx coatings inhibited the adherence and growth of *P. aeruginosa* cells, even during the early development stages. More than that, the AFM data highlighted that the coatings prevented the formation of *P. aeruginosa* biofilms on their surfaces. The adhered bacterial cells retained their characteristic rod-shaped morphology, with lengths of 0.98–2.15 μm and widths of 0.55–0.72 μm. Furthermore, the 2D AFM images showed that the CrHApAx coatings exhibited enhanced antibacterial activity compared to CrHAp alone, suggesting a synergistic effect between chromium ions and amoxicillin. The AFM results emphasized that there was a significant reduction in *P. aeruginosa* adherence within the first 24 h of incubation, with further decrease in bacterial cell attachment over time. After 72 h, only some isolated bacterial cells were observed on the surfaces of the coatings, as emphasized by both 2D AFM micrographs and their 3D representations. These findings showed that the coatings successfully inhibited bacterial colonization and biofilm formation over extended periods. The antibacterial mechanisms of chromium ions and amoxicillin are hypothesized to involve several distinct but interconnected processes that lead to bacterial cell death [64–70].

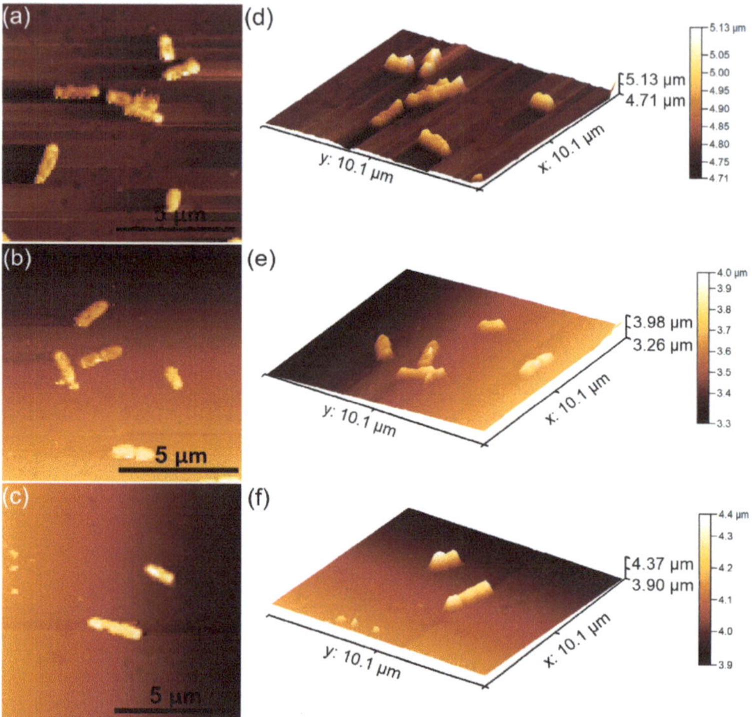

Figure 19. Two-dimensional AFM topography of *Pseudomonas aeruginosa* 27853 ATCC cells attached to the surface of the CrHApAx coatings after a 24 (**a**), 48 (**b**), and 72 h (**c**) incubation period and their 3D representation (**d–f**).

One of the key mechanisms is their ability to disrupt the integrity of the bacterial cell membranes. This interference has the ability to compromise the membrane permeability, provoking an uncontrolled leakage of the vital cellular components like ions, proteins, and nucleotides. This loss weakens the cell's structural integrity and homeostasis and leads to cell lysis. Moreover, besides targeting the bacterial membrane, both chromium ions and amoxicillin are believed to be able to interfere with the fundamental cellular processes such as protein synthesis and DNA replication. By disrupting these essential processes, the agents impair the bacterium's ability to grow and divide, effectively stopping its proliferation. Chromium ions are also reported to exhibit their antibacterial effects through the generation of reactive oxygen species (ROS). These highly reactive molecules can produce damage to the cellular components (lipids, proteins, and DNA), causing oxidative stress and compromising the bacterium's viability. However, the exact pathways and efficacy of Cr^{3+} ions as antibacterial agents require further investigation to be fully understood [56,60,68–71]. Additionally, amoxicillin, which is a beta-lactam antibiotic, exhibits its antimicrobial effects by disrupting bacterial cell wall synthesis. It achieves this by binding to and inhibiting penicillin-binding proteins (PBPs), which play a crucial role in cross-linking peptidoglycan chains—an essential component of bacterial cell walls. This inhibition compromises the integrity of the cell wall, rendering bacteria vulnerable to osmotic pressure and ultimately leading to cell lysis and death. Amoxicillin is renowned for its broad-spectrum activity, being particularly effective against a range of gram-positive and certain gram-negative bacteria, which makes it a widely prescribed antibiotic for treating various bacterial infections [64–70]. The combined synergistic impact of membrane disruption, inhibition of critical biosynthetic processes, and oxidative damage underscores the complex nature of the antibacterial action of chromium ions and amoxicillin. Together, these mechanisms contribute to their effectiveness in combating bacterial infections. These mechanisms align with previously reported studies and provide insightful information for the development of novel antibacterial agents.

This study highlighted the importance of stability in obtaining coatings with potential uses in the medical field. In the case of stable suspensions, the particles are uniformly distributed in the liquid medium. This uniformity is essential to create a consistent and defect-free coating. On the other hand, in stable suspensions, the particles remain suspended and do not settle at the bottom, resulting in uniform layer deposition. This ensures the uniformity of the layer by a uniform deposition of the suspension. Furthermore, stable suspensions prevent the particles from agglomerating, thus avoiding defects that could occur during the coating process, which can cause defects in the coating. In addition, well-dispersed particles can form stronger bonds with the surface, which leads to a more durable coating. The ultrasonic measurements allowed us to evaluate the stability of the two concentrated suspensions. The results of the physicochemical and biological studies align with the stability of the suspensions used to obtain the coatings. SEM and AFM studies showed that the surface of the CrHAp coatings is smoother compared to the CrHApAx coatings. The more uneven surface of the CrHApAx coatings could be attributed to particle agglomerations in the suspension, leading to uneven regions during deposition. It was also observed that coatings resulting from very stable suspensions (CrHAp) have better adhesion. Moreover, the biological studies emphasized the impact of the solution stability on the obtained coatings. Thus, a better cell viability was observed in the case of the CrHAp coatings. Consequently, it can be said that by ensuring the stability of the suspension, high-quality coatings can be obtained that adhere well to the substrate and provide the desired properties.

Therefore, future research should be focused on determining the optimum ratio of chromium ions to amoxicillin to maximize the coating's antibacterial efficacy while maintaining its stability and bioavailability. Nonetheless, the preliminary results obtained in this study emphasized that CrHApAx coatings represent a promising strategy for reducing bacterial contamination and hospital-related infections, having a great potential to improve patient outcomes in clinical practice.

4. Conclusions

The coatings obtained by dip coating technique starting from CrHAp and CrHApAx suspensions showed good stoichiometry. XPS results showed the presence of C, O, Ca, P, and Cr in both samples. In addition, in the CrHApAx sample, the presence of N and S, which are specific to amoxicillin, was shown. On the other hand, the present study demonstrated that the stability of the suspensions plays a very important role in obtaining the coatings. Hydroxyapatite phase as the main crystalline phase was identified by XRD studies in both samples. The amoxicillin-specific peak was also observed in the XRD studies of the CrHApAx sample.

The presence of the functional groups of hydroxyapatite and amoxicillin in the CrHAp and CrHApAx layers was confirmed by the results of the FTIR studies. The results of the SEM analysis revealed the influence of the gels' stability on the layers' surface morphology. On the other hand, it was noticed that the dip coating deposition technique allows the development of continuous layers. The results of the biological assays demonstrated that CrHAp and CrHApAx have potential as biomaterials, combining biocompatibility and antibacterial properties. The MTT assay on MG63 cells highlighted their good biocompatibility, with a high percentage of viable cells, indicating that these materials could support cell proliferation and do not exhibit cytotoxic effects towards MG63 cells. These findings attest their suitability for use in applications for bone tissue engineering and implants, where compatibility with human osteoblast-like cells is critical. Additionally, the in vitro antibacterial activity against *Pseudomonas aeruginosa* has been verified. The results of the in vitro antibacterial assays highlighted an excellent antibacterial activity. More than that, the results showed that the CrHApAx exhibited a higher inhibitory effect against *P. aeruginosa* than CrHAp. Furthermore, the AFM studies revealed the strong antibacterial activity of these materials. This activity is particularly important for preventing infections in biomedical settings, including for implantable devices and wound-healing applications. Together, these results demonstrate that CrHAp and CrHApAx are promising candidates for biomedical use, effectively combining the ability to support healthy cell growth with the ability to combat bacterial pathogens. The present study demonstrated that coatings with physicochemical properties similar to pure hydroxyapatite with high biological properties can be obtained by dip coating technique starting from stable suspensions.

Author Contributions: C.S.C.: conceptualization, methodology, software, validation, formal analysis, investigation, data curation, writing—original draft preparation, writing—review and editing, visualization. D.P.: conceptualization, software, validation, methodology, data curation, formal analysis, investigation, resources, writing—original draft preparation, writing—review and editing, visualization; supervision, project administration, funding acquisition. S.L.I.: methodology, software, validation, formal analysis, investigation, data curation, writing—original draft preparation, writing—review and editing, visualization. K.R.: validation, formal analysis, investigation, writing—original draft preparation, writing—review and editing, visualization. S.R.: validation, formal analysis, investigation, writing—original draft preparation, writing—review and editing, visualization. C.C.N.: validation, formal analysis, investigation, writing—review and editing, visualization. L.G.: validation., formal analysis, investigation, writing—review and editing, visualization. C.B.: validation, formal analysis, investigation, writing—review and editing, visualization. M.V.P.: conceptualization, software, validation, data curation, methodology, formal analysis, investigation, resources, writing—original draft preparation, writing—review and editing, visualization, supervision, project administration, funding acquisition. All authors have read and agreed to the published version of the manuscript.

Funding: This work is funded by the Core Program of the National Institute of Materials Physics, granted by the Romanian Ministry of Research, Innovation and Digitalization through the Project PC1-PN23080101.

Institutional Review Board Statement: Not applicable.

Informed Consent Statement: Not applicable.

Data Availability Statement: The original contributions presented in the study are included in the article. Further inquiries can be directed to the corresponding author/s.

Conflicts of Interest: The authors declare no conflicts of interest. The funders had no role in the design of the study; in the collection, analyses, or interpretation of data; in the writing of the manuscript; or in the decision to publish the results.

References

1. Mohd Pu'ad, N.A.S.; Koshy, P.; Abdullah, H.; Idris, M.I.; Lee, T. Syntheses of hydroxyapatite from natural sources. *Heliyon* **2019**, *5*, e01588. [CrossRef] [PubMed]
2. Gronwald, B.; Kozłowska, L.; Kijak, K.; Lietz-Kijak, D.; Skomro, P.; Gronwald, K.; Gronwald, H. Nanoparticles in Dentistry—Current Literature Review. *Coatings* **2023**, *13*, 102. [CrossRef]
3. Anjaneyulu, U.; Pattanayak, D.K.; Vijayalakshmi, U. Snail Shell Derived Natural Hydroxyapatite: Effects on NIH-3T3 Cells for Orthopedic Applications. *Mater. Manuf. Process.* **2016**, *31*, 206–221. [CrossRef]

4.	Alotaibi, N.H.; Munir, M.U.; Alruwaili, N.K.; Alharbi, K.S.; Ihsan, A.; Almurshedi, A.S.; Khan, I.U.; Bukhari, S.N.; Rehman, M.; Ahmad, N. Synthesis and characterization of antibiotic–loaded biodegradable citrate functionalized mesoporous hydroxyapatite nanocarriers as an alternative treatment for bone infections. *Pharmaceutics* **2022**, *14*, 975. [CrossRef]

5.	Alhasan, H.S.; Yasin, S.A.; Alahmadi, N.; Alkhawaldeh, A.K. The Application of Hydroxyapatite NPs for Adsorption Antibiotic from Aqueous Solutions: Kinetic, Thermodynamic, and Isotherm Studies. *Processes* **2023**, *11*, 749. [CrossRef]

6.	Bandgar, S.S.; Yadav, H.M.; Shirguppikar, S.S.; Shinde, M.A.; Shejawal, R.V.; Kolekar, T.V.; Bamane, S.R. Enhanced Hemolytic Biocompatibility of Hydroxyapatite by Chromium (Cr^{3+}) Doping in Hydroxyapatite Nanoparticles Synthesized by Solution Combustion Method. *J. Korean Ceram. Soc.* **2017**, *54*, 158–166. [CrossRef]

7.	Lima, T.A.R.M.; Brito, N.S.; Peixoto, J.A.; Valerio, M.E.G. The Incorporation of Chromium (III) into Hydroxyapatite Crystals. *Mater. Lett.* **2015**, *140*, 187–191. [CrossRef]

8.	Tautkus, S.; Ishikawa, K.; Ramanauskas, R.; Kareiva, A. Zinc and Chromium Co-Doped Calcium Hydroxyapatite: Sol-Gel Synthesis, Characterization, Behaviour in Simulated Body Fluid and Phase Transformations. *J. Solid State Chem.* **2020**, *284*, 121202. [CrossRef]

9.	Dorozhkin, S.V. A Detailed History of Calcium Orthophosphates from 1770s till 1950. *Mater. Sci. Eng. C* **2013**, *33*, 3085–3110. [CrossRef]

10.	Tanaskovic, D.; Jokic, B.; Socol, G.; Popescu, A.; Mihailescu, I.N.; Petrovic, R.; Janackovic, D. Synthesis of Functionally Graded Bioactive Glass-Apatite Multistructures on Ti Substrates by Pulsed Laser Deposition. *Appl. Surf. Sci.* **2007**, *254*, 1279–1282. [CrossRef]

11.	López, E.O.; Rossi, A.L.; Bernardo, P.L.; Freitas, R.O.; Mello, A.; Rossi, A.M. Multiscale connections between morphology and chemistry in crystalline, zinc-substituted hydroxyapatite nanofilms designed for biomedical applications. *Ceram. Intern.* **2019**, *45*, 793–804. [CrossRef]

12.	Hidalgo-Robatto, B.M.; López-Álvarez, M.; Azevedo, A.S.; Dorado, J.; Serra, J.; Azevedo, N.F.; González, P. Pulsed laser deposition of copper and zinc doped hydroxyapatite coatings for biomedical applications. *Surf. Coat. Technol.* **2018**, *333*, 168–177. [CrossRef]

13.	Gergeroglu, H.; Ebeoglugil, M.F.; Bayrak, S.; Aksu, D.; Azar, Y.T. Systematic investigation and controlled synthesis of Ag/Ti co-doped hydroxyapatite for bone tissue engineering. *Mater. Today Chem.* **2024**, *39*, 102175. [CrossRef]

14.	Uysal, I.; Yilmaz, B.; Evis, Z. Zn-doped hydroxyapatite in biomedical applications. *J. Aust. Ceram. Soc.* **2021**, *57*, 869–897. [CrossRef]

15.	Smiciklas, I.; Onjia, A.; Markovic, J.; Raicevic, S. Comparison of Hydroxyapatite Sorption Properties towards Cadmium, Lead, Zinc and Strontium Ions. *Mater. Sci. Forum* **2005**, *494*, 405–410. [CrossRef]

16.	LeGeros, R.Z. Calcium Phosphate-Based Osteoinductive Materials. *Chem. Rev.* **2008**, *108*, 4742–4753. [CrossRef]

17.	Niinomi, M.; Nakai, M.; Hieda, J. Development of New Metallic Alloys for Biomedical Applications. *Acta Biomater.* **2012**, *8*, 3888–3903. [CrossRef] [PubMed]

18.	Prasanna, A.P.S.; Venkatasubbu, G.D. Sustained Release of Amoxicillin from Hydroxyapatite Nanocomposite for Bone Infections. *Prog. Biomater.* **2018**, *7*, 289–296. [CrossRef] [PubMed]

19.	Rentsch, B.; Bernhardt, A.; Henß, A.; Ray, S.; Rentsch, C.; Schamel, M.; Gbureck, U.; Gelinsky, M.; Rammelt, S.; Lode, A. Trivalent Chromium Incorporated in a Crystalline Calcium Phosphate Matrix Accelerates Materials Degradation and Bone Formation in Vivo. *Acta Biomater.* **2018**, *69*, 332–341. [CrossRef]

20.	Collery, P.; Maymard, Y.; Theophanides, T.; Khassanova, L.; Collery, T. *Metal Ions in Biology*; J. John Libbey: New Barnet, UK, 2008; Volume 10, pp. 739–742.

21.	Mabilleau, G.; Filmon, R.; Petrov, P.K.; Baslé, M.F.; Sabokbar, A.; Chappard, D. Cobalt, Chromium and Nickel Affect Hydroxyapatite Crystal Growth in vitro. *Acta Biomater.* **2010**, *6*, 1555–1560. [CrossRef]

22.	de Araujo, T.S.; Macedo, Z.S.; de Oliveira, P.A.; Valerio, M.E. Production and Characterization of Pure and Cr^{3+}-Doped Hydroxyapatite for Biomedical Applications as Fluorescent Probes. *J. Mater. Sci.* **2007**, *42*, 2236–2243. [CrossRef]

23.	Yadav, H.M.; Kolekar, T.V.; Barge, A.S.; Thorat, N.D.; Delekar, S.D.; Kim, B.M.; Kim, B.J.; Kim, J.S. Enhanced Visible Light Photocatalytic Activity of Cr3+-Doped Anatase TiO2 Nanoparticles Synthesized by Sol–Gel Method. *J. Mater. Sci. Mater. Electron.* **2016**, *27*, 526–534. [CrossRef]

24.	Devoya, J.; Gehin, A.; Muller, S.; Melczer, M.; Remy, A.; Antoine, G.; Sponne, I. Evaluation of Chromium in Red Blood Cells as an Indicator of Exposure to Hexavalent Chromium: An in vitro Study. *Toxicol. Lett.* **2016**, *255*, 63–70. [CrossRef]

25.	Balamurugan, K.; Vasant, C.; Rajaram, R.; Ramasami, T. Hydroxopentaamminechromium(III) promoted phosphorylation of bovine serum albumin: Its potential implications in understanding biotoxicity of chromium. *Biochim Biophys Acta* **1999**, *1427*, 357. [CrossRef]

26.	ATSDR. Agency for Toxic Substances and Disease Registry (2000) Toxicological Profile for Chromium. Syracuse. US Department of Health & Human Services. Available online: https://wwwn.cdc.gov/TSP/ToxProfiles/ToxProfiles.aspx?id=62&tid=17 (accessed on 30 October 2024).

27. Cimpeanu, C.; Predoi, D.; Ciobanu, C.S.; Iconaru, S.L.; Rokosz, K.; Predoi, M.V.; Raaen, S.; Badea, M.L. Development of Novel Biocomposites with Antimicrobial-Activity-Based Magnesium-Doped Hydroxyapatite with Amoxicillin. *Antibiotics* **2024**, *13*, 963. [CrossRef] [PubMed]

28. Predoi, D.; Iconaru, S.L.; Ciobanu, S.C.; Ţălu, Ş.; Predoi, S.A.; Buton, N.; Ramos, G.Q.; da Fonseca Filho, H.D.; Matos, R.S. Synthesis, Characterization, and Antifungal Properties of Chrome-Doped Hydroxyapatite Thin Films. *Mater. Chem. Phys.* **2024**, *324*, 129690. [CrossRef]

29. Predoi, D.; Iconaru, S.L.; Predoi, M.V.; Motelica-Heino, M.; Buton, N.; Megier, C. Obtaining and Characterizing Thin Layers of Magnesium Doped Hydroxyapatite by Dip Coating Procedure. *Coatings* **2020**, *10*, 510. [CrossRef]

30. Predoi, D.; Iconaru, S.L.; Predoi, M.V.; Motelica-Heino, M.; Guegan, R.; Buton, N. Evaluation of Antibacterial Activity of Zinc-Doped Hydroxyapatite Colloids and Dispersion Stability Using Ultrasounds. *Nanomaterials* **2019**, *9*, 515. [CrossRef] [PubMed]

31. Debye, P.; Scherrer, P. Interference of irregularly oriented particles in X-rays. *Phys. Zeit.* **1916**, *17*, 277–283.

32. Debye, P.; Scherrer, P. Interference on inordinate orientated particles in X-ray light. III. *Phys. Zeit.* **1917**, *18*, 291–301.

33. Danilchenko, S.N.; Kukharenko, O.G.; Moseke, C.; Protsenko, I.Y.; Sukhodub, L.F.; Sulkio-Cleff, B. Determination of the bone mineral crystallite size and lattice strain from diffraction line broadening. *Cryst. Res. Technol.* **2002**, *37*, 1234–1240. [CrossRef]

34. Richardson, J.W.; Faber, J., Jr. *Advances in X-ray Analysis*, 2nd ed.; Barrett, C.S., Cohen, J.B., Faber, J., Jr., Jenkins, R., Leyden, D.E., Russ, J.C., Predecki, P.K., Eds.; Plenum Publishing: New York, NY, USA, 1986; Volume 29.

35. Bragg, W.H.; Bragg, W.L. The Reflexion of X-rays by Crystals. *Proc. R. Soc. Lond. A.* **1913**, *88*, 428–438. [CrossRef]

36. Casa Software Ltd. CasaXPS: Processing Software for XPS, AES, SIMS and More. 2009. Available online: www.casaxps.com (accessed on 30 September 2022).

37. Biesinger, M.C.; Lau, L.W.; Gerson, A.R.; Smart, R.S.C. Resolving surface chemical states in XPS analysis of first row transition metals, oxides and hydroxides: Sc, Ti, V, Cu and Zn. *Appl. Surf. Sci.* **2010**, *257*, 887–898. [CrossRef]

38. Wagner, C.D.; Naumkin, A.V.; Kraut-Vass, A.; Allison, J.W.; Powell, C.J.; Rumble, J.R., Jr. NIST Standard Reference Database 20, Version 3.4. 2003. Available online: https://srdata.nist.gov/xps (accessed on 20 November 2024).

39. Gwyddion. Available online: http://gwyddion.net/ (accessed on 1 October 2024).

40. Iconaru, S.L.; Motelica-Heino, M.; Predoi, D. Study on Europium-Doped Hydroxyapatite Nanoparticles by Fourier Transform Infrared Spectroscopy and Their Antimicrobial Properties. *J. Spectrosc.* **2013**, *2013*, 284285. [CrossRef]

41. Iconaru, S.L.; Predoi, D.; Ciobanu, C.S.; Motelica-Heino, M.; Guegan, R.; Bleotu, C. Development of Silver Doped Hydroxyapatite Coatings for Biomedical Applications. *Coatings* **2022**, *12*, 341. [CrossRef]

42. ImageJ. Available online: http://imagej.nih.gov/ij (accessed on 20 November 2024).

43. Ciobanu, C.S.; Iconaru, S.L.; Predoi, D.; Truşcă, R.-D.; Prodan, A.M.; Groza, A.; Chifiriuc, M.C.; Beuran, M. Fabrication of Novel Chitosan–Hydroxyapatite Nanostructured Coatings for Biomedical Applications. *Coatings* **2021**, *11*, 1561. [CrossRef]

44. Kittel, C. Introduction to Solid State Physics. In *Introduction to Solid State Physics*, 7th ed.; Wiley: Brisbane, Australia, 1996.

45. Andrade, A.V.C.; da Silva, J.C.Z.; Paiva, S.C.O.; Weber, C.; Tebchernai, S.M. Synthesis and Crystal Phase Evaluation of Hydroxylapatite Using the Rietveld-Maximum Entropy Method. In Proceedings of the 28th International Conference on Advanced Ceramics and Composites B: Ceramic Engineering and Science Proceedings, Cocoa Beach, FL, USA, 24–29 January 2004; John Wiley & Sons: Hoboken, NJ, USA, 2004; Volume 24, pp. 639–645. [CrossRef]

46. Person, A.; Bocherens, H.; Saliège, J.F.; Paris, F.; Zeitoun, V.; Gérard, M. Early diage netic evolution of bone phosphate: An X-ray diffractometry analysis. *J. Archaeol. Sci.* **1995**, *22*, 211–221. [CrossRef]

47. Manoj, M.; Subbiah, R.; Mangalaraj, D.; Ponpandian, N.; Viswanathan, C.; Park, K. Influence of Growth Parameters on the Formation of Hydroxyapatite (HAp) Nanostructures and Their Cell Viability Studies. *Nanobiomedicine* **2015**, *2*. [CrossRef] [PubMed]

48. Antonakos, A.; Liarokapis, E.; Leventouri, T. Micro-Raman and FTIR Studies of Synthetic and Natural Apatites. *Biomaterials* **2007**, *28*, 3043–3054. [CrossRef] [PubMed]

49. *ISO 10993-1:2018*; Biological Evaluation of Medical Devices–Part 1: Evaluation and Testing within a Risk Management Process. ISO: Geneva, Switzerland, 2018. Available online: https://www.iso.org/standard/68936.html (accessed on 21 December 2024).

50. Wątroba, M.; Bednarczyk, W.; Szewczyk, P.K.; Kawałko, J.; Mech, K.; Grünewald, A.; Unalan, I.; Taccardi, N.; Boelter, G.; Banzhaf, M.; et al. In vitro cytocompatibility and antibacterial studies on biodegradable Zn alloys supplemented by a critical assessment of direct contact cytotoxicity assay. *J. Biomed. Mater. Res. B Appl. Biomater.* **2023**, *111*, 241–260. [CrossRef] [PubMed]

51. Sukumaran, A.; Sweety, V.K.; Vikas, B.; Joseph, B. Cytotoxicity and Cell Viability Assessment of Biomaterials. In *Cytotoxicity-Understanding Cellular Damage and Response*; Sukumaran, A., Mahmoud, A.M., Eds.; Intech Open: London, UK, 2023.

52. Khan, F.; Bai, Z.; Kelly, S.; Skidmore, B.; Dickson, C.; Nunn, A.; Rutledge-Taylor, K.; Wells, G. Effectiveness and Safety of Antibiotic Prophylaxis for Persons Exposed to Cases of Invasive Group a Streptococcal Disease: A Systematic Review. *Open Forum Infect. Dis.* **2022**, *9*, ofac244. [CrossRef] [PubMed]

53. Kaur, S.; Rao, R.; Nanda, S. Amoxicillin: A Broad-Spectrum Antibiotic. *Int. J. Pharm. Sci.* **2011**, *3*, 3.

54. Awasthi, S.; Pandey, S.K.; Arunan, E.; Srivastava, C. A Review on Hydroxyapatite Coatings for Biomedical Applications: Experimental and Theoretical Perspectives. *J. Mater. Chem. B* **2021**, *9*, 228–249. [CrossRef] [PubMed]

55. Cacciotti, I. Multisubstituted Hydroxyapatite Powders and Coatings: The Influence of the Codoping on the Hydroxyapatite Performances. *Int. J. Appl. Ceram. Technol.* **2019**, *16*, 1864–1884. [CrossRef]

56. Fu, J.; Liang, X.; Chen, Y.; Tang, L.; Zhang, Q.H.; Dong, Q. Oxidative Stress as a Component of Chromium-Induced Cytotoxicity in Rat Calvarial Osteoblasts. *Cell Biol. Toxicol.* **2008**, *24*, 201–212. [CrossRef] [PubMed]

57. Predoi, D.; Iconaru, S.L.; Predoi, M.V. Dextran-Coated Zinc-Doped Hydroxyapatite for Biomedical Applications. *Polymers* **2019**, *11*, 886. [CrossRef] [PubMed]

58. Predoi, D.; Iconaru, S.L.; Ciobanu, S.C.; Predoi, S.-A.; Buton, N.; Megier, C.; Beuran, M. Development of Iron-Doped Hydroxyapatite Coatings. *Coatings* **2021**, *11*, 186. [CrossRef]

59. Predoi, S.A.; Ciobanu, S.C.; Chifiriuc, M.C.; Motelica-Heino, M.; Predoi, D.; Iconaru, S.L. Hydroxyapatite Nanopowders for Effective Removal of Strontium Ions from Aqueous Solutions. *Materials* **2023**, *16*, 229. [CrossRef] [PubMed]

60. Nickens, K.P.; Patierno, S.R.; Ceryak, S. Chromium Genotoxicity: A Double-Edged Sword. *Chem. Biol. Interact.* **2010**, *188*, 276–288. [CrossRef] [PubMed]

61. Hancock, R.E.W.; Speert, D.P. Antibiotic resistance in *Pseudomonas aeruginosa*: Mechanisms and impact on treatment. *Drug Resist. Updat.* **2000**, *3*, 247–255. [CrossRef]

62. Poole, K. Resistance to β-lactam antibiotics in *Pseudomonas aeruginosa*. *Clin. Microbiol. Infect.* **2004**, *10*, 16–22. [CrossRef]

63. Livermore, D.M. Multiple Mechanisms of Antimicrobial Resistance in Pseudomonas aeruginosa: Our Worst Nightmare? *Clin. Infect. Dis.* **2002**, *34*, 5–634. [CrossRef] [PubMed]

64. Demurtas, M.; Perry, C.C. Facile one-pot synthesis of amoxicillin-coated gold nanoparticles and their antimicrobial activity. *Gold Bull.* **2014**, *47*, 103–107. [CrossRef]

65. Todd, P.A.; Benfield, P. Amoxicillin/Clavulanic Acid. *Drugs* **1990**, *39*, 264–307. [CrossRef]

66. Bankole, O.M.; Ojubola, K.I.; Adanlawo, O.S.; Adesina, A.O.; Lawal, I.O.; Ogunlaja, A.S.; Achadu, O.J. Amoxicillin Encapsulation on Alginate/Magnetite Composite and Its Antimicrobial Properties Against Gram-Negative and Positive Microbes. *BioNanoScience* **2022**, *12*, 1136–1149. [CrossRef]

67. Weber, D.J.; Tolkoff-Rubin, N.E.; Rubin, R.H. Amoxicillin and Potassium Clavulanate: An Antibiotic Combination Mechanism of Action, Pharmacokinetics, Antimicrobial Spectrum, Clinical Efficacy and Adverse Effects. *Pharmacotherapy* **1984**, *4*, 122–136. [CrossRef] [PubMed]

68. Güncüm, E.; Işıklan, N.; Anlaş, C.; Ünal, N.; Bulut, E.; Bakırel, T. Development and characterization of polymeric-based nanoparticles for sustained release of amoxicillin—An antimicrobial drug. *Artif. Cells Nanomed. Biotechnol.* **2018**, *46*, 964–973. [CrossRef]

69. Elabbasy, M.T.; Algahtani, F.D.; Alshammari, H.F.; Kolsi, L.; Dkhil, M.A.; Abd El-Rahman, G.I.; El-Morsy, M.A.; Menazea, A.A. Improvement of mechanical and antibacterial features of hydroxyapatite/chromium oxide/graphene oxide nanocomposite for biomedical utilizations. *Surf. Coat. Technol.* **2022**, *440*, 128476. [CrossRef]

70. Li, Y.; Ho, J.; Ooi, C.P. Antibacterial efficacy and cytotoxicity studies of copper (II) and titanium (IV) substituted hydroxyapatite nanoparticles. *Mater. Sci. Eng. C* **2010**, *30*, 1137–1144. [CrossRef]

71. Cohen, M.D.; Sisco, M.; Prophete, C.; Yoshida, K.; Chen, L.C.; Zelikoff, J.T.; Ghio, A.J. Effects of metal compounds with distinct physicochemical properties on iron homeostasis and antibacterial activity in the lungs: Chromium and vanadium. *Inhal. Toxicol.* **2010**, *22*, 169–178. [CrossRef] [PubMed]

 *coatings*

Article

An Advanced Surface Treatment Technique for Coating Three-Dimensional-Printed Polyamide 12 by Hydroxyapatite

Abdulaziz Alhotan [1], Saleh Alhijji [1], Sahar Ahmed Abdalbary [2], Rania E. Bayoumi [3], Jukka P. Matinlinna [4], Tamer M. Hamdy [5] and Rasha M. Abdelraouf [6,*]

1   Department of Dental Health, College of Applied Medical Sciences, King Saud University, P.O. Box 10219, Riyadh 12372, Saudi Arabia; aalhotan@ksu.edu.sa (A.A.); smalhijji@ksu.edu.sa (S.A.)
2   Department of Orthopaedic Physical Therapy, Faculty of Physical Therapy, Nahda University, Beni Sueif 62521, Egypt; saharabdalbary@yahoo.com
3   Biomaterials Department, Faculty of Dentistry (Girls), Azhar University, Cairo 11754, Egypt; raniaezzat.26@azhar.edu.eg
4   Biomaterials Science, Division of Dentistry, Faculty of Biology, Medicine and Health, The University of Manchester, Manchester M13 9PL, UK; jukka.matinlinna@manchester.ac.uk
5   Restorative and Dental Materials Department, Oral and Dental Research Institute, National Research Centre (NRC), El Bohouth St., Dokki, Giza 12622, Egypt; tm.hamdy@nrc.sci.eg
6   Biomaterials Department, Faculty of Dentistry, Cairo University, Cairo 11553, Egypt
*   Correspondence: rasha.abdelraouf@dentistry.cu.edu.eg

Citation: Alhotan, A.; Alhijji, S.; Abdalbary, S.A.; Bayoumi, R.E.; Matinlinna, J.P.; Hamdy, T.M.; Abdelraouf, R.M. An Advanced Surface Treatment Technique for Coating Three-Dimensional-Printed Polyamide 12 by Hydroxyapatite. *Coatings* 2024, 14, 1181. https://doi.org/10.3390/coatings14091181

Academic Editor: Emerson Coy

Received: 23 May 2024
Revised: 28 August 2024
Accepted: 3 September 2024
Published: 12 September 2024

Abstract: Polymer 3D printing has is used in a wide range of applications in the medical field. Polyamide 12 (PA12) is a versatile synthetic polymer that has been used to reconstruct bony defects. Coating its surface with calcium phosphate compounds, such as hydroxyapatite (HA), could enhance its bonding with bone. The aim of this study was to coat 3D-printed polyamide 12 specimens with hydroxyapatite by a simple innovative surface treatment using light-cured resin cement. Polyamide 12 powder was printed by selective laser sintering to produce 80 disc-shaped specimens (15 mm diameter × 1.5 mm thickness). The specimens were divided randomly into two main groups: (1) control group (untreated), where the surface of the specimens was left without any modifications; (2) treated group, where the surface of the specimens was coated with hydroxyapatite by a new method using a light-cured dental cement. The coated specimens were characterised by both Fourier transform infrared spectroscopy (FTIR) and Transmission Electron Microscopy (TEM), (n = 10/test). The control and treated groups were further randomly subdivided into two subgroups according to the immersion in phosphate-buffered saline (PBS). The first subgroup was not immersed in PBS and was left as 3D-printed, while the second subgroup was immersed in PBS for 15 days (n = 10/subgroup). The surfaces of the control and treated specimens were examined using an environmental scanning electron microscope (SEM) and energy dispersive X-ray analysis (EDXA) before and after immersion in PBS. Following the standard American Society for Testing and Materials (ASTM D3359), a cross-cut adhesion test was performed. The results of the FTIR spectroscopy of the coated specimens were confirmed the HA bands. The TEM micrograph revealed agglomerated particles in the coat. The SEM micrographs of the control 3D-printed polyamide 12 specimens illustrated the sintered 3D-printed particles with minimal porosity. Their EDXA revealed the presence of carbon, nitrogen, and oxygen as atomic%: 52.1, 23.8, 24.1 respectively. After immersion in PBS, there were no major changes in the control specimens as detected by SEM and EDXA. The microstructure of the coated specimens showed deposited clusters of calcium and phosphorus on the surface, in addition to carbon, nitrogen, and oxygen, with atomic%: 9.5, 5.9, 7.2, 30.9, and 46.5, respectively. This coat was stable after immersion, as observed by SEM and EDXA. The coat adhesion test demonstrated a stable coat with just a few loose coating flakes (area removed <5%) on the surface of the HA-coated specimens. It could be concluded that the 3D-printed polyamide 12 could be coated with hydroxyapatite using light-cured resin cement.

Keywords: polyamide; 3D printing; hydroxyapatite; coating; surface treatment technique

---

1. Introduction

In subtractive manufacturing (SM), or conventional milling, a material is removed from a block until the desired item shape is achieved [1,2]. Despite still being extensively utilised in the creation of several medical and dental restorations, it has some drawbacks, among them the high amount of material waste [3]. This potentially raises material costs and may cause further environmental issues [4]. Additionally, this method is unsuitable for complex geometries. Additionally, in subtractive manufacturing, when milling tools age over time, they need to be replaced and maintained. This will add to the cost and duration of manufacturing [4].

The opposite of subtractive manufacturing is additive manufacturing (AM), sometimes referred to as 3D printing. The AM allows for customisation of complicated forms without the typical moulds and characteristic tools required for conventional milling processes. In addition, this technology enables the direct production of actual real products from virtual 3D images, saving time, material, and money [5].

Computer-aided design and computer-aided manufacturing (CAD/CAM) have been increasingly popular since the turn of the twenty-first century. These days, more and more materials are being printed in 3D. Using CAD software (https://www.autodesk.com/), a 3D digital model is created and sent to a 3D printer, which translates the digital model into a 3D product. At present, 3D printing is extensively utilised in a variety of industries and has advanced significantly [6].

Polymers, ceramics, and metals are used in dental 3D printing. The most utilised materials for 3D printing in dentistry are polymers because of their ease of processing, affordability, unique surface characteristics, mechanical and biological properties. They can be utilised to create functional casts, surgical guides, customised trays, and provisional restorations [7]. Polymethyl methacrylate (PMMA), polyurethane (PU), polyethylene (PE), polycarbonate (PC), polyetheretherketone (PEEK), polyethylene glycol (PEG), polydimethylsiloxane (PDMS), polylactic acid (PLA), poly(e-caprolactone) (PCL), acrylonitrile butadiene styrene (ABS), and polypropylene (PP) are some of the polymers that are commonly used in dental applications [8].

In orthopedics, polymer 3D-printing has also gained wide applications, which may aid in the reconstruction of complex bone structures with minimal waste [9]. Polyamide 12 (PA12) is a member of a large family of polyamides that are recognised for their exceptional toughness, strength, and impact resistance [10]. PA12 has been used in reconstructing bony defects and has given promising results. Its biocompatibility and versatility have been reported in several studies [11]. PA12 has become a successful orthopedic reconstructive material. It was successfully used in reconstructing defects in cranial and zygomatic bones [12,13].

A successful additive manufacturing technique for polyamides is selective laser sintering (SLS). The use of lasers raises the powdered material's temperature to a point at which the particles agglomerate and form products with an exact dimension [14].

Coating its surface with calcium phosphate compounds could enhance biointegration, which is the adhesion of living tissue to the surface of a biomaterial or implant. For orthopedic and dental uses, hydroxyapatite (HA) is commonly used as a synthetic substitute for the calcium phosphate present naturally in the body because of its excellent biocompatibility and bioactive reactivity qualities [15,16]. It adheres to bone and encourages the growth of new bone, which is essential for the integration of prosthetics with bone, e.g., dental titanium implants [17].

Over the past few decades, calcium phosphate bioceramics have become commonly utilised in alloplastic bone transplants in dental applications. The components of calcium phosphate bioceramics include tricalcium phosphate (α- and β-TCP), hydroxyapatite (HA), or biphasic calcium phosphate (BCP), which is a combination of β- and HA-TCP. Bioceramics can be mixed to create composite scaffolds with improved mechanical qualities [18,19]. Among the calcium phosphate ceramics, hydroxyapatite ($Ca_{10}(PO_4)_6(OH)_2$) is one of the most used calcium phosphate compounds due to its structural and chemical similarity to teeth and bones. The composition of human bone is composed of approximately 70% inorganic matter (hydroxyapatite), 25% organic matter, and 5% water [19,20].

Several studies have investigated the effect of adding hydroxyapatite to light-curd dental materials such as glass ionomer and resin composites [20–22]. Its influence on the materials' properties and the bond strength to the tooth structure was examined [20–22]. In these studies, the hydroxyapatite was added as fillers within the matrix of such restorative materials [20–22]. However, in the current study, hydroxyapatite was used as a coat rather than fillers, and the light-cured resin cement was used as an adhesive layer rather than a matrix, upon which the hydroxyapatite coat was applied.

The hydroxyapatite was also added previously as fillers to bone cements for bioactivity [15]. Regarding dental implants, a common technique to increase the bond strength of dental implant materials to bone was to coat them with hydroxyapatite [23]. It has been demonstrated that hydroxyapatite exhibited high cell affinity, which influenced osteoblast adhesion, proliferation, and direct bone integration [24]. It had better biomimetic properties concerning performance, structural, and functional aspects when it interacted with human bones [19,20].

When hydroxyapatite is implanted, a carbonated apatite layer will be precipitated due to the partial solubility of calcium and phosphorous ions in hydroxyapatite. Moreover, the formed layer acts as a platform for osteoblastic cells, promoting their growth and differentiation and impeding the formation of a fibrous capsule around the implant material. The greater production of this apatite-like layer improves the quality of the contact at the bone–implant interface. The primary challenge lies in achieving a proper attachment between hydroxyapatite and the surface of the polymeric material [25].

There are several methods for coating hydroxyapatite on implant surfaces, including electrophoretic deposition, sol–gel methods, biomimetics, and plasma spraying [26]. Yet, most of these methods require several stages, such as a high temperature, specialised equipment, and demanding circumstances [26].

Resin cements are used to bond indirect dental restorations to tooth structures. Resin cements containing 10-methacryloyloxydecyldihydrogenphosphate (10-MDP) have shown superior bonding to the hydroxyapatite of the enamel, dentin, ceramic and some metal surfaces [27]. This laboratory study introduces an innovative surface treatment technique using light-cured resin cement containing 10-MDP to coat 3D-printed polyamide 12 with hydroxyapatite. To simulate the behaviour of materials in a human body, solutions having a composition comparable to that of blood plasma can be utilised, such as phosphate-buffered saline (PBS).

Thus, the aim of the current laboratory study was to examine and assess the microstructure and elemental analysis of 3D-printed polyamide 12 coated with hydroxyapatite using an innovative technique (light-cured resin cement) versus an uncoated one (control). The effect of immersion in phosphate-buffered saline on the previous analysis was examined. The adhesion of the coating to the underlying substructure was also investigated. The null hypothesis postulated was that there was no difference between the control and coated specimens in microstructure and elemental analysis either before or after immersion in PBS.

2. Materials and Methods

2.1. Specimens Preparation

Polyamide 12 powder (Franz Eckert Gmbh, Waldkirch, Germany) was used for printing by selective laser sintering using a 3D-printing laser (Sintratec, Brugg, Switzerland). The dimensions of the specimens were inserted as input data into computerised software attached to the 3D-printed device (computer-aided design). A total of 80 disc-shaped specimens (15 mm in diameter × 1.5 mm in thickness) were fabricated.

2.2. Specimens Grouping

After the 3D-printing process, the obtained specimens were divided randomly into the following two main study groups: (1) the control group (untreated), where the surface of the specimens was left as 3D-printed without any modifications; and (2) the treated group, where the surface of the specimens was coated with hydroxyapatite. Each group was further randomly subdivided into two subgroups according to its immersion in phosphate-buffered saline (PBS). The first subgroup was not immersed in PBS, while the second subgroup was immersed in PBS for 15 days.

The study was approved by the Medical Research Ethical Committee (MREC) of the National Research Centre (NRC), Cairo, Egypt (Ref. number: 1287112022). The G*Power (version 3.1.9.7) sample size calculator was used to determine the sample size based on means and standard deviations [26]. The coated specimens were characterised by both Fourier-Transform Infrared spectroscopy and Transmission Electron Microscopy (n = 10/test). For the microstructure and elemental analysis, the projected sample size for each group was 20 (10/subgroup). The adhesion of the coat to treated specimens was exposed to a coat adhesion test before and after immersion in PBS (10 each). Eighty is the total number of specimens in the study.

2.3. Surface Treatment

Figure 1 represents the steps of the novel surface treatment. One type of dental resin cement was used that was dual cured, i.e., light and chemical cure (Panavia™ F2.0, Kuraray, Japan). The resin cement consisted of two pastes. The first paste contained 10-methacryloyloxydecyldihydrogenphosphate (MDP), dimethacrylate, silica, an initiator, a catalyst, and camphoroquinone. The second paste was made of hydrophobic aromatic and aliphatic dimethacrylate, sodium aromatic sulphinate, N,N-diethanol-p-toluidine, functionalised sodium fluoride, and silanized barium glass [28]. Barium was added to the glass to increase its radiopacity [29].

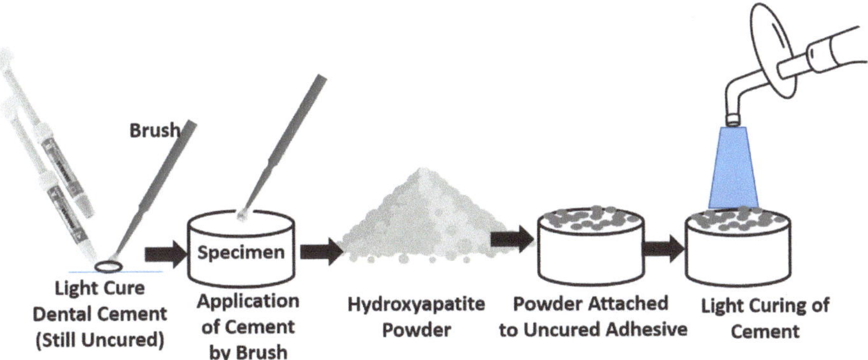

Figure 1. Diagram representing the surface treatment steps.

The resin cement was mixed following the manufacturer's instructions. Being a two-paste system, one drop from each tube was dispensed into a mixing paper pad and mixed immediately using a plastic spatula for 20 s. The mixed resin was applied to the specimens using a disposable brush tip. Then, the hydroxyapatite powder (Ossila, Sheffield, UK) was applied to the uncured resin cement using a spatula. The cement holding the hydroxyapatite powder was light-polymerised for 20 s using a light-curing device (Mini LED, Satelec, Acteon, France). According to the hydroxyapatite manufacturer, its chemical formula was $HCa_5O_{13}P_3$, and its average particle size was 10–20 µm.

Stereomicroscopic images were taken at magnification 12.5× to illustrate the steps of coating (Figures 2 and 3) using a stereomicroscope (Leica, Allendale, NJ, USA). The left-hand side part of Figure 2a shows the untreated 3D-printed polyamide 12, the middle portion (b) illustrates the resin cement, and the right part (c) demonstrates the application of the hydroxyapatite on the resin cement. Figure 3 is a stereomicroscopic image showing coated and uncoated surfaces before immersion in phosphate-buffered saline.

Figure 2. Stereomicroscopic image illustrating the steps a–c of coating. Magnification: 12.5×.

Figure 3. Stereomicroscopic image showing coated and uncoated surfaces (before immersion in PBS). Magnification: 12.5×.

2.4. Characterisation by Fourier-Transform Infrared and Transmission Electron Microscopy

The coated specimens were characterised by Fourier-Transform Infrared (FTIR) spectroscopy (FT/IR-4000 Series Spectrometer, JASCO, Tokyo, Japan) to detect the functional groups on the specimens' surface. In addition, the coated specimens were examined using transmission electron microscopy (TEM) (JEM-2100 HR TEM, JEOL Tokyo, Japan) at a 200 kV accelerating voltage and 22 Å imaging resolution.

2.5. Storage in Phosphate-Buffered Saline

Both the control and treated specimens were next randomly subdivided into two subgroups. Half of the specimens from each group were left without immersion, whereas the other half was immersed in PBS (Dulbecco's Phosphate-Buffered Saline, Lonza, Verviers, Belgium) for 15 days at 37 °C in an incubator (BTC, Cairo, Egypt). The composition of PBS was 0.0095 M PO_4^{3-}. The stereomicroscopic image was taken for the coated and uncoated surfaces after immersion in PBS at a magnification of 12.5×, as seen in Figure 4.

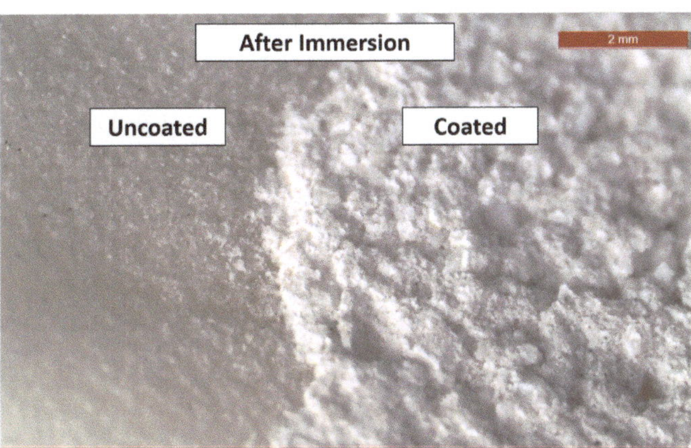

Figure 4. Stereomicroscopic image after immersion in PBS for coated and uncoated surfaces). Magnification: 12.5×.

2.6. Examination of Surface Microstructure and Elemental Analysis

The surfaces of the control and treated specimens were examined using an environmental scanning electron microscope (SEM; JSM-5200, JEOL, Tokyo, Japan) before and after immersion in PBS. A magnification of 200× was used with a 30 kV accelerating voltage. An extra SEM micrograph was taken for the coated specimen (lateral view) to measure the coat thickness with 500× magnification.

Energy dispersive X-ray analysis (EDXA) was performed to detect the elements present on the surfaces of control and treated specimens both before and after immersion in PBS. Energy dispersive X-ray spectroscopy (EDX) (Oxford Inca Energy 350, Oxford Instruments, Abingdon, UK) was used with a 10 mm working distance, 3 nm resolution, and a 30 kV accelerating voltage. The quantification of the elements was presented in atomic%. The Ca/P ratio was calculated in the coated groups to determine the form of the calcium phosphate compound.

2.7. Coat Adhesion Test

Following the standard of the American Society for Testing and Materials (ASTM D3359) [30], the cross-cut adhesion test was performed. A sharp cutting blade (Blades for Adhesion Tester, GLTL, Changzhou, Jiangsu, China) was used to pierce the coating until it reached the substrate. Six uniformly spaced incisions (2 mm apart from each other) were made both vertically and horizontally to create a lattice pattern on the test area's surface, and then an adhesive tape (ASTM D3359 Cross Hatch Adhesion Test Tape, GLTL, Changzhou, Jiangsu, China) was used to remove the coat following the standard steps of the tape test according to ASTM D3359 [30]. Figure 5 and Table 1 display the adhesion scale ranges.

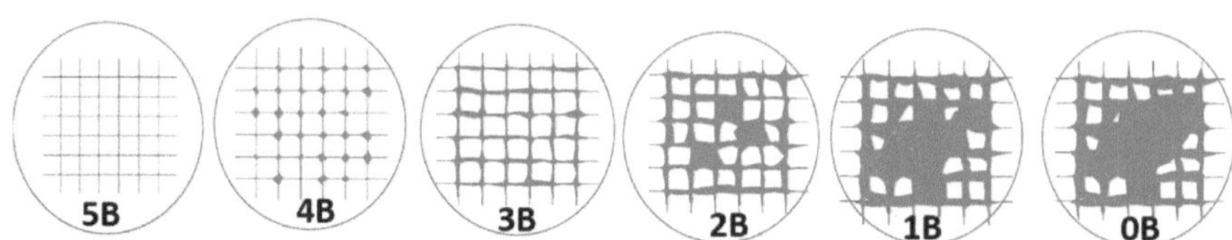

Figure 5. Diagram showing adhesion-scale according to ASTM D3359 standard.

Table 1. Adhesion-scale according to ASTM D3359 standard.

Score	Description and Area% removed
5B	The whole coat is attached, and the margins of the incisions are perfectly smooth (Area removed is zero).
4B	Minor coating flakes detached at lines of intersections (Area removed <5%).
3B	Little coating flakes come off at cut intersections and around edges, (5 to 15% of the lattice).
2B	Parts of the squares and their edges have chipped off from the coat (15–35%).
1B	Whole squares have come away from the coat, and enormous ribbons with cut edges have flaked (35–65%).
0B	Detachment and flaking are worse than in Grade 1 (>65%).

3. Results

3.1. FTIR and TEM

The spectra of the FTIR, seen in Figure 6, revealed the absorption peaks of hydroxyapatite on the coated specimens [31]. The vibration peaks of PO_4^{-3} were shown by the sharp peaks at 566.005 and 601.682 cm^{-1}. The PO_4^{-3} group's stretching mode was indicated by the bands at 823.455 and 1034.62 cm^{-1}. The band that is seen at 1380.78 cm^{-1} indicated CO_3^{-2}. The O-H stretching mode was represented by the band at 1639.2 cm^{-1}. Adsorbed water was indicated by the wide band at 3432.67 cm^{-1}.

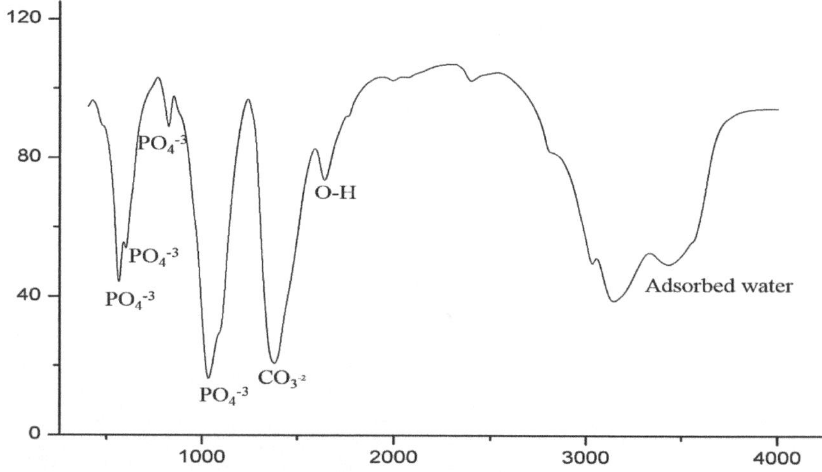

Figure 6. FTIR Spectra of the coated specimen.

The TEM micrograph of the coated specimen is shown in Figure 7. The micrograph revealed non-uniform clusters representing the hydroxyapatite coat.

Figure 7. TEM micrographs of the coated specimen.

3.2. SEM and EDXA

The SEM micrograph of the control 3D-printed polyamide 12 specimens (before immersion in PBS) illustrated the agglomerated 3D-printed particles with minimal porosity (Figure 8; magnification 200×). The EDX analysis of the control specimens (Figure 9) revealed the presence of carbon, nitrogen, and oxygen as atomic%: 52.1 ± 2.4, 23.8 ± 0.8, 24.1 ± 3.2, respectively.

Figure 8. SEM micrograph of control specimens before immersion in phosphate-buffered saline.

The SEM micrograph of the treated specimens (before immersion in PBS) illustrated the deposition of clusters on the surface (Figure 10). The EDX analysis of the coated specimens (Figure 11) showed the presence of calcium and phosphorus in addition to carbon, nitrogen, and oxygen with atomic%: 9.5 ± 3.1, 5.9 ± 1.6, 7.2 ± 3.8, 30.9 ± 2.0 and 46.5 ± 4.6 respectively. The Ca/P ratio was 1.6.

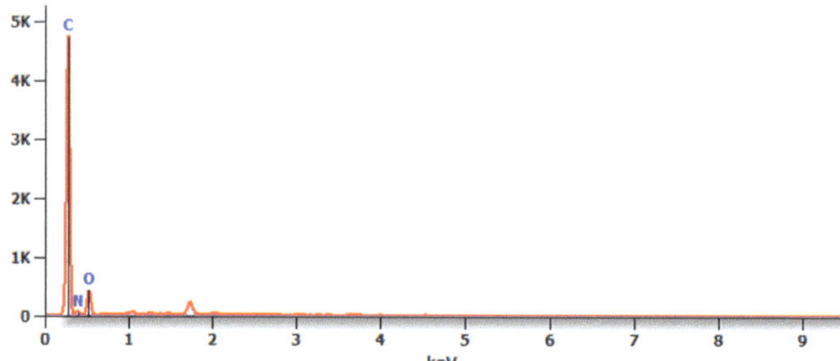

Figure 9. EDXA of control specimens before immersion in phosphate-buffered saline.

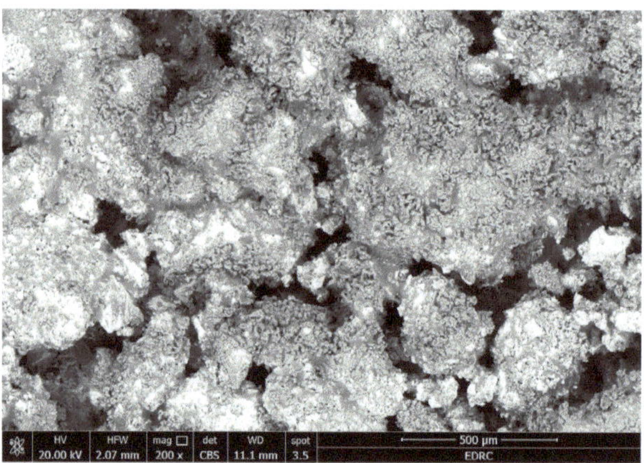

Figure 10. SEM micrograph of treated specimens before immersion in phosphate-buffered saline (magnification: 200×).

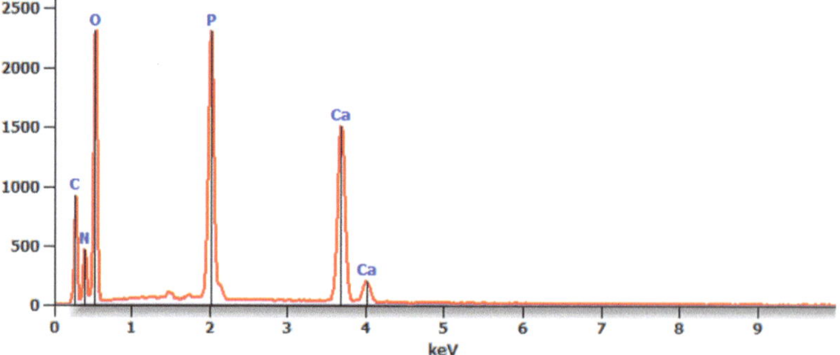

Figure 11. EDXA spectrum of treated specimens before immersion in phosphate-buffered saline.

The SEM micrograph of the control 3D-printed polyamide 12 specimens after immersion in PBS (Figure 12) illustrated nearly the same microstructure as before immersion, which was agglomerated particles. The EDX analysis of the control specimens after immersion in PBS revealed the same elements detected before immersion, namely carbon, nitrogen, and oxygen with atomic%: 54.4 ± 2, 24.6 ± 1 and 20.9 ± 3, respectively (Figure 13).

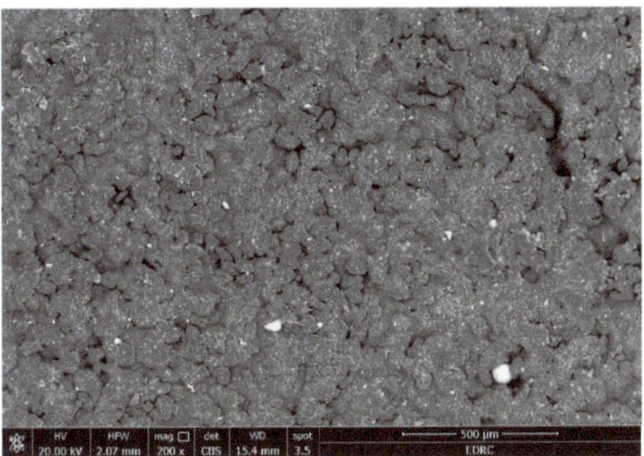

Figure 12. SEM micrograph of control specimens after immersion in phosphate-buffered saline.

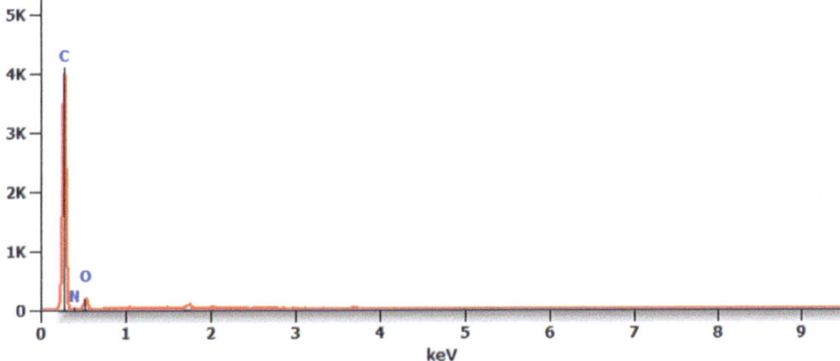

Figure 13. EDXA spectrum of control specimens after immersion in phosphate-buffered saline.

The SEM micrograph of the treated specimens after immersion in PBS illustrated the persistence of the deposited clusters upon the surface (Figure 14). The EDX analysis of the coated specimens (Figure 15) showed calcium and phosphorus in addition to carbon and oxygen with atomic%: 15.7 ± 2, 7.8 ± 1.5, 17.5 ± 3, 59 ± 4, respectively. The Ca/P ratio was 2.

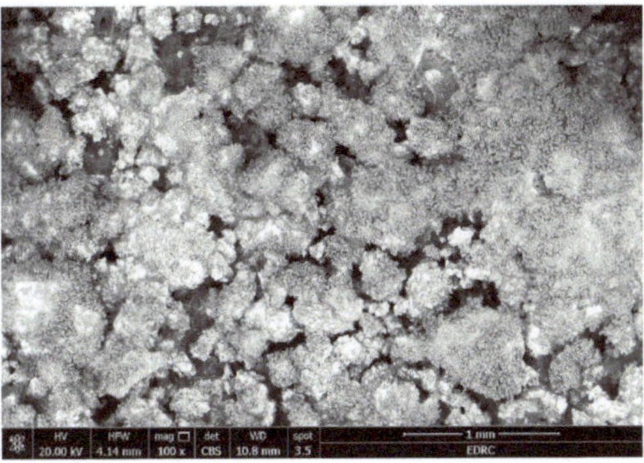

Figure 14. SEM micrograph of treated specimens after immersion in phosphate-buffered saline.

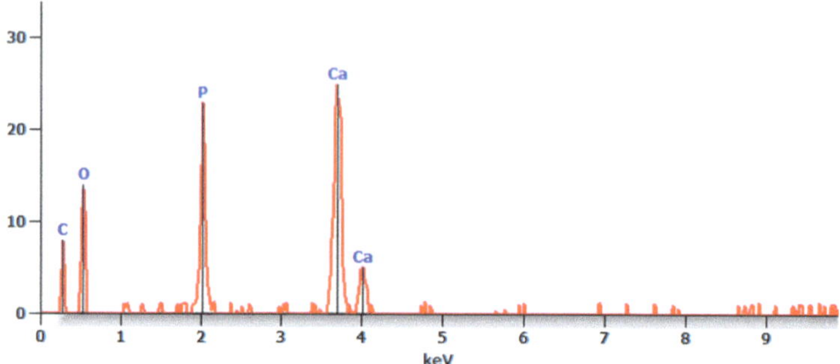

Figure 15. EDXA spectrum of treated specimens after immersion in phosphate-buffered saline.

To determine the coat thickness, the SEM micrograph of the lateral view of the coated specimen (Figure 16) displays that the average thickness of the coat was 100 ± 5 μm, the resin cement layer thickness was 47.2 ± 4.6 μm, and the hydroxyapatite was 52.8 ± 4.9 μm. The resin cement is represented by fillers (of different sizes) and the resinous matrix, and it is located as an intermediary layer between the substructure below and the hydroxyapatite above it.

Figure 16. SEM micrograph of lateral view of the coated specimens (magnification: 500×).

3.3. Coat Adhesion Test

The surface of the coated specimens displayed the score 4B, which indicated minor coating flakes detached at the lines of intersections (area removed <5%) both before and after storage in PBS for 15 days at 37 °C.

4. Discussion

The 3D-printed polyamide 12 is a promising polymeric material that has been used in orthopedic reconstruction with successful results [32]. Its coating with hydroxyapatite (HA) could enhance its bonding with bone [19]. Resin cement containing 10-MDP has shown a durable bond with calcium and phosphate present in tooth structure [33,34]. Yet, it is noteworthy that this cement has not been previously used as an adhesive layer to coat the hydroxyapatite with the polymeric substructure.

The coated specimens were characterised by FTIR spectroscopy. Since each chemical bond absorbs infrared (IR) radiation with a unique wavenumber, FTIR allows the characterisation of many chemical bonds and functional groups found in a substance. The observed FTIR absorption spectrum's shape and intensity matched the bands generated by HA particles. The obtained HA spectra agreed with those reported by previous studies [35–37]. The coated specimens were further characterised by TEM, which provided in-depth examination of the morphology and microstructure of the specimens. The TEM micrograph revealed non-uniform clusters representing the hydroxyapatite coat.

The current study examined the microstructure and elemental analysis of 3D-printed polyamide 12 coated with hydroxyapatite using light-cured resin cement versus an uncoated one (control). Now, the null hypothesis was rejected as there were differences between the control and coated specimens in the microstructure and elemental analysis either before or after immersion in PBS.

The SEM micrographs of the control 3D-printed polyamide 12 specimens illustrated agglomerated particles with minimal porosity due to the sintering process performed by the 3D printing process performed by the selective laser sintering device. The minimal porosity of PA12 in this study after 3D printing was not in agreement with a previous study that reported a non-homogenous surface with several particles as well as multiple pores and attributed this to the wavelength of the laser beam used [38].

The EDXA of the control specimens revealed the presence of carbon, nitrogen, and oxygen, as polyamide 12 is made up of amide group (-CO-NH-) repeating units joined by carbon atoms. Yet, the EDX was unable to detect hydrogen because of its solitary valence electron that makes up the K-shell of the hydrogen atom and participates in chemical bonding [39].

After immersion in PBS, there were no major changes in the control specimens, as detected by SEM and EDXA. This may be attributed to the surface properties of polyamide 12 and its resistance against deterioration as a result of its molecular structure with entangled long chains [40], in addition to its thermal stability due to the amide groups [41]. Moreover, polyamide 12 exhibits favourable characteristics such as excellent dimensional stability and low water sorption [42,43].

The microstructure of the coated specimens showed deposited calcium phosphate clusters on the surface, as the coating is mainly composed of hydroxyapatite. The molecular formula of hydroxyapatite is $Ca_5(PO_4)_3OH$ [44]. Its crystal unit cell contains ten calcium ions (10 Ca^{2+}) and its chemical formula is frequently written as $Ca_{10}(PO_4)_6(OH)_2$ [44]. Its Ca/P atomic ratio is 10/6 or 1.67 in the unit cell [45]. Any deviation from the exact Ca/P ratio destabilises the crystal and enhances the dissolution of the material. In this research, the Ca/P ratio was 1.60, indicating calcium-deficient HA, which is slightly more bioactive than stoichiometric HA with a Ca/P ratio of 1.67 [46].

Several analytical techniques have been established in order to assess the success materials within the human body during their service life. Testing the materials for use in bone applications in a simulated testing medium is one of the most significant of these techniques [47]. It is reported that, to replicate the behaviour of materials, solutions having a composition comparable to that of human blood plasma can be utilised, such as phosphate-buffered saline (PBS) [48].

This current calcium phosphate coat did not show detachment from the surface after immersion, as revealed by SEM and EDXA. In addition, adhesion of the coating showed that the surface of the coated specimens displayed minor coating flakes. This might be due to the durable bond between the 10-MDP monomer and hydroxyapatite with the low rate of dissolution of resin cement in water [49]. After immersion in PBS, the Ca/P ratio was 2, which denoted tetra-calcium-phosphate (TTCP) with the chemical formula $Ca_4(PO_4)_2O$. This conversion could be attributed to the chemical interaction between the hydroxyapatite and phosphate-buffered saline [50,51].

The surface treatment of hydroxyapatite using light-cured resin cement in PA12 could be considered a time-saving process. Contrary to the sol–gel technique used previously, this was a multi-step process. This started with the following preparation of the sol: tetra-ethyl-orthosilicate (4 mL), calcium-nitrate ethanolic solution (1 mL), and 85% phosphoric acid (0.13 mL) to ethanol (32.0 mL) under magnetic stirring. After agitation (30 min), an ammonia-ethanolic solution (2.4 mL) was added to the reaction mixture. This was followed by polyamide coating, where this sol was agitated for two hours to be deposited on PA by dip-coating. The substrates were left in the sols for 20 min; they were then dried for a day at 50 °C, and the resultant coat thickness was less than 3 μm [38].

On the other hand, the introduced coating method using light-cured resin cement was much simpler and seemed promising in this study, yet further studies are needed to assess and verify its interactions with osteoblasts, wettability, and the resultant surface roughness.

Although tribo-mechanical aspects of the coat were not measured in this study, it was expected that the coat could be stable during service due to the following several reasons: (1) the coated specimens were stored in phosphate-buffered saline for 15 days and no detachment was detected (examined by stereomicroscope, SEM and EDXA); (2) a coat adhesion test was performed following the standard American Society for Testing and Materials (ASTM D3359) after storage for 15 days in phosphate-buffered saline, and, in this test, mechanical stress was applied during cutting to penetrate the coat until reaching the substructure; it should be noted that in this study this test resulted a stable coat with just a few loose coating flakes (area removed <5%); (3) the used resin cement was the type containing 10-MDP, which was reported previously in the literature to bond chemically to various substructures and resist water degradation [52]. However, it is recommended to test the coat when subjected to wear or friction in future studies.

5. Conclusions

In summary, the 3D-printed polyamide 12 could be coated with hydroxyapatite using light-cured resin cement containing 10-methacryloyloxydecyldihydrogenphosphate (10-MDP). To check the optimal structure of the coated specimens, Fourier transform infrared spectroscopy (FTIR) and transmission electron microscopy (TEM) were used for characterisation. The stability of the coat was assessed after immersion for 15 days in phosphate-buffered saline (PBS) at 37 °C. The surfaces of the coated specimens were examined using a scanning electron microscope (SEM) and energy dispersive X-ray analysis (EDXA) both before and after immersion in PBS and compared to uncoated control specimens. In addition, a cross-cut adhesion test was performed following the standard American Society for Testing and Materials (ASTM D3359) before and after immersion in 37 °C PBS for 15 days. The main findings were presented as follows:

1. The results of the FTIR spectroscopy of the coated specimens confirmed the HA bands. The TEM micrographs revealed HA-agglomerated particles in the coat.
2. The coat was stable after immersion in PBS, as observed by SEM and EDXA as well as the coat adhesion test, which demonstrated a stable coat with just a few loose coating flakes (area removed <5%) on the surface of the HA-coated specimens.
3. There were no major changes in both the coated and uncoated (control) specimens before and after immersion in PBS. The SEM micrographs of the control 3D-printed polyamide 12 specimens illustrated the sintered 3D-printed particles with minimal porosity. Their EDXA revealed the presence of carbon, nitrogen, and oxygen. The microstructure of the coated specimens showed deposited clusters of calcium and phosphorus on the surface in addition to carbon, nitrogen, and oxygen.

Therefore, the light-cured resin cement resin with 10-MDP functional groups could be used for coating the 3D-printed polyamide 12 with hydroxyapatite, yet biological assessment is the further step to ensure its biocompatibility for biomedical applications.

Author Contributions: Conceptualisation, R.M.A. and T.M.H.; data curation, T.M.H., R.M.A., S.A.A. and A.A.; formal analysis, T.M.H., R.M.A. and A.A.; investigation, R.M.A. and S.A.A.; methodology, T.M.H., R.M.A., A.A. and S.A.A.; visualisation, A.A., S.A., S.A.A., R.E.B. and J.P.M.; resources, A.A., S.A., S.A.A. and R.E.B.; supervision, T.M.H., R.M.A., A.A. and J.P.M.; writing—original draft, R.M.A.,

T.M.H. and S.A.A.; writing—review and editing; T.M.H., R.M.A., J.P.M. and A.A. All authors have read and agreed to the published version of the manuscript.

Funding: This research was funded by Researchers Supporting Project number (RSPD2024R790), King Saud University, Riyadh, Saudi Arabia.

Institutional Review Board Statement: This study is approved by the Medical Research Ethical Committee (MREC) of National Research Centre (NRC), Cairo, Egypt (Ref. number: 1287112022).

Informed Consent Statement: Not applicable.

Data Availability Statement: The data presented in this study are available on request from the corresponding author.

Acknowledgments: The authors are grateful to the Researchers Supporting Project number (RSPD2024R790), King Saud University, Riyadh, Saudi Arabia.

Conflicts of Interest: The authors declare no conflicts of interest.

References

1. Earar, K.; Iliescu, A.A.; Popa, G.; Iliescu, A.; Rudnic, I.; Feier, R.; Voinea-Georgescu, R.N. Additive vs. Subtractive CAD/CAM Procedures in Manufacturing of the PMMA Interim Dental Crowns a Comparative in Vitro Study of Internal Fit. *Rev. Chim.* **2020**, *71*, 405–410. [CrossRef]
2. Hamdy, T.M.; Abdelnabi, A.; Othman, M.S.; Bayoumi, R.E. Alterations in Surface Gloss and Hardness of Direct Dental Resin Composites and Indirect CAD/CAM Composite Block after Single Application of Bifluorid 10 Varnish: An In Vitro Study. *J. Compos. Sci.* **2024**, *8*, 58. [CrossRef]
3. Bhatnagar, A.; Bhardwaj, A.; Verma, S. Additive Manufacturing: A 3-Dimensional Approach in Periodontics. *J. Adv. Med. Med. Res.* **2020**, *32*, 105–117. [CrossRef]
4. Moon, J.-M.; Jeong, C.-S.; Lee, H.-J.; Bae, J.-M.; Choi, E.-J.; Kim, S.-T.; Park, Y.-B.; Oh, S.-H. A Comparative Study of Additive and Subtractive Manufacturing Techniques for a Zirconia Dental Product: An Analysis of the Manufacturing Accuracy and the Bond Strength of Porcelain to Zirconia. *Materials* **2022**, *15*, 5398. [CrossRef]
5. Jeong, M.; Radomski, K.; Lopez, D.; Liu, J.T.; Lee, J.D.; Lee, S.J. Materials and Applications of 3D Printing Technology in Dentistry: An Overview. *Dent. J.* **2023**, *12*, 1. [CrossRef]
6. Cai, H.; Xu, X.; Lu, X.; Zhao, M.; Jia, Q.; Jiang, H.-B.; Kwon, J.-S. Dental Materials Applied to 3D and 4D Printing Technologies: A Review. *Polymers* **2023**, *15*, 2405. [CrossRef]
7. Revilla-León, M.; Sadeghpour, M.; Özcan, M. An Update on Applications of 3D Printing Technologies Used for Processing Polymers Used in Implant Dentistry. *Odontology* **2020**, *108*, 331–338. [CrossRef]
8. Rokaya, D.; Srimaneepong, V.; Sapkota, J.; Qin, J.; Siraleartmukul, K.; Siriwongrungson, V. Polymeric Materials and Films in Dentistry: An Overview. *J. Adv. Res.* **2018**, *14*, 25–34. [CrossRef]
9. Nathanael, A.J.; Oh, T.H. Biopolymer Coatings for Biomedical Applications. *Polymers* **2020**, *12*, 3061. [CrossRef]
10. Vidakis, N.; Petousis, M.; Tzounis, L.; Maniadi, A.; Velidakis, E.; Mountakis, N.; Kechagias, J.D. Sustainable Additive Manufacturing: Mechanical Response of Polyamide 12 over Multiple Recycling Processes. *Materials* **2021**, *14*, 466. [CrossRef]
11. Priyadarshini, B.M.; Kok, W.K.; Dikshit, V.; Feng, S.; Li, K.H.H.; Zhang, Y. 3D Printing Biocompatible Materials with Multi Jet Fusion for Bioreactor Applications. *Int. J. Bioprint.* **2023**, *9*, 14–35. [CrossRef] [PubMed]
12. Arafat, S.W.; Ibrahim, W.H.; Shaker, S.; Aldainy, D.G.; Salama, D.; Shaheen, H.A. Reconstruction of Cranial Bone Defects Using Polyamide 12 Patient-Specific Implant: Long Term Follow Up. *J. Craniofac. Surg.* **2022**, *33*, 1825–1828. [CrossRef] [PubMed]
13. Shaheen, H.A.; Shaker, S.; Ibrahim, W.H.; AlDainy, D.G.; Salama, D.; Emara, A.S. Delayed Complex Fronto-Zygomatico-Orbital Reconstruction Using Patient Specific 3D Printed Implants; A Report of Three Complex Cases—A Short Communication. *J. Plast. Reconstr. Aesthetic Surg.* **2022**, *75*, 3877–3903. [CrossRef] [PubMed]
14. Gomes, P.C.; Piñeiro, O.G.; Alves, A.C.; Carneiro, O.S. On the Reuse of SLS Polyamide 12 Powder. *Materials* **2022**, *15*, 5486. [CrossRef] [PubMed]
15. Hamdy, T.M.; Saniour, S.H.; Sherief, M.A.; Zaki, D.Y. Effect of Incorporation of 20 Wt% Amorphous Nano-Hydroxyapatite Fillers in Poly Methyl Methacrylate Composite on the Compressive Strength. *Res. J. Pharm. Biol. Chem. Sci.* **2015**, *6*, 1136–1141. [CrossRef]
16. Abdelnabi, A.; Hamza, M.K.; El-Borady, O.M.; Hamdy, T.M. Effect of Different Formulations and Application Methods of Coral Calcium on Its Remineralization Ability on Carious Enamel. *Open Access Maced. J. Med. Sci.* **2020**, *8*, 94–99. [CrossRef]
17. Hamdy, T.M.; Mousa, S.M.A.; Sherief, M.A. Effect of Incorporation of Lanthanum and Cerium-Doped Hydroxyapatite on Acrylic Bone Cement Produced from Phosphogypsum Waste. *Egypt. J. Chem.* **2020**, *63*, 1823–1832. [CrossRef]
18. Dutta, S.R.; Passi, D.; Singh, P.; Bhuibhar, A. Ceramic and Non-Ceramic Hydroxyapatite as a Bone Graft Material: A Brief Review. *Ir. J. Med. Sci.* **2015**, *184*, 101–106. [CrossRef]
19. Hamdy, T.M. Dental Biomaterial Scaffolds in Tooth Tissue Engineering: A Review. *Curr. Oral Health Reports* **2023**, *10*, 14–21. [CrossRef]

20. Chae, M.H.; Lee, Y.K.; Kim, K.N.; Lee, J.H.; Choi, B.J.; Choi, H.J.; Park, K.T. The Effect of Hydroxyapatite on Bonding Strength in Light Curing Glass Ionomer Dental Cement. *Key Eng. Mater.* **2006**, *309*, 881–884. [CrossRef]
21. Razali, R.A.C.; Rahim, N.A.; Zainol, I.; Sharif, A.M. Preparation of Dental Composite Using Hydroxyapatite from Natural Sources and Silica. In *Proceedings of the Journal of Physics: Conference Series*; IOP Publishing: Bristol, UK, 2018.
22. Domingo, C.; Arcs, R.W.; Lpez-Macipe, A.; Osorio, R.; Rodrguez-Clemente, R.; Murtra, J.; Fanovich, M.A.; Toledano, M. Dental Composites Reinforced with Hydroxyapatite: Mechanical Behavior and Absorption/Elution Characteristics. *J. Biomed. Mater. Res.* **2001**, *56*, 297–305. [CrossRef]
23. Ong, J.L.; Chan, D.C.N. Hydroxyapatite and Their Use as Coatings in Dental Implants: A Review. *Crit. Rev. Biomed. Eng.* **2000**, *28*, 667–707. [CrossRef] [PubMed]
24. Soleymani, S.; Naghib, S.M. 3D and 4D Printing Hydroxyapatite-Based Scaffolds for Bone Tissue Engineering and Regeneration. *Heliyon* **2023**, *9*, e19363. [CrossRef] [PubMed]
25. Szcześ, A.; Hołysz, L.; Chibowski, E. Synthesis of Hydroxyapatite for Biomedical Applications. *Adv. Colloid Interface Sci.* **2017**, *249*, 321–330. [CrossRef] [PubMed]
26. Abdulghafor, M.A.; Mahmood, M.K.; Tassery, H.; Tardivo, D.; Falguiere, A.; Lan, R. Biomimetic Coatings in Implant Dentistry: A Quick Update. *J. Funct. Biomater.* **2023**, *15*, 15. [CrossRef]
27. Carvalho, P.C.K.; Almeida, C.C.M.S.; Souza, R.O.A.; Tango, R.N. The Effect of a 10-MDP-Based Dentin Adhesive as Alternative for Bonding to Implant Abutment Materials. *Materials* **2022**, *15*, 5449. [CrossRef]
28. Al-Shehri, E.Z.; Al-Zain, A.O.; Sabrah, A.H.; Al-Angari, S.S.; Al Dehailan, L.; Eckert, G.J.; Özcan, M.; Platt, J.A.; Bottino, M.C. Effects of Air-Abrasion Pressure on the Resin Bond Strength to Zirconia: A Combined Cyclic Loading and Thermocycling Aging Study. *Restor. Dent. Endod.* **2017**, *42*, 206–215. [CrossRef]
29. Pekkan, G. Radiopacity of Dental Materials: An Overview. *Avicenna J. Dent. Res.* **2016**, *8*, 8. [CrossRef]
30. *ASTM D3359/D3359M-17*; Standard Test Methods for Measuring Adhesion by Tape Test. American Society for Testing and Materials, ASTM International: West Conshohocken, PA, USA, 2017.
31. Torabinejad, B.; Mohammadi-Rovshandeh, J.; Davachi, S.M.; Zamanian, A. Synthesis and Characterization of Nanocomposite Scaffolds Based on Triblock Copolymer of L-Lactide, ε-Caprolactone and Nano-Hydroxyapatite for Bone Tissue Engineering. *Mater. Sci. Eng. C Mater. Biol. Appl.* **2014**, *42*, 199–210. [CrossRef]
32. Zakręcki, A.; Cieślik, J.; Bazan, A.; Turek, P. Innovative Approaches to 3D Printing of PA12 Forearm Orthoses: A Comprehensive Analysis of Mechanical Properties and Production Efficiency. *Materials* **2024**, *17*, 663. [CrossRef]
33. Pimentel de Oliveira, R.; de Paula, B.L.; Ribeiro, M.E.; Alves, E.; Costi, H.T.; Silva, C. Evaluation of the Bond Strength of Self-Etching Adhesive Systems Containing HEMA and 10-MDP Monomers: Bond Strength of Adhesives Containing HEMA and 10-MDP. *Int. J. Dent.* **2022**, *2022*, 5756649. [CrossRef] [PubMed]
34. Hamdy, T.M. Polymerization Shrinkage in Contemporary Resin-Based Dental Composites: A Review Article. *Egypt. J. Chem.* **2021**, *64*, 3087–3092. [CrossRef]
35. Gheisari, H.; Karamian, E.; Abdellahi, M. A Novel Hydroxyapatite -Hardystonite Nanocomposite Ceramic. *Ceram. Int.* **2015**, *41*, 5967–5975. [CrossRef]
36. Chandrasekar, A.; Sagadevan, S.; Dakshnamoorthy, A. Synthesis and Characterization of Nano-Hydroxyapatite (n-HAP) Using the Wet Chemical Technique. *Int. J. Phys. Sci.* **2013**, *8*, 1639–1645.
37. Bouropoulos, N.; Stampolakis, A.; Mouzakis, D.E. Dynamic Mechanical Properties of Calcium Alginate-Hydroxyapatite Nanocomposite Hydrogels. *Sci. Adv. Mater.* **2010**, *2*, 239–242. [CrossRef]
38. Bandeira, L.C.; Ciuffi, K.J.; Calefi, P.S.; Nassar, E.J.; Silva, J.V.L.; Oliveira, M.; Maia, I.A.; Salvado, I.M.; Fernandes, M.H.V. Effect of Calcium Phosphate Coating on Polyamide Substrate for Biomaterial Applications. *J. Braz. Chem. Soc.* **2012**, *23*, 810–817. [CrossRef]
39. Alhotan, A.; Abdelraouf, R.M.; El-Korashy, S.A.; Labban, N.; Alotaibi, H.; Matinlinna, J.P.; Hamdy, T.M. Effect of Adding Silver-Doped Carbon Nanotube Fillers to Heat-Cured Acrylic Denture Base on Impact Strength, Microhardness, and Antimicrobial Activity: A Preliminary Study. *Polymers* **2023**, *15*, 2976. [CrossRef]
40. Chen, C.-W.; Ranganathan, P.; Mutharani, B.; Shiu, J.-W.; Rwei, S.-P.; Chang, Y.-H.; Chiu, F.-C. Synthesis of High-Value Bio-Based Polyamide 12,36 Microcellular Foams with Excellent Dimensional Stability and Shape Recovery Properties. *Polymers* **2024**, *16*, 159. [CrossRef]
41. Guo, B.; Xu, Z.; Luo, X.; Bai, J. A Detailed Evaluation of Surface, Thermal, and Flammable Properties of Polyamide 12/Glass Beads Composites Fabricated by Multi Jet Fusion. *Virtual Phys. Prototyp.* **2021**, *16*, S39–S52. [CrossRef]
42. Lewandowski, G.; Rytwińska, E.; Milchert, E. Physical Properties and Application of Polyamide 12. *Polimery/Polymers* **2006**. [CrossRef]
43. Morano, C.; Alfano, M.; Pagnotta, L. Effect of Strain Rates and Heat Exposure on Polyamide (PA12) Processed via Selective Laser Sintering. *Materials* **2023**, *16*, 4654. [CrossRef] [PubMed]
44. Shi, H.; Zhou, Z.; Li, W.; Fan, Y.; Li, Z.; Wei, J. Hydroxyapatite Based Materials for Bone Tissue Engineering: A Brief and Comprehensive Introduction. *Crystals* **2021**, *11*, 149. [CrossRef]
45. Vokhidova, N.R.; Ergashev, K.H.; Rashidova, S.S. Hydroxyapatite-Chitosan Bombyx Mori: Synthesis and Physicochemical Properties. *J. Inorg. Organomet. Polym. Mater.* **2020**, *30*, 3357–3368. [CrossRef]

46. Beaufils, S.; Rouillon, T.; Millet, P.; Le Bideau, J.; Weiss, P.; Chopart, J.-P.; Daltin, A.-L. Synthesis of Calcium-Deficient Hydroxyapatite Nanowires and Nanotubes Performed by Template-Assisted Electrodeposition. *Mater. Sci. Eng. C* **2019**, *98*, 333–346. [CrossRef] [PubMed]
47. Baino, F.; Yamaguchi, S. The Use of Simulated Body Fluid (SBF) for Assessing Materials Bioactivity in the Context of Tissue Engineering: Review and Challenges. *Biomimetics* **2020**, *5*, 57. [CrossRef]
48. Yilmaz, B.; Pazarceviren, A.E.; Tezcaner, A.; Evis, Z. Historical Development of Simulated Body Fluids Used in Biomedical Applications: A Review. *Microchem. J.* **2020**, *155*, 104713. [CrossRef]
49. Chen, Y.; Lu, Z.; Qian, M.; Zhang, H.; Chen, C.; Xie, H.; Tay, F.R. Chemical Affinity of 10-Methacryloyloxydecyl Dihydrogen Phosphate to Dental Zirconia: Effects of Molecular Structure and Solvents. *Dent. Mater.* **2017**, *33*, e415–e427. [CrossRef]
50. Chen, C.; Lee, I.S.; Zhang, S.M.; Yang, H.C. Biomimetic Apatite Formation on Calcium Phosphate-Coated Titanium in Dulbecco's Phosphate-Buffered Saline Solution Containing $CaCl_2$ with and without Fibronectin. *Acta Biomater.* **2010**, *6*, 2274–2281. [CrossRef]
51. Cieplik, F.; Rupp, C.M.; Hirsch, S.; Muehler, D.; Enax, J.; Meyer, F.; Hiller, K.-A.; Buchalla, W. Ca^{2+} Release and Buffering Effects of Synthetic Hydroxyapatite Following Bacterial Acid Challenge. *BMC Oral Health* **2020**, *20*, 85. [CrossRef]
52. Shokry, M.; Al-Zordk, W.E.G.; Ghazy, M.H. Influence of Different Primer/Resin Cement Systems on Retention of Monolithic Zirconia Crowns. *Mansoura J. Dent.* **2021**, *8*, 53–57. [CrossRef]

Article

Chemical Bonding of Nanorod Hydroxyapatite to the Surface of Calciumfluoroaluminosilicate Particles for Improving the Histocompatibility of Glass Ionomer Cement

Sohee Kang [1], So Jung Park [2], Sukyoung Kim [3] and Inn-Kyu Kang [2,*]

[1] Department of Dentistry, College of Medicine, Yeungnam University, Daegu 42415, Republic of Korea; kangsh@yu.ac.kr
[2] Department of Polymer Science and Engineering, Kyungpook National University, Daegu 41566, Republic of Korea; sojung90714@naver.com
[3] Materials Science and Engineering, Yeungnam University, Gyeongsan-si 38541, Republic of Korea; sykim@yu.ac.kr
* Correspondence: ikkang@knu.ac.kr

Abstract: Glass ionomer cement (GIC) is composed of anionic polyacrylic acid and a silica-based inorganic powder. GIC is used as a filling material in the decayed cavity of the tooth; therefore, compatibility with the tooth tissue is essential. In the present study, we aimed to improve the histocompatibility of GIC by introducing nano-hydroxyapatite (nHA), a component of teeth, into a silica-based inorganic powder. CFAS-nHA was prepared by chemically bonding nanorod hydroxyapatite (nHA) to the surface of calciumfluoroaluminosilicate (CFAS). The synthesis of CFAS-nHA was confirmed using Fourier transform infrared spectroscopy (FTIR) and scanning electron microscopy (SEM). The prepared CFAS-nHA was mixed with polyacrylic acid and cured to prepare GIC containing nHA (GIC-nHA). Cytocompatibility tests of GIC-nHA and GIC were performed using osteoblasts. Osteoblast activity and bone formation ability were superior after GIC-nHA treatment than after control GIC treatment. This enhanced histocompatibility is believed to be due to the improvement of the biological activity of osteoblasts induced by the HA introduced into the GIC. Therefore, to enhance its compatibility with dental tissues, GIC could be manufactured by chemically bonding nHA to the surface of GI inorganic powder.

Keywords: bioactivity; calciumfluoroaluminosilicate; glass ionomer cement; histocompatibility; hydroxyapatite; osteoblast

Citation: Kang, S.; Park, S.J.; Kim, S.; Kang, I.-K. Chemical Bonding of Nanorod Hydroxyapatite to the Surface of Calciumfluoroaluminosilicate Particles for Improving the Histocompatibility of Glass Ionomer Cement. *Coatings* **2024**, *14*, 893. https://doi.org/10.3390/coatings14070893

Academic Editors: Hicham Benhayoune and Richard Drevet

Received: 21 June 2024
Revised: 13 July 2024
Accepted: 15 July 2024
Published: 17 July 2024

1. Introduction

Glass ionomer cement (GIC) has various applications in dentistry, including use as a bonding agent for restorative materials, liners, fissure sealants, and orthodontic brackets for permanent and primary teeth [1]. Compared with previously used materials, dental restoration approaches using GIC have been developed. The GIC comprises a mixture of silica-based inorganic powders and anionic polyacrylic acid. GIC, which exhibits hydrophilicity owing to the characteristics of silica powder with many hydroxyl groups, can effectively absorb the liquid remaining at the bottom of the fissure compared to composite resins, thus adhering well to the tooth enamel [2]. Although GIC is actively utilized in dental restorations because of its high adhesion to teeth, there is a need for further improvement in terms of its brittleness, abrasion resistance, bending, and tensile strength [3]. It has been reported that the wear resistance or brittleness of GIC can be improved via the introduction of polyalkenoic acid [4] or nanosized bioceramics [5]. Among them, nanosized biomaterials have shown promising potential for improving the strength, gloss, and aesthetics of dental filling materials compared to conventional modifiers [6]. In recent years, there have been many reports focused on improving the mechanical strength of GICs through the introduction of hydroxyapatite (HA) [7,8]. HA has a chemical structure that closely resembles that of human teeth and skeletal systems [9]. Recent advances in HA synthesis technology has enabled the synthesis of HA of various sizes and shapes [10–12], facilitating their application as biocompatible fillers that resemble natural teeth. In addition to exhibiting unique radiopaque properties [13], HA plays an important role in orthopedic surgery because of its excellent biological activity and osteoconductivity. Furthermore, research has been conducted to enhance the antibacterial activity of GIC through the use of HA. Praveen et al. [14] added 8% HA powder to an existing glass ionomer (GC Fuji Type IX gold label, GC Corporation, Tokyo, Japan) and combined it with liquid polyacrylic acid to obtain cylindrical GIC test specimens. They reported that GIC with 8% HA demonstrated higher antibacterial activity than GIC without HA through antibacterial testing using *Streptococcus mutans*. Haider et al. [15] investigated the topographical effects of HA on the physiological activity of osteoblasts by preparing nanocomposites. They added spherical HA (sHA) and nanorod HA (nHA) to the poly(lactic-co-glycolic acid) scaffold. Based on the results of studying the interaction between the nanocomposite and osteoblasts, they concluded that the HA-containing nanorod scaffold further promoted osteoblast bioactivity. In this study, nHA was chemically bonded to the surface of calciumfluoroaluminosilicate (CFAS), a solid component of GIC, using the biocomponents L-glutamic acid (G) and albumin (Alb) as spacers. The progress of the surface reaction was confirmed by attenuated total reflectance Fourier transform infrared (ATR-FTIR) and scanning electron microscopy (SEM). The resulting CFAS-nHA powder was mixed with polyacrylic acid and UV-cured to prepare the disc-shaped GIC-nHA. The cytocompatibility and bone formation ability of the GIC and GIC-nHA discs were investigated using pre-osteoblasts.

2. Materials and Methods

2.1. The Preparation of Materials

The following materials were purchased from Sigma Aldrich Chemical Company (St. Louis, MO, USA): 3-Aminopropyltriethoxysilane (A), L-glutamic acid, N-(3-dimethylaminopropyl)-N′-ethylcarbodiimide hydrochloride (EDC), N-hydroxysuccinimide (NHS), fluorescein isothiocyanate (FITC), and bovine serum albumin. The nHA was synthesized according to previously described procedures [10,15]. The mouse MC3T3-E1 cell line was purchased from the Korea Cell Bank (Seoul, Republic of Korea) and stored in liquid nitrogen until cell seeding. The MC3T3-E1 cell line is a cultured mouse osteoblast cell line derived from mouse embryo pre-osteoblasts; it is capable of differentiating into osteoblasts. A phosphate-buffered saline (PBS) solution (pH 7.4) containing Na_2HPO_4, KH_2PO_4, NaCl, and KCl was acquired from Sigma-Aldrich. CFAS, a solid powder used in the preparation of GIC, was provided by Professor Sukyoung Kim of Yeungnam University, Republic of Korea. The XRD pattern of the CFAS was broad and showed poor crystallinity. The nHA was chemically bonded to the surface of the CFAS for use as a solid component of the GIC. The liquid component of GIC (acquired from GC International, Tokyo, Japan, GC Fuji II LC) was used as the polyacrylic liquid component to prepare the GIC. The MC3T3-E1 cells were cultured in α-minimum essential medium (α-MEM) supplemented with 10% fetal bovine serum and 1.0% penicillin G-streptomycin at a temperature of 37 °C. Under a 5% CO_2 atmosphere, the culture medium was replaced every other day. 3-(4,5-Dimethylthiazol-2 yl)-2,5-diphenyltetrazolium bromide (MTT) was acquired from Sigma-Aldrich (United States). The presence of nHA on the surfaces of the solid particles was observed using SEM (Hitachi S-400, Tokyo, Japan).

2.2. Synthesis and Surface Modification of CFAS

We used the sol–gel method described by Khiri et al. [16] to prepare the CFAS. The compositions of the ingredients used to prepare the CFAS are shown in Table 1. Initially, 23.89 mL of tetraethylorthosilicate was dissolved in 400 mL of ethanol. Subsequently, an aqueous solution (100 mL) containing aluminum nitrate, calcium nitrate, and ammonium dihydrogen phosphate was added dropwise. Finally, rapid addition of fluorosilicic acid followed by stirring the mixture at 80 °C for 4 h resulting in the formation of a gel-like product. The gel was then dried for 24 h at a temperature of 80 °C and then heat-treated for 10 min at a temperature of 750 °C. After heat treatment, the sample was quenched in water, powdered, and sieved to obtain a micrometre-sized sample (Figure 1).

Table 1. Feed ratio of calciumfluoroaluminosilicate microparticles.

Ingredient	Chemical Formula	Molar Ratio	50 mmol (Amount Used in the Experiment)
Tetraethylorthosilicate	$Si(OC_2H_5)_4$	2.133	23.89 mL
Ammonium dihydrogen phosphate	$NH_4H_2PO_4$	0.160	0.92 g
Fluorosilicic acid	H_2SiF_6	0.167	2.95 mL
Aluminum nitrate	$Al(NO_3)_3$	2.200	41.27 g
Calcium nitrate	$Ca(NO_3)_3$	1.000	11.81 g

Figure 1. SEM image of calciumfluoroaluminosilicate (CFAS) microparticles prepared in the present study.

To incorporate the primary amino groups on the surface of the CFAS particles, a mixture of aminopropyltriethoxysilane (A) and distilled water in a 1:9 ratio was prepared, and 0.06 g was added to the mixture. The mixture was subjected to ultrasonication for 30 min. Subsequently, acetic acid was used to set the pH of the reaction solution at 4.5–5.0. The reaction solution was maintained at 90 °C for 2 h under flowing nitrogen. The reactants were then transferred to distilled water and ultrasonicated for 5 min to remove unreacted A. The resulting CFAS with immobilized A (CFAS-A) was dried under reduced pressure for 12 h at a temperature of 25 °C.

2.3. Surface Modification of nHA

Transmission electron microscopy (TEM) analysis of nHA synthesized according to a previously reported method [15] showed that it was rod-shaped with a length of 30–150 nm (Figure 2). nHA particles contain numerous hydroxyl groups on their surfaces [16,17]. In this study, we aimed to chemically bind nHA particles to the surfaces of CFAS microparticles via a chemical reaction. These solid–solid reactions have low reactivity because they occur in heterogeneous systems [18]. To increase the surface reactivity of the nHA particles and improve their dispersibility in aqueous solutions, albumin was chemically coupled to the nHA surface. The carboxyl group present in the side chain of the introduced albumin molecule was reacted with the CFAS particles (Figure 3). For this purpose, nHA-G (nHA into which L-glutamic acid was introduced) was first prepared by reacting L-glutamic acid with hydroxyl groups on the nHA surface [19]. L-glutamic acid was dissolved in an aqueous solution at pH 5. Then, 1-ethyl-3-(3-dimethylaminopropyl) carbodiimide hydrochloride (0.5 g, 0.25 wt%) and N-hydroxysuccinimide (0.5 g, 0.25 wt%) was added and stirred for 4 h at room temperature, resulting in the activation of the carboxyl group of L-glutamic acid. Next, nHA (0.5 g) was added and the mixture was stirred for 24 h. Finally, the reaction mixture was centrifuged to obtain the precipitate. After washing the precipitate three times with deionized water, it was lyophilized to obtain nHA-G [19].

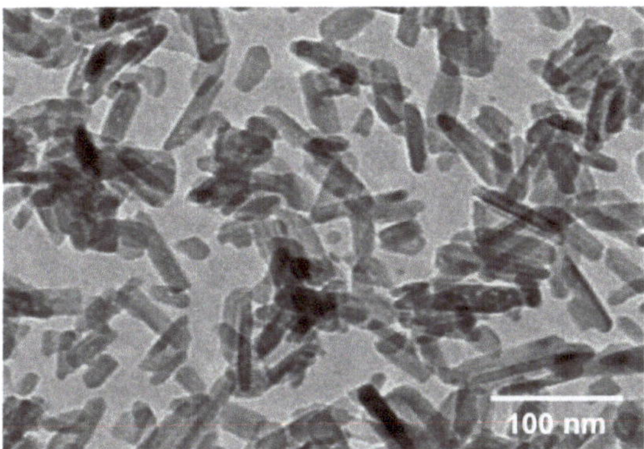

Figure 2. TEM image of nHA used in the present study synthesized by the chemical precipitation method (reference [15] is cited for the picture).

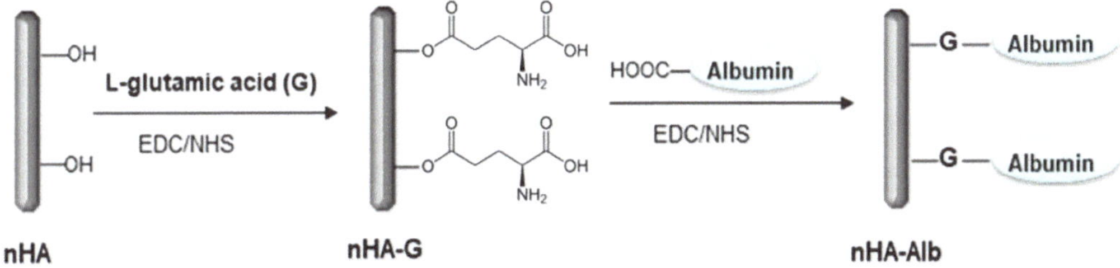

Figure 3. Schematic diagram showing the surface modification of nHA using L-glutamic acid and albumin as linkers.

The resulting nHA-G was mixed with distilled water, stirred for 1 min, and allowed to stand. No precipitate was observed with the naked eye when observing the solution. The nHA particles aggregated in an aqueous solution and formed precipitates. However, when L-glutamic acid was bound to the surface, nHA aggregation did not occur, and the dispersibility of the particles improved, indicating that precipitation did not occur. Despite the improved dispersibility of nHA, to increase its reactivity with microsized CFAS solid particles, the spacer introduced on the nHA surface must be sufficiently long and dynamically mobile in an aqueous solution [20]. Therefore, we attempted to bind water-soluble albumin to the surface of nHA-G as a second spacer [21]. We dissolved 0.06 g of albumin, acting as a chain extender, in 100 mL of deionized water. Thereafter, 0.1 g each of EDC and NHS was added and stirred for 2 h at a temperature of 25 °C, resulting in the activation of the carboxyl group of albumin. The nHA-G particles (0.5 g) were then added to the albumin solution, and the mixture was allowed to react for 24 h. The reaction solution was placed on a semi-permeable membrane (MWCO: 100,000) and dialyzed to separate the unreacted albumin. It was then washed thrice with deionized water. The resulting product was termed as albumin-immobilized nHA (nHA-Alb). Albumin immobilized on the nHA surface in an aqueous solution not only exhibited dynamic motion but also contained several carboxyl groups. Therefore, it is expected that the probability of a chemical bond forming between the primary amino group of A immobilized on the CFAS surface and albumin immobilized on nHA increases. The ATR-FTIR spectra of albumin chemically bound to the nHA particle surface and nHA-Alb chemically bound to the CFAS surface were obtained using a Galaxy 7020A device (Mattson, Fremont, CA, USA).

2.4. Chemical Bonding of nHA to the CFAS Surface

The chemical bonding of nHA-Alb to CFAS-A was performed as follows (Figure 4): nHA-Alb (0.5 g) was added to 100 mL of an aqueous solution containing 0.5 g of EDC (0.25 wt%) and 0.5 g of NHS (0.25 wt%). By stirring at 25 °C for 6 h, the carboxyl group of albumin was activated. CFAS-A was then added to the aqueous solution and stirred mechanically for 24 h. The reaction mixture was then centrifuged at 1200 rpm for 10 min to precipitate CFAS-nHA. Distilled water was added to the precipitate and centrifugation was repeated to remove any unreacted activator. The introduction of nHA into the CFAS was confirmed by ATR-FTIR spectroscopy. Additionally, the surface morphology of CFAS-nHA was examined using FE-SEM on a Hitachi 400 instrument (Tokyo, Japan).

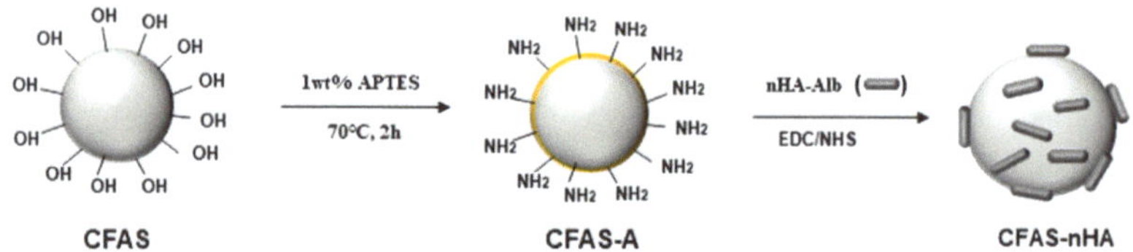

Figure 4. Schematic diagram showing the chemical bonding of nHA on the surface of CFAS microparticles.

2.5. Preparation of GIC-nHA Discs

GIC-nHA discs were prepared as follows (Figure 5): 1 g of CFAS-nHA was placed on a paper mixing pad and a commercially available liquid component (0.5 g) of GIC (GC International, Japan, GC Fuji II LC) was added dropwise and mixed. After mixing for 30 s, the cells were transferred to a single well of a 24-well culture dish. After that, LED Spotlights (470 nm blue) were irradiated for 20 s to proceed with photopolymerization to manufacture a GIC-nHA disc. A control sample was prepared by photopolymerization in the same manner using CFAS (0.5 g) and a liquid component (0.5 g) from the GC Fuji II LC. After preparing the experimental specimens, we left them for 2 days under UV clean benches before the cell culture experiment.

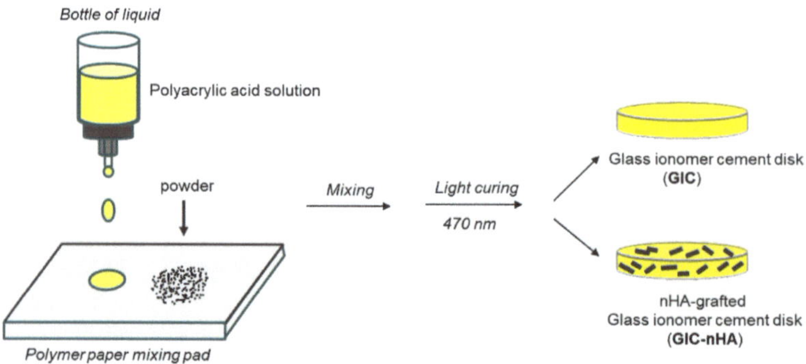

Figure 5. Schematic diagram showing the manufacturing process of glass ionomer disc using polyacrylic acid and surface-modified CFAS powder.

2.6. Cytocompatibility of GIC-nHA

Using a standard protocol, we investigated the adhesion, proliferation, and osteogenic characteristics of MC3T3-E1 cells to confirm the effects of nHA incorporation into GIC.

2.6.1. Cell Adhesion

We evaluated the cellular response of MC3T3-E1 cells (4×10^4 cells/mL) on disc surfaces by seeding them onto the GIC and GIC-nHA discs. The cells were cultured in a humidified atmosphere for 24 h using α-MEM. After removing the supernatant, the cells were washed with PBS and fixed with a 2.5% glutaraldehyde solution for 10 min. The samples were then dehydrated and dried using a critical-point dryer. Before capturing SEM images, the samples were sputter-coated with Au.

2.6.2. MTT Assay

MC3T3-E1 cells were cultured on the GIC and GIC-nHA discs for 3 days. Afterwards, the viability of the cells was measured through MTT assay. This assay relies on the reduction of MTT, a yellow water-soluble tetrazolium dye primarily produced by mitochondrial dehydrogenases, to purple formazan crystals [22]. For MTT analysis, sterilized disc samples were plated in 24-well dishes, 500 µL of non-osteogenic α-MEM was added, and MC3T3-E1 cells were seeded onto each disc at a density of 1.2×10^4 cells/cm^2. After three days of culture, the supernatant was removed and two subsequent washes of the scaffold with PBS solution were conducted sequentially. Cell-seeded scaffolds were cultured in 500 µL of MTT solution (500 µg/mL) at 37 °C for 4 h, followed by the removal of the supernatant. The generated purple formazan crystals were dissolved for 10 min using 250 µL of dimethyl sulfoxide. Wells injected with MC3T3-E1 cells without discs served as positive controls, whereas empty wells without cells served as negative controls. The absorbance of the extract was recorded at 570 nm based on 690 nm for the medium using a Synergy HT multidetection microplate reader (Synergy HT; BioTek, Shoreline, WA, USA). The amount of formazan generated was determined using a microplate reader. Data from negative controls were subtracted from the measurements. The number of viable cells was correlated with optical density, and cell viability was assessed by normalizing the values to those of the positive control wells.

2.6.3. Cytotoxicity

The cytotoxicity of GIC and GIC-nHA was evaluated by seeding cells at a density of 2.0×10^4 cells/mL onto both discs. The discs were then incubated in Dulbecco's Modified Eagle Medium (DMEM) at 37 °C in a 5% CO_2 atmosphere for 1 or 2 days. Afterwards, 5 µL of calcein-AM (4 mM in anhydrous DMSO) and 20 µL of ethidium homodimer III (2 mM in DMSO/H_2O) were added to 10 mL of PBS solution and thoroughly mixed to prepare the staining solution, which was then used to stain the cells [19,23]. After removing the DMEM from the culture wells, both the GIC and GIC-nHA discs were washed twice with PBS solution two times. A large amount of the staining solution was added to fully cover the cells. The discs were then incubated in the dark at room temperature for an additional period. After 30 min, the staining solution was removed and the cells were washed three times with PBS [19]. Finally, the cells were preserved in PBS until observation under a confocal laser scanning microscope. Live and dead cells were simultaneously assessed using calcein-AM (494 nm excitation wavelength) and ethidium homodimer III (530 nm excitation wavelength).

2.6.4. Actin Cytoskeleton Assay

Osteoblasts were seeded onto GIC discs at a concentration of 2×10^4 cells/mL and cultured for 3 days. The cells were then fixed with 4% paraformaldehyde solution. The cells were then washed with PBS containing 0.05% Tween-20. Subsequently, the samples were permeabilized with 0.1% Triton X-100 in PBS solution for 15 min at 25 °C. After permeabilization, the samples were incubated in a mixture of 1% bovine serum albumin (BSA) and PBS for 30 min. After washing the cells 2–3 times with PBS, 5(6)-tetramethylrhodamine isothiocyanate (TRITC)-conjugated phalloidin (Millipore, Burlington, MA, USA, Cat. No. 90228) was added, and the cells were incubated at ambient temperature for 30 min. Finally, a confocal laser scanning microscope (Model 700, Carl Zeiss, Oberkochen, Germany) was used to capture fluorescence images [24].

2.6.5. Von Kossa Assay

Using the von Kossa assay, we assessed the calcium deposition response of MC3T3-E1 cells cultured on the GIC and GIC-nHA discs. MC3T3-E1 cells were seeded in 24-well culture plates at a density of 5×10^4 cells/mL and cultured for 14 days in α-MEM at 37 °C in a humidified atmosphere of 5% CO_2. After gently washing the discs three times with PBS for 5 min each, they were fixed in 10% formaldehyde for 30 min. Afterwards, each disc was washed thrice with distilled water for 10 min each, followed by treatment with 5% $AgNO_3$ solution and exposure to UV irradiation for 5 min. The GIC discs were washed twice with PBS to remove the remaining $AgNO_3$ and soaked in 5% $Na_2S_2O_3$ solution for 5 min. After washing the GIC discs twice with distilled water, digital images of the stained cells were captured using an optical microscope equipped with a camera (Nikon E 4500, Minato, Japan) [22].

2.7. Statistical Analysis

The experiments were performed in triplicates, and all data were expressed as means ± standard deviation. Statistical analyses were conducted using Student's two-tailed test and Scheffe's test for multiple comparisons using IBM SPSS (version 20.0; IBM Corp., Armonk, NY, USA). Differences with p-values < 0.05 were considered statistically significant.

3. Result

3.1. Immobilization of nHA on CFAS Particle Surface

ATR-FTIR is a recommended method for confirming whether biomolecules such as amino acids or proteins have been introduced to the surface of solid particles such as CFAS or nHA [25]. Figure 6 shows the results of the ATR-FTIR measurements after the surface modification of nHA and CFAS. The infrared spectrum of the nHA sample showed typical apatite modes near 1086, 1021, 961, 600, and 562 cm^{-1} [25]. In the ATR-FTIR spectrum of nHA-Alb, obtained by binding L-glutamic acid to nHA, followed by binding to albumin, the carbonyl stretching vibration of the ester formed by the reaction between the hydroxyl group of nHA and the carboxyl group of L-glutamic acid appeared at 1707 cm^{-1}. Furthermore, peaks attributed to amides I and II of the introduced albumin appeared at 1647 cm^{-1} and 1545 cm^{-1}, respectively [26]. The characteristic peak of nHA-Alb appeared in the IR spectrum of CFAS-nHA, indicating that nHA was chemically bonded to the CFAS surface (Figure 6).

Figure 7 shows SEM images of the CFAS surface before (a) and after (b) the chemical bonding of nHA. The CFAS surface showed a nanoporous structure (Figure 7a), whereas the CFAS-nHA surface had nanorod nHA bound to the surface (Figure 7b). As such, nHA bound to the CFAS surface is expected to play a major role in improving histocompatibility when used as a filling material for hard tooth tissues [27].

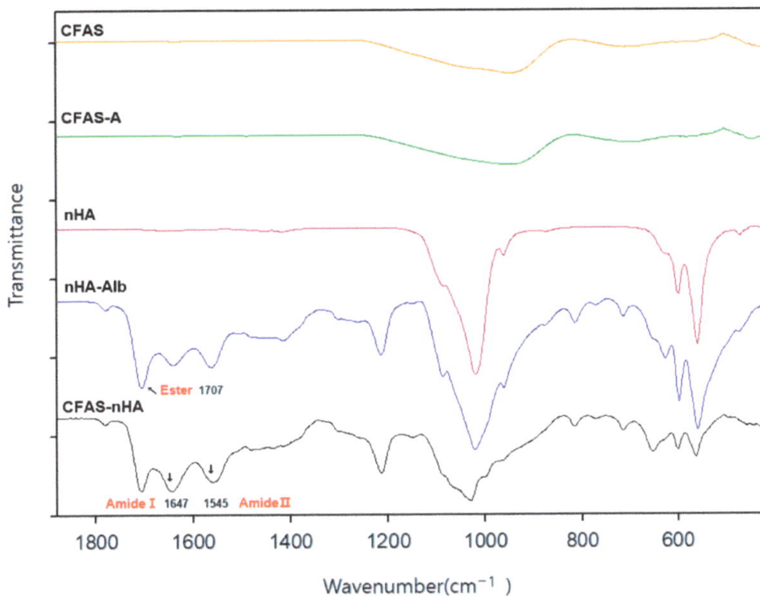

Figure 6. ATR-FTIR spectra of albumin-bound nHA (nHA-Alb) and nHA-bound CFAS (CFAS-nHA).

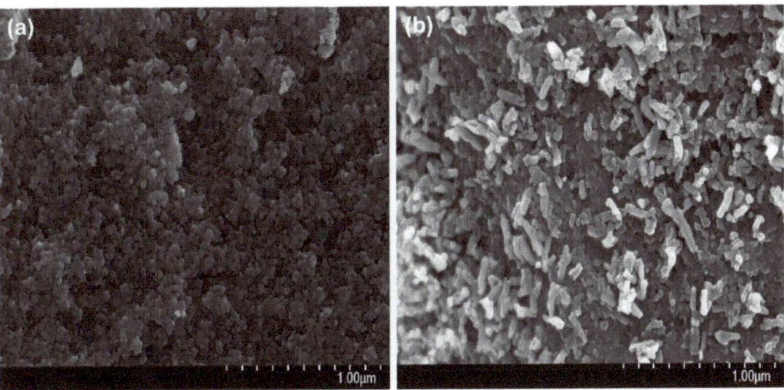

Figure 7. SEM images of CFAS microparticles before (**a**) and after (**b**) the chemical bonding of nHA.

3.2. Cytocompatibility of GIC-nHA

The GIC was prepared using CFAS and polyacrylic acid. The GIC-nHA was prepared from CFAS-nHA and polyacrylic acid. After culturing the osteoblasts on the two samples for 24 h, they were observed by SEM. As shown in Figure 8, the osteoblasts were well spread on the GIC (Figure 8a) and GIC-nHA (Figure 8b) samples after 24 h.

Osteoblasts were cultured in GIC and GIC-nHA samples for 24 and 48 h and then stained using calcein-AM and ethidium homodimer III dye; the results are shown in Figures 9 and 10. No red fluorescence was observed in the GIC or GIC-nHA samples. These results indicated that the GIC and GIC-nHA samples were not toxic. Additionally, as shown in Figure 10, the number of cells attached to GIC-nHA was much higher than that attached to GIC. This is believed to be because the introduction of CFAS-nHA into the GIC promotes cell proliferation.

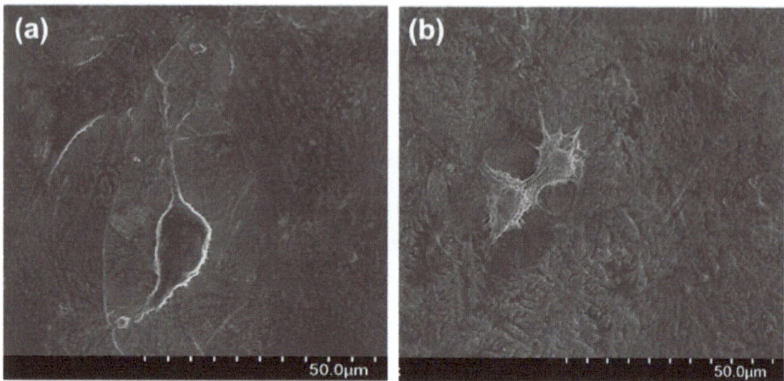

Figure 8. SEM images of osteoblasts cultured for 24 h in GIC (**a**) and GIC-nHA (**b**).

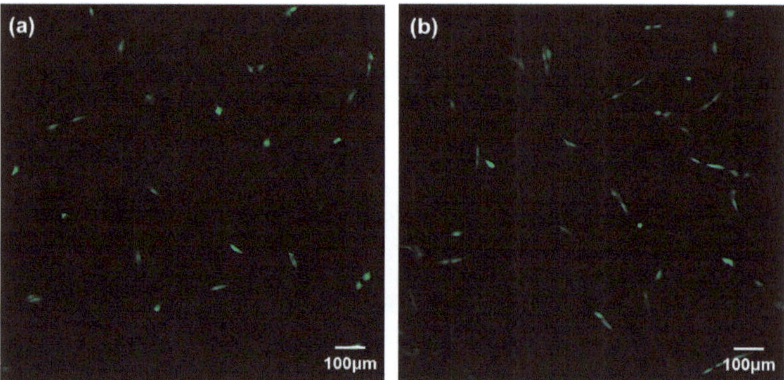

Figure 9. Live and dead assay of MC3T3-E1 cells cultured for 24 h in the presence of GIC (**a**) and the GlC-nHA (**b**). The cells were stained with calcein AM and ethidium homodimer III. Initial cell number = 1.2×10^4/mL.

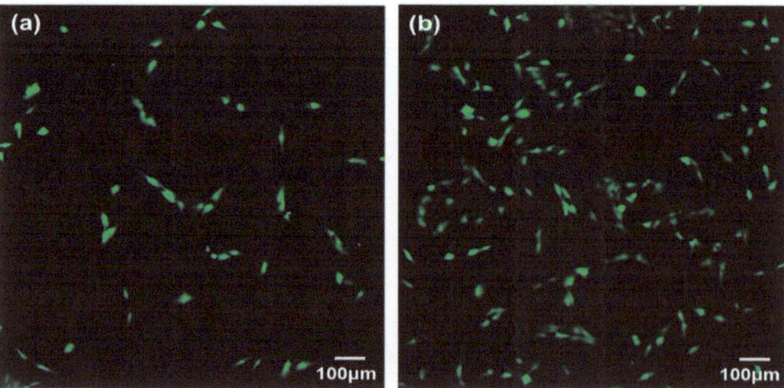

Figure 10. Live and dead assay of MC3T3-E1 cells cultured for 48 h in the presence of GIC (**a**) and the GlC-nHA (**b**). The cells were stained with calcein AM and ethidium homodimer III. Initial cell number = 1.2×10^4/mL.

MC3T3-E1 cell proliferation in GIC and GIC-nHA was assessed using an MTT assay [28]. According to the MTT analysis, the proliferation rate of MC3T3-E1 cells cultured on the GIC-nHA sample for 3 days was significantly higher than that on the GIC (Figure 11). This result indicates that the immobilization of nHA on the surface of CFAS not only enhances adhesion but also provides better opportunities for cell proliferation. This may be because nHA acts as an extracellular matrix.

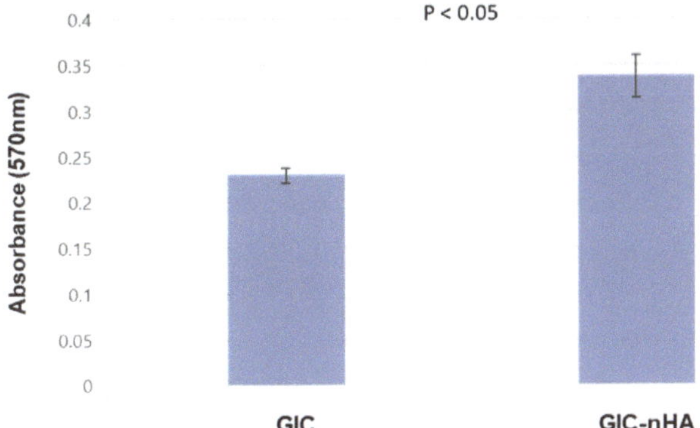

Figure 11. MTT assay of MC3T3-E1 cells cultured for 3 days in the presence of GIC and GIC-nHA in α-MEM. Initial cell concentration: 1.2×10^4/mL.

Actin microfilaments, which constitute the cytoskeleton, play a crucial role in cellular processes, cell shape determination, and cell attachment patterns. When cells attach to the extracellular matrix, they form filopodia. They are guided into their position by actin and interact with the plasma membrane [29]. Figure 12 shows confocal microscopy images of osteoblasts on the GIC (a) and GIC-nHA (b) discs. These results suggested that osteoblasts in both GIC and GIC-nHA exhibited a highly organized cytoskeleton.

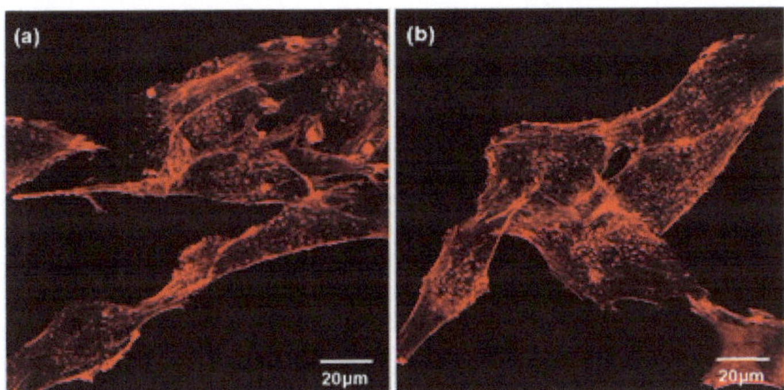

Figure 12. Confocal microscopy images of the actin filaments of osteoblasts cultured for 3 days in the presence of GIC (**a**) and GIC-nHA (**b**) (magnification, ×400).

Figure 13 shows the results of the von Kossa analysis performed after culturing the osteoblasts in the GIC and GIC-nHA samples for 14 days. Bone nodules are considered as specific markers of osteoblast differentiation. In von Kossa analysis, calcified areas were stained as black spots. According to the von Kossa analysis, more bone nodule formation was observed in GIC-nHA (Figure 13b) than in GIC (Figure 13a). These results indicated that HA triggered and accelerated osteoblast differentiation [30].

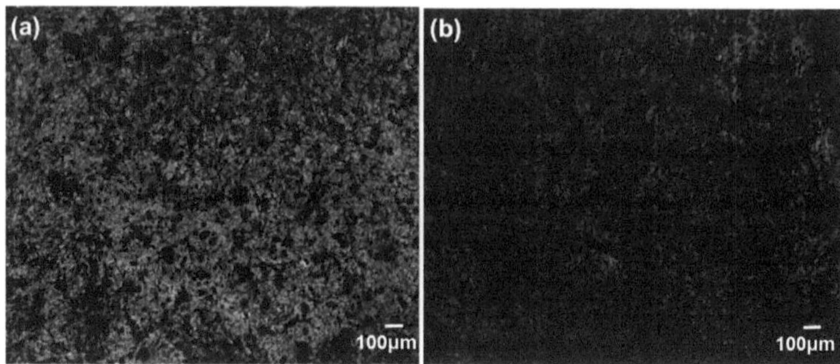

Figure 13. Von Kossa staining of osteoblast cells cultured on GIC (**a**) and GIC-nHA (**b**) for 14 days. The calcium-containing areas are stained in black.

4. Discussion

Glass ionomer cement (GIC) has been recognized as an ideal dentin substitute because of its ability to exhibit anti-cavity properties, form stable ionic bonds, and aid the remineralization process. Initially developed for grade 5 cavity restorations, GIC is now being expanded for use as a permanent adhesive, base, grade 1 and 3 filling, and as core and fissure sealants. To be effective as dental filling materials, GIC must withstand high loads. Light-cured GIC exhibits high mechanical strength and is suitable for areas where occlusal forces are applied. Additionally, light-cured GIC offers advantages such as reduced curing time, which significantly reduces the overall clinical chair time. However, formulations of light-cured material are associated with the cytotoxicity of resin monomers; this limits their clinical use near the dental pulp. In this case, traditional self-curing GICs have been used instead of light-cured GICs. Many studies have aimed at improving the mechanical properties of GIC composed of polyacrylic acid-based resins and silica-based inorganic particles [3,31,32]. However, while numerous studies have examined the mechanical properties of GIC in direct contact with the enamel or dentin, few have reported the tissue compatibility of GIC. As silica-based inorganic powder, a key component of GIC, differs chemically from enamel or dentin, long-term tissue compatibility between GIC and teeth is crucial.

Hydroxyapatite (HA) is an inorganic mineral found in bones and teeth that contributes to structural strength and bone regeneration. It has been reported that HA induces a favourable immune response [33], exerts an angiogenic effect on defective bone tissue [34], and activates osteoblasts and osteoclasts to accelerate the differentiation of these cells, resulting in bone tissue remodelling. [35]. A study was conducted to improve the histocompatibility of GIC by introducing HA into an inorganic solid powder of a glass ionomer. Noorani et al. [36] investigated the cytotoxicity of GICs containing 5% HA-SiO$_2$, Fuji IX GP, and Fuji II LC. Fuji IX GP exhibited the lowest cytotoxicity, HA-SiO$_2$-GIC showed medium toxicity, and Fuji II LC showed the highest toxicity, which was attributed to the residual unreacted monomers [37]. This study suggests that the introduction of 5% HA-SiO$_2$ into GIC was insufficient to effectively suppress toxicity.

CFAS are inorganic particles tens of micrometres in size. nHA is an inorganic particle that is several tens of nanometres in size. These two types of high-density inorganic particles easily precipitate in aqueous solution; therefore, no reaction occurs between the two particles. First, we introduced primary amino groups by reacting 3-aminopropyltriethoxysilane on the CFAS surface. In addition, L-glutamic acid was chemically bonded to the nHA surface to introduce an organic spacer on the CFAS surface. By binding L-glutamic acid to nHA, the water dispersibility of nHA-G was slightly improved, but the water dispersibility of nHA-G was still low [38]; therefore, no reaction occurred between the two particles. After trial and error, albumin was selected as the second linker. When albumin was combined with nHA-G, the water dispersibility of nHA-Alb significantly improved. Thus, the reaction between CFAS-A and nHA-Alb was successful [21]. Chemical bonding between the CFAS surface and nHA was confirmed using ATR-FTIR spectroscopy, and the presence of nHA on the CFAS particle surface was directly confirmed using a scanning electron microscope [39].

GIC can dissolve upon exposure to acidic media. Jaiswal et al. [40] measured the weight of GIC after immersion in drinking water and distilled water, which is similar to the acidic environment of the oral cavity, for a certain period. It has been reported that solubility increases when immersed in acidic beverages. Further research is required on the hydrolysis and abrasion resistance of the GIC-nHA prepared in this study in an oral environment. Kumar et al. [41] investigated the interactions between various HA discs and osteoblasts. Cell adhesion, proliferation, and metabolic activity are significantly increased in HA with charges developed on its surface. Polarizing HA discs and generating positive and negative charges on their surfaces can affect the in vitro response and activity of osteoblasts [41]. It has been reported that when a charge is generated on the HA surface, cell adhesion increases within 4 h and cell proliferation increases over a period of 7 days. From this, it can be seen that adding charge to the HA surface resulted in a significant increase in metabolic activity. Simply physically mixing nHA with CFAS can cause phase separation; therefore, nanosized nHA cannot be uniformly sprayed onto the microsized CFAS surface. In this study, the activity of fibroblasts was improved by nHA-CFAS, which was introduced as the main component of GIC by chemically bonding nHA to the surface of CFAS, an artificial mineral.

Oliva et al. [42] compared the response of cultured human osteoblasts to five commercially used glass ionomer cements (Ketac-Fil Aplicap, Ionocem, GC Fuji II, GC Fuji II LC, and Vitremer 3M). Most glass ionomer cements tested showed biocompatibility, except for Vitremer 3M, and proliferated and expressed biochemical markers of the osteoblast phenotype. As in the above study, Figure 8 shows that the osteoblasts were well spread on both the GIC and GIC-nHA samples after 24 h of culture. After culturing osteoblasts on GIC and GIC-nHA discs for 24 and 48 h, dead and live cells were observed under a confocal laser microscope. Staining with calcein-AM and ethidium homodimer III revealed that neither GIC nor GIC-nHA discs were toxic, as only live (green-stained) cells were observed (Figures 9 and 10). Additionally, more osteoblasts adhered to the GIC-nHA disc than to the GIC disc, likely because of the increased affinity of osteoblasts from the HA for binding to the inorganic powder component of GIC. The MTT assay, performed after culturing osteoblasts on GIC and GIC-nHA discs for 3 days, revealed greater cell proliferation on GIC-nHA than on pristine GIC; this was attributable to nHA enhancing osteoblastic cell growth [41].

Von Kossa staining, used to quantify calcium-like mineralization in cell cultures, showed darker black images for the GIC-nHA discs than for the GIC discs after 14 days of osteoblast culture (Figure 13). This suggests an increased osteoblast affinity for GIC-nHA due to nHA on the CFAS surface [43,44].

In the present study, the complete setting time of the light-cured GIC was not measured post-light curing. This is a limitation, as the setting process continues beyond the initial light curing phase through a chemical setting reaction. According to other studies, the complete chemical setting of light-cured glass ionomer cement typically takes 24 h to 7 days, depending on the specific formulation and environmental conditions [1,45]. For example, Sidhu et al. [1] report that while initial light curing substantially hardens the material, the chemical setting continues and typically completes within 24 h to several days. These findings suggest that researchers should consider a prolonged observation period post-light curing to fully understand the setting characteristics of GICs. Future studies should aim to measure the physical properties of the material at various intervals post-light curing to better understand the setting process. This will help provide clearer guidelines for clinical application and expected performance timelines.

The use of CFAS-nHA in GIC-nHA significantly enhanced its biocompatibility compared to conventional GIC. The introduction of nanohydroxyapatite (nHA) effectively supported cell compatibility and proliferation, as demonstrated by the MTT assays and von Kossa staining. The chemical bonding between CFAS and nHA improves the water dispersibility and ensures uniform particle distribution, indicating potential improvements in the mechanical properties. This study aimed to overcome the long working time and poor physical properties of conventional GIC, as well as the limited use of light-cured GIC due to its lower biocompatibility. Future research should explore the long-term tissue compatibility of GIC-nHA in vivo to validate these findings and optimize the properties of GIC by adjusting the nHA concentration and exploring other potential linkers. The development of GIC formulations tailored for specific dental applications may lead to broader clinical use and improved patient outcomes.

5. Conclusions

In conclusion, in the present study, we synthesized CFAS-nHA by chemically bonding nHA to the surface of CFAS, a silica-based solid particle. GIC-nHA was prepared by mixing CFAS-nHA with polyacrylic acid, a liquid component of Fuji II LC. Additionally, CFAS was mixed with the liquid components of Fuji II LC to prepare the GIC. If nano-HA is combined with the silicate-based inorganic powder of commercialized GIC using the spacer method proposed in this study, nHA-GICs with improved tooth tissue compatibility can be manufactured in large quantities. Culturing osteoblasts on GIC-nHA and GIC discs and examining cell activity and osteogenic potential revealed that GIC-nHA exhibited higher cell activity and osteogenic potential than GIC. This was attributed to the chemical bonded to the surface of GIC's silica-based solid GIC powder, which promoted osteoblast activity. Chemically combining the surface of silica-based solid particles with nHA and using it as a GIC component could improve the histocompatibility of GIC.

Author Contributions: Conceptualization, S.K. (Sohee Kang); data curation, S.J.P. and I.-K.K.; formal analysis, S.J.P.; funding acquisition, S.K. (Sohee Kang); investigation, S.J.P. and S.K. (Sukyoung Kim); methodology, S.K. (Sohee Kang), S.J.P., S.K. (Sukyoung Kim) and I.-K.K.; resources, S.K. (Sukyoung Kim); supervision, I.-K.K.; visualization, S.K. (Sohee Kang) and I.-K.K.; writing—original draft, S.J.P. and I.-K.K.; writing—review and editing, S.K. (Sohee Kang). All authors have read and agreed to the published version of the manuscript.

Funding: This work was supported by the 2023 Yeungnam University research grant (223A580032).

Institutional Review Board Statement: Not applicable.

Informed Consent Statement: Not applicable.

Data Availability Statement: Data is contained within the article.

Conflicts of Interest: The authors declare no conflicts of interest. The funders had no role in the design of the study, such as the collection, analyses, or interpretation of data; writing of the manuscript; or the decision to publish the results.

References

1. Sidhu, S.K.; Nicholson, J.W. A review of glass-ionomer cements for clinical dentistry. *J. Funct. Biomater.* **2016**, *7*, 16. [CrossRef] [PubMed]
2. Farahnaz, S.; Ali, A.A.; Saba, S.; Mina, S. Comparison of shear bond strength of three types of glass ionomer cements containing hydroxyapatite nanoparticles to deep and superficial dentin. *J. Dent.* **2020**, *21*, 132–140.
3. Bilić-Prcić, M.; Brzović Rajić, V.; Ivanišević, A.; Pilipović, A.; Gurgan, S.; Miletić, I. Mechanical properties of glass ionomer cements after incorporation of marine derived hydroxyapatite. *Materials* **2020**, *13*, 3542. [CrossRef] [PubMed]
4. Lathikumari, S.S.; Saraswathy, M. Modifications of Polyalkenoic Acid and Its Effect on Glass Ionomer Cement. *Mater. Adv.* **2024**, *5*, 2719–2735. [CrossRef]
5. Faraji, F.; Heshmat, H.; Banava, S. Effect of Protective Coating on Microhardness of a New Glass Ionomer Cement: Nanofilled Coating Versus Unfilled Resin. *J. Conserv. Dent.* **2017**, *20*, 260–263. [CrossRef] [PubMed]
6. Amin, F.; Rahman, S.; Khurshid, Z.; Zafar, M.S.; Sefat, F.; Kumar, N. Effect of nanostructures on the properties of glass ionomer dental restoratives/cements: A comprehensive narrative review. *Materials* **2021**, *14*, 6260. [CrossRef] [PubMed]
7. Arita, K.; Yamamoto, A.; Shinonaga, Y.; Harada, K.; Abe, Y.; Nakagawa, K.; Sugiyama, S. Hydroxyapatite particle characteristics influence the enhancement of the mechanical and chemical properties of conventional restorative glass ionomer cement. *Dent. Mater. J.* **2011**, *30*, 672–683. [CrossRef] [PubMed]

8. Alatawi, R.A.S.; Elsayed, N.H.; Mohamed, W.S. Influence of Hydroxyapatite Nanoparticles on the Properties of Glass Ionomer Cement. *J. Mater. Res. Technol.* **2019**, *8*, 344–349. [CrossRef]
9. Mohd Pu'ad, N.A.S.; Abdul Haq, R.H.; Mohd, N.; Abdullah, H.Z.; Idris, M.I.; Lee, T.C. Synthesis method of hydroxyapatite: A review. *Mater. Today Proc.* **2020**, *29*, 233–239. [CrossRef]
10. Ramay, H.R.; Zhang, M. Biphasic Calcium Phosphate Nanocomposite Porous Scaffolds for Load-Bearing Bone Tissue Engineering. *Biomaterials* **2004**, *25*, 5171–5180. [CrossRef]
11. Sadat-Shojai, M.; Khorasani, M.T.; Dinpanah-Khoshdargi, E.; Jamshidi, A. Synthesis methods for nanosized hydroxyapatite with diverse structures. *Acta Biomater.* **2013**, *9*, 7591–7621. [CrossRef] [PubMed]
12. Mohandes, F.; Salavati-Niasari, M. Particle Size and Shape Modification of Hydroxyapatite Nanostructures Synthesized via a Complexing Agent-Assisted Route. *Mater. Sci. Eng. C* **2014**, *40*, 288–298. [CrossRef] [PubMed]
13. Kang, I.G.; Park, C.I.; Lee, H.; Kim, H.E.; Lee, S.M. Hydroxyapatite microspheres as an additive to enhance radiopacity, biocompatibility and osteoconductivity of poly(methylmethacrylate) bone cement. *Materials* **2018**, *11*, 258. [CrossRef] [PubMed]
14. Praveen, B.; Attiguppe, R.P.; Nadig, B. An in vitro comparative evaluation of compressive strength and antibacterial activity of conventional GIC and hydroxyapatite reinforced GIC in different storage media. *J. Clin. Diagn. Res.* **2015**, *9*, ZC51–ZC55.
15. Haider, A.; Gupta, K.C.; Kang, I.K. Morphological effects of HA on the cell compatibility of electrospun HA/PLGA composite nanofiber scaffolds. *Biomed. Res. Int.* **2014**, *2014*, 11. [CrossRef]
16. Khiri, M.Z.A.; Matori, K.A.; Zaid, M.H.M.; Abdullah, A.C.; Zainuddin, N.; Jusoh, W.N.W.; Jalil, R.A.; Rahman, N.A.A.; Kul, E.; Wahab, S.A.A.; et al. Soda lime silicate glass and clam shell act as precursor in synthesize calciumfluoroaluminosilicate glass to fabricate glass ionomer cement with different ageing time. *J. Mater. Res. Technol.* **2020**, *9*, 6125–6134. [CrossRef]
17. Jia, S.; Liu, Y.; Ma, Z.; Liu, C.; Chai, J.; Li, Z.; Song, W.; Hu, K. A novel vertical aligned mesoporous silica coated nanohydroxyapatite particle as efficient dexamethasone carrier for potential application in osteogenesis. *Biomed. Mater.* **2021**, *16*, 035030. [CrossRef]
18. Aykol, M.; Montoya, J.H.; Hummelshøj, J. Rational solid-state synthesis routes for inorganic materials. *J. Am. Chem. Soc.* **2021**, *143*, 9244–9259. [CrossRef]
19. Haider, A.; Versace, D.; Gupta, K.C.; Kang, I.K. Pamidronic acid-grafted nHA/PLGA hybrid nanofiber scaffolds suppress osteoclastic cell viability and enhance osteoblastic cell activity. *J. Mater. Chem. B* **2016**, *4*, 7596. [CrossRef]
20. Kamiya, H.; Iijima, M. Surface modification and characterization for dispersion stability of inorganic nanometer-scaled particles in liquid media. *Sci. Technol. Adv. Mater.* **2010**, *11*, 044304. [CrossRef]
21. Park, S.J.; Gupta, K.C.; Kim, H.; Kim, S.; Kang, I.K. Osteoblast behaviours on nanorod hydroxyapatite-grafted glass surfaces. *Biomater. Res.* **2019**, *23*, 28. [CrossRef] [PubMed]
22. Ghasemi, M.; Turnbull, T.; Sebastian, S.; Kempson, I. The MTT Assay: Utility, Limitations, Pitfalls, and Interpretation in Bulk and Single-Cell Analysis. *Int. J. Mol. Sci.* **2021**, *22*, 12827. [CrossRef] [PubMed]
23. Xing, Z.C.; Chae, W.P.; Baek, J.Y.; Choi, M.J.; Jung, Y.; Kang, I.K. In Vitro Assessment of Antibacterial Activity and Cytocompatibility of Silver-Containing PHBV Nanofibrous Scaffolds for Tissue Engineering. *Biomacromolecules* **2010**, *11*, 1248–1253. [CrossRef] [PubMed]
24. Przekora, A.; Benko, A.; Blazewicz, M.; Ginalska, G. Hybrid Chitosan/β-1,3-Glucan Matrix of Bone Scaffold Enhances Osteoblast Adhesion, Spreading and Proliferation via Promotion of Serum Protein Adsorption. *Biomed. Mater.* **2016**, *11*, 045001. [CrossRef]
25. Tiernan, H.; Byrne, B.; Kazarian, S.G. ATR-FTIR spectroscopy and spectroscopic imaging for the analysis of biopharmaceuticals. *Spectrochim. Acta Part A Mol. Biomol. Spectrosc.* **2020**, *241*, 118636. [CrossRef] [PubMed]
26. Alhazmi, H.A. FT-IR spectroscopy for the identification of binding sites and measurements of the binding interactions of important metal ions with bovine serum albumin. *Sci. Pharm.* **2019**, *87*, 5. [CrossRef]
27. Wan Jusoh, W.N.; Matori, K.A.; Mohd Zaid, M.H.; Zainuddin, N.; Ahmad Khiri, M.Z.; Abdul Rahman, N.A.; Abdul Jalil, R.; Kul, E. Incorporation of Hydroxyapatite into Glass Ionomer Cement (GIC) Formulated Based on Alumino-Silicate-Fluoride Glass Ceramics from Waste Materials. *Materials* **2021**, *14*, 954. [CrossRef]
28. Soesilawati, P.; Rizqiawan, A.; Roestamadji, R.I.; Arrosyad, A.R.; Firdauzy, M.A.; Abu Kasim, N.H. In Vitro Cell Proliferation Assay of Demineralized Dentin Material Membrane in Osteoblastic MC3T3-E1 Cells. *Clin. Cosmet. Investig. Dent.* **2021**, *13*, 443–449. [CrossRef]
29. Fletcher, D.A.; Mullins, R.D. Cell mechanics and the cytoskeleton. *Nature* **2010**, *28*, 485–492. [CrossRef]
30. Tayalia, P.; Mooney, D.J. Controlled growth factor delivery for tissue engineering. *Adv. Mater.* **2009**, *21*, 3269–3285. [CrossRef]
31. Moshaverinia, A.; Ansari, S.; Moshaverinia, M.; Roohpour, N.; Darr, J.A.; Rehman, I. Effects of incorporation of hydroxyapatite and fluoroapatite nanobioceramics into conventional glass ionomer cements (GIC). *Acta Biomater.* **2008**, *4*, 432–440. [CrossRef] [PubMed]
32. Barandehfard, F.; Kianpour Rad, M.; Hosseinnia, A.; Khoshroo, K.; Tahriri, M.; Jazayeri, H.E.; Moharamzadeh, K.; Tayebi, L. The addition of synthesized hydroxyapatite and fluorapatite nanoparticles to a glass-ionomer cement for dental restoration and its effects on mechanical properties. *Ceram. Int.* **2016**, *42*, 17866–17875. [CrossRef]
33. Teng, N.C.; Nakamura, S.; Takagi, Y.; Yamashita, Y.; Ohgaki, M.; Yamashita, K. A New Approach to Enhancement of Bone Formation by Electrically Polarized Hydroxyapatite. *J. Dent. Res.* **2001**, *80*, 1925–1929. [CrossRef] [PubMed]
34. Kobayashi, T.; Nakamura, S.; Yamashita, K. Enhanced Osteobonding by Negative Surface Charges of Electrically Polarized Hydroxyapatite. *J. Biomed. Mater. Res.* **2001**, *57*, 477–484. [CrossRef] [PubMed]

35. Royce, B.S.H. Field Induced Transport Mechanism in Hydroxyapatite. *Ann. N. Y. Acad. Sci.* **1974**, *238*, 131–138. [CrossRef] [PubMed]
36. Noorani, T.Y.; Luddin, N.; Ab Rahman, I.; Masudi, S.M. In vitro cytotoxicity evaluation of novel nano-hydroxyapatite-silica incorporated glass ionomer cement. *J. Clin. Diagn. Res.* **2017**, *11*, ZC105–ZC109. [CrossRef] [PubMed]
37. Stanislawski, L.; Daniau, X.; Lautié, A.; Goldberg, M. Factors responsible for pulp cell cytotoxicity induced by resin-modified glass ionomer cements. *J. Biomed. Mater. Res.* **1999**, *48*, 277–288. [CrossRef]
38. Kang, S.; Haider, A.; Gupta, K.C.; Kim, H.; Kang, I.K. Chemical bonding of biomolecules to the surface of nano-hydroxyapatite to enhance its bioactivity. *Coatings* **2022**, *12*, 999. [CrossRef]
39. Kim, H.; Lee, Y.H.; Kim, N.K.; Kang, I.K. Bioactive surface of zirconia implant prepared by nano-hydroxyapatite and type I collagen. *Coatings* **2022**, *12*, 1335. [CrossRef]
40. Jaiswal, S.; Kusumitha, P.; Alla, R.K.; Guduri, V.; AV, R.; Sajjan, S.M.C. Solubility of Glass Ionomer Cement in Various Acidic Beverages at Different Time Intervals: An In Vitro Study. *Int. J. Dent. Mater.* **2022**, *4*, 78–81. [CrossRef]
41. Kumar, D.; Gittings, J.P.; Turner, I.G.; Bowen, C.R.; Bastida-Hidalgo, A.; Cartmell, S.H. Polarization of hydroxyapatite: Influence on osteoblast cell proliferation. *Acta Biomater.* **2010**, *6*, 1549–1554. [CrossRef] [PubMed]
42. Oliva, A.; Della Ragione, F.; Salerno, A.; Riccio, V.; Tartaro, G.; Cozzolino, A.; D'Amato, S.; Pontoni, G.; Zappia, V. Biocompatibility studies on glass ionomer cements by primary cultures of human osteoblasts. *Biomaterials* **1996**, *17*, 1351–1356. [CrossRef]
43. Schneider, M.R. Von Kossa and his staining technique. *Histochem. Cell Biol.* **2021**, *156*, 523–526. [CrossRef] [PubMed]
44. Chen, C.S.; Chang, J.H.; Srimaneepong, V.; Wen, J.Y.; Tung, O.H.; Yang, C.H.; Lin, H.C.; Lee, T.H.; Han, Y.; Huang, H.H. Improving the in vitro cell differentiation and in vivo osseointegration of titanium dental implant through oxygen plasma immersion ion implantation treatment. *Surf. Coat. Technol.* **2020**, *399*, 126125. [CrossRef]
45. Pearson, G.J.; Atkinson, A.S. Long-Term Flexural Strength of Glass Ionomer Cements. *Biomaterials* **1991**, *12*, 658–660. [CrossRef]

Article

Influence of TiO$_2$ on the Microstructure, Mechanical Properties and Corrosion Resistance of Hydroxyapatite HaP + TiO$_2$ Nanocomposites Deposited Using Spray Pyrolysis

Hafedh Dhiflaoui [1,2,*], Sarra Ben Salem [1], Mohamed Salah [1], Youssef Dabaki [1,3], Slah Chayoukhi [4], Bilel Gassoumi [5], Anouar Hajjaji [6], Ahmed Ben Cheikh Larbi [1], Mosbah Amlouk [7] and Hicham Benhayoune [8,*]

[1] Laboratoire de Mécanique, Matériaux et Procédés LR99ES05, Ecole Nationale Supérieure d'Ingénieurs de Tunis, Université de Tunis, Tunis 1008, Tunisia; sarrabensalem475@gmail.com (S.B.S.); bohr2011@gmail.com (M.S.); dabakiyoussef@gmail.com (Y.D.); ahmed.cheikhlarbi@gmail.com (A.B.C.L.)

[2] Institut Supérieur des Sciences Appliquées et de Technologie de Kasserine, Université de Kairouan, Tunis 3100, Tunisia

[3] Laboratoire de Physico-Chimie de l'Atmosphere, Université du Littoral Cote d'Opale, 59140 Dunkerque, France

[4] Laboratoire Mécanique, Productique et Énergétique (LMPE), Ecole Nationale Supérieure d'Ingénieurs de Tunis, Université de Tunis, B.P. 56, Bab Menara, Tunis 1008, Tunisia; slah.chayoukhi@fst.utm.tn

[5] Laboratoire Nanomatériaux, Nanotechnologie et Énergie (L2NE), Faculté des Sciences de Tunis, Université de Tunis El Manar, Tunis 2092, Tunisia; gassoumi.bilel@gmail.com

[6] Laboratoire de Photovoltaïque, Centre de Recherches et des Technologies de l'Energie, Technopole de Borj-Cédria, BP 95, Hammam-Lif 2050, Tunisia

[7] Department de Physique, Unité de Physique des Dispositifs à Semi-Conducteurs, Faculté des Sciences de Tunis, Tunis El Manar University, Tunis 2092, Tunisia; mosbah.amlouk@gmail.com

[8] Institut de Thermique, Mécanique et Matériaux (ITheMM), Université de Reims Champagne-Ardenne (URCA), 51100 Reims, France

* Correspondence: dhafedh@gmail.com (H.D.); hicham.benhayoune@univ-reims.fr (H.B.)

Citation: Dhiflaoui, H.; Ben Salem, S.; Salah, M.; Dabaki, Y.; Chayoukhi, S.; Gassoumi, B.; Hajjaji, A.; Ben Cheikh Larbi, A.; Amlouk, M.; Benhayoune, H. Influence of TiO$_2$ on the Microstructure, Mechanical Properties and Corrosion Resistance of Hydroxyapatite HaP + TiO$_2$ Nanocomposites Deposited Using Spray Pyrolysis. *Coatings* **2023**, *13*, 1283. https://doi.org/10.3390/coatings13071283

Academic Editor: Maria Vittoria Diamanti

Received: 29 May 2023
Revised: 12 July 2023
Accepted: 19 July 2023
Published: 21 July 2023

Abstract: Titanium oxides and their alloys are widely used in medical applications because of their biocompatibility. However, they are characterized by their low resistance to corrosion. The HaP + TiO$_2$ nanocomposites' coating was applied in different experiments, especially on a Ti-6Al-4V substrate with the spray pyrolysis process to deal with such weakness. The TiO$_2$ content effects on the surface morphology and the phase composition were investigated using a scanning electron microscopy, X-ray microanalysis (SEM-EDXS) and X-ray diffraction (XRD). The mechanical properties were determined with nanoindentation. The potentiodynamic polarization, electrochemical impedance spectroscopy (EIS) and simulated body fluid (SBF) solution environment tests were carried out to investigate the corrosion resistance of HaP + TiO$_2$/Ti6Al4V systems. The experimental findings revealed that sprayed thin films possessed uniform morphology. The coatings' nanoindentations proved that the HaP + 20% TiO$_2$ coating hardness (252.77 MPa) and the elastic modulus (52.48 GPa) overtopped those of the pure hydroxyapatite coatings. The corrosion test demonstrated that the corrosion current density of about 36.1 μA cm^{-2} and the corrosion potential of the order of -392.7 mV of HaP + 20% TiO$_2$ was lower compared to the pure HaP coating.

Keywords: HaP nanocomposite; corrosion; nanoindentation; mocrosctructure

1. Introduction

During the past decades, metallic implants (e.g., stainless steel, titanium, alloys, etc.) have been intensively used in different medical applications. The Ti-6Al-4Valloy is generally used as an implant in artificial hip and knee joints due to its low elastic modulus and its good biocompatibility [1,2]. However, the use of Ti-6Al-4V implants and their application in the human body can cause health problems due to the release of toxic aluminum (Al) and vanadium (V) ions that can damage the cells and tissues, and it may even lead to cancer [3–5]. To overcome this deficit, surface modification consists of one of the envisaged solutions to reduce the release of Al and V ions and to improve their corrosive effect on human body fluids. Hydroxyapatite ($Ca_{10}(PO_4)_6(OH)_2$) has also attracted great interest in the medical field due to its excellent bioactivity and biocompatibility. Indeed, this material has become particularly interesting in artificial implants and human tissue restorations since it offers bone growth around the implant [6,7]. Moreover, its morphology an dits structure are similar to those of human hard tissues, especially those of bones and teeth. HaP-based bioactive coatings are the biomaterials that replace tissues in the body. They consist of nanoparticles similar to those of the inorganic molecules of human bones.

However, HaP coatings are generally synthesized using several methods such as physical vapor deposition (PVD) [8], sol–gel [9], micro-arc oxidation [10], plasma spray [11], etc. Among these techniques, spray pyrolysis shows some advantages, including uniformity in the coating's thickness, a high deposition rate, low cost and the ability to coat complex shapes and geometries. In addition, this technique can be used to produce crystalline coatings without applying post-heating [12]. HaP has unfavorable mechanical properties, which makes it unsuitable for load-bearing conditions. As it is brittle and weak, HaP's clinical application in orthopedics and dentistry is remarkably reduced [13]. Thus, several solutions were proposed to improve the long-term performance of this compound. For instance, the use of titanium and its alloys as bearing implants can be mentioned due to their excellent mechanical toughness, strength, biocompatibility and corrosion resistance [14,15]. The mechanical properties of HaP can be enhanced by employing other ceramic materials. This approach presents an alternative to improve the overall durability and strength of HaP in various applications. According to Lee et al. [16], ZrO_2, Al_2O_3 and their composite ceramics exhibit strong biocompatibility, exceptional oxidation resistance and high wear resistance, making them efficiently employed in both matrix and reinforcement phases. Tanaka et al. [17] used ZrO_2, La_2O_3 and TiO_2 as binding agents to improve HaP's bending and tensile strength. Also, Oktar [18] observed that the microstructure and phases formed during the sintering of the composite materials significantly influenced the densification and mechanical properties of the resulting composites. These composites, containing 5 wt% or 10 wt% of TiO_2, were made of HaP obtained from bovine bone. Recently, Vemulapalli et al. [19] have found that the mechanical properties (hardness, Young's modulus, fracture toughness, bending strength and compressive strength) of the composites (HaP/TiO_2), produced using ball milling, improved significantly with the addition of TiO_2 content. On the other hand, Ahmadi et al. [6] showed that TiO_2/HaP + 15 wt% of TiO_2 + 5 wt% of Ag coating had a higher adhesion strength (20.5 MPa), hardness (5.2 GPa) and elastic modulus (110.6 GPa), compared to the pure hydroxyapatite coatings. Indeed, they also showed that these coatings had the lowest corrosion current density (0.22 $\mu A/cm^2$).

To date, there are a few studies that have investigated the characterization of HaP + TiO_2 thin films with various TiO_2 contents in the literature, but they did not combine the study of microstructural properties with that of the nanomechanical and electrochemical characteristics [20,21]. The objective of this investigation is to grow HaP coatings on Ti-6Al-4V substrates, which have a structure most similar to that of human bone. Therefore, a detailed analysis of the mechanical and corrosion resistance of the deposited HaP + TiO_2 thin film seems to be in demand for the assessment of its reliability. The present contribution is in the detailed study of the correlation between the nanomechanical properties, corrosion performance in a simulated body fluid (SBF) medium, coating morphology and TiO_2 content.

2. Materials and Methods

2.1. Sample Preparations

Firstly, HaP + TiO_2 nanocomposites were successfully synthesized using an aqueous solution of 0.042 M of $Ca(NO_3)_2 \cdot 4H_2O$ by adding 25 mL of a 5 mM acetic acid as a catalyst. TiO_2 nano-powder (particle size lower than 21 nm) [22], provided by Sigma-Aldrich, was added as a doping agent with varying amounts of 0%, 5%, 10% and 20% (calculated to total weight of TiO_2). Then, the solution was stirred for 15 min. Cylindrical rods of Ti-6Al-4V of dimensions 12 mm × 4 mm were used as substrates. To accomplish the synthesis of the coatings, the substrates underwent both mechanical polishing with abrasive paper and chemical etching in diluted solution acids consisting of nitric acid (HNO_3, 6% in volume) and hydrofluoric acid (HF, 3% in volume) at room temperature for 30 s. Afterward, the substrates were ultrasonically cleaned with acetone for 15 min and washed with distilled water after each step.

Secondly, the substrates were gradually heated on a heating plate to prevent thermal shock. Once the desired synthesis temperature of 450 °C was reached, the solution was pumped at a fixed rate of 4 mL/min towards the sputtering device positioned at about 27 cm away from the substrates [23]. Nitrogen gas was employed as the carrier gas with a pressure of approximately 0.5 bars. Finally, the resulting coatings were cooled naturally for three hours until attaining room temperature (Figure 1) [24,25].

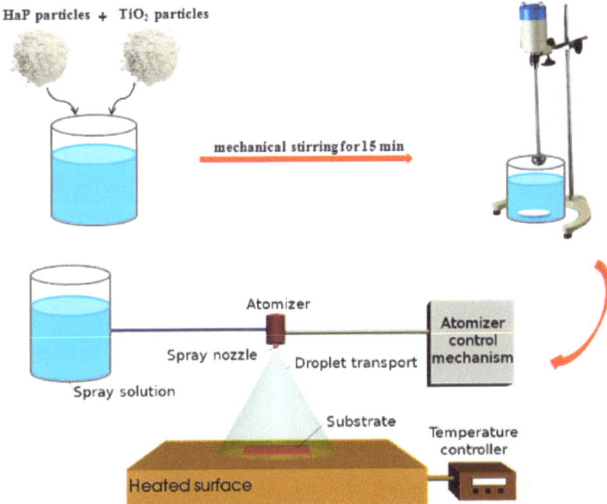

Figure 1. Sample preparations and principle of the spray pyrolysis process.

2.2. Characterization of the Coatings

The coatings morphology was characterized via a LaB6 electron microscope (Virginia Tech Switchboard, Blacksburg, CA, USA) operating at 0–30 kV. The chemical composition of the material was analyzed using both X-ray microanalysis (SEM-EDXS) (IXRF, Austin, TX, USA) and a scanning electron microscope (SEM) (IXRF, Austin, TX, USA). Furthermore, the phase identification and composition analysis were conducted with the help of a Bruker D8 Advance diffractometer (Bruker, Santa Barbara, CA, USA) utilizing X-ray diffraction techniques. The experiment involved adjusting the angle within a range of $2\theta = 20$ to $80°$ using monochromatic CuKα radiation. The increment between each measurement was set to 0.04°. The mechanical properties of the HaP + TiO_2 coatings were specified through nanoindentation. The nanoindentation tests were performed using an Anton Paar NHT^2 instrument in Anton Paar, Buchs, Switzerland). The equipment was fitted with a diamond Berkovich tip that featured a nominal angle of 65.3° and a nominal radius curvature of 20 nm. All samples were subjected to a consistent loading and unloading rate of 160 mN/min during the tests. The tests were restricted to a maximum load of 80 mN. During the tests, a nanoindentation technique was used with a Berkovich diamond tip that featured a nominal angle of 65.3° and a nominal radius with a curvature equal to 20 nm. The nanoindentation process was carried out using a nanoindenter instrument CSM type manufactured by Anton Paar, Buchs, Switzerland). The coatings' elastic modulus

(E_{IT}) and hardness (H_{IT}) were measured as part of the experiment. The measurements were reached by recording the indenter displacement as a function of the applied normal load. During the loading phase, the penetration depth increased until reaching a maximum value (P_{max}). After removing the indenter, the indentation depth declined to the residual indentation depth (h_r). Additionally, the hardness and elasticity modulus were determined by analyzing the discharge curve and applying the Oliver–Pharr model [26] that allowed their calculation, relying on the results presented by the discharge curve. As part of the analysis, the maximum load was divided by the projected area of the indenter in contact with the sample at the point of maximum load. This measurement was used as a factor in determining the hardness of the sample. Figure 2 depicts the typical curve of a nanoindentation test.

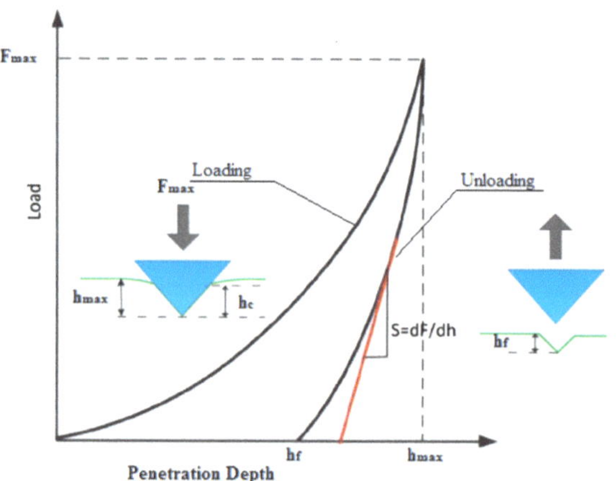

Figure 2. Typical curve of a nano-indentation test.

To assess how TiO_2 impacted the hardness and elastic modulus of HaP coatings, nanoindentation measurements were carried out in control mode. The maximum load in all the experiments was maintained at 80 mN. The loading and unloading rates were set to 160 mN/min. To ensure that the test results were reliable and accurate, each test was conducted five times.

Electrochemical measurements were conducted using the Ec-Lab system (Biologic, Seyssinet-Pariset, France) (version V11.50) made up of a Potentiostat–Galvanostat PGZ301 at room temperature in a three-electrode cell: (i) a saturated calomel reference electrode (SCE) (Hg/Hg_2Cl_2) filled with 1 M KCl and employed to measure the potential; (ii) an auxiliary electrode consisting of a coiled nickel wire; and (iii) a working electrode (test sample). The Tafel graphs are measured in the potential range of [−200 mV; +200 mV] at 1 mV/s. The polarization curves of Ti-6Al-4V and Ti-6Al-4V coatings were generated while in a simulated body fluid solution at a temperature of 37 °C. To ensure their reliability, each corrosion test was conducted three times. The main components of the used solution are listed in Table 1.

To examine the electrochemical behavior of the samples, electrochemical impedance spectra (EIS) measurements were carried out within a frequency span of 10 mHz to 10 kHz. The ZSimpWin software (Gamry Instruments, Warminster, PA, USA) (version 3.5) was used to model and interpret the results via an equivalent circuit. The following Figure 3 summarizes all the test instruments and product materials used in this work.

Table 1. Chemical composition of SBF solution [27].

Component	Amount in 1000 mL
NaCl	8.035 g
$NaHCO_3$	0.355 g
KCl	0.225 g
$MgCl_2.6H_2O$	0.311 g
Na_2SO_4	0.072 g
1.0 M HCl	39 mL
$CaCl_2$	0.292 g
$((HOCH_2)_3CNH_2)$	6.118 g

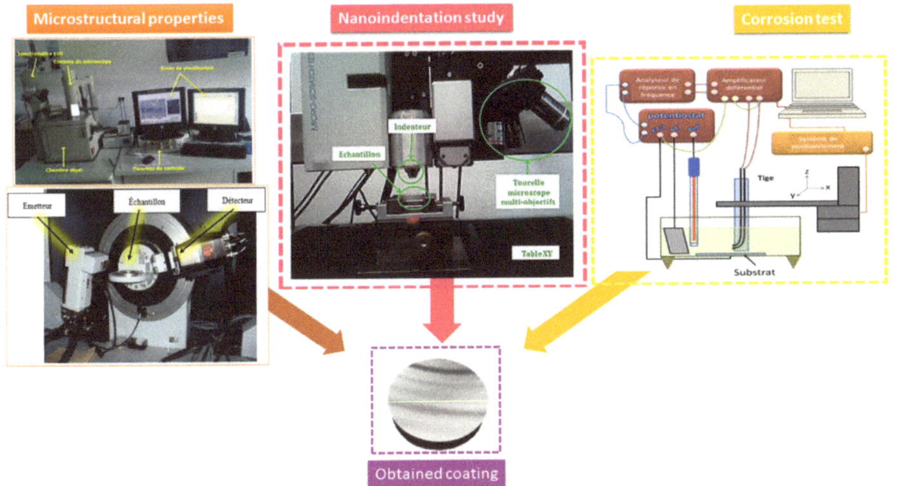

Figure 3. Macro photograph of the used test instruments.

3. Results and Discussion

3.1. Effect of TiO_2 Contents on the Morphologies and Microstructural Properties

The SEM images of HaP + TiO_2 coatings are presented in Figure 4. The experimental results show that the sprayed coatings possessed a homogeneous surface without any cracks or holes. Moreover, the TiO_2 particles were evenly distributed throughout the film. It was also noticed that the shape and uniformity of the grains were affected by the concentration of doping in the TiO_2 material. At a 20% doping concentration, the grains had a more uniform and spherical shape. However, a 10% concentration resulted in non-uniform and irregularly shaped grains. The results show that the addition of TiO_2 played a crucial role in determining the morphology and size of the HaP. For instance, Nathanael et al. [28] reported that the presence of the crystallized TiO_2 anatase phase promoted the molecular ordering of Ca and P ions, which ordinarily occurred in a random distribution resulting in the formation of the apatite structure. They also showed that the anatase phase of TiO_2 stimulated the nucleation and growth of HaP more than the rutile phase due to the better lattice compatibility between anatase TiO_2 and HaP.

EDX proved that, when the content of TiO_2 increased, the obtained Ca/P ratio for the HaP-coated sample was 1.74 and reduced to the range of HaP + 5% TiO_2 (1.73), HaP + 10% TiO_2 (1.71) and HaP + 20% TiO_2 (1.68) when the TiO_2 content increased. The Ca/P ratios between 1.67 and 1.76 are suitable as standard ratios for bioimplant coatings [29].

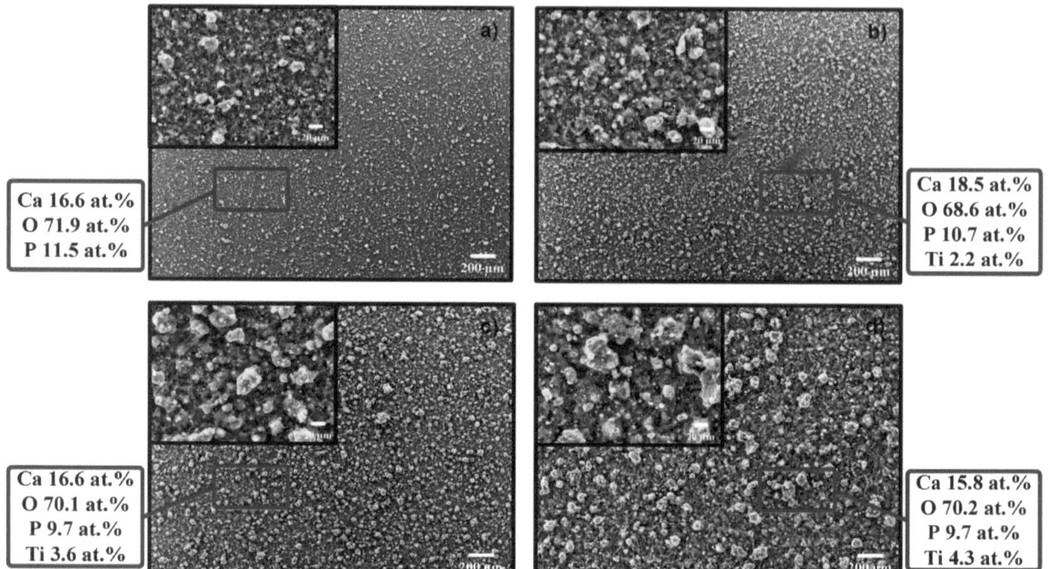

Figure 4. SEM images of HaP and HA/TiO$_2$ coatings. (**a**) HaP, (**b**) HaP + 5% TiO$_2$, (**c**) HaP + 10% TiO$_2$ and (**d**) HaP + 20% TiO$_2$.

The XRD patterns for the samples HaP, HaP + 5% TiO$_2$, HaP + 10% TiO$_2$ and HaP + 20% TiO$_2$ are presented in Figure 5. According to the XRD patterns, the HaP peaks were observed in positions 2θ = 31°, 78°, 51°, 52°, 69° and 74°. The diffraction peaks observed were consistent with the pure hexagonal phase, which arein agreement with the bulk hydroxyapatite crystals and can be easily indexed (JCPDS#9-432). The addition of TiO$_2$ resulted in the appearance of the phase in positions 2θ = 26.1°, 36.8°, 47.55°, 54.1°, 55.61°, 66.3° and 73.4° (JCPDS#84-1286) [30,31]. All X-ray peaks could be indexed along with the Tetragonal phase.

We noticed that the diffraction pattern was almost identical to that of pure HaP, although with a slight decrease in the intensity of HaP peaks. Also, in nanocomposites containing 20% of TiO$_2$, the TiO$_2$ phase was dominated. The introduction of TiO$_2$ in the early stages initiated the nucleation of nanorods and accelerated their growth through a process of heterogeneous nucleation [32,33]. Nathaniel et al. [34] showed that after the addition of 40% TiO$_2$, the excess of Ti(OH)$_4$washydrolyzed, nucleated as spherical particles and deposited on the surface of the HaP nanorods. They found that the lattice parameters for the HaP phase varied, with parameters (a) and (c) decreasing from 9.418 to 9.397 Å and from 6.884 to 6.801 Å, respectively. Likewise, for the TiO$_2$ phase, parameters "a" and "c" increased, respectively, from 3.785 to 3.823 Å and from 9.513 to 9.585 Å. These variations were assigned to the presence of TiO$_2$ in the HaP structure. Hence, as more TiO$_2$ was added and covered the HaP phase's surface, the intensity of the XRD peak of HaP was decreasing [28]. Observations indicated that as the concentration of TiO$_2$ increased, there was a broadening in the XRD peaks, and this resulted in a decrease in the crystallite size decrease. The crystallite size of the composites was estimated from Scherrer's formula (Equation (1)) [35,36]:

$$D = 0.9\lambda/B \cdot \cos\theta \tag{1}$$

where λ is the X-ray wavelength, θ represents the diffraction angle and B corresponds to the full width at the diffraction peak half maximum.

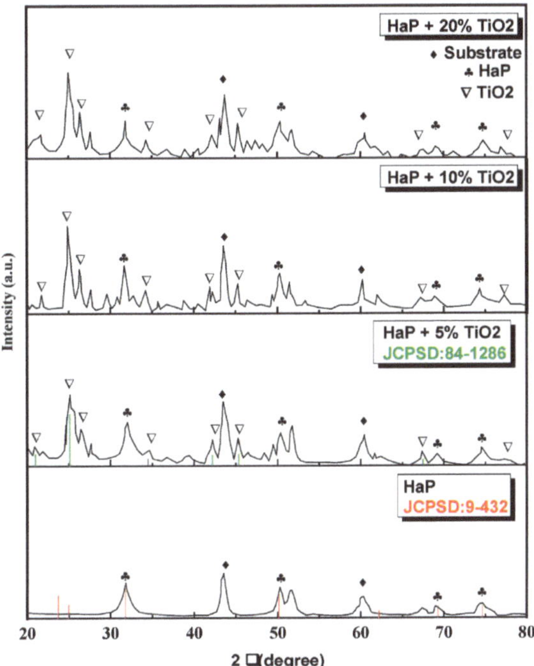

Figure 5. X-ray diffraction pattern of HaP and HaP/TiO$_2$ composites.

The result shows a gradual decrease in the crystallite size from 52 nm (5% TiO$_2$) to 20 nm with a 20% increase in TiO$_2$ amount.

3.2. Effect of TiO$_2$ Contents on the Hardness and the Elastic Modulus

The evolution of the applied load according to the penetration depth provided by the nanoindentation for different coatings is depicted in Figure 6. The experimental findings exhibit a similar trend in which the maximum displacement declined due to the addition of TiO$_2$. The numerical data of the hardness (H) and the elastic modulus (E) are gathered in Table 2.

Table 2 and Figure 6 demonstrate that the pure HaP coating exhibited the lowest hardness (98.98 MPa) and elastic modulus (21.29 GPa), compared to the other coatings. By adding 5% of TiO$_2$, the hardness and elastic modulus rose to 122.86 MPa and 30.13 GPa, respectively. Many researchers proved that the addition of TiO$_2$ to HaP leads to a decrease in the grain size of the used material, as observed in the increase of both the hardness and the elastic modulus. This increase was attributed to the relationship between the nano-hardness, elastic modulus, grain size and crystal size. This relation shows that the decrease in the grain size led to an improvement in the latter properties [4,7,17]. The highest hardness and Young's modulus values of 252.77 MPa and 52.48 GPa were obtained from the 20% TiO$_2$ nanocomposite. From the experimental results, it was found that the addition of TiO$_2$ remarkably enhanced the Young's modulus and the hardness of the HaP/TiO$_2$ composites. Refs. [28,37] show that the microstructural and morphological modifications due to the addition of TiO$_2$ on HaP have a strong influence on the mechanical properties of the composite coating. According to the results regarding the nanoindentation test, it was noted that a progressive increase in the mechanical properties of hardness and Young's modulus was found as the content of TiO$_2$ reinforcement increased. Also, the test depicted that the HaP + 20% TiO$_2$ coating had the highest hardness value, while the pure HaP-coated sample had the lowest hardness.

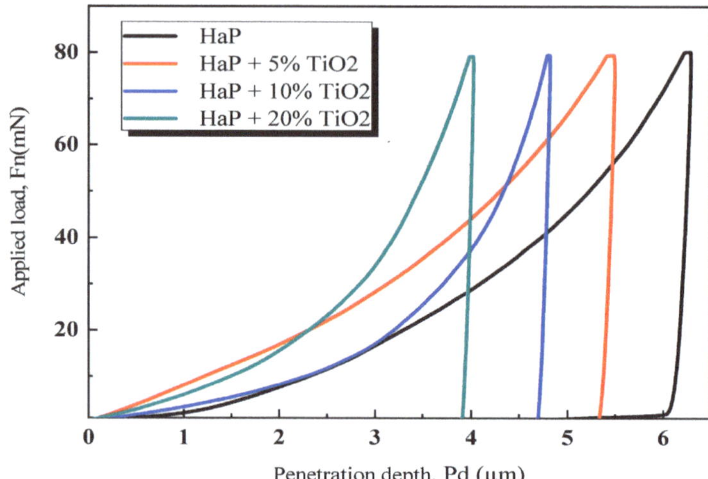

Penetration depth, Pd (μm)

Figure 6. Force–penetration depth curves for HaP, HaP + 5% TiO_2, HaP + 10% TiO_2 and HaP + 20% TiO_2 coatings.

Table 2. Nanoindentation results of the different used coatings.

Sample	H (MPa)	E (GPa)
HaP	98.98 ± 1.81	21.29 ± 0.13
HaP + 5% TiO_2	122.86 ± 2.2	30.13 ± 0.22
HaP + 10% TiO_2	172.18 ± 1.52	42.74 ± 0.34
HaP + 20% TiO_2	252.77 ± 1.63	52.48 ± 0.51

3.3. Effect of TiO_2 Contents on the Corrosion Properties of the Coatings

The Tafel extrapolation technique, demonstrated in the polarization curves, was utilized to provide several corrosion parameters (e.g., Tafel constants (β_a and β_c), corrosion potential (E_{corr}) and corrosion current density (I_{corr})). Afterward, the polarization resistance (R_p) was computed using the typical Stern–Geary equation [21]. Figure 7 presents the Tafel curves given by the polarization tests, while Table 3 shows the obtained values.

$$R_p = \frac{\beta_a \times \beta_c}{2.303(\beta_a + \beta_c)i_{corr}} \tag{2}$$

The obtained results reveal that the coated samples had a superior E_{corr} compared to the uncoated substrate. The corrosion current density I_{corr} of the uncoated substrate was equal to 38.80×10^{-2} (mA cm^{-2}), indicating poor corrosion resistance. Although the introduction of HaP coatings remarkably minimized the corrosion current density to 267 μA cm^{-2}, the appearance ofcracks in the pure HaP coating made it possible to expose the surface of the substrate to the SBF solution under such high current density.

A similar result was obtained by Fathi et al. [38] for HaP coatings on implant materials in simulated body fluids. The authors showed that the corrosion potential E_{corr} value of the coated sample (358.79 mV) shifted to a valuable amount, compared to that of the substrate (−363.96 mV). This result further confirmed that the coated samples had superior passivation than the substrate. Amaravathy et al. [39] found that, in comparison to the uncoated sample, the HaP-coated substrate exhibited a higher corrosion potential and less corrosion current density. As it was observed that the pure HaP coatings had less adhesion behavior, resulting in these coatings breaking easily and losing barrier properties. Such behavior allowed the simulatedbody fluid (SBF) to penetrate the cracks and pores. Byanalyzing the results, it was concluded that the HaP + TiO_2 coatings could provide better corrosion protection than the pure HaP coatings. Moreover, with the rise of TiO_2 in HaP, the I_{corr} value declined, and the HaP + 20% TiO_2-coated sample showed the lowest I_{corr} value. Moskalewicz et al. [40] demonstrated that adding TiO_2 coatings resulted in the corrosion resistance increase of the metallic substrate. This finding was observed because of their homogeneous microstructure and very good adhesion to the substrate. In fact, the corrosion potential rose from ~-0.62 (potential of the cathodicanodic transition of the pristine Co-28Cr-5Mo alloy) to about -0.48 V for the HaP and to -0.59 V for the TiO_2/HaP coatings.

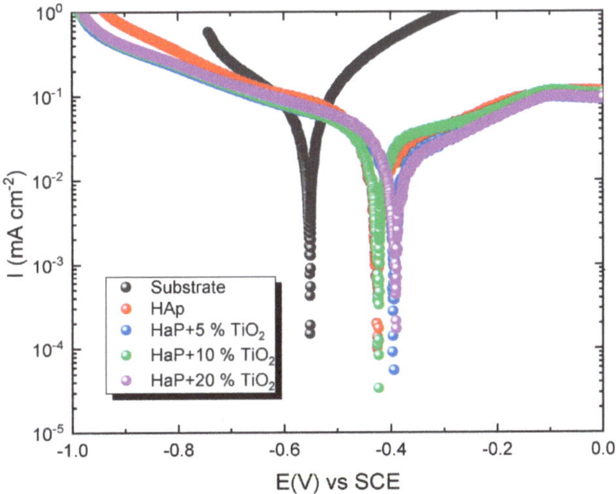

Figure 7. Tafel polarization curves of the uncoated and coated samples.

Table 3. Corrosion parameter values obtained from Tafel curves.

Coating	E_{corr} (mV)	I_{corr} (μA cm^{-2})	β_a	β_c	R_p (kΩ cm^2)
Substrate	-553.3 ± 1.33	204 ± 0.81	0.10	01.80	0.89
HaP	-427.5 ± 1.21	38.9 ± 0.17	0.10	20.31	1.19
HaP + 5% TiO$_2$	-424.9 ± 1.27	388.0 ± 0.88	0.10	01.33	0.10
HaP + 10% TiO$_2$	-392.8 ± 1.36	267 ± 0.93	0.10	02.80	0.15
HaP + 20% TiO$_2$	-392.7 ± 1.53	36.1 ± 0.13	0.10	15.57	1.10

The results obtained in this study reveal that the corrosion resistance of the HaP coating in SBF solution is significantly increased with TiO_2 content. The sample coated with HaP + 20% TiO_2 revealed a lower corrosion current of 36.1 μA cm^{-2} and a corrosion rate of 1.1 c kΩ cm^2, compared to the uncoated Ti-Al6-4V sample.

3.4. Electrochemical Impedance Spectroscopy (EIS)

The electrochemical impedance spectroscopy (EIS) is widely known as an effective technique applied to study the corrosion behavior of the used materials in the interface between the coatings and the electrolytes. Based on EIS, the corrosion resistance property of the elaborated coatings (e.g., the behavior of the dual-layer capacitance and charge transfer resistance) was determined. In Figure 8, the Nyquist plots of the impedance diagrams of both the Ti-6Al-4V substrate and the HaP + TiO_2 coatings are presented. To simulate the experimental data, the EIS results were fitted through an equivalent circuit using the ZSimpWin software. As shown in Figure 9, the equivalent circuit model consists of the following elements: R_s (solution resistance), R_p (coating resistance), R_{ct} (charge transfer resistance at the electrode–electrolyte interface), Q_{dl} (double-layer capacitance) and Q_c (interfacial capacitance). The different value parameters fitting the circuit model are detailed in Table 4.

Table 4. The fitting parameters of the equivalent circuit.

Coating	R_s (Ω cm^2)	CPE_c		R_p (Ω cm^2)	CPE_{dl}		R_{ct} (Ω cm^2)	$\chi^2 \times 10^{-4}$
		Y_c (mF cm^{-2})	n_c		Y_{dl} (mF cm^{-2})	n_{dl}		
Substrate	13.88	02.67	0.58	056.21	76.8	1.00	01.53	1.47
HaP	09.48	11.86	0.45	075.17	78.1	0.86	11.22	9.5
HaP + 5% TiO$_2$	16.71	17.59	0.38	142.90	157	0.88	41.85	1.85
HaP + 10% TiO$_2$	07.68	55.20	0.71	208.70	19.7	0.38	97.94	3.03
HaP + 20% TiO$_2$	17.36	18.06	0.70	214.00	10.9	0.40	697.10	2.36

During the modeling calculation, an iterative approach is adopted herein using the chi-square (χ^2) value for the model along with the percentage error values for each circuit component. This was conducted to ascertain the effectiveness of a particular model in fitting the experimental data. The calculation of the (χ^2) value was based on the expression mentioned below.

$$\chi^2 = \sum_{i=1}^{i=n} [W_i'(Z_{i,exp}' - Z_{i,cal}'(\omega_i, \overline{p})^2 + W_i''(Z_{i,exp}'' - Z_{i,cal}''(\omega_i, \overline{p}))^2] \tag{3}$$

The experimental (calculated) impedance at a particular frequency is represented by its real and imaginary parts, denoted by $Z_{i,exp}'$ ($Z_{i,cal}'$) and $Z_{i,exp}''$ ($Z_{i,cal}''$), respectively. The statistical weights of the data are denoted by these quantities as well. Also, the number of data points 'i' is represented by the variable 'n'.

Achieving the correlation between the experimental and theoretical data was determined by minimizing the value of the (χ^2). This was accomplished with the calculation of the difference between the experimental data and squaring it, which emphasized large variances in the data. The literature [41] reported that a value of one order of magnitude less than or equal to 10^{-3} is generally considered acceptable for a given model.

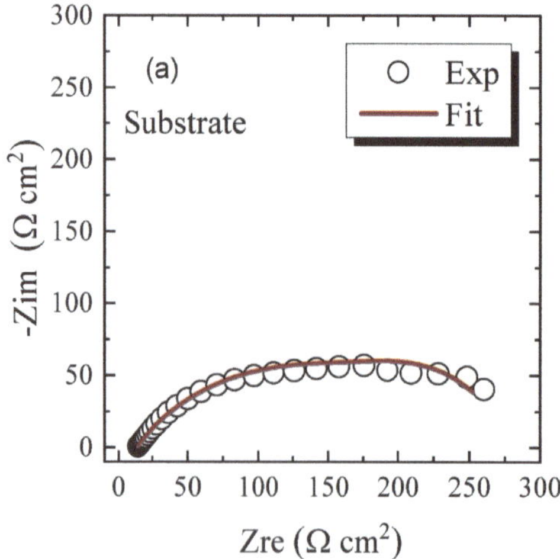

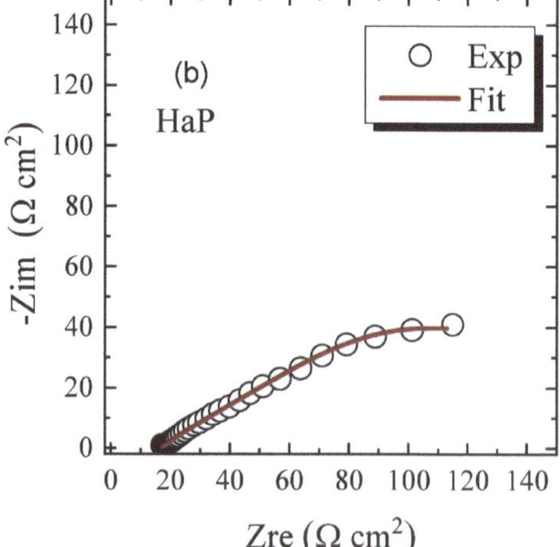

Figure 8. *Cont.*

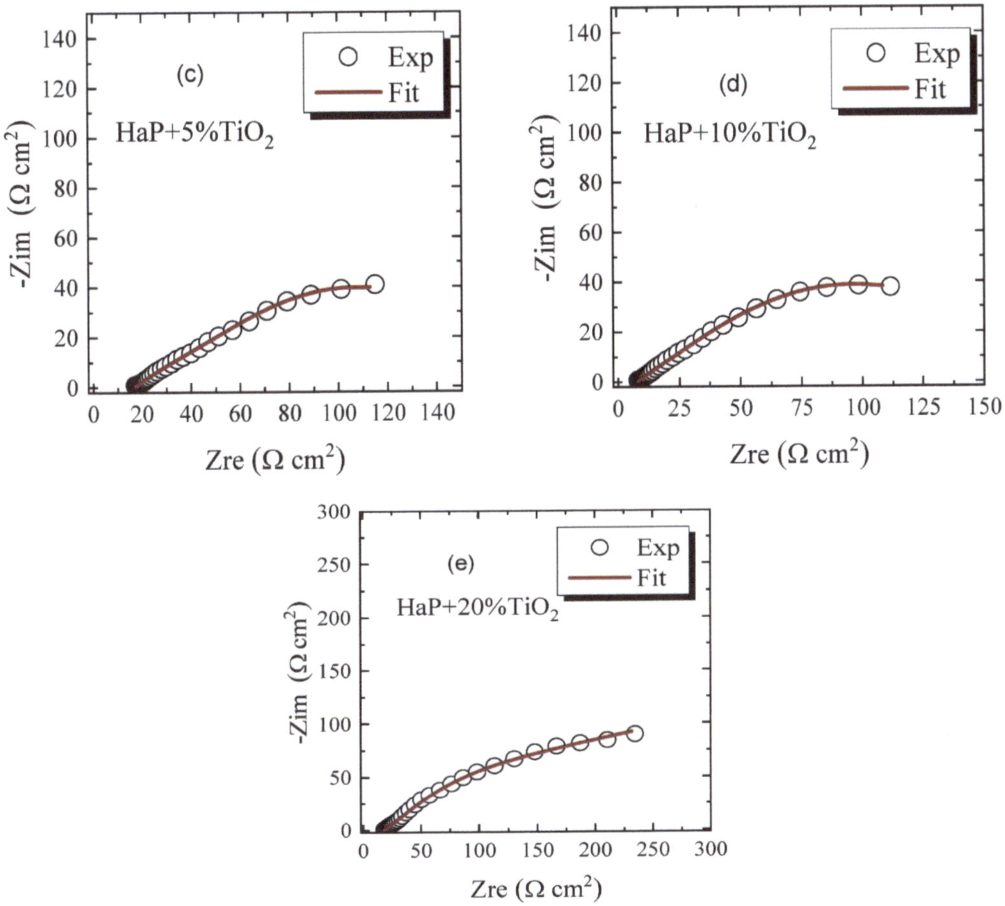

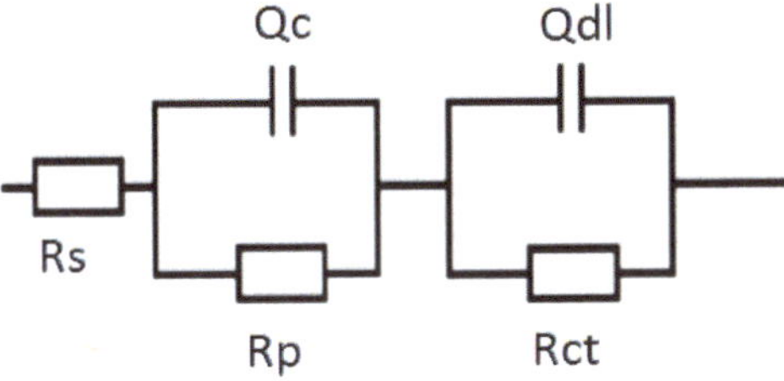

Figure 8. Nyquist experimental and modeling curves. (**a**) Substrate, (**b**) HaP, (**c**) HaP + 5% TiO$_2$, (**d**) HaP + 10% TiO$_2$ and, (**e**) HaP + 20% TiO$_2$.

Figure 9. Equivalent circuits.

As revealed in the latter, the increase in TiO$_2$ content corresponds to the rise in the charge transfer resistance R_{ct} (or corrosion resistance) of the HaP coatings. The values of R_{ct} range from 55.51 to 697.10 Ω cm^2. These findings are consistent with the results obtained by analyzing the Tafel curves. R_p increased from 56.21 to 214.00 Ω cm^2, suggesting the better corrosion resistance of HaP + TiO$_2$ coatings compared to that of the uncoated substrate. The observed behavior provided evidence that the enhanced corrosion resistance of the sprayed coatings was attributed to the formation of a stable structure and a uniformly smooth surface that effectively prevented the penetration of the electrolytes.

As per Ahmadi et al. [6,42], the formation of a HaP layer causes a substantial increase in the diameter of the resistance semicircle (5.3×10^4 Ω cm^2). The polarization outcomes confirmed that an interface layer of TiO_2 leads to a reduction in the corrosion current density. The semicircle diameter for coatings that contained a TiO_2 interlayer was larger than that of the substrate or the pure HaP coating. It was also observed that the sample TiO_2/HaP + 15 wt% of TiO_2 + 5 wt% of Ag exhibited the largest semicircle diameter (6.54×10^5 Ω cm^2), indicating that it offers the highest resistance to corrosion.

To study in detail the interface between the electrolyte and the coated material, it is important to examine the double-layer capacitance (C_{dl}) calculated using the expression writtenbelow:

$$C_{dl} = \frac{(Y_{dl}. \times R_{ct})^{\frac{1}{n_{dl}}}}{R_{ct}} \tag{4}$$

where Y_{dl} and n_{dl} are, respectively, the admittance and the exponent parameters.

In order to determine the surface roughness factor (f), the following expression was used:

$$f = \frac{C_{dl}}{22\mu Fcm^{-2}} \tag{5}$$

The customary value of the capacitance (C_{dl}) for metal electrodes was almost equal to 20 μF cm^{-2} [43].

For the real surface area, S_r, the following expression was utilized:

$$S_r = S \times f \tag{6}$$

where S is the geometric surface area of the coatings (~1 cm^2).

The formula applied to calculate the coating capacitance (C_c) is written below:

$$C_c = \frac{(Y_c \cdot R_c)^{\frac{1}{n_c}}}{R_c} \tag{7}$$

The values of C_{dl}, C_c and S_r are presented in Table 5.

Table 5. The double-layer capacitance (C_{dl}), the film capacitance (C_c) and the real surface area S_r of HaP + TiO_2 coatings.

Coating	C_{dl} (mF cm^{-2})	S_r(cm^2)	C_c (mF cm^{-2})
Substrate	0.076	3.8	0.017
HaP	0.076	2.2	0.37
HaP + 5% TiO_2	0.20	9.1	0.79
HaP + 10% TiO_2	0.057	2.3	0.98
HaP + 20% TiO_2	0.22	10.3	0.18

The results shown in Figure 10 reveal that the double-layer capacitance (C_{dl}) initially decreases. Then, it increases with the rise in the TiO_2 concentration. This observation suggests that the thickness of the corrosive layer formed on the HaP surface was reduced due to changes on the surface which, in turn, enhanced the anti-corrosive properties of the HaP + TiO_2 coatings.

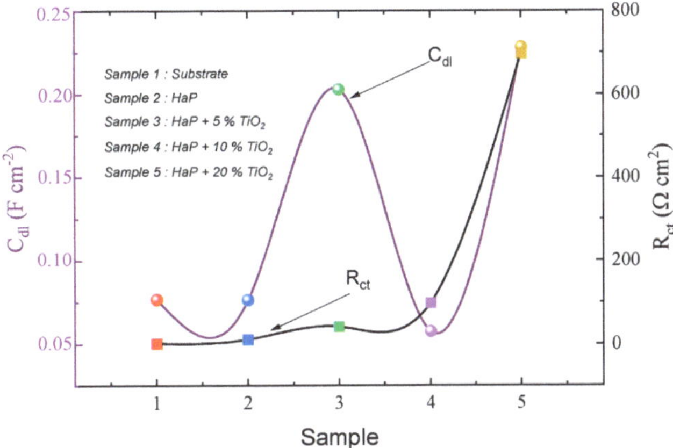

Figure 10. Variation of the charge transfer resistance (*Rct*) and the double-layer capacitance (*Cdl*) with different TiO$_2$ contents.

As displayed in Figure 11, the real surface area is larger than the geometric surface. Indeed, it increases with the rise in the TiO$_2$ concentration. This observation can be attributed to the surface roughness and heterogeneity rather than to its homogeneity.

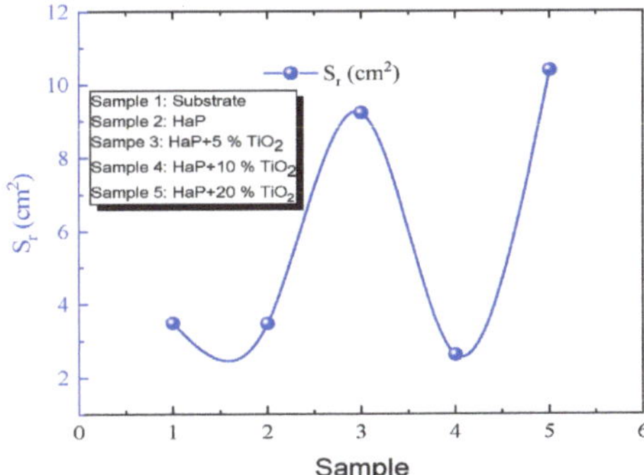

Figure 11. The real surface area, S_r.

The Bode diagrams of impedance and the phase for different coatings in the SBF solution are presented in Figure 12. The Bode plot generally shows that higher values of |Z| at lower frequencies indicate higher levels of corrosion resistance. Based on the result presented in Figure 12a, it can be inferred that the HaP + 20% TiO$_2$ coating has the highest corrosion resistance due to its higher Z modulus at low frequencies. The lowest Z modulus is related to the uncoated substrate. The electrochemical response of the HaP + 20% TiO$_2$ coating, as shown by the slope in the data, suggests that its electric double-layer capacitance is mostly capacitive. This observation implies that the coating has the potential to effectively minimize the corrosion rate of Ti-6Al-4V when exposed to SBF [41]. According to the Bode plots of the phase exposed in Figure 12b, a capacitive behavior results from the maximum phase angle occurring at low frequencies. At high frequencies, the impedance is not affected by frequency, while the phase angle approaches 0°, indicating a resistive behavior that corresponds to the resistance of the SBF solution between the working and reference electrodes. Conversely, in the low-frequency region, the HaP + 20% TiO$_2$ coating demonstrates a phase angle approaching $-35°$. This value indicates the capacitive behavior of the electrode–electrolyte interface.

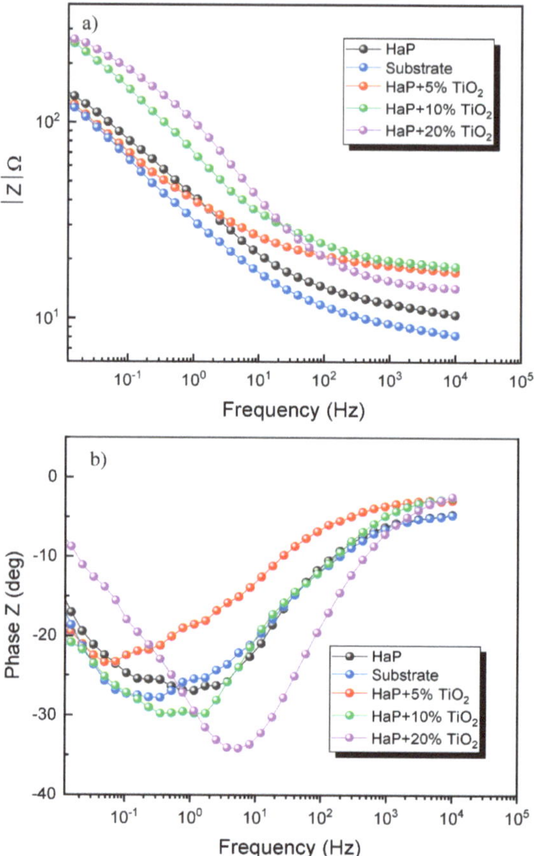

Figure 12. Bode impedance (**a**) and Bode phase diagram (**b**).

4. Conclusions

In summary, the spray technique was used to prepare nanocomposite thin films consisting of hydroxyapatite doped with titanium oxides. To determine the optimal structure of the thin oxide films, we conducted an X-ray diffraction analysis on the various deposits, examining the effect of the doping concentration. We also investigated some mechanical properties related to HaP in terms of TiO_2 content. These findings indicated that TiO_2 doping has a significant effect on the microstructural, morphological, and nanomechanical properties, as well as the corrosion rate, of HaP + TiO_2 coatings. The major obtained findings are presented below:

1. XRD results stand for the presence of HaP and TiO_2 compounds with no additional phases, while SEM micrographs revealed homogeneously distributed grain coatings;
2. The use of the nanoindentation test revealed that TiO_2 doping plays a crucial role in enhancing HaP thin films' mechanical properties, such as hardness and Young's modulus;
3. The corrosion studies using potentiodynamic polarization measurements as well as EIS techniques showed that TiO_2-doped HaP coatings exhibit much-improved corrosion resistance properties when compared to that of pure HaP thin films. This investigation is particularly noteworthy since we used a cost-effective, straightforward spray pyrolysis technique to obtain HaP thin films. Along the same line, these

TiO$_2$-doped films hold considerable potential for various applications, especially in environmental areas like photocatalysis, as well as in protective coating systems.

Author Contributions: Writing—original draft, H.D., S.B.S., M.S., Y.D., S.C., B.G. and H.B.; Supervision, A.H., A.B.C.L., M.A. and H.B. All authors have read and agreed to the published version of the manuscript.

Funding: This research received no external funding.

Institutional Review Board Statement: Not applicable.

Informed Consent Statement: Not applicable.

Data Availability Statement: Not applicable.

Conflicts of Interest: The authors declare no conflict of interest.

References

1. Ling, H.; Ling, P.; Wei, Z.; Zibo, N.; Xuan, C.; Maolin, C.; Qinzhao, Z.; Wenqian, P.; Peng, X.; Yang, L. Thermal corrosion behavior of Yb$_4$Hf$_3$O$_{12}$ ceramics exposed to calcium-ferrum-alumina-silicate (CFAS) at 1400 °C. *J. Eur. Ceram. Soc.* **2023**, *43*, 4114–4123.
2. Ling, P.; Ling, H.; Zibo, N.; Peng, X.; Wei, Z.; Yang, L. Corrosion behavior of ytterbium hafnate exposed to water-vapor with Al(OH)$_3$ impurities. *J. Eur. Ceram. Soc.* **2023**, *43*, 612–620.
3. Tlotleng, M.; Akinlabi, E.; Shukla, M.; Pityana, S. Microstructures, hardness and bioactivity of hydroxyapatite coatings deposited by direct laser melting process. *Mater. Sci. Eng. C* **2014**, *3*, 189–198. [CrossRef]
4. Nicholson, J.W. Titanium alloys for dental implants: A review. *Prosthesis* **2020**, *2*, 11. [CrossRef]
5. Okazaki, Y.; Gotoh, E. Implant applications of highly corrosion-resistant Ti-15Zr-4Nb-4Ta alloy. *Mater. Trans.* **2002**, *43*, 2943–2948. [CrossRef]
6. Ahmadi, R.; Asadpourchallou, N.; Kaleji, B.K. In vitro study: Evaluation of mechanical behavior, corrosion resistance, antibacterial properties and biocompatibility of HAp/TiO$_2$/Ag coating on Ti6Al4V/TiO$_2$ substrate. *Surf. Interfaces* **2021**, *24*, 101072. [CrossRef]
7. Mohan, L.A.; Durgalakshmi, D.; Geetha, M.; Narayanan, T.S.; Asokamani, R. Electrophoretic deposition of nanocomposite (Hap + TiO$_2$) on titanium alloy for biomedical applications. *Ceram. Int.* **2012**, *38*, 3435–3443. [CrossRef]
8. Duta, L. In vivo assessment of synthetic and biological-derived calcium phosphate-based coatings fabricated by pulsed laser deposition: A review. *Coatings* **2021**, *11*, 99. [CrossRef]
9. Singh, N.; Batra, U.; Kumar, K.; Siddiquee, A.N. Evaluating the Electrochemical and In Vitro Degradation of an HA-Titania Nano-Channeled Coating for Effective Corrosion Resistance of Biodegradable Mg Alloy. *Coatings* **2023**, *13*, 30. [CrossRef]
10. Wang, Z.; Ye, F.; Chen, L.; Lv, W.; Zhang, Z.; Zang, Q.; Lu, S. Preparation and degradation characteristics of MAO/APS composite bio-coating in simulated body fluid. *Coatings* **2021**, *11*, 667. [CrossRef]
11. Heimann, R.B. Structural Changes of Hydroxylapatite during Plasma Spraying: Raman and NMR Spectroscopy Results. *Coatings* **2021**, *11*, 987. [CrossRef]
12. Farooq, S.A.; Raina, A.; Mohan, S.; Arvind Singh, R.; Jayalakshmi, S.; Irfan Ul Haq, M. Nanostructured coatings: Review on processing techniques, corrosion behaviour and tribological performance. *Nanomaterials* **2022**, *12*, 1323. [CrossRef]
13. Samanipour, F.; Bayati, M.R.; Zargar, H.R.; Golestani-Fard, F.; Troczynski, T.; Taheri, M. Electrophoretic enhanced micro arc oxidation of ZrO$_2$–HAp–TiO$_2$ nanostructured porous layers. *J. Alloys Compd.* **2011**, *509*, 9351–9355. [CrossRef]
14. Liu, X.; Chu, P.K.; Ding, C. Surface modification of titanium, titanium alloys, and related materials for biomedical applications. *Mater. Sci. Eng. R Rep.* **2004**, *47*, 49–121. [CrossRef]
15. Kunze, J.; Müller, L.; Macak, J.M.; Greil, P.; Schmuki, P.; Müller, F.A. Time-dependent growth of biomimetic apatite on anodic TiO$_2$ nanotubes. *Electrochim. Acta* **2008**, *53*, 6995–7003. [CrossRef]
16. Lee, B.T.; Jang, D.H.; Kang, I.C.; Lee, C.W. Relationship between Microstructures and Material Properties of Novel Fibrous Al$_2$O$_3$–(m-ZrO$_2$)/t-ZrO$_2$ Composites. *J. Am. Ceram. Soc.* **2005**, *88*, 2874–2878. [CrossRef]
17. Tanaka, A.; Nishimura, Y.; Sakaki, T.; Fujita, A.; Shin-Ike, T. Histologic evaluation of tissue response to sintered lanthanum-containing hydroxyapatites subcutaneously implanted in rats. *J. Osaka Dent. Univ.* **1989**, *23*, 111–120. [PubMed]
18. Oktar, F.N. Hydroxyapatite–TiO$_2$ composites. *Mater. Lett.* **2006**, *60*, 2207–2210. [CrossRef]
19. Vemulapalli, A.K.; Penmetsa, M.R.; Nallu, R.; Siriyala, R. HAp/TiO$_2$ nanocomposites: Influence of TiO$_2$ on microstructure and mechanical properties. *J. Compos. Mater.* **2020**, *54*, 765–772. [CrossRef]
20. Patrick, L.; Jonathan, A.; Stephen, M.; Jeroen, J.J.P.; Joanna, W.; Adrian, B.; Brian, J.M. Nanoindentation and nano-scratching of hydroxyapatite coatings for resorbable magnesium alloy bone implant applications. *J. Mech. Behav. Biomed. Mater.* **2022**, *133*, 105306.
21. Ivanova, A.A.; Surmeneva, M.A.; Tyurin, A.I.; Pirozhkova, T.S.; Shuvarin, I.A.; Prymak, O.; Epple, M.; Chaikina, M.V.; Surmenev, R.A. Fabrication and physico-mechanical properties of thin magnetron sputter deposited silver-containing hydroxyapatite films. *Appl. Surf. Sci.* **2016**, *360*, 929–935. [CrossRef]

22. Dhiflaoui, H.; Khlifi, K.; Barhoumi, N.; Ben Cheikh Larbi, A. The tribological and corrosion behavior of TiO_2 coatings deposited by the electrophoretic deposition process. *Proc. Inst. Mech. Eng. C J. Mech. Eng. Sci.* **2020**, *234*, 1231–1238. [CrossRef]

23. Gassoumi, B.; Jaballah, R.; Boukhachem, A.; Kamoun-Turki, N.; Amlouk, M. Simple route deposition and some physical investigations on nanoflower $NiMoO_4$ sprayed thin films. *Bull. Mater. Sci.* **2021**, *44*, 128. [CrossRef]

24. Larbi, T.; Ouni, B.; Boukachem, A.; Boubaker, K.; Amlouk, M. Electrical measurements of dielectric properties of molybdenum-doped zinc oxide thin films. *Mater. Sci. Semicond. Process.* **2014**, *22*, 50–58. [CrossRef]

25. Mrabet, C.; Boukhachem, A.; Amlouk, M.; Manoubi, T. Improvement of the optoelectronic properties of tin oxide transparent conductive thin films through lanthanum doping. *J. Alloys Compd.* **2016**, *666*, 392–405. [CrossRef]

26. Oliver, W.C.; Pharr, G.M. An improved technique for determining hardness and elastic modulus using load and displacement sensing indentation experiments. *J. Mater. Res.* **1992**, *7*, 1564–1583. [CrossRef]

27. Anurag, K.P.; Gautam, R.K.; Behera, C.K. Corrosion and wear behavior of Ti–5Cu-xNb biomedical alloy in simulated body fluid for dental implant applications. *J. Mech. Behav. Biomed. Mater.* **2023**, *137*, 105533.

28. Nathanael, A.J.; Mangalaraj, D.; Ponpandian, N. Controlled growth and investigations on the morphology and mechanical properties of hydroxyapatite/titania nanocomposite thin films. *Compos. Sci. Technol.* **2010**, *70*, 1645–1651. [CrossRef]

29. Singh, B.; Singh, G.; Sidhu, B.S. Investigation of the in vitro corrosion behavior and biocompatibility of niobium (Nb)-reinforced hydroxyapatite (HA) coating on CoCr alloy for medical implants. *J. Mater. Res.* **2019**, *34*, 1678–1691. [CrossRef]

30. Lee, S.W.; Paraguay-Delgado, F.; Arizabalo, R.D.; Gómez, R.; Rodríguez-González, V. Understanding the photophysical and surface properties of $TiO_2–Al_2O_3$ nanocomposites. *Mater. Lett.* **2013**, *107*, 10–13. [CrossRef]

31. Singh, G.; Singh, S.; Prakash, S. Surface characterization of plasma sprayed pure and reinforced hydroxyapatite coating on Ti6Al4V alloy. *Surf. Coat. Technol.* **2011**, *205*, 4814–4820. [CrossRef]

32. So, W.W.; Park, S.B.; Kim, K.J.; Moon, S.J. Phase transformation behavior at low temperature in hydrothermal treatment of stable and unstable titania. *Sol. J. Colloid. Interf. Sci.* **1997**, *191*, 398–406. [CrossRef] [PubMed]

33. Sujaridworakun, P.; Koh, F.; Fujiwara, T.; Pongkao, D.; Ahniyaz, A.; Yoshimura, M. Preparation of anatase nanocrystals deposited on hydroxyapatite by hydrothermal treatment. *Mater. Sci. Eng. C* **2005**, *25*, 87–91. [CrossRef]

34. Nathanael, A.J.; Mangalaraj, D.; Chen, P.C.; Ponpandian, N. Mechanical and photocatalytic properties of hydroxyapatite/titania nanocomposites prepared by combined high gravity and hydrothermal process. *Compo. Sci. Technol.* **2010**, *70*, 419–426. [CrossRef]

35. Dhiflaoui, H.; Ben Jaber, N.; Simescu Lazar, F.; Faure, J.; Ben Cheikh Larbi, A.; Benhayoune, H. Effect of annealing temperature on the structural and mechanical properties of coatings prepared by electrophoretic deposition of TiO_2 nanoparticles. *Thin Solid Film.* **2017**, *638*, 201–212. [CrossRef]

36. Rouchdi, M.; Salmani, E.; Fares, B.; Hassanain, N.; Mzerd, A. Synthesis and characteristics of Mg doped ZnO thin films: Experimental and abinitio study. *Results Phys.* **2017**, *7*, 620–627. [CrossRef]

37. Lee, S.H.; Kim, H.E.; Kim, H.W. Nano-sized hydroxyapatite coatings on Ti substrate by E beam deposition. *J. Am. Ceram. Soc.* **2007**, *90*, 50–56. [CrossRef]

38. Fathi, M.H.; Azam, F. Novel hydroxyapatite/tantalum surface coating for metallic dental implant. *Mater. Lett.* **2007**, *61*, 1238–1241. [CrossRef]

39. Amaravathy, P.; Sathyanarayanan, S.; Sowndarya, S.; Rajendran, N. Bioactive HA/TiO_2 coating on magnesium alloy for biomedical applications. *Ceram. Int.* **2014**, *40*, 6617–6630. [CrossRef]

40. Moskalewicz, T.; Łukaszczyk, A.; Kruk, A.; Kot, M.; Jugowiec, D.; Dubiel, B.; Radziszewska, A. Porous HA and nanocomposite nc-TiO_2/HA coatings to improve the electrochemical corrosion resistance of the Co-28Cr-5Mo alloy. *Mater. Chem. Phys.* **2017**, *199*, 144–158. [CrossRef]

41. Dhiflaoui, H.; Dabaki, Y.; Zayani, W.; Debbich, H.; Faure, J.; Ben Cheikh Larbi, A.; Benhayoune, H. Effects of Hydrogen Peroxide Concentration and Heat Treatment on the Mechanical Characteristics and Corrosion. *J. Mater. Eng. Perform.* **2023**, *in press*. [CrossRef]

42. Ahmadi, R.; Afshar, A. In vitro study: Bond strength, electrochemical and biocompatibility; evaluations of TiO_2/Al_2O_3 reinforced hydroxyapatite sol–gel coatings on 316L SS. *Surf. Coat. Technol.* **2021**, *405*, 126594. [CrossRef]

43. Zayani, W.; Azizi, S.; El-Nasser, K.S.; Othman Ali, I.; Molière, M.; Fenineche, N.; Mathlouthi, H.; Lamloumi, J. Electrochemical behavior of a spinel zinc ferrite alloy obtained by a simple sol-gel route for Ni-MH battery applications. *Int. J. Energy Res.* **2020**, *45*, 5235–5247. [CrossRef]

Article

Powder Synthesized from Aqueous Solution of Calcium Nitrate and Mixed-Anionic Solution of Orthophosphate and Silicate Anions for Bioceramics Production

Daniil Golubchikov [1,*], Tatiana V. Safronova [1,2,*], Elizaveta Nemygina [1], Tatiana B. Shatalova [1,2], Irina N. Tikhomirova [3], Ilya V. Roslyakov [1,4], Dinara Khayrutdinova [5], Vadim Platonov [2], Olga Boytsova [2], Maksim Kaimonov [1], Denis A. Firsov [2] and Konstantin A. Lyssenko [2]

1   Department of Materials Science, Lomonosov Moscow State University, Building, 73, Leninskie Gory, 1, 119991 Moscow, Russia
2   Department of Chemistry, Lomonosov Moscow State University, Building, 3, Leninskie Gory, 1, 119991 Moscow, Russia
3   Department of General Technology of Silicates, Mendeleev University of Chemical Technology, Building, 1, Geroyev Panfilovtsev, 20, 125480 Moscow, Russia
4   Kurnakov Institute of General and Inorganic Chemistry, Russian Academy of Sciences, Leninskii Prosp., 31, 119071 Moscow, Russia
5   A.A. Baikov Institute of Metallurgy and Materials Science, Russian Academy of Sciences, Leninskii Prosp., 49, 119334 Moscow, Russia
*   Correspondence: golubchikovdo@my.msu.ru or dddannn2113@gmail.com (D.G.); safronovatv@my.msu.ru (T.V.S.)

Abstract: Synthesis from mixed-anionic aqueous solutions is a novel approach to obtain active powders for bioceramics production in the $CaO-SiO_2-P_2O_5-Na_2O$ system. In this work, powders were prepared using precipitation from aqueous solutions of the following precursors: $Ca(NO_3)_2$ and Na_2HPO_4 (CaP); $Ca(NO_3)_2$ and Na_2SiO_3 (CaSi); and $Ca(NO_3)_2$, Na_2HPO_4 and Na_2SiO_3 (CaPSi). Phase composition of the CaP powder included brushite $CaHPO_4 \cdot 2H_2O$ and the CaSi powder included calcium silicate hydrate. Phase composition of the CaPSi powder consisted of the amorphous phase (presumably containing hydrated quasi-amorphous calcium phosphate and calcium silicate phase). All synthesized powders contained $NaNO_3$ as a by-product. The total weight loss after heating up to 1000 °C for the CaP sample—28.3%, for the CaSi sample—38.8% and for the CaPSi sample was 29%. Phase composition of the ceramic samples after the heat treatment at 1000 °C based on the CaP powder contained β-$NaCaPO_4$ and β-$Ca_2P_2O_7$, the ceramic samples based on the CaSi powder contained α-$CaSiO_3$ and $Na_2Ca_2Si_2O_7$, while the ceramics obtained from the CaPSi powder contained sodium rhenanite β-$NaCaPO_4$, wollastonite α-$CaSiO_3$ and $Na_3Ca_6(PO_4)_5$. The densest ceramic sample was obtained in $CaO-SiO_2-P_2O_5-Na_2O$ system at 900 °C from the CaP powder (ρ = 2.53 g/cm^3), while the other samples had densities of 0.93 g/cm^3 (CaSi) and 1.22 (CaPSi) at the same temperature. The ceramics prepared in this system contain biocompatible and bioresorbable phases, and can be recommended for use in medicine for bone-defect treatment.

Keywords: mixed-anionic solution; precipitation; brushite; calcium silicate hydrate; amorphous calcium phosphate; sodium nitrate; calcium pyrophosphate; sodium rhenanite; sodium-calcium silicate; wollastonite; bioceramics

Citation: Golubchikov, D.; Safronova, T.V.; Nemygina, E.; Shatalova, T.B.; Tikhomirova, I.N.; Roslyakov, I.V.; Khayrutdinova, D.; Platonov, V.; Boytsova, O.; Kaimonov, M.; et al. Powder Synthesized from Aqueous Solution of Calcium Nitrate and Mixed-Anionic Solution of Orthophosphate and Silicate Anions for Bioceramics Production. *Coatings* **2023**, *13*, 374. https://doi.org/10.3390/coatings13020374

Academic Editor: Richard Drevet

Received: 31 December 2022
Revised: 1 February 2023
Accepted: 2 February 2023
Published: 7 February 2023

1. Introduction

Modern science faces a problem of increasing levels of bone diseases, traumas and cancers. Thereby, the development of bone repair orthopedic approaches as well as enhancement of implemented biomaterials is crucial [1,2]. Bone tissues possess a great regenerative potential and suitable biomaterials are required to support natural regeneration process [3–6]. One of the crucial processes of native bone forming is angiogenesis [7], which is vital for bone healing [8]. The most commonly used biomaterials in the field of bone tissue engineering are polymers [9–11], composites [12], as well as glass and ceramics, obtained in systems containing calcium phosphates and silicates [13–16]. Such materials are also required to provide a sufficient mechanical support and appropriate environment for cell attachment, proliferation, and differentiation [17,18].

Among the first steps in this area were materials in the Na_2O-CaO-SiO_2-P_2O_5 system, suggested by L. L. Hench [19]. This study marked the beginning of calcium-sodium phosphate-silicate glass-ceramics development as well as application of such materials in bone implant production. L.L. Hench et al. [20] demonstrated that synthesized surface-active bioglass-ceramics can be used in vivo without inflammation.

Materials synthesized in the considered system tend to be osteoinductive, since the presence of Si-containing ions is suggested to stimulate the proliferation of human aortic endothelial cells as well as to support the expression of genes encoding the proangiogenic downstream cytokines [21], which is necessary for successful angiogenesis processes [22].

A special part of the Na_2O-CaO-SiO_2-P_2O_5 system is the CaO-SiO_2 system, in which two congruently melting compounds, Ca_2SiO_4 and $CaSiO_3$ (wollastonite), and two incongruently melting compounds, Ca_3SiO_5 (stable from 1250 to 2050 °C), and $Ca_3Si_2O_7$ (stable from low temperature to 1464 °C), can be obtained [23]. Congruently melting compounds are more promising for bone tissue engineering: for example, 3D-printed β-Ca_2SiO_4 scaffolds sintered at a higher temperature stimulated the adhesion, proliferation, ALP activity, and osteogenic-related gene expression of rBMSCs [24]. $CaSiO_3$ ceramics have been extensively researched as biomaterial to replace other materials due to their superior biological activity, for example, compared to hydroxyapatite (HA) [25] showing their potential prospects in in vivo trials [26]. In addition, calcium silicate ceramics can be used for skin healing and cartage regeneration [27]. In [28], silicate ceramics are obtained by solid-phase synthesis from a mixture of powders.

In the CaO-P_2O_5, which is a part of the Na_2O-CaO-SiO_2-P_2O_5 system, the following biocompatible and bioresorbable phases of tricalcium phosphate ($Ca_3(PO_4)_2$) and calcium pyrophosphate ($Ca_2P_2O_7$) [29] can be obtained.

In Na_2O-CaO-P_2O_5 sodium-substituted tricalcium phosphate $Ca_{10}Na(PO_4)_7$, sodium rhenanite β-$NaCaPO_4$, mixed sodium-calcium pyrophosphate ($CaNa_2P_2O_7$) [29] and mixed sodium-calcium phosphate ($Na_3Ca_6(PO_4)_5$) can be obtained.

In the CaO-SiO_2-P_2O_5, as a part of the Na_2O-CaO-SiO_2-P_2O_5 system, two main phases can be obtained: silicocarnotite ($Ca_5(PO_4)_2SiO_4$), with a structure type carnotite and a wide range of solid solutions [30,31], and nagelshmidtite ($Ca_7Si_2P_2O_{16}$) [32]. Both phases provide satisfactory biocompatibility. The addition of sodium to this system leads to the appearance of $Na_2Ca_2Si_3O_9$ and $Na_2CaSi_3O_8$ after the sintering of material above 600 °C [33].

There are several approaches for obtaining ceramics or glass-ceramics in the CaO-SiO_2-P_2O_5-Na_2O system. First of all, the samples can be obtained in the form of glasses with subsequent crystallization during the heat treatment at an appropriate temperature [34]; nevertheless, this method has final phase composition limitations. In this case, the bioactive glass decomposition occurs, which leads to the appearance of additional crystalline phases. The composition of this phases directly depends on the composition of the initial mixture. For example, using the mixture of the initial components with the ratio of SiO_2:NaO_2:CaO = 41.8:26.7:31.5 (mol) makes it possible to obtain the $Na_2Ca_2Si_2O_7$ phase [35]. This phase was also synthesized in [36] from the mixture of reagent grade Na_2CO_3, CaO, bovine bone (P_2O_5 source), and rice husk (SiO_2 source), as additional to the $Na_6Ca_3Si_6O_{18}$ main phase. Nevertheless, the sample containing this phase showed good results in mechanical properties and in in vitro tests.

Another approach to obtain ceramics in the considered CaO-SiO_2-P_2O_5-Na_2O system is based on the use of hydroxyapatite or tricalcium phosphate powders as a filler and an aqueous solution of sodium silicate as a binder [37]. This method leads to the formation of two main phases, which are β-rhenanite (β-$NaCaPO_4$) and sodium calcium phosphate

($Na_3Ca_6(PO_4)_5$) after heat treatment. These phases are biocompatible and are used for the restoration of bone-tissue defects [38,39].

It is also possible to prepare ceramic materials using intermediate phases precipitated from the solution (such as hydroxyapatite and amorphous sodium silicate), which can be converted into final phases (sodium rhenanite, calcium-sodium silicate, and wollastonite) during the heat treatment process [37,40].

In addition, it is possible to print scaffolds from a composite of calcium phosphate powders and 45S5 Bioglass using a cementation reaction during printing [41,42]. In [42], calcium hydrogen phosphate dihydrate ($CaHPO_4 \cdot 2H_2O$) was formed as a result of the cementing reaction. And the final phase composition of material after the heat treatment at 1000 °C contained sodium rhenanite ($NaCaPO_4$) and wollastonite ($CaSiO_3$).

Ceramics in the CaO-SiO_2-P_2O_5-Na_2O system can be obtained from powders synthesized from solutions. To obtain such ceramics, active powders with a highly homogeneous distribution of components are required. Amorphous powders synthesized from mixed-anionic solutions have the necessary homogeneity.

There are some examples of powders synthesized from mix-anionic HPO_4/P_2O_7 [43] P_2O_7/CO_3 [44], HPO_4/CO_3 [45,46], or mix-cationic K/Na [47] aqueous solutions with appropriate homogeneity that were used for the ceramic materials preparation. Preservation of reaction by-product as a component of the powder mixture also can be used as a method of preparation of powder mixtures with high homogeneity [25,48,49].

However, an approach involving synthesis from mixed-anionic solutions is more interesting, since this method leads to a more homogeneous distribution of components in the powder mixture [43–46]. In addition, a chemical approach to the synthesis of mixed-anionic powders makes it possible to obtain more dispersed powders with a larger specific surface area, which are more active, and the sintering process is more effective [48,49]. Additionally, the by-product of the synthesis from mixed-anionic solutions can be kept in the composition to form melts and promote the formation of new phases [48,49]. Thereby, this method can be the most promising approach to obtain ceramics in the CaO-SiO_2-P_2O_5-Na_2O system.

Thus, the aim of the present work consisted in the synthesis of powder, which contains precursors of calcium phosphate and silicate high-temperature phases, as well as the reaction by-product from the mixed-anionic solution, for preparation of composite ceramics in Na_2O-CaO-SiO_2-P_2O_5 system. The obtained powder mixture was implemented as a powder precursor with homogeneous distribution of the components for production of the composite material, while the by-product of synthesis was used both as a participant of the heterogeneous reactions and as a sintering aid.

2. Materials and Methods

2.1. Materials

Powders of calcium nitrate tetrahydrate $Ca(NO_3)_2 \cdot 4H_2O$ (CAS no. 13477-34-4, ACS reagent, Sigma-Aldrich, Mumbai, India), sodium metasilicate pentahydrate $Na_2SiO_3 \cdot 5H_2O$ (CAS no. 10213-79-3, RusKhim, Moscow, Russia), and sodium phosphate dibasic (CAS no. 7558-79-4, BioXtra, Sigma-Aldrich, Gillingham, UK) were used for powder mixture preparation.

2.2. Synthesis of Powders

The synthesis of powders was carried out using the Na_2SiO_3, $Ca(NO_3)_2$, Na_2HPO_4 (Table 1) according to the equations:

$$Ca(NO_3)_2 + Na_2HPO_4 + 2H_2O \rightarrow CaHPO_4 \cdot 2H_2O + 2\,NaNO_3 \tag{1}$$

$$2\,Ca(NO_3)_2 + Na_2HPO_4 + Na_2SiO_3 + xH_2O \rightarrow CaHPO_4 \cdot 2H_2O + CaSiO_3 \cdot xH_2O + 4\,NaNO_3 \tag{2}$$

$$Ca(NO_3)_2 + Na_2SiO_3 + xH_2O \rightarrow CaSiO_3 \cdot xH_2O + 2\,NaNO_3 \tag{3}$$

Table 1. Conditions used for the syntheses of powders from aqueous solutions Na_2SiO_3, Na_2HPO_4, and $Ca(NO_3)_2$.

No.	Labeling	Concentration $\times$ Volume		
		Na_2SiO_3	Na_2HPO_4	$Ca(NO_3)_2$
1	CaP	-	0.5 M $\times$ 0.5 L	0.5 M $\times$ 0.5 L
2	CaPSi	0.5 M $\times$ 0.25 L	0.5 M $\times$ 0.25 L	0.5 M $\times$ 0.5 L
3	CaSi	0.5 M $\times$ 0.5 L	-	0.5 M $\times$ 0.5 L

The starting salts were dissolved in distilled water at a concentration of 0.5 M, then the calcium salt solution was slowly added to the sodium salt solution in the volume ratios corresponding to the reactions (1–3), and the suspension was stirred for 1 h. The synthesis was carried out at a temperature of 37 °C. Then, the precipitate was filtered using a vacuum filter and evenly distributed over a large surface area and left to dry for 1 week.

Further, the obtained powders were disaggregated in acetone medium using a planetary ball mill (Fritch Pulverisette, Bavaria, Germany) for 10 min in zirconia containers with grinding ZrO_2-media (m_{powder}:m_{balls} = 1:5). After that, when the acetone was completely removed each of the powders was sieved through a polyester sieve with a mesh size of 200 μm.

2.3. Preparation of Ceramic Samples

The obtained samples were pressed into the form of simple disks by uniaxial one-sided pressing on a manual press (Carver Laboratory Press model C, Fred S. Carver, Inc., Wabash, IN, USA) using steel die with a diameter of 12 mm. Pressing was carried out at a pressure of 100 MPa for 10 s.

The pressed samples in the form of disks were fired at 800, 900 and 1000 °C. Additionally, the powders were fired at 400 and 600 °C to control the phase transformations. Heating in the furnace was carried out at a speed of 5 °C/min. The holding time at these temperatures was 2 h. The heat treatment of the samples was carried out in order to study the effect of high temperatures on the initial composition, as well as to determine the mass losses and determine the thermal behavior of the materials in the temperature range mentioned. The linear shrinkage and density of the samples after the heat treatment were also calculated.

2.4. Methods of Analysis

The linear shrinkage after the heat treatment and the density of the samples before and after the heat treatment were calculated using Equations (4) and (5), respectively.

$$\Delta D_{rel} = (D_0 - D)/D_0 \times 100, \%, \tag{4}$$

where:

ΔD_{rel}—linear shrinkage of the sample after the heat treatment, %;
D—diameter of the sample after the heat treatment, cm;
D_0—diameter of the sample after pressing, cm.

$$\rho = m/(h \times \pi D^2/4), g/cm^3, \tag{5}$$

where:

ρ—density of the sample, g/cm^3;
m—weight of the sample, g;
h—thickness of the sample, cm;
D—diameter of the sample, cm.

The mass and the linear dimensions of the samples were measured with accuracy of ±0.001 g and ±0.01 mm, respectively, before and after the heat treatment.

Thermal analysis (TA) including thermogrvimetry (TG) and differential thermal analysis (DTA) was performed using an STA 409 PC Luxx thermal analyzer (NETZSCH, Selb, Germany) during heating in air (10 °C/min, 40–1000 °C), the specimen mass being at least 10 mg. The gas-phase composition was monitored by a Netzsch QMS 403C Aëolos quadrupole mass spectrometer (NETZSCH, Selb, Germany) coupled with a Netzsch STA 409 PC Luxx thermal analyzer (NETZSCH, Selb, Germany). The mass spectra were registered for the following m/Z values: 18 (H_2O); 30 (NO).

The phase composition of the powders obtained after the synthesis was determined by X-ray powder diffraction (XRD) analysis using Rigaku D/Max-2500 diffractometer (Rigaku Corporation, Tokyo, Japan) with a rotating anode (Cu–Ka radiation), angle interval 2Θ: from 2° to 70° (step 2Θ − 0.02°). XRD analysis of the ceramic composition was also implemented using a Rigaku Miniflex 600 diffractometer (CuKα radiation, Kβ filter, and D/teX Ultra detector) in Bragg–Brentano geometry (Rigaku Corporation, Tokyo, Japan) with an angle interval 2Θ from 3° to 70° (step 2Θ − 0.02°). Phase analysis was performed using the ICDD PDF2 database and Match software (version https://www.crystalimpact.com/, 25 December 2022).

Scanning electron microscopy (SEM) images of the synthesized powder and powder mixtures were characterized by SEM on an NVision 40 microscope (Carl Zeiss, Jena, Germany), and SEM images of ceramic samples were taken with Tescan Vega II (Tescan, Brno, Czech Republic) at accelerating voltages from 1 to 20 kV in secondary electron imaging mode (SE2 detector). A chromium/gold layers ($\leq$10 nm in thickness) on the surface of the ceramic sample was applied to the samples (Quorum Technologies spraying plant, Q150T ES, Great Britain, London, UK).

3. Results and Discussion

Table 2 shows weights of the prepared powders after synthesis and drying, weights of by-products calculated from the reactions (1–3) and weights of by-products isolated from the mother liquor by drying. The mass of prepared powders (column 3, Table 2) may include the mass of target minerals and the mass of by-product. In the row of powders CaP, CaPSi, CaSi the weight of the by-product collected from the mother liquor became lower. It was the lowest for CaSi, therefore, the amount of the by-product adsorbed and occluded by powder was the largest. The mass of adsorbed and occluded by-products increases from CaP to CaSi powder. The amount of the isolated reaction by-product apparently can be interpreted as an indirect confirmation of the presence of the largest active surface for CaSi and CaPSi powders. These values can apparently correlate with the surface area, which makes it possible to expect the highest activity for these powders.

Table 2. Weights of prepared powders and by-products.

No.	Labeling	Weight of Prepared Powders, g	Calculated Weight of By-Product, g	Weight of Collected By-Product, g	Difference in Weight of By-Product, g
1	CaP	41.97	42.44	40.01	2.44
2	CaPSi	47.89	42.44	32.89	9.55
3	CaSi	62.76	42.44	22.05	20.39

According to XRD (Figure S1) the phase composition of by-products of all reactions (1–3) isolated from the mother liquors by drying contained only $NaNO_3$.

The phase composition of all synthesized powders after disaggregation in acetone is presented in Figure 1. It should be noted that XRD patterns for powders after synthesis and after disaggregation were the same for each powder. The phase composition of the synthesized powders after the disaggregation process was not changed.

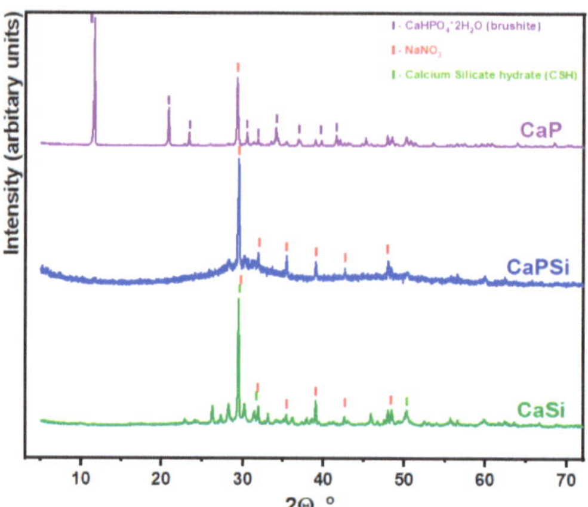

Figure 1. XRD data of powders after synthesis and disaggregation in acetone (CaP, CaPSi, and CaSi). (PDF card 9-77)—brushite, (PDF card 36-1474)—sodium nitrate, (PDF card 33-306)—calcium silicate hydrate.

According to the XRD data (Figure 1), the powder synthesized from $Ca(NO_3)_2$ and Na_2HPO_4 solutions (CaP) after disaggregation contained brushite (dicalcium phosphate dihydrate, $CaHPO_4 \cdot 2H_2O$), as well as the reaction by-product of sodium nitrate ($NaNO_3$).

According to the XRD analysis (Figure 1), the CaSi powder contained crystalline phases of calcium silicate hydrate (CSH) [50] and $NaNO_3$. Under considered synthesis conditions (37 °C and maturation of precipitates during 1 h in the mother liquors), the formation of well-crystallized calcium silicate hydrates (CSH — $Ca_{1.5}SiO_{3.5} \cdot xH_2O$ [PDF card 33-306]) was not expected, since their formation occurs in the process of high-temperature synthesis under autoclave conditions as it is known from the literature [50]. Low-basic hydrosilicates (Ca/Si = 0.8–1.5) have similar peaks on the X-ray diffraction pattern, and weak-crystallized products have 3–4 peaks, which lead to the ambiguous interpretations of the chemical composition. Peaks with interplanar spacing of 3.02A (29.54°) and 1.81A (50.45°) possibly belong to calcium silicate hydrate (CSH), which has a variable Ca/Si ratio from 0.8 to 1.5. The main peaks of CSH are 29.54, 33.25 and 50.45° [50]. And peaks 29.54, 33.25 of CSH overlaps with sodium nitrate peaks. The formation of tobermorite-like gel is also possible [50]. Amorphous calcium phosphate can be stabilized by the presence of silicate ions, as well as calcium silicate hydrate with a Ca/Si ratio from 0.8 to 1.5 [16].

Apparently, the CaPSi powder contained the only crystalline phase of $NaNO_3$. Other phases in the CaPSi powder were calcium silicate hydrate and amorphous calcium phosphate. Taking into account the presence of an amorphous halo near 30° on the XRD pattern of the CaPSi sample, this sample also possibly contained amorphous calcium phosphate (ACP), which could be effectively stabilized in the amorphous state by the presence of a silicate anion [16].

The morphology of the obtained powders after the synthesis and after the disaggregation process is presented in Figure 2.

Figure 2. SEM images of powder samples CaP (**a**,**b**), CaPSi (**c**,**d**) and CaSi (**e**,**f**) after synthesis (**a**,**c**,**e**) and disaggregation in acetone (**b**,**d**,**f**).

Brushite crystallizes in the specific shape of large thin bars (longer than 10 μm) characteristic for brushite, which are observed in Figure 2a. This is consistent with the XRD data for CaP powder. After disaggregation, elongated lamellar particles with a size of 1–2 μm remained (Figure 2b).

Large spheroid aggregates with dimensions 4–8 μm are presented in the SEM images of the CaSi powder sample before and after disaggregation (Figure 2e,f). The dimensions of particles in aggregates before and after disaggregation were not bigger than 50 nm. Apparently, this morphology corresponds to the presence of amorphous calcium silicate hydrate.

The CaPSi sample of synthesized powder is also consisted of aggregates (4–8 μm) of spheroid particles (~50 nm), as it is shown by the SEM image (Figure 2c). These spheroid particles may consist of amorphous phase of both hydrated ACP and calcium silicate hydrate. After disaggregation, the powders contain mesoporous aggregates (2–4 μm) of spherical particles of submicron (~50 nm) size (Figure 2d). Thereby, implementation of synthesis via precipitation from mixed-anion solution (CaPSi sample) leads to the production of powders consisted of submicron particles, which are smaller than those reported in previous studies, where the powders were obtained using another method [51].

According to the TG data, the final mass loss of the CaP powder after heating to 1000 °C was 28.3% (Figure 3a). At the first stage (up to 160 °C), there was a gradual decomposition of structurally unbonded water, accompanied by a slight endoeffect. At the next stage (up to 200 °C), the transition of brushite to monetite (DCPD → DCPA (dicalcium phosphate anhydrous)) occurred, accompanied by a significant endoeffect with a maximum at 190 °C (Figure 3b). The further process is the transformation of monetite to γ-pyrophosphate (DCPA → γ-CPP), generally takes place at 400 °C. But the process of transformation of DCPA to CPP in powder CaP under investigation tooks place at lower temperature. The mass loss takes place with an endoeffect at 295 °C (Figure 3b). This endoeffect may also be associated with the melting of sodium nitrate (melting point – 308 °C). So, we can state that the presence of monetite can make the melting point of sodium nitrate lower and consequently the presence of the melt of sodium nitrate leads to the thermal transformation (DCPA → CPP) at lower temperature. According to XRD data (Figure 4,a, Table 3) after heart treatment at 400 °C the phase composition of CaP powder included β-$Ca_2P_2O_7$ and $NaNO_3$. So, the transition of the γ-phase to the β-phase (γ-CPP → β-CPP) also was facilitated and made a contribution in the appearance of endoeffect in interval 250–350 °C. The next step of mass loss with endoeffect at 503 °C can be explained by sodium rhenanite formation from calcium pyrophosphate and sodium nitrate as it was confirmed by XRD data for powder CaP after heat treatment at 600 °C (Figure 4a, Table 3). The maximum endoeffect at the temperature of 503 °C corresponds to the maximum at 500 °C in interval 400–560 °C at mass-spectroscopy graph (Figure S2) by NO ion current (NO (M = 30) release) and the formation of the sodium rhenanite phase according to the formal reaction (6), which is not reflect the complicity of the thermal decomposition of nitrates.

$$Ca_2P_2O_7 + 2NaNO_3 \rightarrow 2NaCaPO_4 + N_2O_5 \tag{6}$$

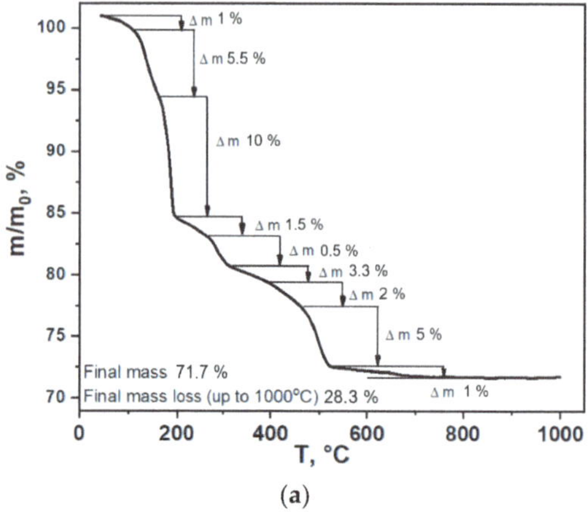
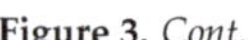

(a)

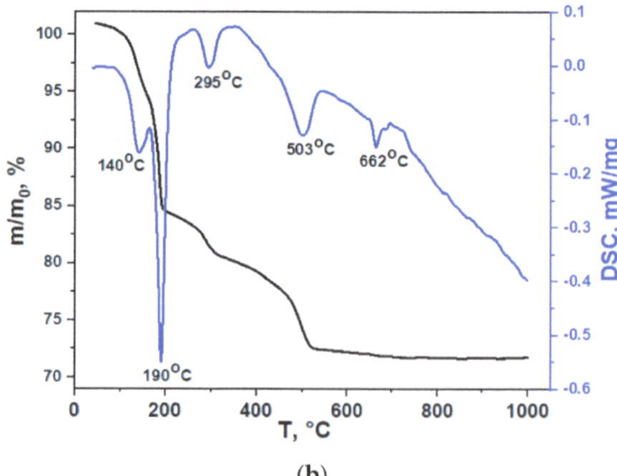

(b)

Figure 3. *Cont.*

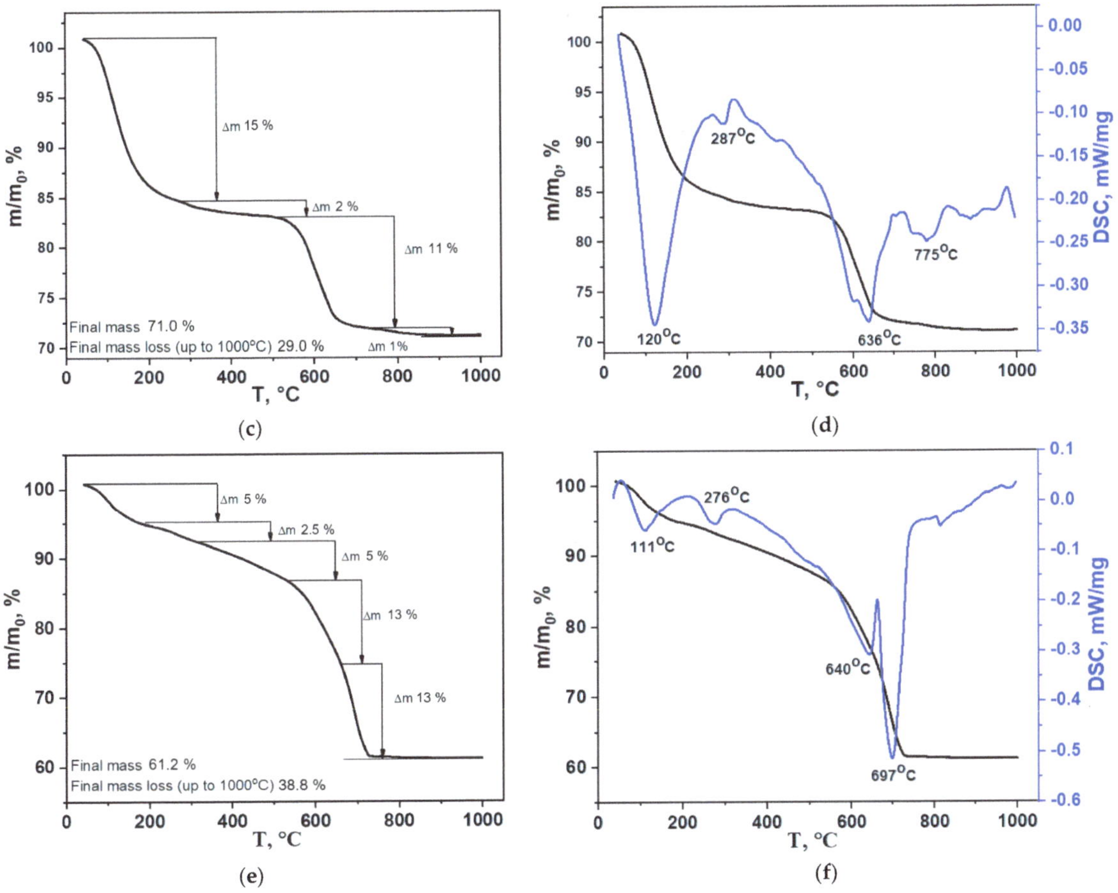

Figure 3. (**a**) step change in mass, and (**b**) TG/DTA curves obtained for the CaP synthesized powders; (**c**) step change in mass, and (**d**) TG/DTA curves obtained for the CaPSi synthesized powders, and (**e**) step change in mass, and (**f**) TG/DTA curves obtained for the CaSi synthesized powders.

Table 3. Transformation of the phase composition of preceramic and ceramic samples after heat treatment at specified temperature for 2 h.

Sample	Heat Treatment Temperature, °C				
	400	600	800	900	1000
CaP	β-$Ca_2P_2O_7$ $NaNO_3$	β-$Ca_2P_2O_7$ β-$NaCaPO_4$	β-$Ca_2P_2O_7$ β-$NaCaPO_4$	β-$Ca_2P_2O_7$ β-$NaCaPO_4$	β-$Ca_2P_2O_7$ β-$NaCaPO_4$
CaPSi	$Ca_5(PO_4)_3OH$ $NaNO_3$ α-$CaSiO_3$	β-$NaCaPO_4$ $Ca_5(PO_4)_3OH$ α-$CaSiO_3$	β-$NaCaPO_4$ $Ca_5(PO_4)_3OH$ α-$CaSiO_3$ $Na_3Ca_6(PO_4)_5$	β-$NaCaPO_4$ α-$CaSiO_3$ $Na_3Ca_6(PO_4)_5$	β-$NaCaPO_4$ α-$CaSiO_3$ $Na_3Ca_6(PO_4)_5$
CaSi	CSH $NaNO_3$	CSH $NaNO_3$ $Ca_5Si_6O_{16}(OH)_2$ $Na_6Ca_3Si_6O_{18}$	α-$CaSiO_3$ $Na_2Ca_2Si_2O_7$	α-$CaSiO_3$ $Na_2Ca_2Si_2O_7$	α-$CaSiO_3$ $Na_2Ca_2Si_2O_7$

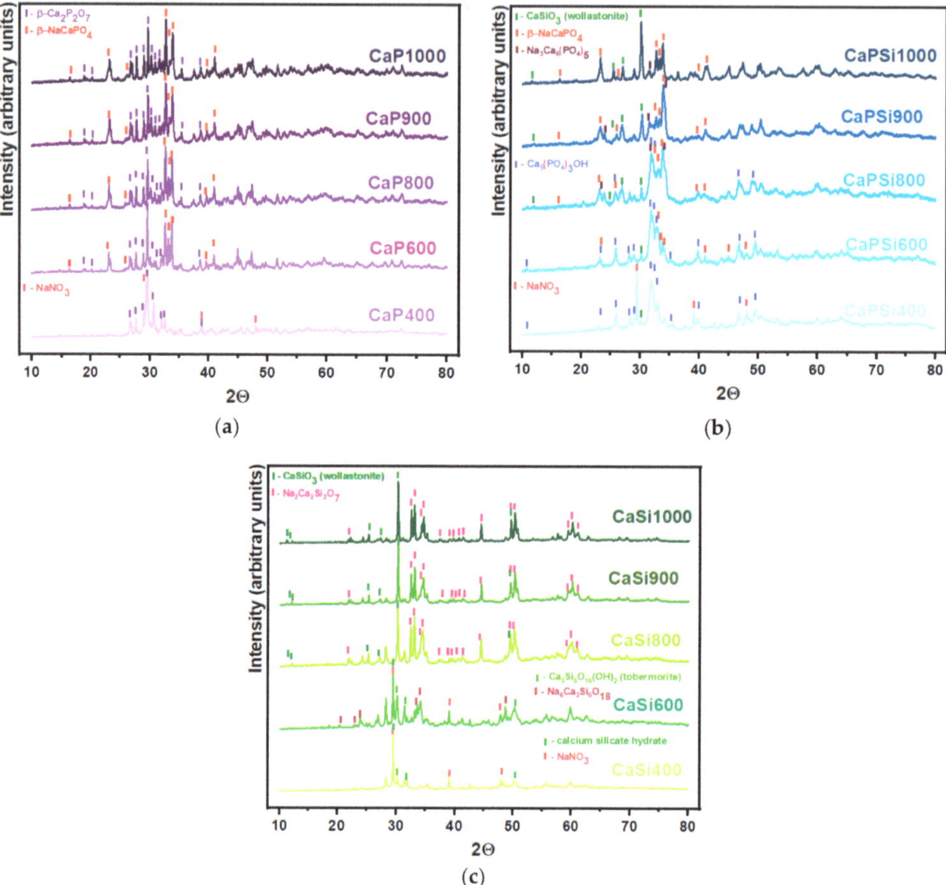

Figure 4. XRD patterns of CaP (**a**), CaSi (**b**), and CaPSi (**c**) samples after heat treatment at 1000 °C; CaPSi samples after heat treatment at 800, 900, and 1000 °C. (9-346)—$Ca_2P_2O_7$—purple, (29-1193)—$NaCaPO_4$—red, (76-186)—α-$CaSiO_3$—green, (10-16)—$Na_2Ca_2Si_2O_7$—pink, (11-236)—$Na_3Ca_6(PO_4)_5$—dark red.

For the CaSi sample, the final mass loss after heating to 1000 °C was 38.8% (Figure 3d). The first endoeffect (Figure 3e) at a temperature of 100–120 °C was apparently due to the elimination of structurally unbonded water. The endoeffect at the temperature of 300–310 °C was associated with the melting of sodium nitrate (melting point - 308 °C). Apparently, then there was a gradual crystallization of the amorphous phase with the formation of α-$CaSiO_3$ (wollastonite), including the possible formation of the intermediate tobermarite phase at 400–500 °C, up to 650 °C (with a significant endoeffect on the DSC graph). Confirmation of the gradual formation of wollastonite is presented in Figure 4c. The formation of wollastonite apparently continues up to 800 °C (Table 3). The maximum endoeffect at a temperature of 700 °C can be associated with the complete decomposition of sodium nitrate, which was also observed at the corresponding mass-spectroscopy graphs (Figure S2) by NO ion current (NO (M = 30) release) and the formation of the $Na_2Ca_2Si_2O_7$ phase (Figure 3f) and the sodium rhenanite phase (Figure 3b,d). The temperature of NO release depends on the composition of the sample. This indirectly indicates the surface area of powder that possibly holds the by-product. Thus, CaSi has the largest surface area, the maximum content of the by-product (Table 2) and a higher temperature, which is required to detach the nitrate. The ratio of intensity values of ion current for $m/Z = 30$ (NO) also indicates an increase in the content of sodium nitrate in the samples in the series CaP-CaPSi-CaSi (correspondingly 0.70×10^{-12}, 1.06×10^{-12}, 2.57×10^{-12}). The NO release peak for this series shifts to higher temperatures. Thus, the reactivity of CaP powder is higher than that of CaSi powder.

The total mass loss of the CaPSi sample after heating to 1000 °C was 29.0% (Figure 3b). Two steps of mass loss (17% and 12%) can be observed at the TG curve. According to MS data one step is connected with H_2O release up to 420 °C. And the second step is observed due to the processes connected with the decomposition of sodium nitrate or with heterphase reactions with sodium nitrate in the interval 450–750 °C. DTA graph (Figure 3c) clearly shows that two significant endoeffects corresponds to these steps of mass loss. According to XRD data for CaPSi sample after heat treatment at 400 and 600 °C hydroxyapatite, sodium nitrate and wollastonite was found. Up to 250 °C, the formation of the hydroxyapatite phase occurs from the amorphous calcium phosphate [52], as well as the beginning of the wollastonite phase forming, which apparently lasts up to 800 °C and also causes a 2% mass change in the temperature range 250–500 °C [25]. After 500 °C, the formation of the $Na_3Ca_6(PO_4)_5$ phase, and the sodium rhenanite (β-NaCaPO$_4$) phase occurs (Figure 3d).

As discussed above, heat treatment of the CaP powder at 600 °C and the ceramic samples at the range of temperatures from 800 to 1000 °C leads to the formation of a two-phase powder and ceramic samples containing calcium pyrophosphate (β-Ca$_2$P$_2$O$_7$) and β-rhenanite (β-NaCaPO$_4$). No other phases were found in the considered quasi-binary system [29].

In the case of the CaSi sample after the thermal treatment at 400 °C, there were no significant changes in phase composition, in comparison with the phase composition before the thermal treatment. After the thermal treatment at 600 °C, the formation of $Ca_5Si_6O_{16}(OH)_2$ and $Na_6Ca_3Si_6O_{18}$ occurred. Two-phase composite ceramics including wollastonite (α-CaSiO3) and $Na_2Ca_2Si_2O_7$ (Table 3, Figure 4; ICDD PDF2 database [53]) were also obtained after heat treatment of the CaSi samples at the range from 800 to 1000 °C.

After the heat treatment at 400 °C, the formation of hydroxyapatite and wollastonite occurred for the CaPSi sample. Presence of H_2O at high temperature give us an opportunity to guess that hydroxyapatite was formed under hydrothermal conditions. At 600 °C, the formation of hydroxyapatite from the amorphous phase, accompanied with the formation of sodium rhenanite from the amorphous phase and $NaNO_3$, occurred. The heat treatment at 800 °C led to the formation of tetraphasic ceramics, containing sodium rhenanite, hydroxyapatite, wollastonite and the $Na_3Ca_6(PO_4)_5$ phase. The phase composition of the CaPSi ceramic samples after the heat treatment in the range of temperatures from 900 to 1000 °C (Figure 4b, Table 3) remained almost unchanged. These samples after heat treatments at 900 and 1000 °C contained wollastonite (α-CaSiO$_3$), sodium rhenanite (NaCaPO$_4$), and mixed calcium-sodium phosphate ($Na_3Ca_6(PO_4)_5$). Nevertheless, the phase ratios varied depending on the heat treatment temperature. With an increase in the heat treatment temperature the phases tended to become more crystallized. The reflexes of the wollastonite (α-CaSiO$_3$) phase increased in the range of temperatures from 800 to 1000 °C, while the hydroxyapatite phase disappeared. After this temperature, heterophase reactions between hydroxyapatite and sodium rhenanite occurred, leading to the formation of $Na_3Ca_6(PO_4)_5$.

The SEM images of the surface of ceramic samples sintered at 1000 °C are shown in Figure 5.

With an increase in the heat treatment temperature, the sintering of the samples proceeds more efficiently due to the elimination of pores and the growth of grains, the size of which is 8–10 μm when fired at 1000 °C (CaP), while the grain sizes for the other samples are 2–3 μm (CaSi) and 2–4 μm (CaPSi).

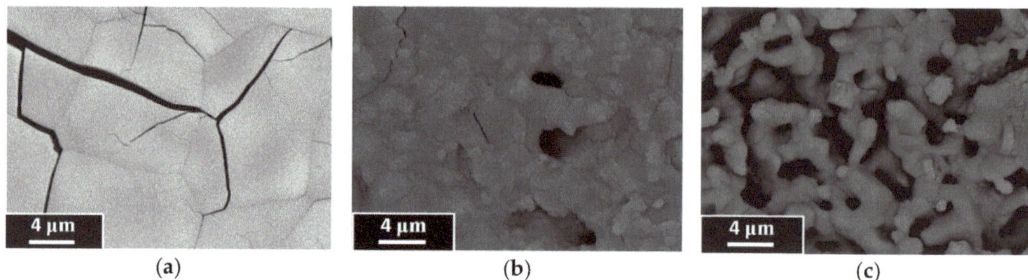

Figure 5. SEM images of ceramics obtained from powders (**a**) CaP; (**b**) CaPSi; (**c**) CaSi after heat treatment at 1000 °C.

Apparently, the occurrence of heterophase reactions in the CaSi and CaPSi samples inhibits grain growth in ceramics. The formation of sodium rhenanite and pyrophosphate is completed already after the heat treatment at 600 °C. During the heat treatment at higher temperatures, grain growth is possible since there are no conditions for heterophase reactions in the quasi-binary calcium pyrophosphate-sodium rhenanite system.

With an increase in the heat treatment temperature (and in the content of wollastonite), the ceramics became denser, which was accompanied by a significant grain growth up to 2–4 μm from submicron size. Microporosity of CaPSi also disappeared and a small number of pores with a diameter of 2–3 μm were formed. The ceramics based on the CaSi powder were highly porous, which was consistent with the literature data [54].

The shrinkage density of the samples CaP, CaPSi, and CaSi during the heat treatment is presented in Figure 6. The initial density of all the powder pre-ceramic samples in the form of disks was approximately equal and was 1.30 g/cm^3 for CaP and CaSi and 1.20 g/cm^3 for CaPSi.

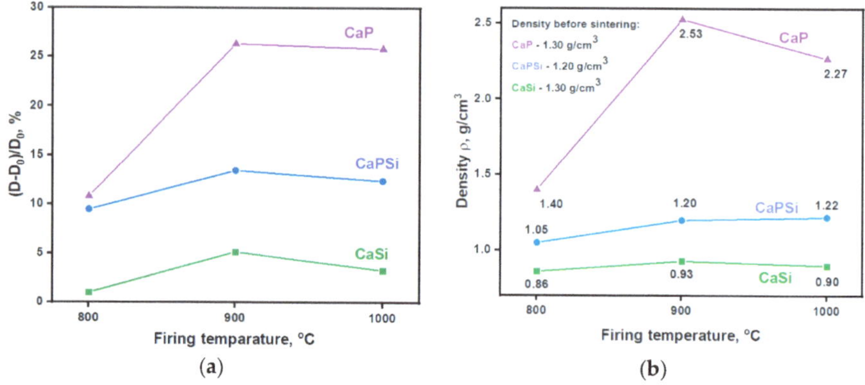

Figure 6. (**a**) Shrinkage during heat treatment, and (**b**) density of samples CaP, CaSi, and CaSi.

The CaP sample had the largest linear shrinkage (up to 26% at 900 and 1000 °C), and the CaSi sample had the smallest linear shrinkage (up to 5%) (Figure 6a). Linear shrinkage for the CaPSi sample ranged from 9 to 13% at 800 to 1000 °C, respectively. At the same time, fluctuations in density changes depending on the temperature for the CaPSi samples are minor, since the obtained components have approximately the same density (1.05–1.22 g/cm^3).

The apparent densities of the obtained ceramics are consistent with the scanning electron microscopy data (Figure 6b). The CaP sample fired at 900 and 1000 °C had the highest density. The uniform distribution of cracks visible on the surface of the CaP ceramic sample was apparently caused by the phase transformation in sodium rhenanite.

4. Conclusions

A novel method of multi-component active powders production was proposed in this work. The powders were prepared using a synthesis from aqueous solutions of $CaNO_3$ and a mixed-anion solution containing Na_2HPO_4 and Na_2SiO_3. For comparison, the powders were synthesized from solutions of $CaNO_3$ and solutions containing these sodium salts separately.

Phase composition of the CaP powder included brushite $CaHPO_4 \cdot 2H_2O$ and the CaSi powder included calcium silicate hydrate. Phase composition of the CaPSi powder consisted of the amorphous phase (presumably containing hydrated quasi-amorphous calcium phosphate and calcium silicate phase). All synthesized powders contained $NaNO_3$ as a by-product. The quasi-amorphous phases of CaPSi powder, obtained by precipitation from mixed-anionic solution, as it can be assumed, stabilized each other. A significant amount of by-product $NaNO_3$ was intentionally kept in the composition of all obtained powders for the synthesis of the ceramics of the Na_2O-CaO-SiO_2-P_2O_5 system during heat treatment of samples. This component acts both as a sintering aid and as a participant of the heterophase reactions of the formation of the ceramics based on the prepared powders (CaP. CaPSi, CaSi). The obtained powders were used to produce ceramics by the heat treatment at 800, 900, and 1000 °C. The main phases obtained for the ceramics based on the CaPSi powder were β-rhenanite β-$NaCaPO_4$, wollastonite α-$CaSiO_3$ and $Na_3Ca_6(PO_4)_5$. The density of the CaPSi ceramics increased with the heat treatment temperature from 1.47 g/cm^3 (800 °C) to 1.72 g/cm^3 (1000 °C). The ceramics based on the CaP powder had the highest density (2.53 g/cm^3) after the heat treatment at 900 °C. The ceramics based on the CaSi powder were significantly porous and had the lowest density (1.52 g/cm^3) after the heat treatment at 900 °C. The ceramics prepared in this work in the CaO-SiO_2-P_2O_5-Na_2O system and containing the biocompatible and bioresorbable phases can be recommended for use in medicine for bone defect treatment. The obtained materials are expected to support cell proliferation and differentiation. Osteoconductive 3D forms also can be manufactured from obtained powders.

Supplementary Materials: The following supporting information can be downloaded at: https://www.mdpi.com/article/10.3390/coatings13020374/s1, Figure S1: XRD data of by-product powders dried from solutions, collected after filtration (mother liquors) (PDF card 36-1474)–sodium nitrate, Figure S2: Mass-spectroscopy data of (a) CaP, (b) CaPSi, (c) CaSi powders.

Author Contributions: Conceptualization, D.G. and T.V.S.; methodology, D.G. and T.V.S.; validation, D.G., E.N. and T.V.S.; investigation, D.G., E.N., T.V.S., T.B.S., I.N.T., I.V.R., D.K., V.P., O.B., M.K., D.A.F. and K.A.L.; resources, T.B.S., I.V.R., O.B. and K.A.L.; data curation, D.G. and E.N.; writing—original draft preparation, D.G. and E.N.; writing—review and editing, D.G., E.N. and T.V.S.; visualization, D.G., E.N., T.B.S., I.V.R., D.K., V.P., O.B. and M.K.; supervision, T.V.S.; funding acquisition, T.V.S. All authors have read and agreed to the published version of the manuscript.

Funding: This work was carried out with financial support from the Russian Science Foundation (grant no. 22-19-00219).

Institutional Review Board Statement: Not applicable.

Informed Consent Statement: Not applicable.

Data Availability Statement: Not applicable.

Acknowledgments: This research was carried out using the equipment of the MSU Shared Research Equipment Center "Technologies for obtaining new nanostructured materials and their complex study" and purchased by MSU in the frame of the Equipment Renovation Program (National Project "Science"), and in the frame of the MSU Program of Development. The authors would like to thank Elena Golubchikova for the assistance with draft preparation. Some of the SEM images were recorded using scientific equipment at the Joint Research Center for Physical Methods of Research, located in the Kurnakov Institute of General and Inorganic Chemistry RAS.

Conflicts of Interest: The authors declare no conflict of interest.

References

1. Stevens, M.M. Biomaterials for Bone Tissue Engineering. *Mater. Today* **2008**, *11*, 18–25. [CrossRef]
2. Ansari, M. Bone Tissue Regeneration: Biology, Strategies and Interface Studies. *Prog. Biomater.* **2019**, *8*, 223–237. [CrossRef]
3. Ashammakhi, N.; Hasan, A.; Kaarela, O.; Byambaa, B.; Sheikhi, A.; Gaharwar, A.K.; Khademhosseini, A. Advancing Frontiers in Bone Bioprinting. *Adv. Healthc. Mater.* **2019**, *8*, 1801048. [CrossRef] [PubMed]
4. Amiryaghoubi, N.; Fathi, M.; Pesyan, N.N.; Samiei, M.; Barar, J.; Omidi, Y. Bioactive Polymeric Scaffolds for Osteogenic Repair and Bone Regenerative Medicine. *Med. Res. Rev.* **2020**, *40*, 1833–1870. [CrossRef] [PubMed]
5. Shuai, C.; Yang, W.; Feng, P.; Peng, S.; Pan, H. Accelerated Degradation of HAP/PLLA Bone Scaffold by PGA Blending Facilitates Bioactivity and Osteoconductivity. *Bioact. Mater.* **2021**, *6*, 490–502. [CrossRef]
6. Yamada, Y.; Inui, T.; Kinoshita, Y.; Shigemitsu, Y.; Honda, M.; Nakano, K.; Matsunari, H.; Nagaya, M.; Nagashima, H.; Aizawa, M. Silicon-Containing Apatite Fiber Scaffolds with Enhanced Mechanical Property Express Osteoinductivity and High Osteoconductivity. *J. Asian Ceram. Soc.* **2019**, *7*, 101–108. [CrossRef]
7. Malhotra, A.; Habibovic, P. Calcium Phosphates and Angiogenesis: Implications and Advances for Bone Regeneration. *Trends Biotechnol.* **2016**, *34*, 983–992. [CrossRef] [PubMed]
8. Schmidt-Bleek, K.; Schell, H.; Schulz, N.; Hoff, P.; Perka, C.; Buttgereit, F.; Volk, H.-D.; Lienau, J.; Duda, G.N. Inflammatory Phase of Bone Healing Initiates the Regenerative Healing Cascade. *Cell Tissue Res.* **2012**, *347*, 567–573. [CrossRef] [PubMed]
9. Ligon, S.C.; Liska, R.; Stampfl, J.; Gurr, M.; Mülhaupt, R. Polymers for 3D Printing and Customized Additive Manufacturing. *Chem. Rev.* **2017**, *117*, 10212–10290. [CrossRef]
10. Mao, M.; He, J.; Li, X.; Zhang, B.; Lei, Q.; Liu, Y.; Li, D. The Emerging Frontiers and Applications of High-Resolution 3D Printing. *Micromachines* **2017**, *8*, 113. [CrossRef]
11. Pina, S.; Ribeiro, V.P.; Marques, C.F.; Maia, F.R.; Silva, T.H.; Reis, R.L.; Oliveira, J.M. Scaffolding Strategies for Tissue Engineering and Regenerative Medicine Applications. *Materials* **2019**, *12*, 1824. [CrossRef] [PubMed]
12. Yunus Basha, R.; Kumar, T.S.S.; Doble, M. Design of Biocomposite Materials for Bone Tissue Regeneration. *Mater. Sci. Eng. C* **2015**, *57*, 452–463. [CrossRef] [PubMed]
13. Guzzo, C.M.; Nychka, J.A. Bone 'Spackling' Paste: Mechanical Properties and In Vitro Response of a Porous Ceramic Composite Bone Tissue Scaffold. *J. Mech. Behav. Biomed. Mater.* **2020**, *112*, 103958. [CrossRef] [PubMed]
14. Guzzo, C.M.; Nychka, J. Fabrication of a Porous and Formable Ceramic Composite Bone Tissue Scaffold at Ambient Temperature. *Metall. Mater. Trans. A* **2020**, *51*, 6110–6126. [CrossRef]
15. Kazakova, G.; Safronova, T.; Golubchikov, D.; Shevtsova, O.; Rau, J.V. Resorbable Mg^{2+}-Containing Phosphates for Bone Tissue Repair. *Materials* **2021**, *14*, 4857. [CrossRef]
16. Zuev, D.M.; Golubchikov, D.O.; Evdokimov, P.V.; Putlyaev, V.I. Synthesis of Amorphous Calcium Phosphate Powders for Production of Bioceramics and Composites by 3D Printing. *Russ. J. Inorg. Chem.* **2022**, *67*, 940–951. [CrossRef]
17. Koons, G.L.; Diba, M.; Mikos, A.G. Materials Design for Bone-Tissue Engineering. *Nat. Rev. Mater.* **2020**, *5*, 584–603. [CrossRef]
18. Hu, X.; Wang, Y.; Tan, Y.; Wang, J.; Liu, H.; Wang, Y.; Yang, S.; Shi, M.; Zhao, S.; Zhang, Y.; et al. A Difunctional Regeneration Scaffold for Knee Repair Based on Aptamer-Directed Cell Recruitment. *Adv. Mater.* **2017**, *29*, 1605235. [CrossRef]
19. Hench, L.L.; Splinter, R.J.; Allen, W.C.; Greenlee, T.K. Bonding Mechanisms at the Interface of Ceramic Prosthetic Materials. *J. Biomed. Mater. Res.* **1971**, *5*, 117–141. [CrossRef]
20. Hench, L.L.; Paschall, H.A. Direct Chemical Bond of Bioactive Glass-Ceramic Materials to Bone and Muscle. *J. Biomed. Mater. Res.* **1973**, *7*, 25–42. [CrossRef]
21. Li, H.; Xue, K.; Kong, N.; Liu, K.; Chang, J. Silicate Bioceramics Enhanced Vascularization and Osteogenesis through Stimulating Interactions between Endothelia Cells and Bone Marrow Stromal Cells. *Biomaterials* **2014**, *35*, 3803–3818. [CrossRef]
22. Kaully, T.; Kaufman-Francis, K.; Lesman, A.; Levenberg, S. Vascularization—The Conduit to Viable Engineered Tissues. *Tissue Eng. Part B Rev.* **2009**, *15*, 159–169. [CrossRef] [PubMed]
23. Hillert, M.; Sundman, B.; Wang, X. An Assessment of the CaO-SiO_2 System. *Metall. Mater. Trans. B* **1990**, *21*, 303–312. [CrossRef]
24. Fu, S.; Liu, W.; Liu, S.; Zhao, S.; Zhu, Y. 3D Printed Porous β-Ca_2SiO_4 Scaffolds Derived from Preceramic Resin and Their Physicochemical and Biological Properties. *Sci. Technol. Adv. Mater.* **2018**, *19*, 495–506. [CrossRef]
25. Pan, Y.; Yin, J.; Yao, D.; Zuo, K.; Xia, Y.; Liang, H.; Zeng, Y. Effects of Silica Sol on the Microstructure and Mechanical Properties of $CaSiO3$ Bioceramics. *Mater. Sci. Eng. C* **2016**, *64*, 336–340. [CrossRef]
26. De Aza, P.N.; Luklinska, Z.B.; Martinez, A.; Anseau, M.R.; Guitian, F.; De Aza, S. Morphological and Structural Study of Pseudowollastonite Implants in Bone. *J. Microsc.* **2000**, *197*, 60–67. [CrossRef]
27. Yu, Q.; Chang, J.; Wu, C. Silicate Bioceramics: From Soft Tissue Regeneration to Tumor Therapy. *J. Mater. Chem. B* **2019**, *7*, 5449–5460. [CrossRef]
28. Hu, Y.; Xiao, Z.; Wang, H.; Ye, C.; Wu, Y.; Xu, S. Fabrication and Characterization of Porous $CaSiO_3$ Ceramics. *Ceram. Int.* **2019**, *45*, 3710–3714. [CrossRef]
29. Filippov, Y.; Murashko, A.; Evdokimov, P.; Safronova, T.; Putlayev, V. Stereolithography 3D printed calcium pyrophosphate macroporous ceramics for bone grafting. *Open Ceram.* **2021**, *8*, 100185. [CrossRef]
30. Serena, S.; Sainz, M.; Caballero, A. Single-phase silicocarnotite synthesis in the subsystem $Ca_3(PO_4)_2$–Ca_2SiO_4. *Ceram. Int.* **2014**, *40*, 8245–8252. [CrossRef]

31. Ros-Tarraga, P.; Mazon, P.; Meseguer-Olmo, L.; De Aza, P. Revising the Subsystem Nurse's A-Phase-Silicocarnotite within the System Ca$_3$(PO$_4$)$_2$–Ca$_2$SiO$_4$. *Materials* **2016**, *9*, 322. [CrossRef] [PubMed]

32. Xu, M.; Zhai, D.; Chang, J.; Wu, C. In Vitro assessment of three-dimensionally plotted nagelschmidtite bioceramic scaffolds with varied macropore morphologies. *Acta Biomater.* **2014**, *10*, 463–476. [CrossRef]

33. Anand, V.; Singh, K.; Kaur, K. Investigation of Mg and Zn doped 45S5 bioactive materials by XRD, FTIR and SEM techniques. *AIP Conf. Proc.* **2014**, *1591*, 745. [CrossRef]

34. Zhang, X.; Guo, X.; Zhang, J.; Fan, X.; Chen, M.; Yang, H. Nucleation, Crystallization and Biological Activity of Na$_2$O-CaO-P$_2$O$_5$-SiO$_2$ Bioactive Glass. *J. Non Cryst. Solids* **2021**, *568*, 120929. [CrossRef]

35. Kahlenberg, V.; Hösch, A. The Crystal Structure of Na$_2$Ca$_2$Si$_2$O$_7$—A Mixed Anion Silicate with Defect Perovskite Characteristics. *Z. Krist. Cryst. Mater.* **2002**, *217*, 155–163. [CrossRef]

36. Leenakul, W.; Pisitpipathsin, N.; Kantha, P.; Tawichai, N.; Tigunta, S.; Eitssayeam, S.; Rujijanagul, G.; Pengpat, K.; Munpakdee, A. Characteristics of 45S5 Bioglass-Ceramics Using Natural Raw Materials. *AMR Adv. Mater. Res.* **2012**, *506*, 174–177. [CrossRef]

37. Kaimonov, M.; Safronova, T.; Shatalova, T.; Filippov, Y.; Tikhomirova, I.; Sergeev, N. Composite Ceramics in the Na$_2$O–CaO–SiO$_2$–P2O$_5$ System Obtained from Pastes Including Hydroxyapatite and an Aqueous Solution of Sodium Silicate. *Ceramics* **2022**, *5*, 550–561. [CrossRef]

38. Safronova, T.V. Inorganic Materials for Regenerative Medicine. *Inorg. Mater.* **2021**, *57*, 443–474. [CrossRef]

39. Demirkiran, H.; Mohandas, A.; Dohi, M.; Fuentes, A.; Nguyen, K.; Aswath, P. Bioactivity and Mineralization of Hydroxyapatite with Bioglass as Sintering Aid and Bioceramics with Na$_3$Ca$_6$(PO$_4$)$_5$ and Ca$_5$(PO$_4$)$_2$SiO$_4$ in a Silicate Matrix. *Mater. Sci. Eng. C* **2010**, *30*, 263–272. [CrossRef]

40. Lin, K.; Zhai, W.; Ni, S.; Chang, J.; Zeng, Y.; Qian, W. Study of the Mechanical Property and In Vitro Biocompatibility of CaSiO$_3$ Ceramics. *Ceram. Int.* **2005**, *31*, 323–326. [CrossRef]

41. Gmeiner, R.; Deisinger, U.; Schönherr, J.; Lechner, B.; Detsch, R.; Boccaccini, A.R.; Stampfl, J. Additive Manufacturing of Bioactive Glasses and Silicate Bioceramics. *J. Ceram. Sci. Technol.* **2015**, *6*, 75–86. [CrossRef]

42. Bergmann, C.; Lindner, M.; Zhang, W.; Koczur, K.; Kirsten, A.; Telle, R.; Fischer, H. 3D Printing of Bone Substitute Implants Using Calcium Phosphate and Bioactive Glasses. *J. Eur. Ceram. Soc.* **2010**, *30*, 2563–2567. [CrossRef]

43. Safronova, T.V.; Knot'ko, A.V.; Shatalova, T.B.; Evdokimov, P.V.; Putlyaev, V.I.; Kostin, M.S. Calcium phosphate ceramic based on powder synthesized from a mixed-anionic solution. *Glass Ceram.* **2016**, *73*, 25–31. [CrossRef]

44. Peranidze, K.; Safronova, T.V.; Filippov, Y.; Kazakova, G.; Shatalova, T.; Rau, J.V. Powders Based on Ca$_2$P$_2$O$_7$-CaCO$_3$-H$_2$O System as Model Objects for the Development of Bioceramics. *Ceramics* **2022**, *5*, 423–434. [CrossRef]

45. Safronova, T.V.; Putlyaev, V.I.; Filippov, Y.Y.; Knot'Ko, A.V.; Klimashina, E.S.; Peranidze, K.K.; Evdokimov, P.V.; Vladimirova, S.A. Powders Synthesized from Calcium Acetate and Mixed-Anionic Solutions, Containing Orthophosphate and Carbonate Ions, for Obtaining Bioceramic. *Glass Ceram.* **2018**, *75*, 118–123. [CrossRef]

46. Joksa, A.A.; Komarovska, L.; Ubele-Kalnina, D.; Viksna, A.; Gross, K.A. Role of carbonate on the crystallization and processing of amorphous calcium phosphates. *Materialia* **2023**, *27*, 101672. [CrossRef]

47. Safronova, T.V.; Putlyaev, V.I.; Filippov, Y.; Shatalova, T.B.; Fatin, D.S. Ceramics based on brushite powder synthesized from calcium nitrate and disodium and dipotassium hydrogen phosphates. *Inorg. Mater.* **2018**, *54*, 195–207. [CrossRef]

48. Safronova, T.V. Phase Composition of Ceramic Based on Calcium Hydroxyapatite Powders Containing Byproducts of the Synthesis Reaction. *Glass Ceram.* **2009**, *66*, 136–139. [CrossRef]

49. Safronova, T.V.; Putlyaev, V.I.; Knot'ko, A.V.; Shatalova, T.B.; Artemov, M.V.; Filippov, Y.Y. Properties of Calcium Phosphate Powder Synthesized from Calcium Chloride and Potassium Pyrophosphate. *Inorg. Mater. Appl. Res.* **2020**, *11*, 44–49. [CrossRef]

50. Gorshkov, V.; Timashev, V.; Saveliev, V. Identification characteristics of compounds containing water. In *Methods for Physical and Chemical Analysis of Binders*; Gaidzhurov, P., Nekrasov, K., Eds.; Vyshaya Shkola: Moscow, Russia, 1981; pp. 292–294. (In Russian)

51. Solonenko, A.P.; Blesman, A.I.; Polonyankin, D.A.; Gorbunov, V.A. Synthesis of Calcium Phosphate and Calcium Silicate Composites. *Russ. J. Inorg. Chem.* **2018**, *63*, 993–1000. [CrossRef]

52. Vecstaudza, J.; Gasik, M.; Locs, J. Amorphous Calcium Phosphate Materials: Formation, Structure and Thermal Behaviour. *J. Eur. Ceram. Soc.* **2019**, *39*, 1642–1649. [CrossRef]

53. ICDD. *PDF-4+ Database*; Kabekkodu, S., Ed.; International Centre for Diffraction Data: Newtown Square, PA, USA, 2010. Available online: https://www.icdd.com/pdf-2/ (accessed on 20 February 2022).

54. Li, L.; Hu, H.; Zhu, Y.; Zhu, M.; Liu, Z. 3D-Printed Ternary SiO$_2$CaO P$_2$O$_5$ Bioglass-Ceramic Scaffolds with Tunable Compositions and Properties for Bone Regeneration. *Ceram. Int.* **2019**, *45*, 10997–11005. [CrossRef]

 *coatings*

Article

The Impact of Graphene Oxide on Polycaprolactone PCL Surfaces: Antimicrobial Activity and Osteogenic Differentiation of Mesenchymal Stem Cell

Letizia Ferroni [1], Chiara Gardin [1], Federica Rigoni [2], Eleonora Balliana [3], Federica Zanotti [4], Marco Scatto [2,5], Pietro Riello [2] and Barbara Zavan [4,*]

[1] Maria Cecilia Hospital, GVM Care & Research, 48033 Cotignola, Ravenna , Italy; lferroni@gvmenet.it (L.F.); gardinc@gvmnet.it (C.G.)

[2] Department of Molecular Sciences and Nanosystems, Ca' Foscari University of Venice, Via Torino 155, 30172 Venezia Mestre, Venice, Italy; rigonii@unive.it (F.R.); scattom@gmail.it (M.S.); riellop@unive.it (P.R.)

[3] Department of Environmental Sciences and Statistics, Ca' Foscari University of Venice, Via Torino 155, 30172 Venezia Mestre, Venice, Italy; eballiana@unive.it

[4] Department of Translational Medicine, University of Ferrara, Via Fossato di Mortara 70, 44121 Ferrara, Ferrara, Italy; zanottiff@unife.it

[5] Nadir S.r.l., Via Torino 155b, c/o Department of Molecular Sciences and Nanosystems, Ca' Foscari University of Venice, Via Torino 155, 30172 Venezia Mestre, Venice, Italy

* Correspondence: barbara.zavan@unife.it

Citation: Ferroni, L.; Gardin, C.; Rigoni, F.; Balliana, E.; Zanotti, F.; Scatto, M.; Riello, P.; Zavan, B. The Impact of Graphene Oxide on Polycaprolactone PCL Surfaces: Antimicrobial Activity and Osteogenic Differentiation of Mesenchymal Stem Cell. *Coatings* **2022**, *12*, 799. https://doi.org/10.3390/coatings12060799

Academic Editors: Nileshkumar Dubey and Vinicius Rosa

Received: 27 April 2022
Accepted: 29 May 2022
Published: 8 June 2022

Abstract: In dentistry, bone regeneration requires osteoinductive biomaterial with antibacterial properties. Polycaprolactone (PCL) may be combined with different nanofillers including reduced graphene oxide (rGO). Here, the amount of rGO filler was defined to obtain a biocompatible and antibacterial PCL-based surface supporting the adhesion and differentiation of human mesenchymal stem cells (MSCs). Compounds carrying three different percentages of rGO were tested. Among all, the 5% rGO-PCL compound is the most bacteriostatic against Gram-positive bacteria. All scaffolds are biocompatible. MSCs adhere and proliferate on all scaffolds; however, 5% rGO-PCL surface supports the growth of cells and implements the expression of extracellular matrix components necessary to anchor the cells to the surface itself. Moreover, the 5% rGO-PCL surface has superior osteoinductive properties confirmed by the improved alkaline phosphatase activity, mineral matrix deposition, and osteogenic markers expression. These results suggest that 5% rGO-PCL has useful properties for bone tissue engineering purposes.

Keywords: graphene; polycaprolactone; antimicrobial surfaces; mesenchymal stem cell; adhesion; differentiation; bone

1. Introduction

In dentistry, bone regeneration is crucial in cases of trauma, infections, and tumors, and it is sometimes required to stabilize dental implants [1]. De facto, implant therapy is effective when an adequate vertical and horizontal bone volume is present at the implant site [2]. Bone augmentation usually requires bone blocks or granulated bone particulates and membrane or titanium meshwork. This can be technically challenging, and the success of the treatment will depend on the space maintenance of the regenerated defect, wound stability, and absence of infection [3]. When bacterial activity occurs in bone defects, bone regeneration can be limited or even hindered. For this reason, novel biomaterials for bone regeneration are still extensively investigated. Currently, most commercial biomaterials are slightly osteoinductive and with poor antibacterial properties [1].

Polycaprolactone (PCL) is a biocompatible, semicrystalline, water-insoluble polymer, which has gained much attention in scaffolding and tissue engineering due to its adaptability and tailorable properties [4]. Thanks to its good ability of plasticity under a certain stress without fracturing, PCL represents a perfect chosen material in dental and orthopedic regenerative medicine [5,6]. In addition, PCL can be combined with a low concentration of specific nanomaterials for the fabrication of scaffolds with definite features for cell and tissue growth. The mechanical properties of nanocomposite polymers depend on the properties of both pure polymer and filler [7]. In recent years, a large body of investigation has been dealing with nanofiller as a reinforcement to improve or introduce new properties at the nano- and macro-scale. Graphene-related materials, including graphene oxide (GO), reduced graphene oxide (rGO), or graphene nanoplates, were embedded in PCL to enhance the crystallization and orientation of polymer matrix or to make PCL conductive [8–10].

Nowadays, stem cells are offering revolutionary therapies for many diseases with limited treatment options. They play a crucial role in tissue repair, as they possess the unique quality to differentiate into specialized cell types in response to biochemical, biophysical, and biomechanical cues. In fact, stem cells adapt themselves to the surrounding environment and to the specific features of the scaffold, including composition, surface topography, ligand availability, and mechanical properties changing migration, proliferation, differentiation, and viability [11,12].

In this work, PCL/rGO composites have been investigated for regenerative medicine purposes as substrates for stem cell proliferation and differentiation. Pure PCL was mixed with three different percentage of rGO (1.6%, 3%, and 5%) to manufacture by extrusion rGO-PCL compounds. First chemical composition, antibacterial activity, and biocompatibility of rGO-PCL compounds were screened. Second, the regenerative properties of the composites were investigated in vitro with human mesenchymal stem cells (HMSCs). All compounds were biocompatible in conformity with the international standard for medical devices and bacteriostatic to Gram-positive bacteria in rGO-dose dependent manner. The percentage of rGO also regulated the adhesion, morphology, and differentiation of the stem cells. rGO impacted the osteogenic differentiation by ensuring the activity of alkaline phosphatase, the deposition of the mineral matrix, and the expression of proteins crucial for the differentiation in osteoblasts. These results suggest that 5% rGO-PCL has useful properties for bone tissue engineering purposes.

2. Materials and Methods

2.1. Compounds Preparation

Pure PCL (Sigma-Aldrich, St. Louis, MO, USA) was mixed with rGO (Abalonyx AS, Oslo, Norway) at the percentages of 1.6% *w/w*, 3% *w/w*, or 5% *w/w* through melt compounding strategy assisted by a twin screw extruder (Themo Fisher Scientific, Waltham, MA, USA) by Nadir s.r.l. PCL pellets were placed in the extruder's pellet feeder, whereas the rGO were placed in the extruder's filler feeder. Next, during the extrusion process, all the materials were fed at the same time to produce a compounded wire that was immediately ground to form composite pellets. After drying the composite pellets at 50 °C, injection molding was performed to manufacture disks. All samples were sterilized with ethylene oxide.

2.2. Chemical Characterization

Raman spectra were recovered using a dispersive Raman system Thermo Scientific™ DXR3™ (Waltham, MA, USA) equipped with an Olympus microscope. The experimental parameters were laser wavelengths 532 nm, power laser of 5 mW, and 50–3350 cm^{-1} full range grating. A 10× objective and a 25 μm-slit aperture were used to obtain more representative spectra from the samples. Total acquisition time for each spectrum was 120 s. Thermo Scientific™ OMNIC™ software 34.3 (Version 3, Waltham, MA, USA) was used to operate the DXR3™ Raman Microscope and to collect the spectra. Raman spectra were compared with the internal library of instrument and with literature for peak assignment.

2.3. Antibacterial Activity

Antibacterial activity of rGO-PCL compounds was evaluated against Streptococcus pyogenes, Staphylococcus aureus, Escherichia coli, and Pseudomonas aeruginosa (all purchased from ATCC, Manassas, VA, USA). Antibacterial assays were performed according to ISO 22196:2011 (Measurement of antibacterial activity on plastics and other non-porous surfaces). The agar diffusion method and the quantification of colony forming units (CFU) were performed. *P. aeruginosa* virulence was evaluated by quantification of pyocyanin production. *E. coli* adhesion was monitored by fluorescent miscopy. A suspension of recombinant *E. coli* bacteria expressing green fluorescent protein GFP (E. coli-GFP; ATCC) was placed in contact with samples at 37 °C with shaking for 24 h and then observed at microscope.

2.4. In Vitro Cytotoxicity

The in vitro cytotoxicity of rGO-PCL compounds was evaluated following the ISO 10993-5 standard test method (Biological evaluation of medical devices—Part 5: Tests for in vitro cytotoxicity) [13]. Briefly, NCTC clone 929 (mouse fibroblast cell line; ATCC) were seeded at the density of 2×10^4 cell/cm^2 in Dulbecco's Modified Eagle Medium (DMEM, EuroClone, Rome, Italy) supplemented with 10% Fetal Bovine Serum (FBS, EuroClone) and 1% Penicillin-Streptomycin (PS, EuroClone). After 24 h, rGO-PCL compounds were placed in direct contact with cells for 48 h. Then, cell viability was determined by incubation with 0.5 mg/mL MTT (Sigma-Aldrich) solution for 3 h. After removing the MTT solution, extraction solution (isopropanol:dimethyl sulfoxide 9:1) was incubated for 30 min at 37 °C, and absorbance at 570 nm was recorded. NCTC in direct contact with sterile titanium was used as control condition. Results were expressed as the percentage of viable cells [14].

2.5. Stem Cell Seeding on rGO-PCL Surfaces

rGO-PCL disks were positioned on the bottom of 24-well plates and 2×10^4 human Adipose-derived Mesenchymal Stem Cells (HMSCs, Sciencell Research Laboratories, New York, NY, USA) were seeded on each of them. Cultures have been maintained in osteogenic differentiation medium consisting in DMEM supplemented with 10% FBS, 1% PS, 10 ng/mL Fibroblast Growth Factor 2 (FGF-2, ProSpec, New York, NY, USA), 10 mM β-glycerophosphate (Sigma-Aldrich, St. Louis, MO, USA), and 10 nM dexamethasone (Sigma-Aldrich, St. Louis, MO, USA). All cultures were incubated at 37 °C and 5% CO$_2$ for up to 21 days, and culture medium was changed three times a week [15]. These cultures were investigated for cell adhesion, mechanical properties, morphology, cell viability, alkaline phosphatase (ALP) activity, mineral matrix deposition, and gene expression.

2.6. Cell Adhesion, Morphology, and Proliferation

The adhesion and morphology of HMSCs on rGO-PCL surfaces were evaluated using the ZEISS SIGMA VP Field Emission Scanning Electron Microscope (FE-SEM, ZEISS, Jena, Germany). Samples were fixed with 2% glutaraldehyde (Sigma-Aldrich, St. Louis, MO, USA) in 0.1 M HEPES buffer (Sigma-Aldrich, St. Louis, MO, USA), dehydrated in ethanol, and chemical dried with hexamethyldisilane (Sigma-Aldrich, St. Louis, MO, USA) [16]. The SEM images of the sample surfaces were recorded in high vacuum (10^{-6} mbar) after the surface metallization with a gold thin film by using a Polaron SC7620 gold sputter coater.

MTS Assay (Abcam, Cambridge, UK) quantified the viability and proliferation of HMSCs on each surface. Briefly, cells were incubated with MTS compound for 4 h at 37 °C and 5% CO$_2$. Then, the produced formazan was quantified by measuring the absorbance at 490 nm.

2.7. ALP Activity

ALP activity was detected by Alkaline phosphates kit (Abcam) that uses p-nitrophenyl phosphate (pNPP) as phosphatase substrate and adsorbed at 405 nm when dephosphorylated by ALP. According to the manufacturer protocol, cells were lysed with ALP Assay

Buffer [17]. Samples were incubated with pNPP for 60 min at 25 °C, and then the absorbance at 405 nm was measured. ALP activity in test samples was calculated as follows:

$$\text{ALP activity (U/L)} = (B/\Delta T \times V) \times D \times 1000 \tag{1}$$

where B is the amount of pNP in sample well calculated from standard curve (μmol), ΔT is the reaction time (minutes), V is the original sample volume added into the reaction well (mL), D is the sample dilution factor [18].

2.8. Quantification of Mineral Matrix Deposition

The deposit of mineral matrix was evaluated by Alizarin Red S (ARS) quantification. After fixation with 10% formalin (Sigma), the staining with 40 mM ARS solution (Sigma-Aldrich) was performed for 20 min with gentle shaking. After washing with ddH2O, ARS staining was extracted with 0.5 mL of 10% acid acetic solution for 20 min with gentle agitation. The extracts were measured at 570 nm in a plate reader (Victor 3 Perkin Elmer).

2.9. Gene Expression Profile

Total RNA was extracted with Total RNA Purification Plus Kit (Norgen Biotek Corporation, Thorold, ON, Canada), according to the manufacture procedures. The RT2 Profiler PCR Array Human Extracellular Matrix and Adhesion Molecules (Qiagen, Hilden, Germany) were performed by reverse transcribing 200 ng of total RNA, following the manufacture procedures. The Ct values of target genes were normalized to the geometric mean Ct values of five housekeeping genes (ACTB: actin, beta; B2M: beta-2-microglobulin; GAPDH: glyceraldehyde-3-phosphate dehydrogenase; HPRT1: hypoxanthine phosphoribosyltransferase 1; RPLP0: ribosomal protein, large, P0). The expression of osteogenesis-related genes including osteocalcin (OC), osteopontin (OPN), runt-related transcription factor 2 (RUNX2), osterix (OSX), receptor activator of NF-kB ligand (RANKL), and alpha-1 type I collagen (COL1A1) was investigated at day 7, 14, and 21. Total RNA was reverse transcribed with SensiFAST cDNA Synthesis kit (Bioline GmbH, Luckenwalde, Germany) following the manufacture conditions. Real-time PCR was performed using SensiFAST SYBR No-ROX mix (Bioline GmbH) with 400 nM primers (selected by Primer 3 software3, Bioline, NY, USA). Ct values of target genes were normalized to the geometric mean Ct values of two housekeeping genes (B2M; GAPDH). Data analysis was performed using the $2\Delta\Delta$Ct method [18]. Results were reported as fold change of target genes on rGO/PCL composites compared with PCL (control).

2.10. Statistical Analysis

All data were expressed as mean and standard deviation (mean ± SD), one-way analysis of variance (ANOVA), and two-tailed t-test were utilized to assess the statistical significance with p-value < 0.05.

3. Results

3.1. Chemical Characterization of rGO-PCL Compounds

The different rGO-PCL compounds were manufactured through the dispersion of low percentage of rGO in order to do not change the bulk macroscopic properties of the substrate such as mechanical and rheological properties. Three percentages of rGO powder (1.6% *w/w*, 3% *w/w* or 5% *w/w*) were well dispersed into melted PCL using a lab-scale twin-screw extruder at temperatures ranging between 55 and 60 °C. Raman spectra of rGO-PCL compounds were compared to pure PCL and pure rGO (Figure 1). The spectrum of pure PCL (Figure 1a) is characterized by two intense bands for CH_2 antisymmetric and symmetric stretching at 2870 cm^{-1} and 2920 cm^{-1}, respectively. The bands at 1728 cm^{-1} (νC=O), at 1110 cm^{-1} (νCOC), and at about 1728 cm^{-1} reflect the crystalline domains of PCL. The other narrow peaks at 916 cm^{-1} (νC–COO), between the spectral range 1046–1110 cm^{-1} (νCOC), and 1298–1419 cm^{-1} (δCH2) are also related to PCL crystalline nature. The remaining minor peaks, such as the broad band at about 860 cm^{-1}, are attributable to the amorphous part of PCL, here present mainly as residues [19–21]. Raman spectrum of pure rGO (Figure 1e) is characterized by two main bands at ~1350 cm^{-1} and at ~1593 cm^{-1} associated to the well-known D-band and the G-band, respectively [22,23]. In 1.6% rGO-PCL (Figure 1b), 3% rGO-PCL (Figure 1c), and 5% rGO-PCL (Figure 1d) the D- and G-band are also present. In particular, 1.6% rGO-PCL sample presents the peculiar peaks and bands of both PCL and rGO. Whereas in 3% rGO-PCL and 5% rGO-PCL samples the D- and G-band are the dominant features due to the cover effect of rGO on PCL film. However, although less intense, the PCL CH$_2$ related bands at 2870 cm^{-1} and 2920 cm^{-1} are also detectable.

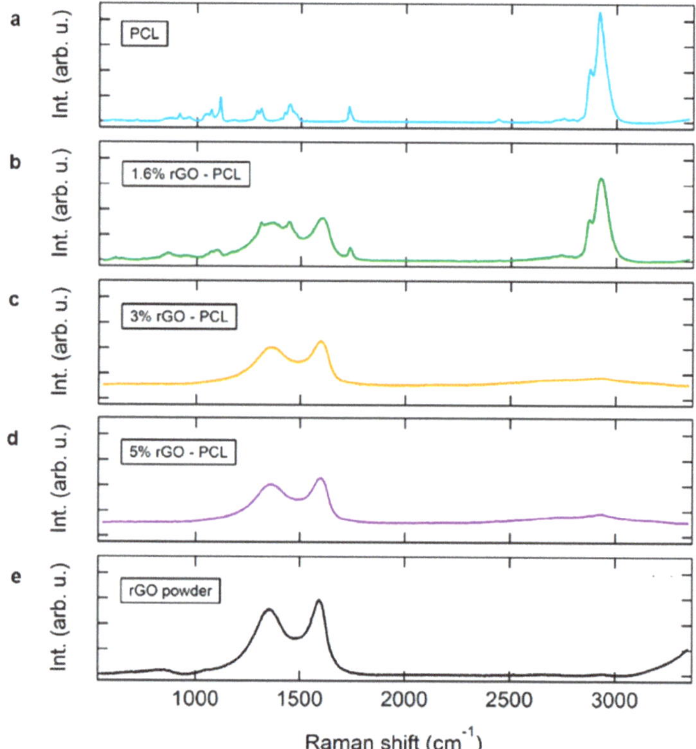

Figure 1. Raman spectra: PCL (**a**), 1.6% rGO-PCL (**b**), 3% rGO-PCL (**c**), 5% rGO-PCL (**d**), and rGO powder (**e**).

3.2. Antibacterial Activity and Cytotoxicity Evaluation of rGO-PCL Compounds

The antibacterial property of each rGO-PCL compound was investigated by means of different approaches, including the agar diffusion method, the count of CFU after treatment or seeding with Gram-positive (*S. pyogenes* and *S. aureus*) or Gram-negative (*E. coli* and *P. aeruginosa*) bacteria, the production of pyocyanin, and the adhesion of E. coli-GFP.

For agar diffusion method, samples were placed in Mueller–Hinton plates seeded with *S. pyogenes*, *S. aureus*, *E. coli,* or *P. aeruginosa*. After 24 h, the inhibition zone around the samples were evaluated. Inhibition zones (diameter 1 mm) were present around 1.6% rGO-PCL, 3% rGO-PCL, and 5% rGO-PCL in *S. aureus* cultures. In case of *S. pyogenes* culture, inhibition region of 1 mm was observed around 1.6% rGO-PCL and 3% rGO-PCL, and inhibition zone of 2 mm around 5% rGO-PCL. Suspensions of each bacterium were incubated with the samples for 1, 4, and 24 h in order to quantify the CFU of bacteria after 24 h of incubation (Table 1). *E. coli* and *P. aeruginosa* growth was not affected by any tested sample. Concentrations of 1.6% rGO-PCL and 3% rGO-PCL have slight bacteriostatic effect against *S. aureus* after 24 h of contact, whereas the increase in contact time yielded an increased bacteriostatic action for 5% rGO-PCL against *S. aureus*. The *S. pyogenes* growth is slightly affected by 5% rGO-PCL after 24 h of treatment. Moreover, the proliferation of 100 bacteria was evaluated after the direct seeding on the samples for 1, 3, and 24 h, and subsequently buffered on a plate (Table 2). Among tested samples, 5% rGO-PCL have bacteriostatic activity on *E. coli*, *S. aureus*, and *S. pyogenes*. The effect of the samples against Gram-negative bacteria was also detected by monitoring the adhesion of *E. coli*, and the production of pyocyanin in *P. aeruginosa*. The *E. coli* adhesion was greater on 1.6% rGO-PCL than on 3% rGO-PCL and 5% rGO-PCL. Instead, the production of pyocyanin in *P. aeruginosa* has been reduced by 27.7% for 1.6% rGO-PCL, 23.34% for 3% rGO-PCL, and 17.58% for 5% rGO-PCL.

Table 1. Colony forming units (CFU) of *E. coli*, *P. aeruginosa*, *S. aureus*, and *S. pyogenes* after treatment with 1.6% rGO-PCL, 3% rGO-PCL, and 5% rGO-PCL for 1, 4, and 24 h.

	Log (CFU/cm^2)			
Time	**1.6% rGO-PCL**	**3% rGO-PCL**	**5% rGO-PCL**	**Control**
	E. coli			
1 h	1.30 (r = −0.02)	1.38 (r = −0.1)	1.23 (r = 0.05)	1.28
4 h	1.76 (r = −0.03)	1.80 (r = −0.07)	1.69 (r = 0.04)	1.73
24 h	2.20 (r = −0.1)	2.20 (r = −0.1)	2.05 (r = 0.05)	2.10
	P. aeruginosa			
1 h	1.58 (r = −0.54)	1.38 (r = −0.34)	1.41 (r = −0.37)	1.04
4 h	1.96 (r = −0.38)	1.82 (r = −0.24)	1.85 (r = −0.27)	1.58
24 h	2.19 (r = 0.01)	2.21 (r = −0.01)	2.23 (r = −0.03)	2.20
	S. aureus			
1 h	1.34 (r = −0.08)	1.38 (r = −0.12)	0.95 (r = 0.31)	1.26
4 h	1.69 (r = −0.11)	1.70 (r = −0.12)	1.23 (r = 0.35)	1.58
24 h	1.90 (r = 0.33)	1.98 (r = 0.25)	1.62 (r = 0.61)	2.23
	S. pyogenes			
1 h	0.84 (r = 0.0)	0.9 (r = −0.06)	0.78 (r = 0.06)	0.84
4 h	1.15 (r = −0.04)	1.18 (r = −0.07)	1.11 (r = 0.0)	1.11
24 h	1.41 (r = 0.07)	1.5 (r = −0.02)	1.34 (r = 0.14)	1.48

r: correlation coefficient.

Table 2. Colony forming units (CFU) of *E. coli*, *P. aeruginosa*, *S. aureus*, and *S. pyogenes* after seeding on 1.6% rGO-PCL, 3% rGO-PCL, and 5% rGO-PCL for 1, 4, and 24 h.

	Log (CFU/cm^2)			
Time	**1.6% rGO-PCL**	**3% rGO-PCL**	**5% rGO-PCL**	**Control**
	E. coli			
1 h	1.18 (r = 0.0)	1.15 (r = 0.03)	1.00 (r = 0.18)	1.18
4 h	1.54 (r = −0.01)	1.50 (r = 0.03)	1.40 (r = 0.13)	1.53
24 h	2.04 (r = 0.0)	2.02 (r = 0.05)	1.91 (r = 0.15)	2.04
	P. aeruginosa			
1 h	1.15 (r = −0.07)	1.08 (r = 0.0)	1.11 (r = −0.03)	1.08
4 h	1.59 (r = −0.02)	1.59 (r = −0.02)	1.56 (r = 0.01)	1.57
24 h	2.18 (r = −0.01)	2.21 (r = −0.04)	2.19 (r = −0.02)	2.17
	S. aureus			
1 h	1.34 (r = 0.0)	1.30 (r = 0.04)	0.95 (r = 0.39)	1.34
4 h	1.62 (r = 0.04)	1.61 (r = 0.05)	1.28 (r = 0.38)	1.66
24 h	1.87 (r = 0.21)	1.99 (r = 0.09)	1.83 (r = 0.25)	2.08
	S. pyogenes			
1 h	1.00 (r = 0.08)	0.9 (r = 0.18)	0.78 (r = 0.30)	1.08
4 h	1.15 (r = 0.13)	1.18 (r = 0.10)	1.08 (r = 0.20)	1.28
24 h	1.36 (r = 0.05)	1.38 (r = 0.03)	1.30 (r = 0.11)	1.41

r: correlation coefficient.

The biocompatibility of rGO-PCL was assessed by measuring the viability of murine fibroblasts (NCTC clone 929) according to the International Organization for Standardization (ISO 10993-5) that assess the in vitro cytotoxicity of medical devices. The viability of the fibroblasts incubated with rGO-PCL compounds ranged between ~81 and ~90% (Figure 2). According to ISO 10993-5, a reduction in cell viability by more than 30% is considered a cytotoxic effect. Therefore, all rGO-PCL compounds exhibited no toxicity against NCTC clone 929 cells.

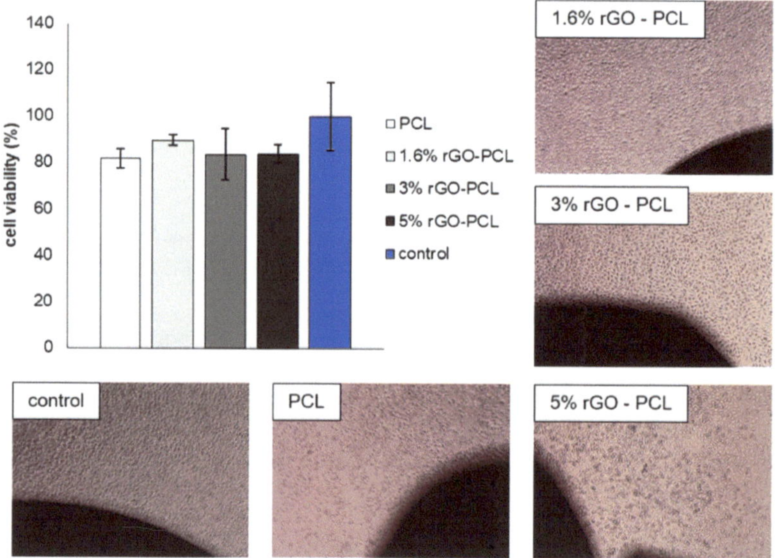

Figure 2. In vitro cytotoxicity evaluation of rGO-PCL compounds. Cell viability percentage of murine fibroblasts (NCTC clone 929) in contact with rGO-PCL compounds for 48 h. Sterile titanium was used as control. Data are expressed as mean ± SD. Multi comparison statistical analyses were performed by using one-way ANOVA. A T test was performed for all pairwise comparisons between group means. Representative images show fibroblasts after 48 h of culture with each sample.

3.3. Stem Cells Adhesion on rGO-PCL Surfaces

HMSCs were seeded and maintained on each rGO-PCL surface for 7 days. In general, the cells have adhered and migrated over the substrates, as shown by the representative SEM images in Figure 3a–d. The cells on rGO-PCL surfaces appeared flat (Figure 3b–d); on the contrary, on pure PCL cells were fusiform (Figure 3a). The gene expression of molecules and proteins of extracellular matrix (ECM) was investigated by real time PCR array (Figure 3e). On rGO-PCL, with more extension on the 5% rGO-PCL surface, the mRNA encoding CD44, COL4A2, CTNNA1, CTNNB1, ECM1, ITGA2, ITGA4, LAMA1, MMP11, SGCE, and TIMP1 genes were upregulated, whereas ITGAM, ITGA1, LAMB3, and VCAM1 were under-expressed genes.

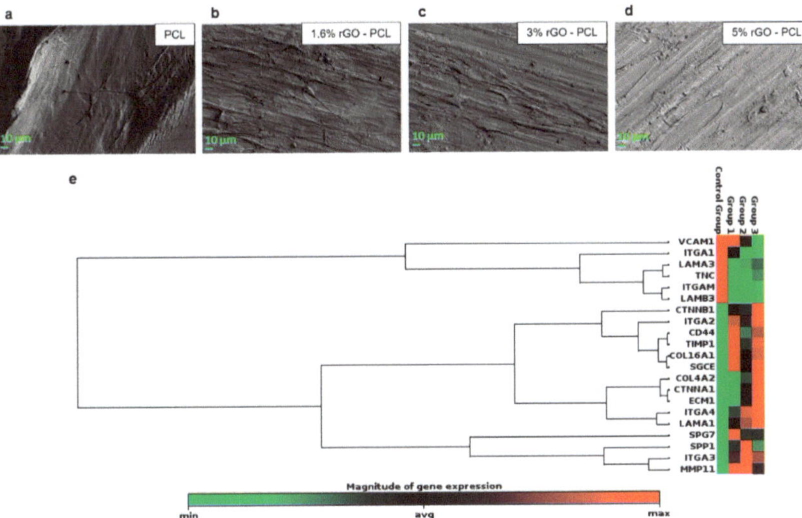

Figure 3. Stem cells adhesion on rGO-PCL surfaces. SEM images of the cells on pure PCL (**a**), 1.6% rGO-PCL (**b**), 3% rGO-PCL (**c**), 5% rGO-PCL at 1 xK magnification (**d**). (**e**) Expression profile of extracellular matrix and adhesion molecules in HMSCs maintained on rGO-PCL surfaces for 7 days. Heat map reporting under expressed (green) and overexpressed (red) genes in 1.6% rGO-PCL (Group 1), 3% rGO-PCL (Group 2), and 5% rGO-PCL (Group 3) compared with pure PCL (Control Group).

3.4. Osteogenic Differentiation of Stem Cell on rGO-PCL Surfaces

In order to investigate rGO-PCL surfaces as promising platform for osteogenic differentiation, cultures of HMSCs were monitored up to 21 days. Firstly, the adhesion and proliferation of HMSCs were evaluated by microscope and enzymatic analyses (Figure 4). As shown by SEM images at 7 days, HMSCs were able to adhere to surfaces of rGO-PCL samples by numerous filopodia (Figure 4a–d). After 14 days of culture, HMSCs appeared more spread out, and an increased cell coverage on each rGO-PCL surface was observed (Figure 4e–h). After 21 day of culture, the number of HMSCs attached to the surfaces was indistinguishable, which implied that the cells were evenly distributed on rGO-PCL surfaces (Figure 4i–l). These observations were in agreement with MTS assay results (Figure 4m–p). In general, the magnitude of the MTS reduction provides information on the number of viable cells indicating if materials support the adhesion and proliferation of the cells. Compared with 7 days of culture, cell proliferation showed a significant increase at the 14th ($p < 0.05$ for PCL and 5% rGO-PCL, $p < 0.01$ for 1.6 %rGO-PCL and 3% rGO-PCL) and 21st day ($p < 0.05$ for PCL, $p < 0.01$ for 1.6 %rGO-PCL, 3% rGO-PCL and 5% rGO-PCL). Furthermore, cell proliferation exhibited a significant increase ($p < 0.05$) on 1.6% rGO-PCL and 3% rGO-PCL by comparing 21 days of culture with 14 days.

The differentiation of HMSCs towards osteoblast phenotype on rGO-PCL surfaces was investigated by the quantification of ALP activity and mineral matrix deposition, and by gene expression profile of osteogenic markers (Figure 5). Compared to 7 days of culture, ALP activity was increased at 14 days. Nevertheless, a significant ($p < 0.05$) increase was observe only in HMSCs seeded on 5% rGO-PCL. At 21 days, ALP activity remained stable on 1.6% rGO-PCL as well as on PCL, and significantly ($p < 0.05$) increased on 3% rGO-PCL. On 5% rGO-PCL, ALP activity at 21 day is less compared to 14 day, but significantly ($p < 0.05$) greater than 7 days (Figure 5a). The Alizarin Red S (ARS) quantification performed on rGO-PCL samples confirmed the progression of osteogenic differentiation up to 21 days. In particular, the matrix deposition was significantly ($p < 0.001$) boosted at 14 and 21 days, and largely on 5% rGO-PCL surface (Figure 5b). In addition, gene expression for both early and late osteogenic differentiation markers osteocalcin (OC), osteonectin (ON), osteopontin (OPN), runt-related transcription factor 2 (RUNX2), osterix (OSX), and receptor activator of NF-kB ligand (RANKL) was assessed (Figure 5c–h). The mRNA expression of the osteogenic markers on each rGO-PCL surface was compared with that on PCL at each time point. The comparison shown that on rGO-PCL the gene expression of matrix proteins OC and ON is similar to those on PCL. Further, 1.6% rGO-PCL, 3% rGO-PCL, and 5% rGO-PCL showed significantly ($p < 0.05$) enhanced mRNA expression level of OPN, which were 2.4 ± 0.22, 1.7 ± 0.04, and 2.3 ± 0.17 fold higher than the PCL group at 14 days, respectively. The transcription factors RUNX2 and OSX were significantly ($p < 0.05$) up regulated on 5% rGO-PCL at 14 days with a fold change of 1.3 ± 0.2 and 5.1 ± 0.4, respectively. In addition, a significant ($p < 0.01$) upregulation of RANKL is also observed on 5% rGO-PCL compared to control condition.

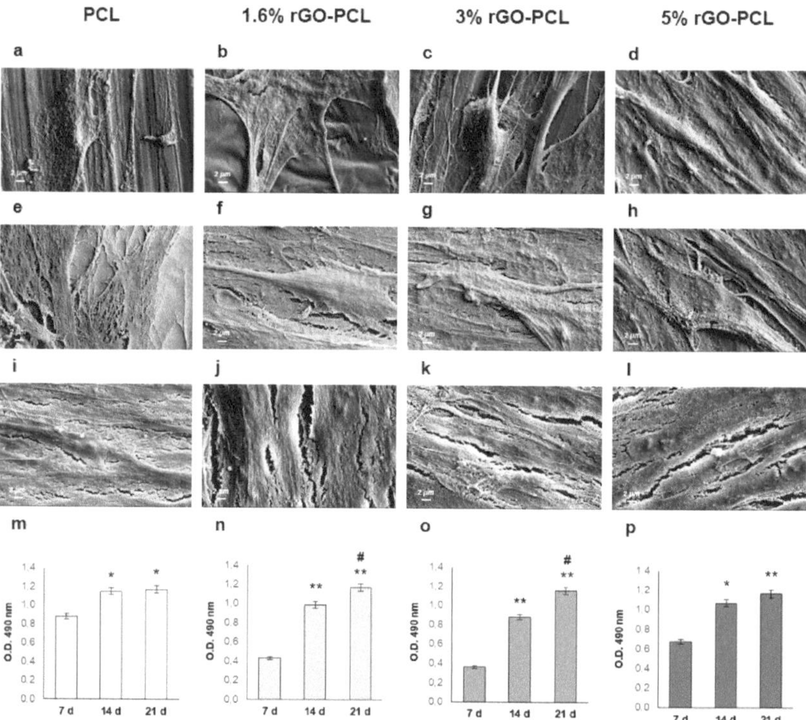

Figure 4. Adhesion and proliferation of HMSCs on rGO-PCL surfaces in osteogenic differentiation medium. SEM images (5 ox magnification) after 7 (**a–d**), 14 (**e–h**), and 21 days (**i–l**) of culture. MTS assay at 7, 14, and 21 day (**m–p**). Data are expressed as mean $\pm$ SD. Multi comparison statistical analysis were performed by using one-way ANOVA. T test was to perform all pairwise comparisons between group means. Note: * $p < 0.05$ and ** $p < 0.01$ mark significant changes compared to 7 days of culture. # $p < 0.05$ marks significant changes compared to 14 days of culture.

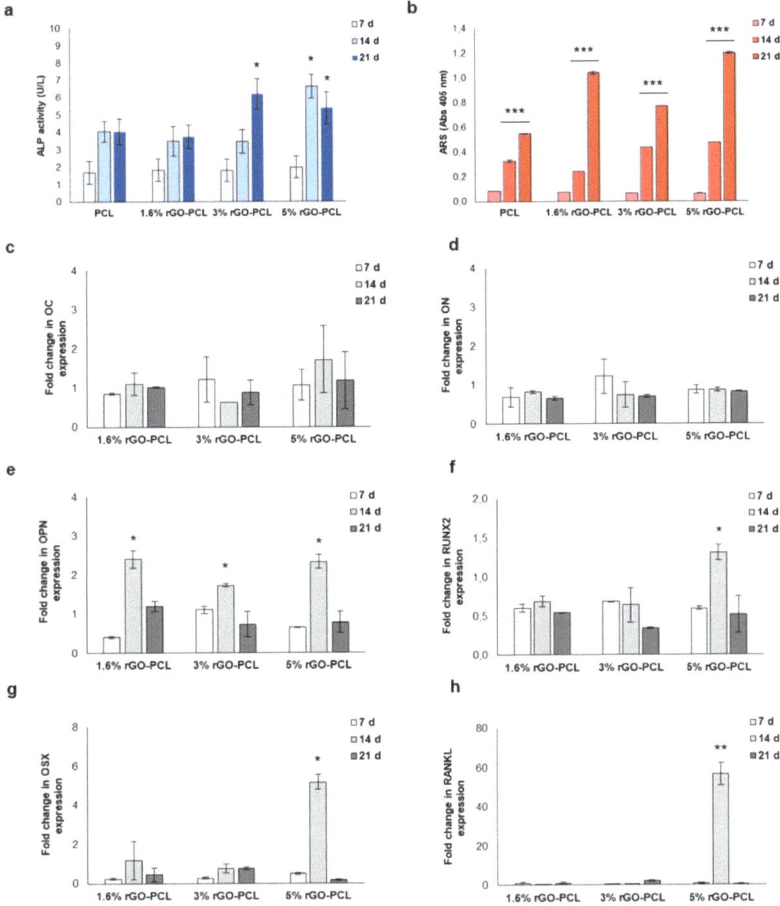

Figure 5. Differentiation of HMSCs towards osteoblast phenotype on rGO-PCL scaffolds up to 21 days. ALP activity (**a**). Quantification of mineral matrix deposition (**b**). Gene expression profile of osteogenic markers: osteocalcin (**c**), osteonectin (**d**), osteopontin (**e**), runt-related transcription factor 2 (**f**), osterix (**g**), receptor activator of NF-kB ligand (**h**). Data are expressed as mean ± SD. Multi comparison statistical analysis were performed by using one-way ANOVA. T test was to perform all pairwise comparisons between group means. Note: * $p < 0.05$, ** $p < 0.01$ and *** $p < 0.001$ mark significant changes compared to pure PCL at each time point.

4. Discussion

Different rGO-PCL compounds were manufactured through the fine dispersion of low percentage of rGO powder in molten PCL. The most suitable combinations to obtain a better dispersion of rGO in molten PCL was obtained with a filler content less than 5% by weight. The 1.6%rGO-PCL, 3%rGO-PCL, and 5%rGO-PCL composites were prepared through extrusion process followed by injection molding. Chemical analysis demonstrated the effective incorporation of rGO into the PLC matrix. The Raman spectra of the composites shown typical characteristics of both PCL and rGO [19,22]. In 1.6% rGO-PCL, the peculiar peaks and bands of PCL and rGO were present, whereas in 3% rGO-PCL and 5%rGO-PCL the bands of rGO were dominant.

After sterilization, antibacterial activity and biocompatibility of rGO-PCL compounds were investigated. By various approaches, the antibacterial activity of each compound on

Gram-negative and Gram-positive bacteria was analyzed. The bacteriostatic properties against Gram-positive bacteria were rGO dose-dependent, and 5% rGO-PCL compound was the best in the prevention of *S. aureus* and *S. pyogenes* growth. Instead, no bacteriostatic effect was shown towards the Gram-negative bacteria *E. coli* and *P. aeruginosa*. Although a reduction in *E. coli* adhesion was detected as the rGO percentage increases, a minor reduction in pyocyanin production in *P. aeruginosa* was observed. All compounds were biocompatible as demonstrate by incubation in direct contact with cultures of murine fibroblasts for 48h. Our results shown cell viability between 81% and 90% for rGO-PCL compounds. According to in vitro cytotoxicity standard, a percentage reduction in viability greater than 30% compared to the control is to be considered a cytotoxic effect [14]. Thus, the rGO-PCL compounds exhibited no cytotoxicity, confirming that rGO-PCL, as pure PCL, has potential biomedical applications.

The adhesion and migration of stem cells were investigated by microscopy and molecular analyses. Stem cells exhibited flat morphology on tested surfaces. Cell physiological processes including growth, division, differentiation, and migration are affected by the adhesion to extracellular components. Via adhesion molecules, cells bind themselves to ECM substratum defining tissue shape, structure, and function. The cell–matrix and cell–cell interactions on rGO-PCL surfaces were also investigated by gene expression profile. In particular, basement membrane constituents, collagens, integrins, selectins, catenins, cell-adhesion molecule family members, and metalloproteinases (MMPs) as well as their inhibitors were analyzed. Twenty-five percent of the genes investigated showed up- or down-regulation compared to pure PCL; the remaining genes conversely were equally expressed. Among the rGO-PCL formulations tested, the one with the highest percentage of rGO resulted in high-level expression of integrins, catenins, and other proteins of ECM. Integrins are heterodimeric proteins composed of alpha and beta subunits that function as transmembrane linkers between the ECM and the actin cytoskeleton of cells. It is known that the chemistry of the substrate can affect both the types of protein adsorbed onto the surface and the conformation of self-same proteins. This in turn induces changes in integrin expression on cell surface [24]. Integrins bind laminins, glycoproteins composed of three different subunits (α-chain, β-chain, and γ-chain) combined and expressed in at least 16 different isoforms in human. By binding cell-surface receptors, laminins regulate vital cellular responses, such as phenotype maintenance, migration, proliferation, and differentiation [25]. Catenins are cell-adhesion molecules that fine control HMSC functions by the intensity of Wnt/β-catenin signals. For instance, proliferation and self-renewal of HMSCs were promoted only under low levels of Wnt/β-catenin, whereas osteogenic differentiation was promoted in a manner proportional to the intensity of the Wnt signals. Therefore, β-catenin functions as a distinct transactivation molecule depending on its level of accumulation to induce different transcriptomic changes in the HMSCs [26]. Moreover, the expression of CD44 gene was rGO dependent. The CD44 protein binds various components of ECM, such as hyaluronan, and through these interactions, it influences cellular behavior, adhesion, and migration. In addition, CD44 proteins functions as specialized platform for growth factors and MMPs. Precisely, the binding CD44-hyaluronan triggers cell migration; instead, the cleavage of CD44 by MMPs hinders migration [27]. The MMPs are key factors of ECM remodeling, the essential prerequisite for the early stages of cell adhesion on biomaterials [28]. Among the MMPs investigated, the MMP11 and the inhibitor TIMP1 were upregulated on all rGO-PCL surfaces. Moreover, on the rGO-PCL surfaces the homeostasis of ECM is favored through the up-regulation of the transmembrane molecule SGCE, that regulates cell-matrix adhesion, and the up-regulation of ECM1 that organizes a variety of extracellular and structural proteins of ECM. The mechanical properties and the expression profile of HMSCs maintained on rGO-PCL surfaces seem to be influenced by the amount of rGO into PCL matrix. In particular, cells maintained on 5% rGO-PCL for 7 days show a gene expression profile for adhesion molecules and extracellular proteins referable to the early stages of osteogenic differentiation. Based on these findings, rGO-PCL surfaces were valued as substrates for the osteogenic differentiation of HMSCs. By microscopy and enzymatic analyses, cell adhesion and proliferation were monitored over three weeks. After 21 day of culture, cells were metabolically active and broadly distributed on rGO-PCL surfaces. However, differences in gene expression profile, in ALP activity, and in mineral matrix deposition were observed between the rGO-PCL surfaces. Generally, ALP is considered an early marker of osteoblast differentiation. As ALP is a byproduct of osteoblast activity, increased level of ALP refers to active bone formation. Conversely, the calcium deposition in matrix is considered a late marker for osteogenesis as matrix mineralization begins after the initial accumulation of osteoblast proteins, such as ALP [29]. Among the tested combinations, only 5% rGO-PCL compound showed maximum ALP activity at 14 days followed by maximum mineralized matrix deposition at 21 days. Furthermore, the expression of the main genes involved in osteogenic differentiation, including OC, ON, OPN, OSX, RUNX2, and RANKL gene was investigated. The gene expression of OC and ON did not change over time in any of the conditions tested. Conversely, OPN gene expression is significantly improved on all surfaces. Instead, the expression of RUNX2, OSX, and RANKL merely increased on 5% rGO-PCL surface. OC, ON, and OPN are major non-collagenous proteins secreted by osteoblasts and pre-osteoblast. They are involved in bone matrix deposition and organization [30]. RUNX2 is the major transcription factor necessary for osteoblast differentiation, as it regulates many bone-related genes. OSX is another osteoblast-specific transcription factor downstream gene of RUNX2 involved in the osteoblast differentiation pathway [31]. RANKL is produced and secreted by osteoblasts during the late stage of differentiation [32–40]. Morphological, mechanical, and gene expression results indicate that the osteogenic differentiation of stem cells is improved on rGO-PCL surfaces in rGO dose-dependent manner. Although cell proliferation and migration are similar on all scaffolds, merely 5% rGO-PCL surface shown ALP activity, mineral matrix deposition, and RUNX2, OSX, OPN, and RANKL gene expression compatible with an effective osteogenic differentiation.

5. Conclusions

In conclusion, this study demonstrated the perfect suitability of rGO-PCL for the adhesion, proliferation, migration, and osteogenic differentiation of HMSCs. Firstly, PCL-based compounds with 1.6%, 3%, and 5% rGO were manufactured by extrusion. After sterilization, the biocompatibility of all compounds was demonstrated by measuring the cell viability according to the international standard for medical devices. Surprisingly, rGO-PCL compounds showed bacteriostatic properties to Gram-positive bacteria in rGO-dependent manner. The percentage of rGO also regulated the adhesion, morphology and differentiation of the stem cells. The 5% rGO-PCL surface supported the growth of cells and implemented the expression of the ECM components necessary to anchor the cells to itself. Furthermore, it promoted the osteogenic differentiation by ensuring the activity of alkaline phosphatase, the deposition of the mineral matrix, and the expression of proteins crucial for the differentiation in osteoblasts. These results suggest that 5% rGO-PCL has useful properties for bone tissue engineering purposes.

Author Contributions: Conceptualization, B.Z. and L.F.; methodology, M.S. and L.F.; validation, L.F. and F.R.; formal analysis, C.G.; investigation, F.Z. and E.B.; data curation, L.F., C.G., F.R., E.B. and F.Z.; writing—original draft preparation, L.F. and F.R.; writing—review and editing, L.F.; visualization, B.Z.; supervision, B.Z. and P.R.; project administration, B.Z. All authors have read and agreed to the published version of the manuscript.

Funding: This research received no external funding.

Institutional Review Board Statement: Not applicable.

Informed Consent Statement: Not applicable.

Data Availability Statement: Not applicable.

Conflicts of Interest: The authors declare no conflict of interest.

References

1. Liu, L.-N.; Zhang, X.-H.; Liu, H.-H.; Li, K.-H.; Wu, Q.-H.; Liu, Y.; Luo, E. Osteogenesis Differences Around Titanium Implant and in Bone Defect Between Jaw Bones and Long Bones. *J. Craniofacial Surg.* **2020**, *31*, 2193–2198. [CrossRef] [PubMed]
2. Javed, F.; Ahmed, H.B.; Crespi, R.; Romanos, G.E. Role of primary stability for successful osseointegration of dental implants: Factors of influence and evaluation. *Interv. Med. Appl. Sci.* **2013**, *5*, 162–167. [CrossRef] [PubMed]
3. Xie, Y.; Li, S.; Zhang, T.; Wang, C.; Cai, X. Titanium mesh for bone augmentation in oral implantology: Current application and progress. *Int. J. Oral Sci.* **2020**, *12*, 37. [CrossRef] [PubMed]
4. Yasin, M.; Tighe, B. Polymers for biodegradable medical devices: VIII. Hydroxybutyrate-hydroxyvalerate copolymers: Physical and degradative properties of blends with polycaprolactone. *Biomaterials* **1992**, *13*, 9–16. [CrossRef]
5. Prabha, R.D.; Kraft, D.C.E.; Harkness, L.; Melsen, B.; Varma, H.; Nair, P.D.; Kjems, J.; Kassem, M. Bioactive nano-fibrous scaffold for vascularized craniofacial bone regeneration. *J. Tissue Eng. Regen. Med.* **2017**, *12*, e1537–e1548. [CrossRef]
6. Kim, S.E.; Yun, Y.-P.; Shim, K.-S.; Kim, H.J.; Park, K.; Song, H.-R. 3D printed alendronate-releasing poly(caprolactone) porous scaffolds enhance osteogenic differentiation and bone formation in rat tibial defects. *Biomed. Mater.* **2016**, *11*, 055005. [CrossRef]
7. Duan, T.; Xu, H.; Tang, Y.; Jin, J.; Wang, Z. Effect of epitaxial crystallization on the structural evolution of PCL/RGO nanocomposites during stretching by in-situ synchrotron radiation. *Polymer* **2018**, *159*, 106–115. [CrossRef]
8. Miao, W.; Wu, F.; Zhou, S.; Yao, G.; Li, Y.; Wang, Z. Epitaxial Crystallization of Poly(ε-caprolactone) on Reduced Graphene Oxide at a Low Shear Rate by In Situ SAXS/WAXD Methods. *ACS Omega* **2020**, *5*, 31535–31542. [CrossRef]
9. Babaie, A.; Rezaei, M.; Sofla, R.L.M. Investigation of the effects of polycaprolactone molecular weight and graphene content on crystallinity, mechanical properties and shape memory behavior of polyurethane/graphene nanocomposites. *J. Mech. Behav. Biomed. Mater.* **2019**, *96*, 53–68. [CrossRef]
10. Damonte, G.; Vallin, A.; Fina, A.; Monticelli, O. On the Development of an Effective Method to Produce Conductive PCL Film. *Nanomaterials* **2021**, *11*, 1385. [CrossRef]
11. Spradling, A.C.; Drummond-Barbosa, D.; Kai, T. Stem cells find their niche. *Nature* **2001**, *414*, 98–104. [CrossRef] [PubMed]
12. Naqvi, S.M.; McNamara, L.M. Stem Cell Mechanobiology and the Role of Biomaterials in Governing Mechanotransduction and Matrix Production for Tissue Regeneration. *Front. Bioeng. Biotechnol.* **2020**, *8*, 597661. [CrossRef] [PubMed]
13. Gardin, C.; Ricci, S.; Ferroni, L.; Guazzo, R.; Sbricoli, L.; De Benedictis, G.; Finotti, L.; Isola, M.; Bressan, E.; Zavan, B. Decellularization and Delipidation Protocols of Bovine Bone and Pericardium for Bone Grafting and Guided Bone Regeneration Procedures. *PLoS ONE* **2015**, *10*, e0132344. [CrossRef] [PubMed]
14. Pankongadisak, P.; Suwantong, O. Enhanced properties of injectable chitosan-based thermogelling hydrogels by silk fibroin and longan seed extract for bone tissue engineering. *Int. J. Biol. Macromol.* **2019**, *138*, 412–424. [CrossRef]
15. Ferroni, L.; Tocco, I.; De Pieri, A.; Menarin, M.; Fermi, E.; Piattelli, A.; Gardin, C.; Zavan, B. Pulsed magnetic therapy increases osteogenic differentiation of mesenchymal stem cells only if they are pre-committed. *Life Sci.* **2016**, *152*, 44–51. [CrossRef] [PubMed]
16. Gardin, C.; Ferroni, L.; Erdoğan, Y.K.; Zanotti, F.; De Francesco, F.; Trentini, M.; Brunello, G.; Ercan, B.; Zavan, B. Nanostructured Modifications of Titanium Surfaces Improve Vascular Regenerative Properties of Exosomes Derived from Mesenchymal Stem Cells: Preliminary In Vitro Results. *Nanomaterials* **2021**, *11*, 3452. [CrossRef]
17. Cecchinato, F.; Karlsson, J.; Ferroni, L.; Gardin, C.; Galli, S.; Wennerberg, A.; Zavan, B.; Andersson, M.; Jimbo, R. Osteogenic potential of human adipose-derived stromal cells on 3-dimensional mesoporous TiO2 coating with magnesium impregnation. *Mater. Sci. Eng. C* **2015**, *52*, 225–234. [CrossRef]
18. Pfaffl, M.W. A new mathematical model for relative quantification in real-time RT-PCR. *Nucleic Acids Res.* **2001**, *29*, e45. [CrossRef]
19. Wesełucha-Birczyńska, A.; Świętek, M.; Sołtysiak, E.; Galiński, P.; Płachta, L.; Piekara, K.; Błażewicz, M. Raman spectroscopy and the material study of nanocomposite membranes from poly(ε-caprolactone) with biocompatibility testing in osteoblast-like cells. *Analyst* **2015**, *140*, 2311–2320. [CrossRef]
20. Baranowska-Korczyc, A.; Warowicka, A.; Jasiurkowska-Delaporte, M.; Grześkowiak, B.; Jarek, M.; Maciejewska, B.M.; Jurga-Stopa, J.; Jurga, S. Antimicrobial electrospun poly(ε-caprolactone) scaffolds for gingival fibroblast growth. *RSC Adv.* **2016**, *6*, 19647–19656. [CrossRef]
21. Vijayavenkataraman, S.; Kannan, S.; Cao, T.; Fuh, J.Y.H.; Sriram, G.; Lu, W.F. 3D-Printed PCL/PPy Conductive Scaffolds as Three-Dimensional Porous Nerve Guide Conduits (NGCs) for Peripheral Nerve Injury Repair. *Front. Bioeng. Biotechnol.* **2019**, *7*, 266. [CrossRef] [PubMed]
22. Wu, J.-B.; Lin, M.-L.; Cong, X.; Liu, H.-N.; Tan, P.-H. Raman spectroscopy of graphene-based materials and its applications in related devices. *Chem. Soc. Rev.* **2018**, *47*, 1822–1873. [CrossRef] [PubMed]
23. Muzyka, R.; Drewniak, S.; Pustelny, T.; Chrubasik, M.; Gryglewicz, G. Characterization of Graphite Oxide and Reduced Graphene Oxide Obtained from Different Graphite Precursors and Oxidized by Different Methods Using Raman Spectroscopy. *Materials* **2018**, *11*, 1050. [CrossRef]
24. Olivares-Navarrete, R.; Rodil, S.E.; Hyzy, S.L.; Dunn, G.R.; Almaguer-Flores, A.; Schwartz, Z.; Boyan, B.D. Role of integrin subunits in mesenchymal stem cell differentiation and osteoblast maturation on graphitic carbon-coated microstructured surfaces. *Biomaterials* **2015**, *51*, 69–79. [CrossRef] [PubMed]
25. Hagbard, L.; Cameron, K.; August, P.; Penton, C.; Parmar, M.; Hay, D.C.; Kallur, T. Developing defined substrates for stem cell culture and differentiation. *Philos. Trans. R. Soc. B Biol. Sci.* **2018**, *373*, 20170230. [CrossRef] [PubMed]

26. Kim, J.-A.; Choi, H.-K.; Kim, T.-M.; Leem, S.-H.; Oh, I.-H. Regulation of mesenchymal stromal cells through fine tuning of canonical Wnt signaling. *Stem Cell Res.* **2015**, *14*, 356–368. [CrossRef]

27. Ponta, H.; Sherman, L.S.; Herrlich, P.A. CD44: From adhesion molecules to signalling regulators. *Nat. Rev. Mol. Cell Biol.* **2003**, *4*, 33–45. [CrossRef]

28. Albano, C.S.; Gomes, A.M.; Feltran, G.D.S.; Fernandes, C.J.D.C.; Trino, L.D.; Zambuzzi, W.F.; Lisboa-Filho, P.N. Bisphosphonate-based surface biofunctionalization improves titanium biocompatibility. *J. Mater. Sci. Mater. Med.* **2020**, *31*, 109. [CrossRef]

29. Lee, J.-M.; Kim, M.-G.; Byun, J.-H.; Kim, G.-C.; Ro, J.-H.; Hwang, D.-S.; Choi, B.-B.; Park, G.-C.; Kim, U.-K. The effect of biomechanical stimulation on osteoblast differentiation of human jaw periosteum-derived stem cells. *Maxillofac. Plast. Reconstr. Surg.* **2017**, *39*, 7. [CrossRef]

30. Bailey, S.; Karsenty, G.; Gundberg, C.; Vashishth, D. Osteocalcin and osteopontin influence bone morphology and mechanical properties. *Ann. N. Y. Acad. Sci.* **2017**, *1409*, 79–84. [CrossRef]

31. Chen, S.; Gluhak-Heinrich, J.; Wang, Y.; Wu, Y.; Chuang, H.H.; Chen, L.; Yuan, G.; Dong, J.; Gay, I.; MacDougall, M. *Runx2, Osx,* and *Dspp* in Tooth Development. *J. Dent. Res.* **2009**, *88*, 904–909. [CrossRef] [PubMed]

32. Tobeiha, M.; Moghadasian, M.H.; Amin, N.; Jafarnejad, S. RANKL/RANK/OPG Pathway: A Mechanism Involved in Exercise-Induced Bone Remodeling. *BioMed Res. Int.* **2020**, *2020*, 6910312. [CrossRef] [PubMed]

33. Brun, P.; Cortivo, R.; Zavan, B.; Vecchiato, N.; Abatangelo, G. In vitro reconstructed tissues on hyaluronan-based temporary scaffolding. *J. Mater Sci. Mater Med.* **1999**, *10*, 683–688. [CrossRef]

34. Figallo, E.; Flaibani, M.; Zavan, B.; Abatangelo, G.; Elvassore, N. Micropatterned Biopolymer 3D Scaffold for Static and Dynamic Culture of Human Fibroblasts. *Biotechnol. Prog.* **2007**, *23*, 210–216. [CrossRef]

35. Gardin, C.; Bressan, E.; Ferroni, L.; Nalesso, E.; Vindigni, V.; Stellini, E.; Pinton, P.; Sivolella, S.; Zavan, B. In vitro concurrent endothelial and osteogenic commitment of adipose-derived stem cells and their genomical analyses through comparative genomic hybridization array: Novel strategies to increase the successful engraftment of tissue-engineered bone grafts. *Stem Cells Dev.* **2012**, *5*, 767–777. [CrossRef]

36. Azzena, B.; Mazzoleni, F.; Abatangelo, G.; Zavan, B.; Vindigni, V. Autologous platelet-rich plasma as an adipocyte in vivo delivery system: Case report. *Aesthetic Plast Surg.* **2008**, 155–158. [CrossRef] [PubMed]

37. Ettorre, V.; De Marco, P.; Zara, S.; Perrotti, V.; Scarano, A.; Di Crescenzo, A.; Petrini, M.; Hadad, C.; Bosco, D.; Zavan, B.; et al. In vitro and in vivo characterization of graphene oxide coated porcine bone granules. *Carbon* **2016**, *103*, 291–298. [CrossRef]

38. Joshian, K.M.; Shelar, A.; Kasabe, U.; Nikam, L.K.; Pawar, R.A.; Sangshetti, J.; Kale, B.B.; Singh, A.V.; Patil, R.; Chaskar, M.G. Biofilm inhibition in *Candida albicans* with biogenic hierarchical zinc-oxide nanoparticles. *Mater Sci. Eng. C Mater Biol. Appl.* **2021**, *134*, 35527134.

39. Zhang, Y.; Cui, K.; Fu, T.; Wang, J.; Shen, F.; Zhang, X.; Yu, L. Formation of MoSi2 and Si/MoSi2 coatings on TZM (Mo–0.5Ti–0.1Zr–0.02C) alloy by hot dip silicon-plating method. *Ceram. Int.* **2021**, *47*, 23053–23065. [CrossRef]

40. Maharjan, R.S.; Singh, A.V.; Hanif, J.; Rosenkranz, D.; Haidar, R.; Shelar, A.; Singh, S.P.; Dey, A.; Patil, R.; Zamboni, P.; et al. Investigation of the Associations between a Nanomaterial's Microrheology and Toxicology. *ACS Omega* **2022**, *7*, 13985–13997. [CrossRef]

Article

Cytotoxicity and Genotoxicity of Metal Oxide Nanoparticles in Human Pluripotent Stem Cell-Derived Fibroblasts

Harish K Handral [1,2], **C. Ashajyothi** [3], **Gopu Sriram** [1], **Chandrakanth R. Kelmani** [4], **Nileshkumar Dubey** [1,5,*] and **Tong Cao** [1,*]

[1] Faculty of Dentistry, National University of Singapore, Singapore 119085, Singapore; mpehkh@nus.edu.sg (H.K.H.); dengs@nus.edu.sg (G.S.)
[2] Department of Mechanical Engineering, National University of Singapore, Singapore 117575, Singapore
[3] Department of Biotechnology, Vijayanagara Sri Krishnadevaraya University, Ballari, Karnataka 583105, India; drashajyothic@vskub.ac.in
[4] Department of Biotechnology, Gulbarga University, Gulbarga, Karnataka 585106, India; ckelmani@gmail.com
[5] Department of Cariology, Restorative Sciences and Endodontics, School of Dentistry, University of Michigan, Ann Arbor, MI 48104, USA
* Correspondence: nileshd@umich.edu (N.D.); dencaot@nus.edu.sg (T.C.)

Abstract: Advances in the use of nanoparticles (NPs) has created promising progress in biotechnology and consumer-care based industry. This has created an increasing need for testing their safety and toxicity profiles. Hence, efforts to understand the cellular responses towards nanomaterials are needed. However, current methods using animal and cancer-derived cell lines raise questions on physiological relevance. In this aspect, in the current study, we investigated the use of pluripotent human embryonic stem cell- (hESCs) derived fibroblasts (hESC-Fib) as a closer representative of the in vivo response as well as to encourage the 3Rs (replacement, reduction and refinement) concept for evaluating the cytotoxic and genotoxic effects of zinc oxide (ZnO), titanium dioxide (TiO$_2$) and silicon-dioxide (SiO$_2$) NPs. Cytotoxicity assays demonstrated that the adverse effects of respective NPs were observed in hESC-Fib beyond concentrations of 200 µg/mL (SiO$_2$ NPs), 30 µg/mL (TiO$_2$ NPs) and 20 µg/mL (ZnO NPs). Flow cytometry results correlated with increased apoptosis upon increase in NP concentration. Subsequently, scratch wound assays showed ZnO (10 µg/mL) and TiO$_2$ (20 µg/mL) NPs inhibit the rate of wound coverage. DNA damage assays confirmed TiO$_2$ and ZnO NPs are genotoxic. In summary, hESC-Fib could be used as an alternative platform to understand toxicity profiles of metal oxide NPs.

Keywords: human pluripotent stem cells; fibroblasts; nanoparticles; cytotoxicity; genotoxicity

Citation: Handral, H.K; Ashajyothi, C.; Sriram, G.; Kelmani, C.R.; Dubey, N.; Cao, T. Cytotoxicity and Genotoxicity of Metal Oxide Nanoparticles in Human Pluripotent Stem Cell-Derived Fibroblasts. *Coatings* **2021**, *11*, 107. https://doi.org/10.3390/coatings11010107

Received: 10 December 2020
Accepted: 12 January 2021
Published: 19 January 2021

Publisher's Note: MDPI stays neutral with regard to jurisdictional claims in published maps and institutional affiliations.

1. Introduction

Advances in nanotechnology are proceeding at a striking rate and a diverse number of industrial, commercial, consumer, and health-care products have expanded by incorporating nanomaterials into these products [1,2]. Due to the large number, diversity, and use of NPs (nanoparticles), toxicology studies cannot keep pace. Since most of the past and current models largely depend on mammalian animals for testing, such time- and resource-intensive studies, albeit informative, cannot assess the avalanche of NP-enabled technology. Owing to differences in the toxicity profiles of the native and nano-sized counterparts, risk-assessment frameworks specific to NPs that incorporate alternative testing models have been proposed [3]. Secondly, reduction and replacement of animal use has been promoted to adopt 3Rs in (Organisation for Economic Co-operation and Development) OECD guidelines and has led to calls for tiered and integrated test strategies that were used to screen and assess NPs along their chemical life cycle by in silico, in vitro and other alternative models [4].

Due to high volumes of testing and limited availability of primary human cells, cell lines of animal-origin, immortalized, and/or cancer-derived sources are commonly used for high-throughput toxicology screening studies. Since cellular immortality leads to epigenetic changes, tumorigenesis, chromosomal and genetic aberrations, the physiological relevance and reliability of cancer-derived and immortalized cell lines is limited [5,6]. Further, though primary human cells are a good platform to rely on procuring large amounts of primary cells typically needed for high-throughput assays is challenging. On the other hand, pluripotent stem cells such as human embryonic stem cells (hESCs), due to their long-term self-renewal and pluripotent ability to derive differentiated lineages, are a potential source for obtaining unlimited amounts of somatic cells [3]. Additionally, hESCs have been shown to be genetically and karyotypically normal, thus considering hESCs as a representative source of how the normal cell acts in vivo. There are some similar reports toxicity of various ZnO, TiO_2, and SiO_2 NPs. The comparative toxic effect of this nanoparticle of hESC-derived fibroblast is nascent and yet to be explored and requires further investigation [7]. Thus, in this study, we explored the potential use of human embryonic stem cell-derived fibroblasts (hESC-Fibs) to evaluate the cellular toxicology profiles of most common metal oxide NPs.

2. Materials and Methods

2.1. Nanoparticle Stock Preparation and Characterization

Nanoparticles TiO_2 (product number: 791326, <100 nm), SiO_2 (product number: 637238, 10–20 nm), and ZnO (product number: 721077, <100 nm) of sizes were purchased from Sigma Aldrich (St. Louis, MO, USA). A stock dispersion of as received nanoparticles was weighed on an analytical mass balance, suspended in deionized water at a concentration of 50 mg/mL, and then probe sonicated for 2 min at 35–40 W to assist in mixing and forming a homogeneous dispersion. For further use, the stock concentration was diluted to prepare working concentrations in DMEM-high glucose media with 10% FBS and 1% pen/strep. NPs were spread on a silicon wafer to characterize using Scanning Electron Microscopy (SEM) (Verios XHR SEM, FEI, Waltham, MA, USA) and Transmission Electron Microscopy (TEM) (Tecnai 20 G2, CSIR-CECRI, Karaikudi, India).

2.2. Cell Culture and Differentiation

H1 hESCs were purchased from WiCell Research Institute, (Madison, WI, USA) and cultured using mTESR1^{TM} culture media (Stem Cell Technologies^{TM}, Vancouver, BC, Canada) by using Matrigel^{TM} coated 6-well tissue culture plates as previously described [8–10]. Our group's established protocol was adapted to obtain hESC-Fib [9–13]. Briefly, hESC colonies were dissociated using dispase (1 mg/mL, Stem Cell Technologies) and plated on ultra-low attachment plates to form embryoid bodies, during which low-glucose DMEM (Biowest, Riverside, MO, USA) culture media was fed for 10 days. On day 11, floating embryoid bodies were seeded on 0.1% gelatin-coated plates, and cells were nourished with DMEM-high glucose (Biowest, Riverside, MO, USA) culture media supplemented with 10% fetal bovine serum (FBS) (Biowest, Riverside, MO, USA). Cells were cultured for 2 weeks until the fibroblast-like outgrowths have emerged, after which, cells were passaged and cultured in high-glucose medium for two more passages before using them for experiments.

Characterization studies for hESC-Fib in comparison with commercially available human primary fibroblasts (PromoCell, Heidelberg, Germany) for morphological features, RT-PCR-based gene expression as well as protein level expression by immunofluorescence were performed.

2.3. Assessment of Cellular Viability

The effect of NPs on viability of cells was accessed by MTS assay (CellTiter 96$^{®}$ AQueous One Solution Cell Proliferation Assay, Promega, Madison, WI, USA) and trypan blue dye exclusion method. In MTS assay, cells were cultured in 96-well plates (1000 cells in 100 µL media/well) until they reach 80% confluency and cells were treated with desired concentrations of metal oxide NPs. After 24 h of NP treatment, 20 µL of MTS reagent in 100 µL of media was added to each group and incubated for 4 h. Culture medium without

NPs was considered as negative control. DMSO was used as the reference toxicant and considered as positive control. Absorbance was read at 490 nm using plate reader (M200 Infinite, Tecan, Switzerland). Cell viability was represented as percentage of viable cells with respect to no treatment control.

For trypan blue assay, cells were treated with respective concentrations of metal oxide NPs for 24 h and stained with trypan blue dye. The number of live and dead cells were manually counted using hemocytometer and cellular viability was expressed as percentage of viable cells.

2.4. Dose-Response Curves

From the MTS assay results, dose-response curves were plotted to investigate no observable adverse effect concentration (NOAEC), inhibitory concentration (IC_{50}) and total lethal concentration (TLC) values [14]. To calculate these values, normalized cell viability (viability of cells in normal culture medium = 1) was plotted as a function of logarithm of the particle. Theoretical dose-response curves were extrapolated using the following equation:

$$y = 1 - 1/1 - e^{a(b-x)} \tag{1}$$

where, a = curve slope, $b = IC_{50}$ value (in µg/mL) and x is the absorbance values expressed in percentage of cell viability.

NOEAC values were determined considering y value as 95%, maximum cell viability in normal cell culture medium. Equation is as follows:

$$NOEAC = b - \ln(19)/a \tag{2}$$

Similarly, TLC value was determined with y being 5% of the maximum response on the normalized viability. Equation is as follows:

$$TLC = b - \ln(1/19)/a \tag{3}$$

2.5. Assessment of Membrane Potential Using Lactate Dehydrogenase Assay (LDH)

CytoTox 96® Non-Radioactive Cytotoxicity Assay kit (Promega, Madison, WI, USA) was used to access the membrane integrity of cells by measuring the release of LDH. Briefly, cells cultured in 96-well plates were treated with respective concentrations of NPs for 24 h. The 100 µL of cell-free supernatant were subsequently transferred from each well into a 96-well plate, and the LDH reaction mixture was added to each well in triplicates. After incubating for 3 h, absorbance values of supernatant were obtained at 490 nm wavelength using plate reader (M200 Infinite, Tecan, Seestrasse, Männedorf, Switzerland).

2.6. Detection of Apoptotic Cells

Flow cytometry analysis was used to detect apoptotic/necrotic profile of NP-treated cells. Double staining of APC Annexin V/Propidium iodide (PI) was done by following manufacturer's instructions of Annexin-PI kit (Invitrogen, ThermoFisher, Waltham, MA, USA). Briefly, hESC-Fib cells treated with respective concentrations of metal oxide NPs were trypsinized after 24 h of NPs' treatment and cell pellet was stained with 4 µL of AnnexinV and incubated in dark at 4 °C for 15 min. Later, cells were washed with 200 µL of FACS buffer and centrifuged to remove an unbound stain; after which, cells were suspended in 500 µL of FACS buffer and stained with 5 µL of PI (1 mg/mL). Stained cells were taken for flow cytometry analysis (CyAn Analyzer, Brea, CA, USA). Based on previously published reports, percentage of apoptotic cells was calculated [15,16].

2.7. Assessment of Cellular Proliferation Using Scratch Wound Assay

Scratch wound assay was performed to assess the ability of metal oxide NPs to promote or inhibit cell proliferation. The hESC-Fib were cultured as confluent monolayers in 12-well plates and wound scratch was made with sterile 100 µL-micropipette tips.

Culture plates were rinsed with PBS before treating the cells with NPs. Cells with no NP treatment were considered as control. After 24 h of NPs' treatment, mean distance between wound edges was measured by imaging under an inverted phase contrast microscope (Olympus IX70 fluorescence microscope, Shinjuku City, Tokyo, Japan). Images were analyzed by using commercially available image analysis software (TScratch software, developed by the group of Dr. Koumoutsakos (CSE Lab), at the Eidgenössische Technische Hochschule (ETH), Zurich, Switzerland [17].

2.8. DNA Damage Assay for Genotoxic Profile

HCS DNA damage assay kit (Invitrogen Carlsbad, CA, USA) was used to analyze genotoxic nature of NPs on hESC-Fib. The hESC-Fib were cultured till 80% confluency. Cells were treated with respective metal oxide NPs for 24 h. We have considered mild and IC_{50} concentrations of all three NPs. After NP treatment, cells were fixed with 4% paraformaldehyde and stained using HCS DNA damage kit as per manufacturer's instructions. Mechanism of the staining works was by phosphorylated γH2AX which is conjugated with Alexa fluor 555; foci formed at the damaged DNA site in the nucleus is measured by specific antibody-based detection and Hoechst 33,342 (a DNA binding dye) stains nucleus to highlight the changes in nuclear morphology of healthy as well as damaged cells.

3. Statistical Analysis

For the cytotoxicity evaluation, concentration response curves were fitted. Statistics were performed with One-way ANOVA and Tukey test ($\alpha = 0.05$).

4. Results

4.1. Characterization of Nanoparticles (NPs)

SEM and TEM analysis demonstrated spherical shape of SiO_2 and TiO_2 NPs with size ranging from 10–20 nm and <100 nm, respectively; with marginal variation particles showed poly-aggregation. Electron microscopy analysis report confirmed the features of ZnO NPs (size < 100 nm) with rod-shaped morphology. Figures 1 and 2, highlight the dimension and surface morphology of the nanoparticle samples examined by SEM and TEM.

Figure 1. SEM (scanning electron microscopy) photomicrographs of SiO_2 NPs (nanoparticles) deposited on the substrate formed the agglomerates consisting of the tangled chains with the particle size ranging from 10–20 nm, spherical shape of TiO_2 NPs with an average crystallite size of about 20 nm and ZnO NPs with an average size of 100 nm. Scale bar: 500 µm.

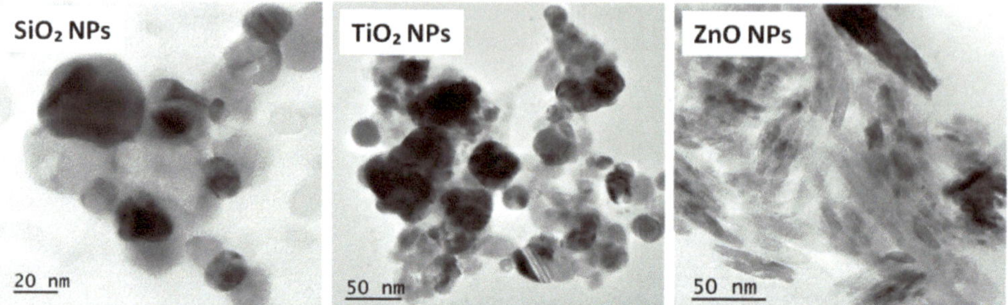

Figure 2. TEM photomicrographs of SiO$_2$, TiO$_2$ and ZnO NPs. Highlighting the spherical shape in case of SiO$_2$ and TiO$_2$ NPs. ZnO NPs showing rod-shaped structures.

4.2. Characterization of hESC-Fib

The hESC-Fib cells were characterized for morphological features in comparison with primary fibroblasts (Figure 3a). Vimentin expression was confirmed by immunofluorescence staining (Figure 3b) and RT-PCR studies (Figure 3c).

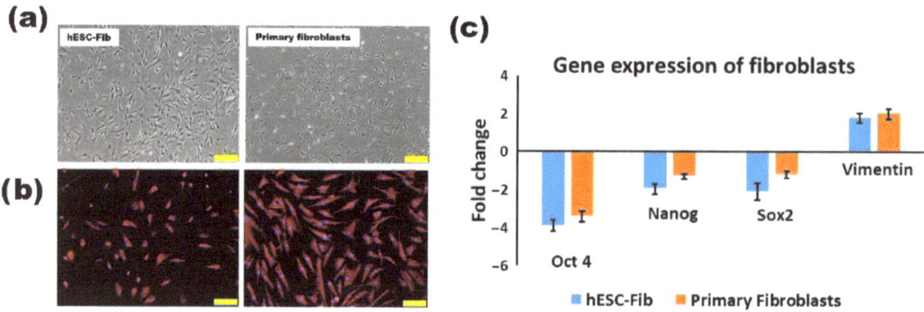

Figure 3. Characterization analysis of hESC-Fib (human embryonic stem cell-derived fibroblast): (**a**) Phase contrast microscopic image of hESC-Fib and primary fibroblasts cultured on gelatin-coated tissue culture plate. Phase contrast images show the spindle-shaped morphology of cells; (**b**) Immunofluorescence-stained photomicrographs of hESC-Fib and primary fibroblasts expressing the presence of Vimentin in red and DAPI stained nucleus in blue; (**c**) Real-time PCR of hESC-Fib compared to primary fibroblasts and both cells were also checked for expression of pluripotent genes. Levels of expression were normalized by corresponding GAPDH values, an internal control. Scale bar: 200 μm.

4.3. Cellular Viability in Presence of NPs

The SiO$_2$, TiO$_2$ and ZnO NP-treated cells displayed morphology close to that of untreated fibroblasts at a concentration of 50, 10, and 5 μg/mL. However, detrimental effect on cell morphology and even cell death was observed with higher NP concentrations (i.e., 1250, 250 and 50 μg/mL SiO$_2$, TiO$_2$ and ZnO NPs respectively). Phase contrast images showed that higher concentrations of SiO$_2$ and TiO$_2$ NPs have disrupted the cell's morphology, whereas ZnO NPs have led to cell death. This implies SiO$_2$ and TiO$_2$ NPs could be cytotoxic at higher concentrations, however, ZnO NPs have shown cytotoxicity at lower concentrations (Figure 4).

As per trypan blue dye exclusion assay, ZnO NPs at and above 30 μg/mL showed sudden decrease in cell viability. In the case of cells treated with TiO$_2$ and SiO$_2$ NPs, there has been steady decline in cell viability with a significant increase in NP concentrations (Figure 5). Similarly, MTS and LDH assay results co-relate with results of trypan blue assays. Cells treated with SiO$_2$ and TiO$_2$ NPs show less LDH release and percentage of toxicity at lower concentrations, however, ZnO NPs at lower concentration (5 μg/mL) have shown 40% cellular toxicity. LDH assay provided a better view of cytotoxic nature of NPs affecting cell membrane integrity (Figure 5). In addition, the optimal dose level for each NP was also calculated (Table 1).

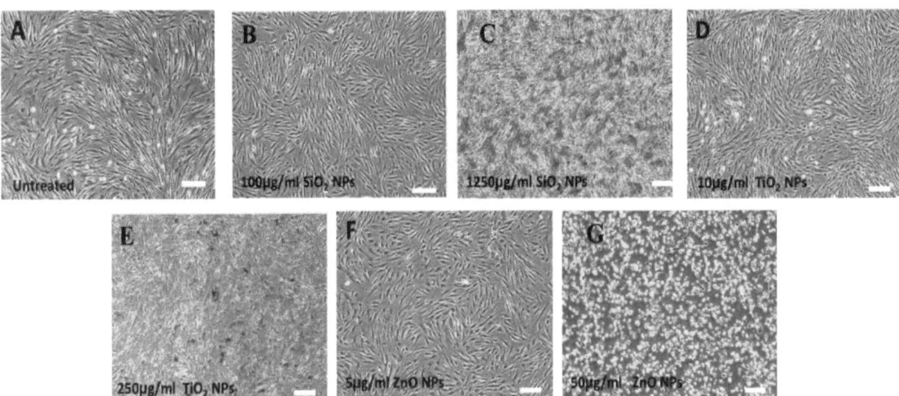

Figure 4. Phase contrast images of hESC-Fib cells taken after 24 h treatment of NPs, where (**A**) was considered as control i.e., cells with no NP treatment (untreated group). This figure also highlights the cell disruptions upon treatment of lower and higher concentrations of SiO_2 (**B,C**), TiO_2 (**D,E**) and ZnO (**F,G**) NPs. Scale bar: 100 μm.

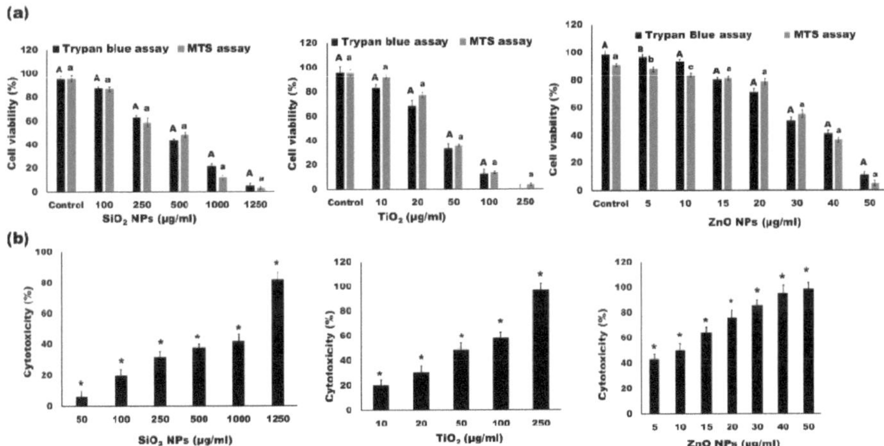

Figure 5. Represents the cytotoxic profiles of hESC-Fib in presence of NPs: (**a**) Trypan blue assay and MTS assay highlights the cell survival. The same letters indicate a significant difference compared with the control. "A" denotes significance at $p < 0.05$ for Trypan blue assay and "a" denotes significance at $p < 0.05$ for MTS assay; (**b**) LDH assay depicts the membrane potential of hESC-Fib cells after NP treatment. (* represents statistical significance compared to control).

Table 1. NOAEC (no observable adverse effect concentration), IC_{50} (inhibitory concentration), and TLC (total lethal concentration) values for hESC-Fib cells cultivated with SiO_2, TiO_2 and ZnO NPs.

NPs	NOAEC	IC_{50}	TLC
TiO_2 (μg/mL)	49.705 ± 2.4	50 ± 3.1	52.39 ± 4
ZnO (μg/mL)	39.412 ± 2	40 ± 3.2	42.39 ± 1.3
SiO_2 (μg/mL)	245.98 ± 6	250 ± 4.3	255.39 ± 7

4.4. Apoptosis Assay

The percentage of cells undergoing apoptosis/necrosis was investigated using flow cytometric analysis (Figure 6). Annexin-V and Propidium Iodide staining kit (Alexa Fluor®, ThermoFisher Scientific, Waltham, MA, USA) was used to analyze the cell apoptotic profile. Cells treated with ZnO NPs have shown apoptosis at 5 µg/mL. Approximately 20%–30% of cells were viable under 20 µg/mL concentration group, whereas cells treated with TiO_2 and SiO_2 NPs have shown lesser viability only in higher concentrations, i.e., 100 µg/mL and 500 µg/mL respectively. Figure 6 depicts the gradual increase in apoptotic cell population with gradual increase in NP concentration. Flow cytometry analysis graphs are mentioned in Supplementary Materials Figure S1. All the tested NPs at higher concentrations have induced apoptosis in hESC-Fib cells ($p \leq 0.05$ and $p \leq 0.001$). Thus, ZnO NP-treated cells showed strong apoptotic response as compared to TiO_2 and SiO_2 NPs.

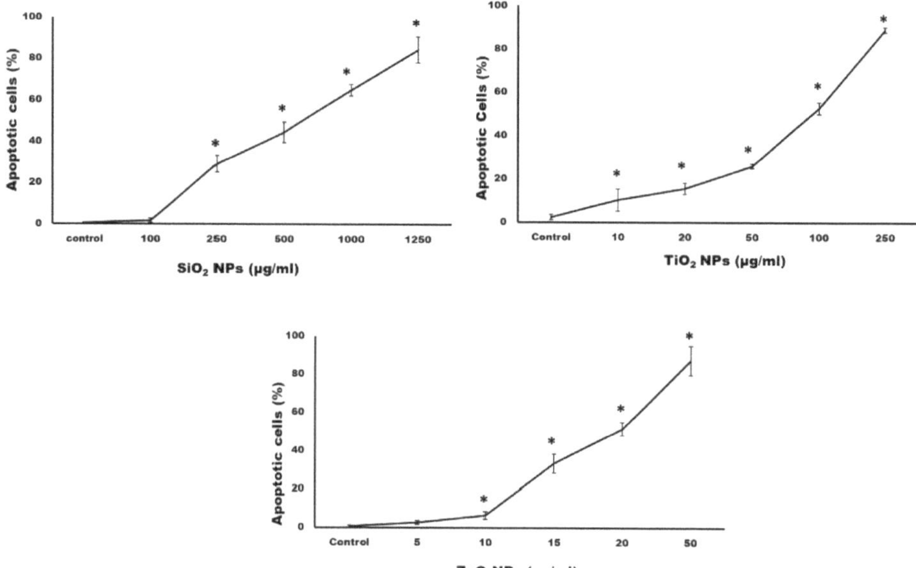

Figure 6. Effect of nanoparticles on the apoptosis of hESC-derived fibroblasts exposed to different concentrations in culture after 24 h. (* represents statistical significance compared to control).

4.5. Wound Healing Scratch Assay

Scratch assay was performed to address the impact of metal oxide nanoparticles on wound healing and collective migratory behavior of hESC-Fib cells. From Figure 7, TiO_2 and ZnO NPs at 20 µg/mL and 15 µg/mL showed approximately 50% of wound closure, respectively. Surprisingly, SiO_2 NPs did not show any cell proliferation beyond 250 µg/mL. The percentage of the wound covered over a period was calculated using TScratch software [17].

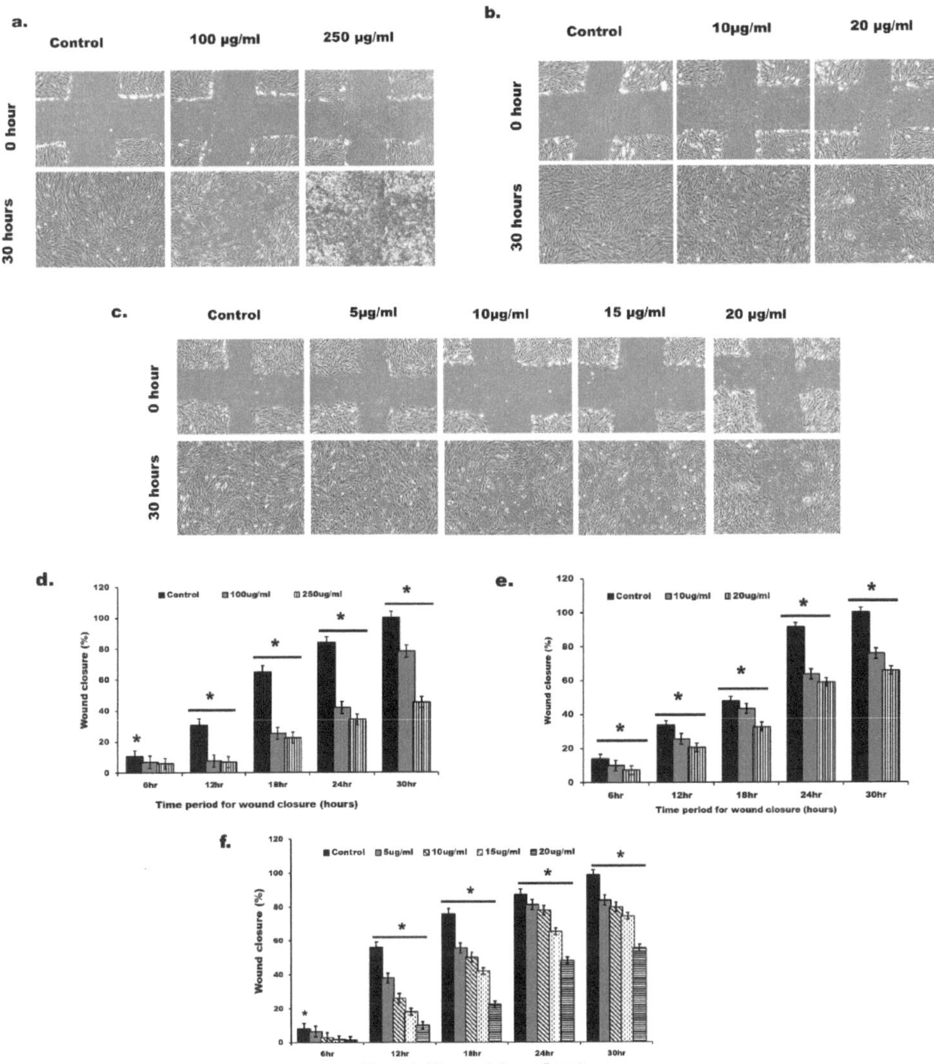

Figure 7. Wound healing scratch assay showing the cell proliferation and migration capability when treated with SiO_2 NPs (**a**,**d**); TiO_2 NPs (**b**,**e**) and ZnO NPs (**c**,**f**). (* represents statistical significance compared to control).

4.6. Genotoxicity Assay

Given the cell viability results in Figures 4 and 5 and dose response concentrations in Table 1, we investigated genotoxic effects of metal oxide NPs on hESC-Fib at NOEAC and IC_{50} concentrations. Figure 8 shows the fluorescent images of nuclear/DNA damages in the NP-treated hESC-Fib cells. Subsequent increase in NP concentration changed the nuclear morphology, which is distinctly visible in ZnO NP-treated cells at 20 µg/mL as compared to TiO_2- (50 µg/mL) and SiO_2- (250 µg/mL) treated cells. ZnO and TiO_2 NPs at IC_{50} concentration have shown pH2AX (TRITC/Alexa Fluor 555) signals, but only ZnO NP-treated hESC-Fib cells showed a shrunken nucleus. In summary, HCS DNA damage assay highlights the genotoxic response of ZnO NPs at IC_{50} concentration. However, TiO_2 and SiO_2 NPs have not shown any major change.

5. Discussion

For evaluating NP-induced mechanism, cell-based in vitro toxicity testing is a very important alternative testing tool that could reduce animal model experimentation in most of the toxicology studies [5]. However, the responses to NPs are known to be dependable on cell type, cell's physiological action and NP size, because diverse physicochemical properties of NPs are known to trigger certain cellular changes [18]. This study aimed to use hESC-derived cells as an alternative cell source to evaluate the toxic effects of commonly used metal oxide NPs. NPs were dispersed in serum containing cell culture media and used to check the effects of NPs on cells. We differentiated hESCs to fibroblasts under in vitro conditions. NP toxicity was assessed by MTS, trypan blue dye exclusion assay and LDH assay for cytotoxic response.

The SiO_2, TiO_2, and ZnO NPs are known for their salient features such as porosity, specific surface characteristics, surface functionalization property and biocompatibility [19]; hence are widely used for biomedical applications, especially in drug delivery, gene delivery, nucleic acid extraction and imaging [20]. From previous reports, silica NPs show toxic effects on human and mouse cells at concentrations beyond 100 μg/mL after 24 h of exposure by indicating the cell lysis after 24 h of incubation, and also cellular agglomeration was observed [21,22]. Similarly, our studies revealed hESC-Fib cells treated with SiO_2 NPs for 24 h showed approximately 80% of viability at 100 μg/mL (Figure 5).

ZnO and TiO_2 NPs are more toxic than SiO_2 NPs. ZnO NPs were reported to initiate cytotoxicity from the range of 5–100 μg/mL in epithelial cells [23]. The rod-shaped ZnO nanoparticles have shown to have higher toxicity compared to the corresponding spherical ones [24]. Thus, nano-rod ZnO particles were used to assess the toxicity response on hESC-derived fibroblasts compared to spherical-shaped SiO_2 or TiO_2 to induce significant toxicity. In the current study, hESC-Fib cells treated with ZnO and TiO_2 NPs showed similar cytotoxic effects at the approximately same concentrations. Likewise, small-sized SiO_2 nanoparticles are widely used compared to bigger size and have been proved more toxic than bigger SiO_2 nanoparticles [25]. Cell viability results were confirmed by trypan blue assay, MTS assay as well as LDH assay (Figure 5). Trypan blue dye exclusion assay, the simplest assay to determine percentage of cell viability and assay, emphasizes the live cells having an intact cell membrane which excludes dyes/stains to enter, whereas dead cells take up the stain [26]. Similarly, MTS assay measures cell viability based on biological reduction of MTS tetrazolium compound and conversion into a formazan product. Amount of formazan produced is directly proportional to the number of live cells, which was measured by colorimetric analysis using a plate reader [27]. MTS assay showed approximately 50% of viability in the cells treated with 250 μg/mL of SiO_2 NPs. Followed by MTS assay, dose response concentrations were calculated for each nanoparticle (Table 1). Previously, the effects of NPs were evaluated also by the release of enzyme lactate dehydrogenase (LDH). LDH, a stable cytosolic enzyme that releases upon cell lysis. LDH assay represented the damages in cell membrane induced by NPs [28]. In our study, LDH assay confirms the toxic concentrations of metal oxide NPs to hESC-Fib cells. ZnO NPs are shown to be more toxic among the three NPs at lower concentration (15 μg/mL) as compared to TiO_2 and SiO_2 NPs. Cytotoxicity assays also explain the speed of toxic response by ZnO NPs. ZnO NPs at 10 μg/mL has damaged nearly 50% of cells in 24 h of exposure, whereas TiO_2 NPs are five times slower and SiO_2 NPs are 50 times. TiO_2 NPs are one of the most commonly used NPs in biomedical industry either in cosmetics or as Ti implants. Interestingly, only 100 μg/mL of TiO_2 NPs upon 24 h of exposure have shown 75–80% of cytotoxic effects in hESC-Fib cells. Based on the cytotoxic response, IC_{50} and TLC values of NPs were calculated. IC_{50} concentrations of NPs were used in further analysis. Various physical factors of NPs such as shapes and geometries including spheres, ellipsoids, cylinders, sheets, cubes, spikes and rods considerably induce toxicity to the cells. In relation to this, we have shown that rod-shaped ZnO NPs at 50 μg/mL have more capacity to induce physical damage to cells leading to higher necrosis than spherical-shaped TiO_2 and SiO_2 NPs at 50 μg/mL. Likewise, in cases of size-dependent nanotoxicity studies, it has been known that distribution of NPs among cells/tissues in an in vivo scenario, that significant difference in distribution of NPs has been reported. Lower-size NPs have been found at most cellular sites, but higher-size NPs at limited locations are one of the main reasons why small-size NPs have been chosen for drug delivery applications. Considering this, toxicity evaluation of small-sized NPs is important to highlight the impact of size of NPs on cytocompatibilty.

Apoptosis is a molecular event characterized by changes in specific morphological features, plasma membrane blebbing, loss in membrane integrity, condensation of cytoplasm, inter-nucleosomal cleavage of DNA, etc. [29]. NPs induce apoptosis in human cells, which is generally due to long-term NP exposure to mitochondria leading to apoptotic pathway independent of caspase-8/t-Bid pathway [30]. Previously, NP-mediated toxicity has shown to induce apoptotic changes in cells [31]. A study on ZnO NP-induced toxicity reported apoptosis in human dermal fibroblasts and was further confirmed by AnnexinV staining [32]. During apoptosis, loss of membrane potential induces the translocation of membrane phospholipid phosphatidylserine (PS) from inner cellular environment to extra cellular spaces [33]. Annexin V has high affinity towards PS, thus binds to apoptotic cell's surface [15,34]. We investigated apoptotic profiles of hESC-Fib cells treated with SiO_2, TiO_2 and ZnO NPs for a desired range of concentration. As shown in Figure 6, all three NPs showed a gradual change in apoptosis along with an increase in concentration. SiO_2 NPs showed apoptotic response at higher concentrations as compared to TiO_2 and ZnO NPs. Among ZnO NP- treated cells, we observed a huge shift in apoptotic population from 10 µg/mL to 20 µg/mL group, percentage of apoptosis rises from 5% to 50% just by doubling the NP concentration. Hereby, we could conclude ZnO NPs are prone to induce drastic changes in apoptosis just by minor changes in NP concentrations.

NPs used in cosmetics, pharmaceuticals, implants, and other biomedical devices, come in contact with proliferating cells and could release into blood/serum leading to long-term toxicity [2,35]. Studies highlighted the adverse effects of NPs on cell migration properties. Destabilization of cellular migration by NPs was found due to changes which occurred in the microtubule network which lead to modification of magnitude, spatiotemporal distribution of cell movement and, ultimately, hampering wound healing capability [36]. Rise in intracellular tension due to NPs have resulted in retarding cellular migration. Live imaging of human lung carcinoma cells in presence of SiO_2 NPs has evidenced the changes in the dynamics of microtubule leading to decrease in cell motility [37,38]. Wound healing scratch assay showed NPs have decreased migration capacity of the cells. Among all three NPs, ZnO NPs at only 20 µg/mL have hindered 60% of cell migration. Cell migration was hindered most by ZnO > TiO_2 > SiO_2 NPs over a period of 30 h (see Figure 7). Interestingly, ZnO NPs at 10 µg/mL have induced 50% of cytotoxic effects on cell membrane integrity (confirmed by LDH assay); similarly, ZnO NPs at the same concentration have inhibited nearly 40% of wound closure. TiO_2 NPs at 20 µg/mL showed moderate cytotoxic effects and 50% inhibition on cellular motility.

Furthermore, NPs have been tested for genotoxic effects such as chromosomal aberration, DNA strand breaks, oxidative DNA damage, mutations, and some commonly used NPs are proven to be genotoxic at certain concentrations [39]. We performed DNA damage assay under in vitro conditions on NP-treated hESC-Fib cells. As presented in Figure 8, ZnO NPs at 20 µg/mL have shown changes in nuclear morphology and prove to be a distinct genotoxic response. In the DNA damage assay, genotoxic response is generally measured by TRITC signals and nuclear changes observed among the treated cells. Among three NPs, SiO_2 NPs showed negligible TRITC signals at 250 µg/mL upon 24 h treatment. However, TiO_2 and ZnO NPs showed strong TRITC signals. In the near future, studies will be planned to evaluate the sub-cellular changes in hESC-Fib cells by TEM imaging and for genotoxic response by real time PCR and Western blotting.

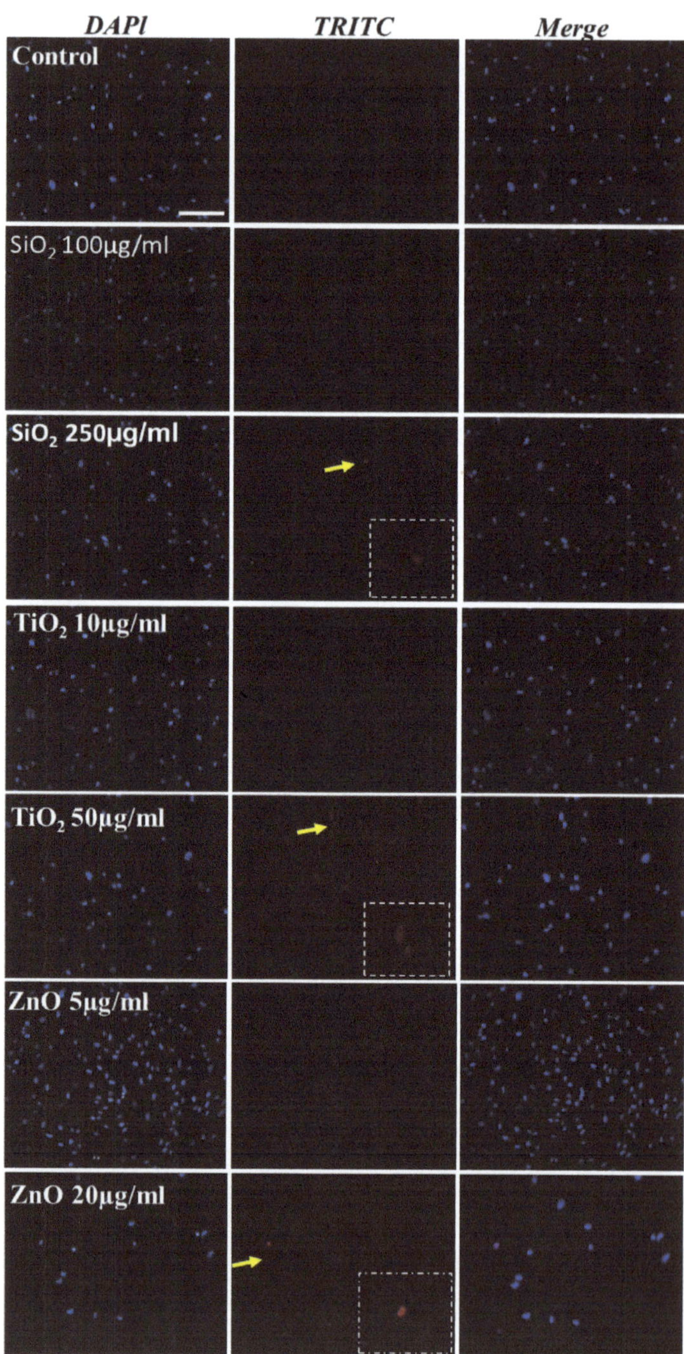

Figure 8. Represents the DNA damage caused by NPs on hESC-Fib cells. NP-treated hESC-Fib cells showing no genotoxic response by SiO_2 NPs, moderate genotoxicity from TiO_2 NPs and strong genotoxicity by ZnO NPs at respective IC_{50} concentration. Dead cells are highlighted in yellow. Scale bar: 100 μm.

6. Conclusions

In this study, we have shown the concentration-dependent toxicity of ZnO, SiO_2 and TiO_2 NPs, which vary considerably, can be due to difference in size, shape, and active surface area of the nanoparticle. Wound healing scratch and MTS assay showed evidence on anti-migratory effects of NPs on hESC-fibroblasts. The membrane damage and apoptosis were mainly responsible for cell death caused by these nanoparticles. Taking into account the ability of hESC to differentiate into an unlimited number of cells that can represent how cells behave in vivo, it should be considered a promising alternative human model for predictive toxicology study of nanomaterials. This will not only lead to less use of animals in research but would also significantly reduce the drug production costs by fostering the 3Rs concept. Therefore, hESCs are a valuable source of cells which would lead to a first step towards intelligent and comprehensive in vitro testing of metal oxide nanoparticles, and in various scientific fields such as in regenerative medicine and drug screening.

Supplementary Materials: The following are available online at https://www.mdpi.com/2079-6412/11/1/107/s1, Figure S1: Flow cytometry graphs showing percentage of live and apoptotic cells treated with (a) SiO_2, (b) TiO_2 and (c) ZnO NPs were mentioned.

Author Contributions: H.K.H.: Conceptualization, Investigation, Methodology, Writing—original draft. C.A.: Formal analysis, Methodology, writing-original draft. G.S.: Conceptualization, Project administration, Writing—review & editing. C.R.K.: Conceptualization, Funding acquisition, Methodology, Project administration, Resources, Supervision. N.D.: Validation, data curation, Investigation, Methodology, Writing—review & editing. T.C.: Conceptualization, Funding acquisition, Methodology, Project administration, Resources, Supervision, Writing—review & editing. All authors have read and agreed to the published version of the manuscript.

Funding: This work was partially supported by grants from National University of Singapore and Singapore Ministry of Education (R221000023112, R221000026112) and National University Health System, Singapore (R221000085515, R221000085733).

Institutional Review Board Statement: Not applicable.

Informed Consent Statement: Not applicable.

Data Availability Statement: Data is contained within the article or supplementary material.

Conflicts of Interest: The authors declare no conflict of interest.

Abbreviations

hESCs	human embryonic stem cells
hESC-Fib	human embryonic stem cell-derived fibroblasts
3Rs	Replacement: Reduction and Refinement
SiO_2	silicon dioxide
TiO_2	titanium dioxide
ZnO	zinc oxide
NPs	nanoparticles
SEM	Scanning Electron Microscope
FBS	fetal bovine serum
MTS	3-(4:5-dimethylthiazol-2-yl)-5-(3-carboxymethoxyphenyl)-2-(4-sulfophenyl)-2H-tetrazolium
LDH	lactate dehydrogenase
OECD	Organization for Economic Co-operation and Development
ROS	reactive oxygen species
DMEM	Dulbecco's Modified Eagle Medium
NOAEC	no observable adverse effect concentration
IC_{50}	50% inhibitory concentration
TLC	total lethal concentration
FACS	fluorescence-assisted cell sorting

References

1. Kaul, S.; Gulati, N.; Verma, D.; Mukherjee, S.; Nagaich, U. Role of nanotechnology in cosmeceuticals: A review of recent advances. *J. Pharm.* **2018**, *2018*. [CrossRef] [PubMed]
2. Gupta, R.; Xie, H. Nanoparticles in Daily Life: Applications, toxicity and regulations. *J. Environ. Pathol. Toxicol. Oncol.* **2018**, *37*, 209–230. [CrossRef] [PubMed]
3. Handral, H.K.; Tong, H.J.; Islam, I.; Sriram, G.; Rosa, V.; Cao, T. Pluripotent stem cells: An in vitro model for nanotoxicity assessments. *J. Appl. Toxicol.* **2016**, *36*, 1250–1258. [CrossRef] [PubMed]
4. Rasmussen, K.; Rauscher, H.; Kearns, P.; González, M.; Riego Sintes, J. Developing OECD test guidelines for regulatory testing of nanomaterials to ensure mutual acceptance of test data. *Regul. Toxicol. Pharmacol.* **2019**, *104*, 74–83. [CrossRef]
5. Fröhlich, E. Comparison of conventional and advanced in vitro models in the toxicity testing of nanoparticles. *Artif. Cells Nanomed. Biotechnol.* **2018**, *46*, 1091–1107. [CrossRef]
6. Drasler, B.; Sayre, P.; Steinhäuser, K.G.; Petri-Fink, A.; Rothen-Rutishauser, B. In vitro approaches to assess the hazard of nanomaterials. *NanoImpact* **2017**, *8*, 99–116. [CrossRef]
7. Zhou, X.; Yuan, L.; Wu, C.; Chen, C.; Luo, G.; Deng, J.; Mao, Z. Recent review of the effect of nanomaterials on stem cells. *RSC Adv.* **2018**, *8*, 17656–17676. [CrossRef]
8. Sriram, G.; Tan, J.Y.; Islam, I.; Rufaihah, A.J.; Cao, T. Efficient differentiation of human embryonic stem cells to arterial and venous endothelial cells under feeder- and serum-free conditions. *Stem Cell Res. Ther.* **2015**, *6*, 261. [CrossRef]
9. Tan, J.Y.; Sriram, G.; Rufaihah, A.J.; Neoh, K.G.; Cao, T. Efficient derivation of lateral plate and paraxial mesoderm subtypes from human embryonic stem cells through GSKi-mediated differentiation. *Stem Cells Dev.* **2013**, *22*, 1893–1906. [CrossRef]
10. Fu, X.; Toh, W.S.; Liu, H.; Lu, K.; Li, M.; Hande, M.P.; Cao, T. Autologous feeder cells from embryoid body outgrowth support the long-term growth of human embryonic stem cells more effectively than those from direct differentiation. *Tissue Eng. Part. C Methods* **2010**, *16*, 719–733. [CrossRef]
11. Sriram, G.; Natu, V.P.; Islam, I.; Fu, X.; Seneviratne, C.J.; Tan, K.S.; Cao, T. Innate Immune Response of Human Embryonic Stem Cell-Derived Fibroblasts and Mesenchymal Stem Cells to Periodontopathogens. *Stem Cells Int.* **2016**, *2016*. [CrossRef] [PubMed]
12. Vinoth, K.J.; Manikandan, J.; Sethu, S.; Balakrishnan, L.; Heng, A.; Lu, K.; Poonepalli, A.; Hande, M.P.; Cao, T. Differential resistance of human embryonic stem cells and somatic cell types to hydrogen peroxide-induced genotoxicity may be dependent on innate basal intracellular ROS levels. *Folia Histochem. Cytobiol.* **2015**, *53*, 169–174. [CrossRef] [PubMed]
13. Cao, T.; Lu, K.; Fu, X.; Heng, B.C. Differentiated fibroblastic progenies of human embryonic stem cells for toxicology screening. *Cloning Stem Cells* **2008**, *10*, 1–10. [CrossRef] [PubMed]
14. Sambale, F.; Wagner, S.; Stahl, F.; Khaydarov, R.R.; Scheper, T.; Bahnemann, D. Investigations of the toxic effect of silver nanoparticles on mammalian cell lines. *J. Nanomater.* **2015**, *2015*. [CrossRef]
15. Van Engeland, M.; Ramaekers, F.C.S.; Schutte, B.; Reutelingsperger, C.P.M. A novel assay to measure loss of plasma membrane asymmetry during apoptosis of adherent cells in culture. *Cytom. J. Int. Soc. Anal. Cytol.* **1996**, *24*, 131–139. [CrossRef]
16. Hanley, C.; Layne, J.; Punnoose, A.; Reddy, K.M.; Coombs, I.; Coombs, A.; Feris, K.; Wingett, D. Preferential killing of cancer cells and activated human T cells using ZnO nanoparticles. *Nanotechnology* **2008**, *19*, 295103. [CrossRef]
17. Gebäck, T.; Schulz, M.M.; Koumoutsakos, P.; Detmar, M. TScratch: A novel and simple software tool for automated analysis of monolayer wound healing assays. *Biotechniques* **2009**, *46*, 265–274. [CrossRef]
18. Akter, M.; Sikder, M.T.; Rahman, M.M.; Ullah, A.K.M.A.; Hossain, K.F.B.; Banik, S.; Hosokawa, T.; Saito, T.; Kurasaki, M. A systematic review on silver nanoparticles-induced cytotoxicity: Physicochemical properties and perspectives. *J. Adv. Res.* **2018**, *9*, 1–16. [CrossRef]
19. Murthy, S.; Effiong, P.; Fei, C.C. 11—Metal oxide nanoparticles in biomedical applications. In *Metal Oxides, Metal Oxide Powder Technologies*; Al-Douri, Y., Ed.; Elsevier: Amsterdam, The Netherlands, 2020. [CrossRef]
20. Nikolova, M.P.; Chavali, M.S. Metal oxide nanoparticles as biomedical materials. *Biomimetics* **2020**, *5*, 27. [CrossRef]
21. Yu, Y.; Duan, J.; Yu, Y.; Li, Y.; Liu, X.; Zhou, X.; Ho, K.-f.; Tian, L.; Sun, Z. Silica nanoparticles induce autophagy and autophagic cell death in HepG2 cells triggered by reactive oxygen species. *J. Hazard. Mater.* **2014**, *270*, 176–186. [CrossRef]
22. Yu, T.; Malugin, A.; Ghandehari, H. Impact of silica nanoparticle design on cellular toxicity and hemolytic activity. *ACS Nano* **2011**, *5*, 5717–5728. [CrossRef] [PubMed]
23. Gu, T.; Yao, C.; Zhang, K.; Li, C.; Ding, L.; Huang, Y.; Wu, M.; Wang, Y. Toxic effects of zinc oxide nanoparticles combined with vitamin C and casein phosphopeptides on gastric epithelium cells and the intestinal absorption of mice. *RSC Adv.* **2018**, *8*, 26078–26088. [CrossRef]
24. Bhattacharya, D.; Santra, C.R.; Ghosh, A.N.; Karmakar, P. Differential toxicity of rod and spherical zinc oxide nanoparticles on human peripheral blood mononuclear cells. *J. Biomed. Nanotechnol.* **2014**, *10*, 707–716. [CrossRef] [PubMed]
25. Murugadoss, S.; Lison, D.; Godderis, L.; Van Den Brule, S.; Mast, J.; Brassinne, F.; Sebaihi, N.; Hoet, P.H. Toxicology of silica nanoparticles: An update. *Arch. Toxicol.* **2017**, *91*, 2967–3010. [CrossRef]
26. Liesche, J.; Marek, M.; Günther-Pomorski, T. Cell wall staining with Trypan blue enables quantitative analysis of morphological changes in yeast cells. *Front. Microbiol.* **2015**, *6*, 107. [CrossRef]
27. Aslantürk, Ö.S. In vitro cytotoxicity and cell viability assays: Principles, advantages, and disadvantages. In *Genotoxicity—A Predictable Risk to Our Actual World*; Larramendy, M.L., Soloneski, S., Eds.; IntechOpen: London, UK, 2018. [CrossRef]

28. Han, X.; Gelein, R.; Corson, N.; Wade-Mercer, P.; Jiang, J.; Biswas, P.; Finkelstein, J.N.; Elder, A.; Oberdörster, G. Validation of an LDH assay for assessing nanoparticle toxicity. *Toxicology* **2011**, *287*, 99–104. [CrossRef]
29. Savitskaya, M.A.; Onishchenko, G.E. Mechanisms of apoptosis. *Biochemistry (Mosc.)* **2015**, *80*, 1393–1405. [CrossRef]
30. Shukla, R.K.; Kumar, A.; Gurbani, D.; Pandey, A.K.; Singh, S.; Dhawan, A. TiO_2 nanoparticles induce oxidative DNA damage and apoptosis in human liver cells. *Nanotoxicology* **2013**, *7*, 48–60. [CrossRef]
31. Khanehzar, A.; Fraire, J.C.; Xi, M.; Feizpour, A.; Xu, F.; Wu, L.; Coronado, E.A.; Reinhard, B.M. Nanoparticle–cell interactions induced apoptosis: A case study with nanoconjugated epidermal growth factor. *Nanoscale* **2018**, *10*, 6712–6723. [CrossRef]
32. Zhang, J.; Qin, X.; Wang, B.; Xu, G.; Qin, Z.; Wang, J.; Wu, L.; Ju, X.; Bose, D.D.; Qiu, F.; et al. Zinc oxide nanoparticles harness autophagy to induce cell death in lung epithelial cells. *Cell Death Dis.* **2017**, *8*, e2954. [CrossRef]
33. Casciola-Rosen, L.; Rosen, A.; Petri, M.; Schlissel, M. Surface blebs on apoptotic cells are sites of enhanced procoagulant activity: Implications for coagulation events and antigenic spread in systemic lupus erythematosus. *Proc. Natl. Acad. Sci. USA* **1996**, *93*, 1624–1629. [CrossRef] [PubMed]
34. Vermes, I.; Haanen, C.; Steffens-Nakken, H.; Reutellingsperger, C. A novel assay for apoptosis Flow cytometric detection of phosphatidylserine expression on early apoptotic cells using fluorescein labelled Annexin V. *J. Immunol. Methods* **1995**, *184*, 39–51. [CrossRef]
35. Grant, C.A.; Twigg, P.C.; Baker, R.; Tobin, D.J. Tattoo ink nanoparticles in skin tissue and fibroblasts. *Beilstein J. Nanotechnol.* **2015**, *6*, 1183–1191. [CrossRef] [PubMed]
36. Tay, C.Y.; Cai, P.; Setyawati, M.I.; Fang, W.; Tan, L.P.; Hong, C.H.L.; Chen, X.; Leong, D.T. Nanoparticles strengthen intracellular tension and retard cellular migration. *Nano Lett.* **2014**, *14*, 83–88. [CrossRef]
37. Sun, Q.; Kanehira, K.; Taniguchi, A. PEGylated TiO_2 nanoparticles mediated inhibition of cell migration via integrin beta 1. *Sci. Technol. Adv. Mater.* **2018**, *19*, 271–281. [CrossRef]
38. Gonzalez, L.; De Santis Puzzonia, M.; Ricci, R.; Aureli, F.; Guarguaglini, G.; Cubadda, F.; Leyns, L.; Cundari, E.; Kirsch-Volders, M. Amorphous silica nanoparticles alter microtubule dynamics and cell migration. *Nanotoxicology* **2015**, *9*, 729–736. [CrossRef]
39. Nallanthighal, S.; Chan, C.; Murray, T.M.; Mosier, A.P.; Cady, N.C.; Reliene, R. Differential effects of silver nanoparticles on DNA damage and DNA repair gene expression in Ogg1-deficient and wild type mice. *Nanotoxicology* **2017**, *11*, 996–1011. [CrossRef]

Review

Electrophoretic Deposition of Bioactive Glass Coatings for Bone Implant Applications: A Review

Richard Drevet [1,*], Joël Fauré [2] and Hicham Benhayoune [2]

[1] Department of Plasma Physics and Technology, Masaryk University, Kotlářská 2, CZ-61137 Brno, Czech Republic

[2] Institut de Thermique, Mécanique et Matériaux (ITheMM), EA 7548, Université de Reims Champagne-Ardenne (URCA), Bât. 6, Moulin de la Housse, BP 1039, 51687 Reims CEDEX 2, France; joel.faure@univ-reims.fr (J.F.); hicham.benhayoune@univ-reims.fr (H.B.)

* Correspondence: richarddrevet@yahoo.fr

Abstract: This literature review deals with the electrophoretic deposition of bioactive glass coatings on metallic substrates to produce bone implants. Biocompatible metallic materials, such as titanium alloys or stainless steels, are commonly used to replace hard tissue functions because their mechanical properties are appropriate for load-bearing applications. However, metallic materials barely react in the body. They need a bioactive surface coating to trigger beneficial biological and chemical reactions in the physiological environment. Bioactive coatings aim to improve bone bonding, shorten the healing process after implantation, and extend the lifespan of the implant. Bioactive glasses, such as 45S5, 58S, S53P4, 13-93, or 70S30C, are amorphous materials made of a mixture of oxides that are accepted by the human body. They are used as coatings to improve the surface reactivity of metallic bone implants. Their high bioactivity in the physiological environment induces the formation of strong chemical bonding at the interface between the metallic implant and the surrounding bone tissue. Electrophoretic deposition is one of the most effective solutions to deposit uniform bioactive glass coatings at low temperatures. This article begins with a review of the different compositions of bioactive glasses described in the scientific literature for their ability to support hard tissue repair. The second part details the different stages of the bioactivity process occurring at the surface of bioactive glasses immersed in a physiological environment. Then, the mechanisms involved in the electrophoretic deposition of bioactive glass coatings on metallic bone implants are described. The last part of the article details the current developments in the process of improving the properties of bioactive glass coatings by adding biocompatible elements to the glassy structure.

Keywords: electrophoretic deposition (EPD); bioactive glass; coating; bone implant; biocompatibility; bioactivity; hard tissue repair

Citation: Drevet, R.; Fauré, J.; Benhayoune, H. Electrophoretic Deposition of Bioactive Glass Coatings for Bone Implant Applications: A Review. *Coatings* **2024**, *14*, 1084. https://doi.org/10.3390/coatings14091084

Academic Editor: Jun-Beom Park

Received: 16 July 2024
Revised: 15 August 2024
Accepted: 21 August 2024
Published: 23 August 2024

1. Introduction

The clinical demand for skeletal repair is constantly increasing due to the aging of the world population [1–8]. Metallic bone implants are commonly used to replace deficient hard tissues of the body because their mechanical properties are appropriate for load-bearing applications [9]. The main metallic materials used to produce bone implants are titanium alloys [10–13], iron-based alloys and stainless steels [14–21], tantalum [22–30], and cobalt chromium alloys [31–38]. They are biocompatible with the human body, meaning they do not cause any adverse effects when they are implanted [39–45]. The mechanical properties of these metallic materials make them relevant in replacing hard tissues, but they barely react with the physiological environment. They need a coating to make their surface bioactive, i.e., able to trigger biological activity and chemical reactions, promoting the formation of an adherent and strong bond with bone tissue (Figure 1). Bioactive coatings deposited on the surface of metallic implants promote long-term stability in the body and prevent or delay revision surgery [46–52].

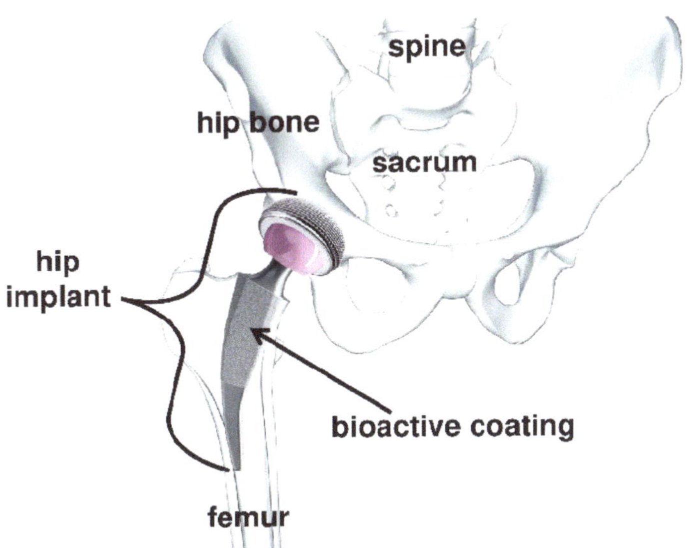

Figure 1. Bioactive coating deposited on titanium hip implant. Reprinted with permission from [52].

Bioactive glasses are the most reactive materials in a physiological environment. The idea to use them for hard tissue repair was introduced first by Larry Hench in 1969, who produced the famous 45S5 Bioglass® by mixing the four oxides SiO_2, CaO, Na_2O, and P_2O_5 [53–55]. In the following years, the ability of 45S5 Bioglass® to bond to bone tissues was demonstrated in vivo with several animal experiments [56–59]. For more than five decades, the bioactivities of glasses of other compositions have been deeply studied, and the idea of using them as coatings deposited on metallic bone implants has grown rapidly. Bioactive glass coatings are commonly described as an alternative to calcium phosphate coatings [60–70] because of their high reactivity in physiological environments.

Several methods have been explored to produce bioactive glass coatings on bone implants, such as plasma spraying [71–76], laser cladding [77–83], magnetron sputtering [84–94], sol–gel process combined with dip or spin coating [95–103], or electrophoretic deposition [104–110]. Among these deposition methods, electrophoretic deposition is attracting increasing attention from industrial and academic researchers because the process produces uniform coatings at low temperatures using quite inexpensive and easy-to-implement equipment. The process requires a stable colloidal suspension of bioactive glass particles, the surface of which is electrically charged when it is in contact with the solution. Under the influence of an electric field, the particles move through the liquid toward an oppositely charged electrode, where they progressively accumulate to form a coating [111–113]. To better understand the mechanisms involved in the process, this article reviews the scientific literature dealing with the electrophoretic deposition of bioactive glass coatings for bone implant applications. The main bioactive glass materials typically used for bone repair are presented first (Table 1). The second section describes the different stages of the bioactivity process occurring when a coated bone implant is immersed in a physiological environment. The third section provides a detailed description of the mechanisms involved in the electrophoretic deposition process. The last section reviews the current developments in the process of improving the properties of bioactive glass coatings by adding biocompatible elements to the glassy structure.

Table 1. Chemical composition of bioactive glasses.

	SiO_2	CaO	Na_2O	P_2O_5	K_2O	MgO	B_2O_3	Reference
45S5 Bioglass®	45 wt.%	24.5 wt.%	24.5 wt.%	6 wt.%	-	-	-	[53–59]
58S	58 wt.%	33 wt.%	-	9 wt.%	-	-	-	[114–117]
77S	77 wt.%	14 wt.%	–	9 wt.%	-	-	-	[118–121]
S53P4	53 wt.%	20 wt.%	23 wt.%	4 wt.%	-	-	-	[122–124]
13-93	53 wt.%	20 wt.%	6 wt.%	4 wt.%	12 wt.%	5 wt.%	-	[125–127]
13-93B1	34 wt.%	20 wt.%	6 wt.%	4 wt.%	12 wt.%	5 wt.%	19 wt.%	[128,129]
13-93B3	-	20 wt.%	6 wt.%	4 wt.%	12 wt.%	5 wt.%	53 wt.%	[129,130]
70S30C	71 wt.%	29 wt.%	-	-	-	-	-	[131–134]

2. Bioactive Glasses

2.1. Bioglass®

Larry Hench's 45S5 Bioglass® was the first bioactive glass produced in the history of biomaterials. This compound was synthesized for the first time in 1969 using the melt-quenching technique at temperatures above 1300 °C [53–55]. The 45S5 Bioglass® chemical composition was obtained by mixing 45 wt.% SiO_2, 24.5 wt.% CaO, 24.5 wt.% Na_2O, and 6 wt.% P_2O_5. In the 1990s, an alternative method was developed to produce 45S5 Bioglass® at low temperatures, the sol–gel process [135–142]. This synthesis method uses a solution of alkoxide precursors, such as tetraethyl orthosilicate (TEOS), typically used to obtain SiO_2. Under appropriate pH conditions, the solution is converted to a sol (colloidal solution) and then to a gel (solid network expended by a liquid phase) via hydrolysis and polycondensation reactions, respectively. 45S5 Bioglass® powder is obtained by drying the gel. Sol–gel 45S5 Bioglass® has a higher porosity and surface area, whereas melt-quenched 45S5 Bioglass® has higher mechanical properties.

The chemical composition of Larry Hench's 45S5 Bioglass® is still the gold standard today. The first in vivo experiments after the discovery showed strong bonding to rat femurs after 6 weeks of implantation. The implanted pieces of 45S5 Bioglass® had become impossible to move or remove from the rat bone [56–59]. These experimental results were rapidly confirmed in other animals, showing the ability of 45S5 Bioglass® to bond both to hard and soft tissues [47,143,144]. Several concentrations of the four oxides were studied, but their bonding abilities to bone and soft tissues were limited at specific regions represented in Hench's diagram (Figure 2) [145–147]. The diagram gathers all the compositions of quaternary glasses containing different concentrations of SiO_2, CaO, and Na_2O, with the concentration of P_2O_5 being kept constant at 6 wt.%. The bone-bonding ability of bioactive glasses is observed in region A. The soft tissue bonding ability of bioactive glasses is observed in region S. The composition of 45S5 Bioglass® corresponds to region E. Glasses produced with compositions in regions B, C, and D do not bond to bone [128,148,149].

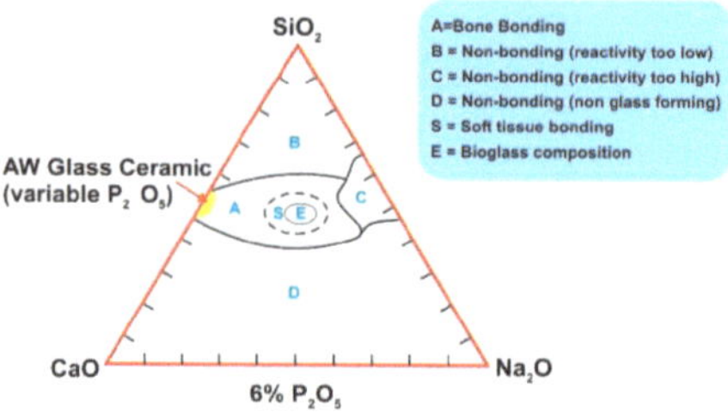

Figure 2. Hench's diagram. Compositional dependence (in weight percent) of glasses for bone bonding and soft tissue bonding. Reprinted with permission from [145].

Larry Hench's 45S5 Bioglass® is a phosphosilicate glass in which SiO_2 and P_2O_5 are network formers and CaO and Na_2O are network modifiers (Figure 3). Each silicon atom of the glassy network is covalently bonded to four oxygen atoms in a tetrahedron structure [150,151]. The whole SiO_2 structure is disrupted by CaO and Na_2O, two network modifiers. They introduce non-bridging oxygen bonds that promote the dissolution of 45S5 Bioglass® in aqueous environments [152–154]. The network former P_2O_5 is not mandatory to make the glass bioactive, but it acts as a nucleation site for the crystallization of apatite during the bioactivity process (see Section 3).

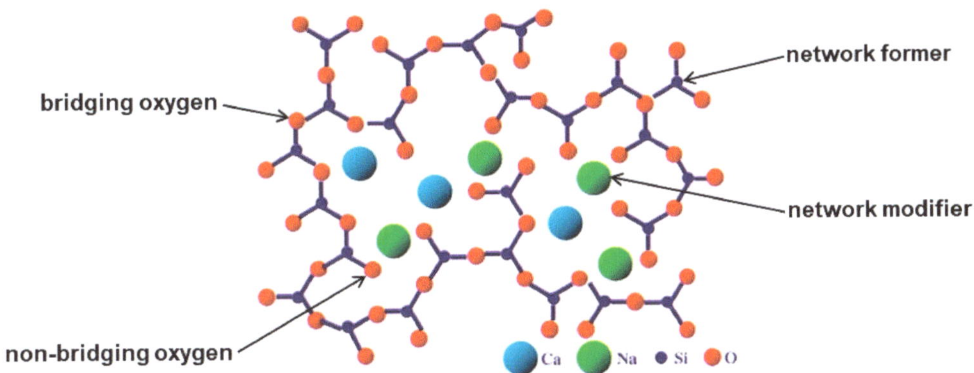

Figure 3. Structure of glass in SiO_2-CaO-Na_2O system. Reprinted and adapted with permission from [150].

2.2. Other Bioactive Glasses

In addition to Larry Hench's 45S5 Bioglass®, the compositions of other phosphosilicate glasses have been explored to evaluate their bioactive properties [155–161]. At a higher SiO_2 concentration, 58S is the second-most studied bioactive glass in academic research [114–117]. In comparison to 45S5 Bioglass®, there is no Na_2O in 58S, meaning that CaO is the only oxide acting as a network modifier of the structure. Another major bioactive glass is S53P4, widely used clinically for more than three decades. S53P4 supports bone repair and simultaneously inhibits bacterial growth [122–124]. A more complex compound is the bioactive glass 13-93, which is made of six different oxides. In addition to the four main oxides used to produce 45S5 Bioglass®, 13-93 includes K_2O and MgO, which are both network modifiers of the glassy structure [125–127]. Several articles also describe bioactive phosphosilicate glasses of less usual compositions such as 60S, 65S, 75S, 77S, 80S, and 91S [118–121].

The addition of boric oxide (B_2O_3) to bioactive glasses is also well documented in the literature [129]. B_2O_3 acts as a network former in the glassy structure as SiO_2 or P_2O_5. Several studies have shown that boron possesses anti-inflammatory properties and stimulates angiogenesis in the body environment. Angiogenesis is the physiological process of producing new blood vessels that supply nutrients to bone cells during bone tissue repair [162,163]. The most usual compositions of borate bioactive glasses are 13-93B1, 13-93B3, 45S5B5, and HB5 [130,164–167].

A simpler bioactive glass containing only two compounds is 70S30C [131–134]. This bioactive glass contains 70 mol.% SiO_2 as a network former of the glassy structure, and 30 mol.% CaO as a network modifier, corresponding to 71 wt.% SiO_2 and 29 wt.% CaO.

3. Bioactivity of Glasses

Discovering bioactive glasses in 1969, Larry Hench introduced the concept of bioactivity and defined it as follows: "A bioactive material is one that elicits a specific biological response at the interface of the material which results in the formation of a bond between the tissues and the material" [53,146]. The bone-bonding ability of bioactive glasses results from the formation of a hydroxycarbonate apatite (HCA) layer on the surface of the glass. The precipitation of this layer is triggered by local supersaturation due to ionic releases from the partial dissolution of the glass under physiological conditions. This chemical process occurs in five stages [47,145,147]:

- Stage 1: Rapid ion exchange between the glass network modifiers (Na^+ and Ca^{2+}) with H^+ ions (or H_3O^+) from the solution leads to the hydrolysis of the silica groups and the creation of silanol (Si–OH) groups on the glass surface:

$$Si\text{–}O\text{–}Na^+ + H^+ + OH^- \rightarrow Si\text{–}OH^+ + Na^+(aq.) + OH^- \tag{1}$$

The pH of the solution increases due to the consumption of H^+ ions.

- Stage 2: The increase in the pH (or OH^- concentration) leads to the attack of the SiO_2 glass network, the dissolution of silica, in the form of silicic acid $Si(OH)_4$ into the solution, and the continuous formation of Si–OH groups on the glass surface:

$$Si\text{–}O\text{–}Si + H_2O \rightarrow Si\text{–}OH + OH\text{–}Si \tag{2}$$

- Stage 3: Condensation and repolymerization of an amorphous SiO_2-rich layer (typically 1 or 2 µm thick) occur on the surface of the glass depleted in Na^+ and Ca^{2+} by leaching:

$$Si\text{–}OH + OH\text{–}Si \rightarrow Si\text{–}O\text{–}Si + H_2O \tag{3}$$

- Stage 4: The migration of Ca^{2+} and PO_4^{3-} ions from the glass through the SiO_2-rich layer and from the solution leads to the formation of an amorphous calcium phosphate (ACP) layer on the surface of the SiO_2-rich layer.
- Stage 5: The amorphous calcium phosphate (APC) layer incorporates OH^- and CO_3^{2-} from the solution and crystallizes as an HCA layer.

Peitl et al. proposed the illustration in Figure 4 to schematically summarize the five chemical stages of the bioactivity process. This reaction sequence occurs when a bioactive phosphosilicate glass, like 45S5 Bioglass®, is immersed in simulated body fluid (SBF) [168].

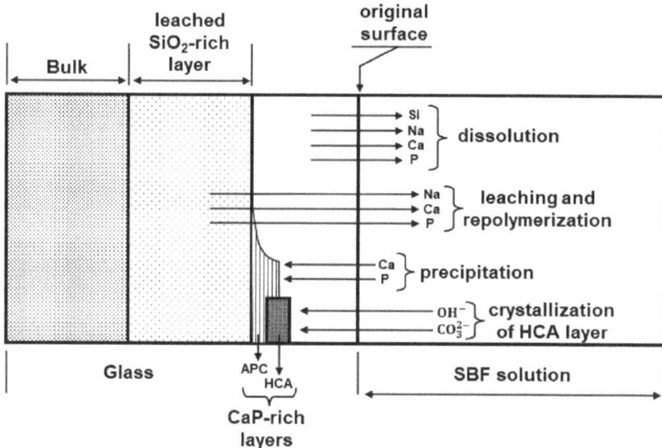

Figure 4. Schematic illustration of stages 1 to 5 of the bioactivity of a bioactive phosphosilicate glass immersed in simulated body fluid (SBF). Reproduced and adapted with permission from [168].

In addition to these five chemical stages resulting in the formation of the HCA layer, Larry Hench defined seven biological stages leading to the formation and growth of new bone tissue (Figure 5).

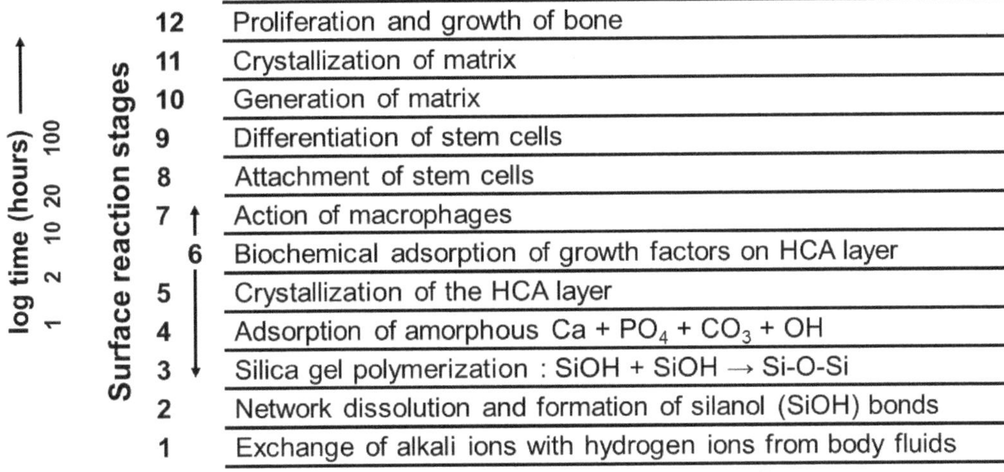

Figure 5. Sequence of interfacial reactions involved in forming a bond between tissue and bioactive glasses. Reprinted with permission from [47].

In stage 6, growth factors are adsorbed on the HCA layer where they activate M2 macrophages. Stage 7 corresponds to the action of M2 macrophages triggering the migration of mesenchymal stem cells and osteoprogenitor cells toward the bioactive glass surface, where they attach to the HCA layer (stage 8). Stem cells and osteoprogenitor cells differentiate in stage 9 to become osteogenic cells, which are precursors of osteoblast cells, like those present in bone tissue. The attachment and differentiation of osteoblasts generate extracellular matrix (ECM) components like type I collagen, enzymes, and glycoproteins (stage 10). Type I collagen is the most abundant protein in bones, representing more than 90% of the organic bone matrix [169].

In stage 11, the ECM is mineralized by hydroxyapatite nanocrystals, forming a three-dimensional network structure at the surface of the implant. The ECM provides structural and biochemical support to the surrounding bone cells, promoting their proliferation (stage 12). Bone growth is triggered at the implant surface, and the bioactivity process progressively continues degrading the bioactive glass coating, forming a greater ECM and supporting the development of new bone cells. The bioactivity process results in the osseointegration (or osteointegration) of the implant by creating an intimate connection with the surrounding bone tissue [170].

4. Electrophoretic Deposition (EPD)

Electrophoretic deposition of bioactive glass coatings on a metallic bone implant requires a stable colloidal suspension of bioactive glass particles in a solvent, typically water, ethanol, or a mixture of both [171,172]. When in contact with the solution, the surface of the particles becomes electrically charged due to electrostatic interactions with the ionic species of the solution. This surface charge maintains the stability of the colloidal suspension due to electrostatic interactions between the particles [173,174]. If two conductive electrodes connected to a generator are immersed in this colloidal suspension, the particles can be set in motion by applying an electric field between the electrodes. The particles move through the liquid in the direction of the oppositely charged electrode, where they progressively agglomerate to form a coating. If the surface of the particles is positively charged, they move toward and agglomerate on the cathode (cathodic EPD in Figure 6a). If the surface of the particles is negatively charged, they move toward and agglomerate on the anode (anodic EPD in Figure 6b).

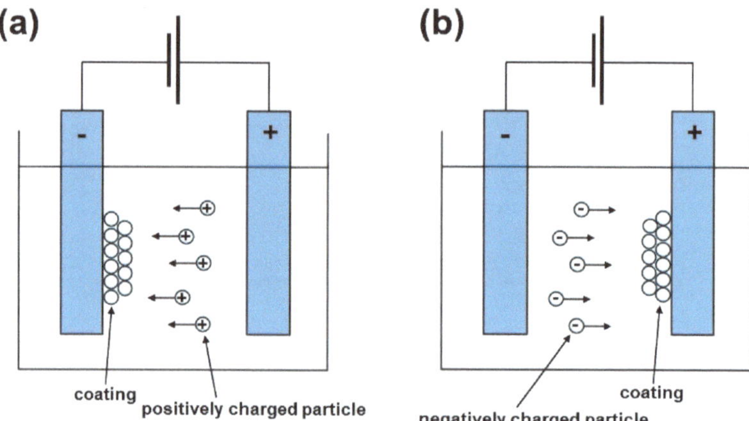

Figure 6. Schematics of (**a**) cathodic and (**b**) anodic electrophoretic deposition.

After deposition, thermal annealing is necessary to evaporate the solvent and improve the cohesive and adhesive properties of the bioactive glass coating [175]. The experimental parameters influencing the electrophoretic deposition of bioactive glass coatings are the pH and stability of the colloidal suspension, the dielectric constant (ε) and viscosity (η) of the solvent, the average particle size, substrate conductivity, electric field and distance between the two electrodes, and the deposition time.

4.1. Surface Charge Formation

When bioactive glass powder is immersed into an aqueous solution, chemical interactions occur on the surface of each powder particle and induce the formation of a net surface charge [176,177]. The main surface phenomena involved in the process are the chemisorption of water molecules, the formation of silanol groups, protonation or deprotonation of the silanol groups, and ionic adsorption.

In an aqueous solution, chemisorbed water readily hydrolyzes the silica groups at the surface of the bioactive glass and creates silanol (Si–OH) groups (Figure 7) similarly to the descriptions of stages 1 and 2 of the bioactivity process (see Section 3).

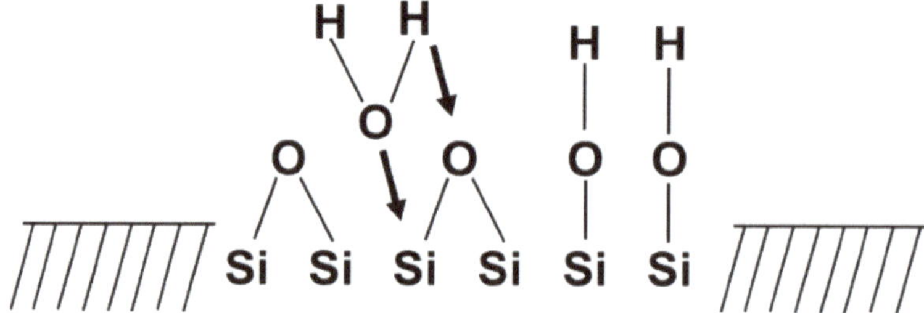

Figure 7. Chemisorption of water and formation of silanol groups at the surface of a bioactive glass particle. Reprinted and adapted with permission from [176].

Water molecules of the aqueous solution react with the silanol groups, resulting in either protonation (Reaction 4) or deprotonation (Reaction 5).

$$\text{Si–OH} + \text{H}_2\text{O} \rightarrow \text{Si–OH}_2^+ + \text{OH}^- \tag{4}$$

$$Si–OH + H_2O \rightarrow Si–O^- + H_3O^+ \tag{5}$$

Lowe et al. deeply described the acid-base dissociation mechanisms and energetics at the silica-water interface [178]. They used ab initio calculations to demonstrate that the deprotonation of the silanol groups (Reaction 5) is the main reaction at pH values above 2. The protonation of the silanol groups according to Reaction 4 occurs only at a very low pH value (below 2), which is an unusual experimental condition for electrophoretic deposition. Consequently, in regular experimental conditions, the surface of a bioactive glass powder immersed in water becomes negatively charged due to the deprotonation of silanol groups.

This negatively charged surface influences the ionic distribution in the vicinity of the powder particle immersed in a polar medium like water. Positive ions (cations) of the solution are attracted toward the particle surface and remain attached to it (Figure 8).

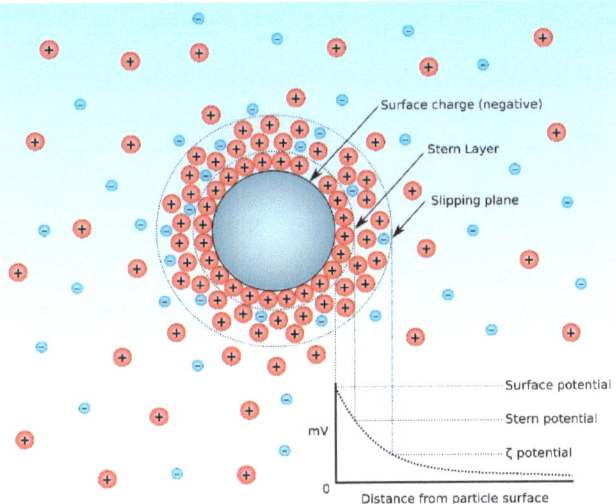

Figure 8. Schematic of the ionic distribution around a negatively charged bioactive glass particle suspended in water. Reprinted with permission from [179]. Copyright Mjones1984.

These ions form the Stern layer of the electrical double-layer model in which positive counterions remain immobile on the particle surface [180]. The second layer of the double-layer model is called the "diffuse layer" and contains cations and anions in electrostatic interaction with the whole system. The diffuse layer ends with the shear slipping plane, the border with the bulk solution from which the charged particle has no more electrostatic influence. The electrical potential at this slipping plane is called the zeta potential (ζ), which is characteristic of the strength of all the charges carried by the particle [181].

4.2. Suspension Stability

The stability of a colloidal suspension depends on the strength of the electrostatic repulsions between the particles. The value of the zeta potential (ζ) must be high enough to prevent coagulation or flocculation of the particles [182]. Zeta potential values higher than +30 mV or lower than −30 mV are generally considered appropriate to obtain a stable suspension [183]. However, the zeta potential of the particles is highly influenced by the pH of the colloidal suspension (Figure 9). The highest values are observed at a low or high pH, whereas the value is zero at the isoelectric point, where the stability of the suspension is at the lowest.

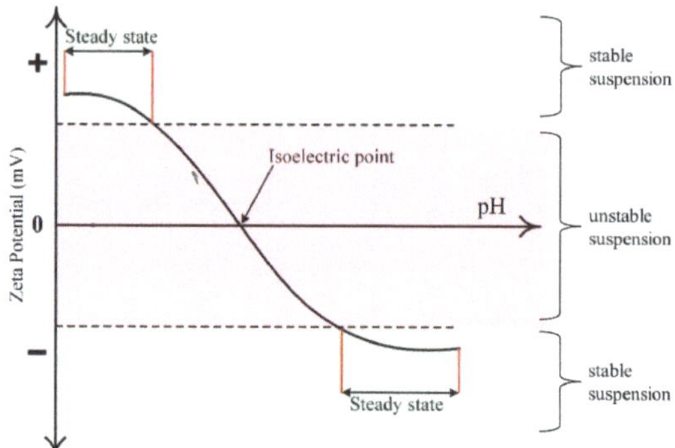

Figure 9. Zeta potential variation as a function of pH in a colloidal suspension. Reprinted and adapted with permission from [183].

4.3. Deposition Mechanism of Bioactive Glass Coating

Thanks to their zeta potential (ζ), the bioactive glass particles of the colloidal suspension can be set in motion by the electric field generated between two conductive electrodes. The particles move toward the oppositely charged electrode [184]. Under the influence of the electric field, the diffuse layer surrounding each bioactive glass particle is distorted and thinned (Figure 10). Some of the counterions of the diffuse layer leave the ion cloud because they are attracted by the other electrode in the opposite direction. Consequently, the zeta potential of each particle progressively decreases on their way toward the electrode surface, and the electrostatic repulsions between the particles are reduced [185]. Reaching the electrode at a high speed, the electric charges of the particles are neutralized, and they progressively accumulate via coagulation to form a uniform bioactive glass coating.

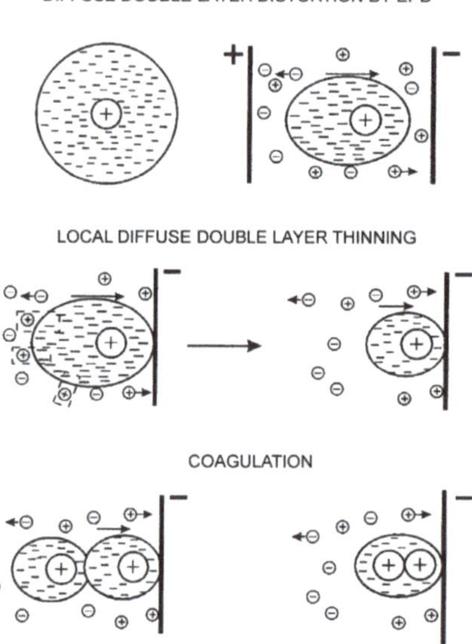

Figure 10. Mechanisms of electrophoretic deposition proposed by Sarkar et al.: distortion and thinning of the double layer surrounding the particle and coagulation of the particles on the electrode surface. Reprinted with permission from [186].

Zhang et al. proposed the following relation to calculate the deposition rate of a coating via electrophoretic deposition [187]:

$$w = \frac{m_0 \, \mu_e SE}{V} \times e^{\left(\frac{-\mu_e SE}{V} t\right)}$$

where w ($kg \cdot s^{-1}$) is the deposition rate; m_0 (kg) is the initial mass of the powder in the suspension; V (m^3) is the volume of the colloidal suspension; μ_e ($m^2 \cdot V^{-1} \cdot s^{-1}$) is the electrophoretic mobility of the particles; S (m^2) is the surface of the coating; and E ($V \cdot m^{-1}$) is the electric field between the two electrodes.

5. Current Developments in the Process and Perspectives

The most recent developments in the process explore solutions to improve the properties of bioactive glass coatings by adding biocompatible elements to the glassy structure. Several strategies are studied, including ionic substitution, deposition of composite coatings, and drug loading.

5.1. Ionic Substitution

As described in Section 2, the most common elements used to produce bioactive glasses are silicon, phosphorus, calcium, and sodium. Their biological properties can be enhanced by adding other ions to the glassy structure of the powder used in the electrophoretic deposition process. The purpose is to take advantage of the dissolution stage of the bioactivity process (see Section 3) to release biologically active ions into the physiological environment after implantation. The most usual ionic substitutions described in the literature involve potassium (K^+), magnesium (Mg^{2+}), zinc (Zn^{2+}), silver (Ag^+), strontium (Sr^{2+}), fluorine (F^-), cobalt (Co^{2+}), copper (Cu^{2+}), and boron (B^{3+}). These ionic substitutions aim to provide osteogenic, angiogenic, anti-inflammatory, or antibacterial properties to the bioactive glass material [188,189]. The quantity of substituting element is generally low, typically a few percent of the whole mixture. Multi-substitution can be used to cumulate several biological enhancements of the bone implant [190,191]. More recently, less common elements have been used to substitute bioactive glass materials. They have been comprehensively reviewed by Pantulap et al., who thoroughly described the effect of these ions on the biological and physical properties of bioactive glasses [192]. They enhance hard and soft tissue repairs, and some of them provide extra functionalities such as anti-inflammatory, antibacterial, or anticancer properties, fluorescence, luminescence, and radiation shielding. Examples of substituting ions and their biological, physical, or chemical effects are reported in Table 2.

Table 2. Ionic substitution in bioactive glasses.

Ions	Biological/Physical Effect	References
Monovalent Ions		
Ag^+	Antibacterial activity	[193–196]
Cl^-	Osteogenesis	[197–199]
F^-	Osteogenesis	[200–202]
K^+	Osteogenesis	[203–205]
Li^+	Antibacterial/osteogenesis	[206–209]
Rb^+	Antibacterial/drug carrier/osteogenesis	[210,211]
Divalent Ions		
Ba^{2+}	Osteogenesis/radiation shielding	[212–215]

Table 2. *Cont.*

Ions	Biological/Physical Effect	References
Co^{2+}	Angiogenesis	[216–219]
Cu^{2+}	Antibacterial activity	[220–223]
Ge^{2+}	Osteogenesis/radiation shielding	[224,225]
Mg^{2+}	Osteogenesis	[226–229]
Mn^{2+}	Antibacterial/osteogenesis	[230–233]
Ni^{2+}	Drug carrier/osteogenesis/radiation shielding	[234–236]
Sr^{2+}	Osteogenesis	[237–241]
Zn^{2+}	Antibacterial/anti-inflammatory/osteogenesis	[242–245]
Trivalent Ions		
Bi^{3+}	Antibacterial/anticancer/osteogenesis	[246–248]
Ce^{3+}	Osteogenesis/antibacterial	[249–252]
Er^{3+}	Osteogenesis/photoluminescence	[253,254]
Eu^{3+}	Drug carrier/photoluminescence	[255–257]
Fe^{3+}	Antibacterial/anticancer/osteogenesis	[258–261]
Ga^{3+}	Antibacterial/anticancer/osteogenesis	[262–265]
Ho^{3+}	Brachytherapy/osteogenesis	[266,267]
La^{3+}	Osteogenesis	[268–270]
N^{3-}	Osteogenesis	[271–273]
Sm^{3+}	Drug carrier/osteogenesis/photoluminescence	[274,275]
Tb^{3+}	Osteogenesis/photoluminescence	[276,277]
Y^{3+}	Anticancer/osteogenesis	[278–280]
Yb^{3+}	Osteogenesis/photoluminescence	[254,281]
Tetravalent Ions		
Se^{4+}	Anticancer/drug carrier/osteogenesis	[282–284]
Te^{4+}	Antibacterial/anticancer/osteogenesis	[285,286]
Zr^{4+}	Antibacterial/osteogenesis	[287–289]
Pentavalent Ions		
Nb^{5+}	Osteogenesis	[290–293]
Ta^{5+}	Antibacterial/osteogenesis	[294–297]
V^{5+}	Osteogenesis/photoluminescence/radiation shielding	[298–300]
Hexavalent Ions		
Mo^{6+}	Drug carrier/osteogenesis	[301,302]

5.2. Composite Coatings

Electrophoretic deposition is a very versatile method. Any powder material added to the colloidal suspension can be simultaneously deposited with bioactive glass to produce a composite coating with improved properties [303]. The only requirement is the sign of all the zeta potentials be the same to move all the materials in the same direction and to deposit them on the same electrode. Similar zeta potential values are necessary to maintain the stability of the colloidal suspension. The main materials used to deposit composite coatings with bioactive glasses are bioceramics, polymers, and carbon nanotubes.

5.2.1. Bioactive Glass and Bioceramics

Several studies have reported on the simultaneous electrophoretic deposition of bioceramics with bioactive glasses. Bioceramics are biocompatible materials that are accepted by the human body without producing any adverse effect [39–41]. Some of them are bioinert, meaning they do not trigger any biological or chemical reaction with the surrounding environment [45,46]. They are typically combined with bioactive glass to improve the mechanical properties of the coating, such as its hardness, Young's modulus, fracture toughness, bending strength, compressive strength, and wear resistance [304]. The most common bioinert ceramics used in composite coatings are Al_2O_3 [305], TiO_2 [306], ZrO_2 [307], ZnO [308], and Si_3N_4 [309]. Other bioceramics are bioactive or bioresorbable materials, meaning they trigger chemical and biological reactions in the physiological environment that promote bone growth and repair [46–48]. The main bioactive or bioresorbable bioceramics are calcium phosphates (hydroxyapatite, β-tricalcium phosphate, octacalium phosphate, and dicalcium phosphate dihydrate) [310], wollastonite ($CaSiO_3$) [311], calcium sulfate ($CaSO_4$) [312], and calcium carbonate ($CaCO_3$) [313]. Their bioactive behavior is linked to their solubility in physiological solutions, which is typically lower than the bioactivity of bioactive glasses. The kinetics of the bioactivity process can be fully controlled by combining two bioactive materials of different solubilities.

5.2.2. Bioactive Glass and Polymers

Polymers are used to improve the mechanical properties of bioactive glass coatings, specifically their hardness and adhesion to the bone implant [314]. They make the coating composition and structure closer to that of bone tissue, which is a composite material of collagen and apatite nanocrystals (calcium phosphate) [5,303]. Typical biocompatible polymers used to produce composite coatings are Polyether ether ketone (PEEK) [315], Poly(lactic-co-glycolic acid) (PLGA) [316], Polycaprolactone (PCL) [317], Poly(vinyl alcohol) (PVA) [318], Polymethyl methacrylate (PMMA) [319], and Poly(styrene-alt-maleic acid) (PSM) [320]. Natural biopolymers such as collagen [321], hyaluronic acid [322], alginate [323], chondroitin sulfate [324], zein [325], and chitosan [326] have also been described to produce composite coatings with improved mechanical and biological properties.

5.2.3. Bioactive Glass and Carbon Nanotubes (CNTs)

Carbon nanotubes are biocompatible, and they have high mechanical strength and flexibility [327]. They are combined with bioactive glass in the electrophoretic deposition process to produce composite coatings with improved mechanical properties and bioactivity [328–330].

5.3. Drug-Loaded Bioactive Glass Coatings

Drugs can be incorporated into bioactive glass coatings produced via electrophoretic deposition to achieve specific biological effects. The partial dissolution of the coating during the bioactivity process can be used to release active substances in the body after implantation. Drugs providing anti-inflammatory and antibacterial properties are mainly used. Several research studies succeeded in adding heparin [331], gentamicin [332], vancomycin [333], ibuprofen [334], ferulic acid [335], tetracycline hydrochloride [336], doxycycline [337], ampicillin [338], or lawsone [339].

6. Conclusions

This article reviewed the scientific literature dealing the electrophoretic deposition of bioactive glass coatings on metallic substrates to produce bone implants. After chemical and structural descriptions of Larry Hench's 45S5 Bioglass® and other bioactive glasses, the different stages of the bioactivity process were explained. According to the model proposed by Larry Hench, bioactive glass coatings immersed in a physiological environment trigger five chemical stages followed by seven biological stages, resulting in the formation and growth of new bone tissue. Then, the mechanisms involved in the electrophoretic deposition process were detailed. The surface of bioactive glass particles of a colloidal suspension becomes electrically charged when it is in contact with the solution. Under the influence of an electric field between two electrodes, the particles are set in motion and move toward an oppositely charged electrode, where they progressively accumulate to form a coating. Finally, the last part of this article reviewed the current developments of the process. The most recent solutions to improve the properties of bioactive glass coatings include the addition of biocompatible elements to the glassy structure, including ionic substitution, deposition of composite coatings, and drug loading.

Author Contributions: Conceptualization, R.D., J.F. and H.B.; validation, R.D., J.F. and H.B.; resources, R.D., J.F. and H.B.; writing—original draft preparation, R.D., J.F. and H.B.; writing—review and editing, R.D., J.F. and H.B. All authors have read and agreed to the published version of the manuscript.

Funding: This research received no external funding.

Institutional Review Board Statement: Not applicable.

Informed Consent Statement: Not applicable.

Data Availability Statement: Not applicable.

Conflicts of Interest: The authors declare no conflicts of interest.

References

1. Ageing and Health. Available online: https://www.who.int/news-room/fact-sheets/detail/ageing-and-health (accessed on 1 July 2024).
2. Demontiero, O.; Vidal, C.; Duque, G. Aging and bone loss: New insights for the clinician. *Ther. Adv. Musculoskel. Dis.* **2012**, *4*, 61–76. [CrossRef] [PubMed]
3. Gheno, R.; Cepparo, J.M.; Rosca, C.E.; Cotton, A. Musculoskeletal Disorders in the Elderly. *J. Clin. Imaging Sci.* **2012**, *2*, 39. [CrossRef]
4. Li, G.; Thabane, L.; Papaioannou, A.; Ioannidis, G.; Levine, M.A.H.; Adachi, J.D. An overview of osteoporosis and frailty in the elderly. *BMC Musculoskelet. Disord.* **2017**, *18*, 46. [CrossRef]
5. Boskey, A.L.; Coleman, R. Aging and Bone. *J. Dent. Res.* **2010**, *89*, 1333–1348. [CrossRef] [PubMed]
6. Montoya, C.; Du, Y.; Gianforcaro, A.L.; Orrego, S.; Yang, M.; Lelkes, P.I. On the road to smart biomaterials for bone research: Definitions, concepts, advances, and outlook. *Bone Res.* **2021**, *9*, 12. [CrossRef] [PubMed]
7. Marie, P.J. Bone Cell Senescence: Mechanisms and Perspectives. *J. Bone Miner. Res.* **2014**, *29*, 1311–1321. [CrossRef]
8. Drevet, R.; Benhayoune, H. Biomaterials Design for Human Body Repair. *Designs* **2024**, *8*, 65. [CrossRef]
9. Chen, Q.; Thouas, G.A. Metallic implant biomaterials. *Mater. Sci. Eng. R* **2015**, *87*, 1–57. [CrossRef]
10. He, G.; Hagiwara, M. Ti alloy design strategy for biomedical applications. *Mater. Sci. Eng. C* **2006**, *26*, 14–19. [CrossRef]
11. Geetha, M.; Singh, A.K.; Asokamani, R.; Gogia, A.K. Ti based biomaterials, the ultimate choice for orthopaedic implants—A review. *Prog. Mater. Sci.* **2009**, *54*, 397–425. [CrossRef]
12. Ijaz, M.F.; Laillé, D.; Héraud, L.; Gordin, D.M.; Castany, P.; Gloriant, T. Design of a novel superelastic Ti-23Hf-3Mo-4Sn biomedical alloy combining low modulus, high strength and large recovery strain. *Mater. Lett.* **2016**, *177*, 39–41. [CrossRef]
13. Sheremetyev, V.; Lukashevich, L.; Kreitcberg, A.; Kudryashova, A.; Tsaturyants, M.; Galkin, G.; Andreev, V.; Prokoshkin, S.; Brailovski, V. Optimization of a thermomechanical treatment of superelastic Ti-Zr-Nb alloys for the production of bar stock for orthopedic implants. *J. Alloys Compd.* **2022**, *928*, 167143. [CrossRef]
14. Drevet, R.; Zhukova, Y.; Malikova, P.; Dubinskiy, S.; Korotitskiy, A.; Pustov, Y.; Prokoshkin, S. Martensitic Transformations and Mechanical and Corrosion Properties of Fe-Mn-Si Alloys for Biodegradable Medical Implants. *Metall. Mater. Trans. A* **2018**, *49*, 1006–1013. [CrossRef]
15. Drevet, R.; Zhukova, Y.; Kadirov, P.; Dubinskiy, S.; Kazakbiev, A.; Pustov, Y.; Prokoshkin, S. Tunable corrosion behavior of calcium phosphate coated Fe-Mn-Si alloys for bone implant applications. *Metall. Mater. Trans. A* **2018**, *49*, 6553–6560. [CrossRef]

16. Kadirov, P.; Zhukova, Y.; Pustov, Y.; Karavaeva, M.; Sheremetyev, V.; Korotitskiy, A.; Shcherbakova, E.; Baranova, A.; Komarov, V.; Prokoshkin, S. Effect of Plastic Deformation in Various Temperature-Rate Conditions on Structure and Mechanical Properties of Biodegradable Fe-30Mn-5Si Alloy. *Metall. Mater. Trans. A* **2024**, *55*, 895–909. [CrossRef]
17. Koumya, Y.; Ait Salam, Y.; Khadiri, M.E.; Benzakour, J.; Romane, A.; Abouelfida, A.; Benyaich, A. Pitting corrosion behavior of SS-316L in simulated body fluid and electrochemically assisted deposition of hydroxyapatite coating. *Chem. Pap.* **2021**, *75*, 2667–2682. [CrossRef]
18. Trincă, L.C.; Burtan, L.; Mareci, D.; Fernández-Pérez, B.M.; Stoleriu, I.; Stanciu, T.; Stanciu, S.; Solcan, C.; Izquierdo, J.; Souto, R.M. Evaluation of in vitro corrosion resistance and in vivo osseointegration properties of a FeMnSiCa alloy as potential degradable implant biomaterial. *Mater. Sci. Eng. C* **2021**, *118*, 111436. [CrossRef]
19. Sultana, N.; Nishina, Y.; Nizami, M.Z.I. Surface Modifications of Medical Grade Stainless Steel. *Coatings* **2024**, *14*, 248. [CrossRef]
20. Bekmurzayeva, A.; Duncanson, W.J.; Azevedo, H.S.; Kanayeva, D. Surface modification of stainless steel for biomedical applications: Revisiting a century-old material. *Mater. Sci. Eng. C* **2018**, *93*, 1073–1089. [CrossRef]
21. Xiao, M.; Chen, Y.M.; Biao, M.N.; Zhang, X.D.; Yang, B.C. Bio-functionalization of biomedical metals. *Mater. Sci. Eng. C* **2017**, *70*, 1057–1070. [CrossRef]
22. Putrantyo, I.; Anilbhai, N.; Vanjani, R.; De Vega, B. Tantalum as a novel biomaterial for bone implant: A literature review. *J. Biomim. Biomater. Biomed. Eng.* **2021**, *52*, 55–65. [CrossRef]
23. Gao, H.; Yang, J.; Jin, X.; Qu, X.; Zhang, F.; Zhang, D.; Chen, H.; Wei, H.; Zhang, S.; Jia, W.; et al. Porous tantalum scaffolds: Fabrication, structure, properties, and orthopedic applications. *Mater. Des.* **2021**, *210*, 110095. [CrossRef]
24. Mahmoodi, M.; Hydari, M.H.; Mahmoodi, L.; Gazanfari, L.; Mirhaj, M. Electrophoretic deposition of graphene oxide reinforced hydroxyapatite on the tantalum substrate for bone implant applications: In vitro corrosion and bio-tribological behavior. *Surf. Coat. Technol.* **2021**, *424*, 127642. [CrossRef]
25. Rupérez, E.; Manero, J.M.; Riccardi, K.; Li, Y.; Aparicio, C.; Gil, F.J. Development of tantalum scaffold for orthopedic applications produced by space-holder method. *Mater. Des.* **2015**, *83*, 112–119. [CrossRef]
26. Luo, C.; Wang, C.; Wu, X.; Xie, X.; Wang, C.; Zhao, C.; Zou, C.; Lv, F.; Huang, W.; Liao, J. Influence of porous tantalum scaffold pore size on osteogenesis and osteointegration: A comprehensive study based on 3D-printing technology. *Mater. Sci. Eng. C* **2021**, *129*, 112382. [CrossRef] [PubMed]
27. Huang, G.; Pan, S.T.; Qiu, J.X. The Clinical Application of Porous Tantalum and Its New Development for Bone Tissue Engineering. *Materials* **2021**, *14*, 2647. [CrossRef] [PubMed]
28. Wang, X.; Ning, B.; Pei, X. Tantalum and its derivatives in orthopedic and dental implants: Osteogenesis and antibacterial properties. *Colloids Surf. B Biointerfaces* **2021**, *208*, 112055. [CrossRef]
29. Balla, V.K.; Bodhak, S.; Bose, S.; Bandyopadhyay, A. Porous tantalum structures for bone implants: Fabrication, mechanical and in vitro biological properties. *Acta Biomater.* **2010**, *6*, 3349–3359. [CrossRef]
30. Wang, X.; Zhu, Z.; Xiao, H.; Luo, C.; Luo, X.; Lv, F.; Liao, J.; Huang, W. Three-Dimensional, MultiScale, and Interconnected Trabecular Bone Mimic Porous Tantalum Scaffold for Bone Tissue Engineering. *ACS Omega* **2020**, *5*, 22520–22528. [CrossRef]
31. Tchana Nkonta, D.V.; Simescu-Lazar, F.; Drevet, R.; Aabouvi, O.; Fauré, J.; Retraint, D.; Benhayoune, H. Influence of the surface mechanical attrition treatment (SMAT) on the corrosion behavior of Co28Cr6Mo alloy in Ringer's solution. *J. Solid State Electrochem.* **2018**, *22*, 1091–1098. [CrossRef]
32. Chen, Y.; Li, Y.; Kurosu, S.; Yamanaka, K.; Tang, N.; Koizumi, Y.; Chiba, A. Effects of sigma phase and carbide on the wear behavior of CoCrMo alloys in Hanks' solution. *Wear* **2014**, *310*, 51–62. [CrossRef]
33. Tchana Nkonta, D.V.; Drevet, R.; Fauré, J.; Benhayoune, H. Effect of surface mechanical attrition treatment on the microstructure of cobalt-chromium-molybdenum biomedical alloy. *Microsc. Res. Tech.* **2021**, *84*, 238–245. [CrossRef] [PubMed]
34. AlMangour, B.; Luqman, M.; Grzesiak, D.; Al-Harbi, H.; Ijaz, F. Effect of processing parameters on the microstructure and mechanical properties of Co–Cr–Mo alloy fabricated by selective laser melting. *Mater. Sci. Eng. A* **2020**, *792*, 139456. [CrossRef]
35. Corona-Gomez, J.; Jack, T.A.; Feng, R.; Yang, Q. Wear and corrosion characteristics of nano-crystalline tantalum nitride coatings deposited on CoCrMo alloy for hip joint applications. *Mater. Charact.* **2021**, *182*, 111516. [CrossRef]
36. Yamanaka, K.; Mori, M.; Kurosu, S.; Matsumoto, H.; Chiba, A. Ultrafine grain refinement of biomedical Co-29Cr-6Mo alloy during conventional hot-compression deformation. *Metall. Mater. Trans. A* **2009**, *40*, 1980–1994. [CrossRef]
37. İbrahim Coşkun, M.; Karahan, İ.H.; Yücel, Y.; Golden, T.D. Optimization of electrochemical step deposition for bioceramic hydroxyapatite coatings on CoCrMo implants. *Surf. Coat. Technol.* **2016**, *301*, 42–53. [CrossRef]
38. Coşkun, M.I.; Karahan, I.H.; Yücel, Y. Optimized electrodeposition concentrations for hydroxyapatite coatings on CoCrMo biomedical alloys by computational techniques. *Electrochim. Acta* **2014**, *150*, 46–54. [CrossRef]
39. Ghasemi-Mobarakeh, L.; Kolahreez, D.; Ramakrishna, S.; Williams, D. Key terminology in biomaterials and biocompatibility. *Curr. Opin. Biomed. Eng.* **2019**, *10*, 45–50. [CrossRef]
40. Williams, D. Revisiting the definition of biocompatibility. *Med. Device Technol.* **2003**, *14*, 10–13. [PubMed]
41. Williams, D.F. On the mechanisms of biocompatibility. *Biomaterials* **2008**, *29*, 2941–2953. [CrossRef]
42. Barrère, F.; Mahmood, T.A.; de Groot, K.; van Blitterswijk, C.A. Advanced biomaterials for skeletal tissue regeneration: Instructive and smart functions. *Mater. Sci. Eng. R* **2008**, *59*, 38–71. [CrossRef]
43. Moniruzzaman, M.d.; O'Neal, C.; Bhuiyan, A.; Egan, P.F. Design and Mechanical Testing of 3D Printed Hierarchical Lattices Using Biocompatible Stereolithography. *Designs* **2020**, *4*, 22. [CrossRef]

44. Nuswantoro, N.F.; Rahandi Lubis, M.A.; Juliadmi, D.; Mardawati, E.; Antov, P.; Kristak, L.; Seng Hua, L. Bio-Based Adhesives for Orthopedic Applications: Sources, Preparation, Characterization, Challenges, and Future Perspectives. *Designs* **2022**, *6*, 96. [CrossRef]

45. Williams, D.F. Biocompatibility pathways and mechanisms for bioactive materials: The bioactivity zone. *Bioact. Mater.* **2022**, *10*, 306–322. [CrossRef]

46. Williams, D.F. On the nature of biomaterials. *Biomaterials* **2009**, *30*, 5897–5909. [CrossRef] [PubMed]

47. Cao, W.; Hench, L.L. Bioactive materials. *Ceram. Int.* **1996**, *22*, 493–507. [CrossRef]

48. Albrektsson, T.; Johansson, C. Osteoinduction, osteoconduction and osseointegration. *Eur. Spine J.* **2001**, *10*, S96–S101. [CrossRef]

49. Andreucci, C.A.; Fonseca, E.M.M.; Jorge, R.N. Bio-Lubricant Properties Analysis of Drilling an Innovative Design of Bioactive Kinetic Screw into Bone. *Designs* **2023**, *7*, 21. [CrossRef]

50. Shaikh, M.S.; Fareed, M.A.; Zafar, M.S. Bioactive Glass Applications in Different Periodontal Lesions: A Narrative Review. *Coatings* **2023**, *13*, 716. [CrossRef]

51. Ducheyne, P.; Qiu, Q. Bioactive ceramics: The effect of surface reactivity on bone formation and bone cell function. *Biomaterials* **1999**, *20*, 2287–2303. [CrossRef]

52. Drevet, R.; Benhayoune, H. Electrodeposition of Calcium Phosphate Coatings on Metallic Substrates for Bone Implant Applications: A Review. *Coatings* **2022**, *12*, 539. [CrossRef]

53. Hench, L.L. The story of Bioglass®. *J. Mater. Sci. Mater. Med.* **2006**, *17*, 967–978. [CrossRef]

54. Hench, L.L. Semiconducting glass-ceramics. *J. Non-Cryst. Solids* **1970**, *2*, 250–277. [CrossRef]

55. Hench, L.L.; Clark, A.E.; Schaake, H.F. Effects of microstructure on the radiation stability of amorphous semiconductors. *J. Non-Cryst. Solids* **1972**, *8–10*, 837–843. [CrossRef]

56. Hench, L.L.; Splinter, R.J.; Allen, W.C.; Greenlee, T.K. Bonding mechanisms at the interface of ceramic prosthetic materials. *J. Biomed. Mater. Res.* **1971**, *5*, 117–141. [CrossRef]

57. Hench, L.L.; Paschall, H.A. Direct chemical bond of bioactive glass-ceramic materials to bone and muscle. *J. Biomed. Mater. Res.* **1973**, *7*, 25–42. [CrossRef] [PubMed]

58. Hench, L.L.; Paschall, H.A. Histochemical responses at a biomaterial's interface. *J. Biomed. Mater. Res.* **1974**, *8*, 49–64. [CrossRef]

59. Piotrowski, G.; Hench, L.L.; Allen, W.C.; Miller, G.J. Mechanical studies of the bone bioglass interfacial bond. *J. Biomed. Mater. Res.* **1975**, *9*, 47–61. [CrossRef]

60. Paital, S.R.; Dahotre, N.B. Calcium phosphate coatings for bio-implant applications: Materials, performance factors, and methodologies. *Mater. Sci. Eng. R* **2009**, *66*, 1–70. [CrossRef]

61. Kharissova, O.V.; Nikolaev, A.L.; Kharisov, B.I.; Dorozhkin, S.V.; López, I.; Peña Méndez, Y.; Gómez de la Fuente, I. Enzymatic synthesis of calcium phosphates: A review. *Nano-Struct. Nano-Objects* **2024**, *39*, 101214. [CrossRef]

62. Vallet-Regí, M.; González-Calbet, J.M. Calcium phosphates as substitution of bone tissues. *Prog. Solid State Chem.* **2004**, *32*, 1–31. [CrossRef]

63. Dorozhkin, S.V. Calcium orthophosphate deposits: Preparation, properties and biomedical applications. *Mater. Sci. Eng. C* **2015**, *55*, 272–326. [CrossRef]

64. Heimann, R.B. Structural Changes of Hydroxylapatite during Plasma Spraying: Raman and NMR Spectroscopy Results. *Coatings* **2021**, *11*, 987. [CrossRef]

65. Dorozhkin, S.V. Calcium Orthophosphate (CaPO$_4$)-Based Bioceramics: Preparation, Properties, and Applications. *Coatings* **2022**, *12*, 1380. [CrossRef]

66. Drevet, R.; Fauré, J.; Benhayoune, H. Bioactive Calcium Phosphate Coatings for Bone Implant Applications: A Review. *Coatings* **2023**, *13*, 1091. [CrossRef]

67. Drevet, R.; Fauré, J.; Benhayoune, H. Calcium Phosphates and Bioactive Glasses for Bone Implant Applications. *Coatings* **2023**, *13*, 1217. [CrossRef]

68. Heimann, R.B. Plasma-Sprayed Osseoconductive Hydroxylapatite Coatings for Endoprosthetic Hip Implants: Phase Composition, Microstructure, Properties, and Biomedical Functions. *Coatings* **2024**, *14*, 787. [CrossRef]

69. Dorozhkin, S.V. There Are over 60Ways to Produce Biocompatible Calcium Orthophosphate (CaPO$_4$) Deposits on Various Substrates. *J. Compos. Sci.* **2023**, *7*, 273. [CrossRef]

70. LeGeros, R.Z. Calcium Phosphate-Based Osteoinductive Materials. *Chem. Rev.* **2008**, *108*, 4742–4753. [CrossRef]

71. Cañas, E.; Orts, M.J.; Boccaccini, A.R.; Sánchez, E. Microstructural and in vitro characterization of 45S5 bioactive glass coatings deposited by solution precursor plasma spraying (SPPS). *Surf. Coat. Technol.* **2019**, *371*, 151–160. [CrossRef]

72. Bellucci, D.; Bolelli, G.; Cannillo, V.; Gadow, R.; Killinger, A.; Lusvarghi, L.; Sola, A.; Stiegler, N. High velocity suspension flame sprayed (HVSFS) potassium-based bioactive glass coatings with and without TiO$_2$ bond coat. *Surf. Coat. Technol.* **2012**, *206*, 3857–3868. [CrossRef]

73. Cañas, E.; Orts, M.J.; Sánchez, E.; Bellucci, D.; Cannillo, V. Deposition of bioactive glass coatings based on a novel composition containing strontium and magnesium. *J. Eur. Ceram. Soc.* **2022**, *42*, 6213–6221. [CrossRef]

74. Cañas, E.; Rojas, O.; Orts, M.J.; Ageorges, H.; Sánchez, E. Effect of feedstock and plasma gun on the microstructure and bioactivity of plasma sprayed bioactive glass coatings. *Surf. Coat. Technol.* **2021**, *406*, 126704. [CrossRef]

75. Cannillo, V.; Sola, A. Different approaches to produce coatings with bioactive glasses: Enamelling vs. plasma spraying. *J. Eur. Ceram. Soc.* **2010**, *30*, 2031–2039. [CrossRef]

76. Monsalve, M.; Ageorges, H.; Lopez, E.; Vargas, F.; Bolivar, F. Bioactivity and mechanical properties of plasma-sprayed coatings of bioglass powders. *Surf. Coat. Technol.* **2013**, *220*, 60–66. [CrossRef]

77. Moritz, N.; Vedel, E.; Ylänen, H.; Jokinen, M.; Hupa, M.; Yli-Urpo, A. Characterisation of bioactive glass coatings on titanium substrates produced using a CO_2 laser. *J. Mater. Sci. Mater. Med.* **2004**, *15*, 787–794. [CrossRef] [PubMed]

78. Kuo, P.H.; Joshi, S.S.; Lu, K.; Ho, Y.H.; Xiang, Y.; Dahotre, N.B.; Du, J. Laser coating of bioactive glasses on bioimplant titanium alloys. *Int. J. Appl. Glass Sci.* **2019**, *10*, 307–320. [CrossRef]

79. Krzyzanowski, M.; Bajda, S.; Liu, Y.; Triantaphyllou, A.; Rainforth, W.M.; Glendenning, M. 3D analysis of thermal and stress evolution during laser cladding of bioactive glass coatings. *J. Mech. Behav. Biomed. Mater.* **2016**, *59*, 404–417. [CrossRef]

80. Bajda, S.; Cholewa-Kowalska, K.; Krzyzanowski, M.; Dziadek, M.; Kopyscianski, M.; Liu, Y.; Rai, A. Laser-directed energy deposition of bioactive glass on Ti-6Al-7Nb titanium alloy substrate with highly refined grain structure. *Surf. Coat. Technol.* **2024**, *485*, 130904. [CrossRef]

81. Bajda, S.; Liu, Y.; Tosi, R.; Cholewa-Kowalska, K.; Krzyzanowski, M.; Dziadek, M.; Kopyscianski, M.; Dymek, S.; Polyakov, A.V.; Semenova, I.P.; et al. Laser cladding of bioactive glass coating on pure titanium substrate with highly refined grain structure. *J. Mech. Behav. Biomed. Mater.* **2021**, *119*, 104519. [CrossRef]

82. Baino, F.; Montealegre, M.A.; Orlygsson, G.; Novajra, G.; Vitale-Brovarone, C. Bioactive glass coatings fabricated by laser cladding on ceramic acetabular cups: A proof-of-concept study. *J. Mater. Sci.* **2017**, *52*, 9115–9128. [CrossRef]

83. Comesaña, R.; Quintero, F.; Lusquiños, F.; Pascual, M.J.; Boutinguiza, M.; Durán, A.; Pou, J. Laser cladding of bioactive glass coatings. *Acta Biomater.* **2010**, *6*, 953–961. [CrossRef] [PubMed]

84. Berbecaru, C.; Stan, G.E.; Pina, S.; Tulyaganov, D.U.; Ferreira, J.M.F. The bioactivity mechanism of magnetron sputtered bioglass thin films. *Appl. Surf. Sci.* **2012**, *258*, 9840–9848. [CrossRef]

85. Stan, G.E.; Montazerian, M.; Shearer, A.; Stuart, B.W.; Baino, F.; Mauro, J.C.; Ferreira, J.M.F. Critical advances in the field of magnetron sputtered bioactive glass thin-films: An analytical review. *Appl. Surf. Sci.* **2024**, *646*, 158760. [CrossRef]

86. Stuart, B.W.; Gimeno-Fabra, M.; Segal, J.; Ahmed, I.; Grant, D.M. Mechanical, structural and dissolution properties of heat treated thin-film phosphate based glasses. *Appl. Surf. Sci.* **2017**, *416*, 605–617. [CrossRef]

87. Popa, A.C.; Stan, G.E.; Besleaga, C.; Ion, L.; Maraloiu, V.A.; Tulyaganov, D.U.; Ferreira, J.M.F. Submicrometer Hollow Bioglass Cones Deposited by Radio Frequency Magnetron Sputtering: Formation Mechanism, Properties, and Prospective Biomedical Applications. *ACS Appl. Mater. Interfaces* **2016**, *8*, 4357–4367. [CrossRef]

88. Stuart, B.; Gimeno-Fabra, M.; Segal, J.; Ahmed, I.; Grant, D.M. Preferential sputtering in phosphate glass systems for the processing of bioactive coatings. *Thin Solid Film.* **2015**, *589*, 534–542. [CrossRef]

89. Stan, G.E.; Popa, A.C.; Galca, A.C.; Aldica, G.; Ferreira, J.M.F. Strong bonding between sputtered bioglass-ceramic films and Ti-substrate implants induced by atomic inter-diffusion post-deposition heat-treatments. *Appl. Surf. Sci.* **2013**, *280*, 530–538. [CrossRef]

90. Stan, G.E.; Morosanu, C.O.; Marcov, D.A.; Pasuk, I.; Miculescu, F.; Reumont, G. Effect of annealing upon the structure and adhesion properties of sputtered bio-glass/titanium coatings. *Appl. Surf. Sci.* **2009**, *255*, 9132–9138. [CrossRef]

91. Stan, G.E.; Marcov, D.A.; Pasuk, I.; Miculescu, F.; Pina, S.; Tulyaganov, D.U.; Ferreira, J.M.F. Bioactive glass thin films deposited by magnetron sputtering technique: The role of working pressure. *Appl. Surf. Sci.* **2010**, *256*, 7102–7110. [CrossRef]

92. Tite, T.; Popa, A.C.; Chirica, I.M.; Stuart, B.W.; Galca, A.C.; Balescu, L.M.; Popescu-Pelin, G.; Grant, D.M.; Ferreira, J.M.F.; Stan, G.E. Phosphate bioglass thin-films: Cross-area uniformity, structure and biological performance tailored by the simple modification of magnetron sputtering gas pressure. *Appl. Surf. Sci.* **2021**, *541*, 148640. [CrossRef]

93. Popa, A.C.; Marques, V.M.F.; Stan, G.E.; Husanu, M.A.; Galca, A.C.; Ghica, C.; Tulyaganov, D.U.; Lemos, A.F.; Ferreira, J.M.F. Nanomechanical characterization of bioglass films synthesized by magnetron sputtering. *Thin Solid Films* **2014**, *553*, 166–172. [CrossRef]

94. Stan, G.E.; Pina, S.; Tulyaganov, D.U.; Ferreira, J.M.F.; Pasuk, I.; Morosanu, C.O. Biomineralization capability of adherent bio-glass films prepared by magnetron sputtering. *J. Mater. Sci. Mater. Med.* **2010**, *21*, 1047–1055. [CrossRef]

95. Durán, A.; Conde, A.; Gómez Coedo, A.; Dorado, T.; García, C.; Ceré, S. Sol-gel coatings for protection and bioactivation of metals used in orthopaedic devices. *J. Mater. Chem.* **2004**, *14*, 2282–2290. [CrossRef]

96. Huang, K.; Cai, S.; Xu, G.; Ren, M.; Wang, X.; Zhang, R.; Niu, S.; Zhao, H. Sol-gel derived mesoporous 58S bioactive glass coatings on AZ31 magnesium alloy and in vitro degradation behavior. *Surf. Coat. Technol.* **2014**, *240*, 137–144. [CrossRef]

97. Niu, S.; Cai, S.; Liu, T.; Zhao, H.; Wang, X.; Ren, M.; Huang, K.; Wu, X. 45S5 bioactive glass-ceramic coated magnesium alloy with strong interfacial bonding strength by "superplasticity diffusion bonding". *Mater. Lett.* **2015**, *141*, 96–99. [CrossRef]

98. Huang, K.; Cai, S.; Xu, G.; Ye, X.; Dou, Y.; Ren, M.; Wang, X. Preparation and characterization of mesoporous 45S5 bioactive glass-ceramic coatings on magnesium alloy for corrosion protection. *J. Alloys Compd.* **2013**, *580*, 290–297. [CrossRef]

99. Omar, S.A.; Ballarre, J.; Castro, Y.; Martinez Campos, E.; Schreiner, W.; Durán, A.; Cere, S.M. 58S and 68S sol-gel glass-like bioactive coatings for enhancing the implant performance of AZ91D magnesium alloy. *Surf. Coat. Technol.* **2020**, *400*, 126224. [CrossRef]

100. Soule, L.D.; Pajares Chomorro, N.; Chuong, K.; Mellott, N.; Hammer, N.; Hankenson, K.D.; Chatzistavrou, X. Sol-Gel-Derived Bioactive and Antibacterial Multi-Component Thin Films by the Spin-Coating Technique. *ACS Biomater. Sci. Eng.* **2020**, *6*, 5549–5562. [CrossRef]

101. Ye, X.; Leeflang, S.; Wu, C.; Chang, J.; Zhou, J.; Huan, Z. Mesoporous Bioactive Glass Functionalized 3D Ti-6Al-4V Scaffolds with Improved Surface Bioactivity. *Materials* **2017**, *10*, 1244. [CrossRef]
102. Farag, M.M.; Ahmed, H.Y.; Al-Rashidy, Z.M. Improving the Corrosion Resistance of Magnesium Alloy by Magnesium Phosphate/Glass Composite Coatings Using Sol-Gel Method. *Silicon* **2023**, *15*, 3841–3854. [CrossRef]
103. Wang, Y.C.; Lin, S.H.; Chien, C.S.; Kung, J.C.; Shih, C.J. In Vitro Bioactivity and Antibacterial Effects of a Silver-Containing Mesoporous Bioactive Glass Film on the Surface of Titanium Implants. *Int. J. Mol. Sci.* **2022**, *23*, 9291. [CrossRef] [PubMed]
104. Krause, D.; Thomas, B.; Leinenbach, C.; Eifler, D.; Minay, E.J.; Boccaccini, A.R. The electrophoretic deposition of Bioglass® particles on stainless steel and Nitinol substrates. *Surf. Coat. Technol.* **2006**, *200*, 4835–4845. [CrossRef]
105. Rojaee, R.; Fathi, M.; Raeissi, K.; Taherian, M. Electrophoretic deposition of bioactive glass nanopowders on magnesium based alloy for biomedical applications. *Ceram. Int.* **2014**, *40*, 7879–7888. [CrossRef]
106. Fiorilli, S.; Baino, F.; Cauda, V.; Crepaldi, M.; Vitale-Brovarone, C.; Demarchi, D.; Onida, B. Electrophoretic deposition of mesoporous bioactive glass on glass-ceramic foam scaffolds for bone tissue engineering. *J. Mater. Sci. Mater. Med.* **2015**, *26*, 21. [CrossRef] [PubMed]
107. Virk, R.S.; Atiq Ur Rehman, M.; Munawar, M.A.; Schubert, D.W.; Goldmann, W.H.; Dusza, J.; Boccaccini, A.R. Curcumin-Containing Orthopedic Implant Coatings Deposited on Poly-Ether-Ether-Ketone/Bioactive Glass/Hexagonal Boron Nitride Layers by Electrophoretic Deposition. *Coatings* **2019**, *9*, 572. [CrossRef]
108. Azzouz, I.; Faure, J.; Khlifi, K.; Cheikh Larbi, A.; Benhayoune, H. Electrophoretic Deposition of 45S5 Bioglass® Coatings on the Ti6Al4V Prosthetic Alloy with Improved Mechanical Properties. *Coatings* **2020**, *10*, 1192. [CrossRef]
109. Ammam, M. Electrophoretic deposition under modulated electric fields: A review. *RSC Adv.* **2012**, *2*, 7633–7646. [CrossRef]
110. Mehdipour, M.; Afshar, A.; Mohebali, M. Electrophoretic deposition of bioactive glass coating on 316L stainless steel and electrochemical behavior study. *Appl. Surf. Sci.* **2012**, *258*, 9832–9839. [CrossRef]
111. Sergi, R.; Bellucci, D.; Cannillo, V. A Comprehensive Review of Bioactive Glass Coatings: State of the Art, Challenges and Future Perspectives. *Coatings* **2020**, *10*, 757. [CrossRef]
112. Maximov, M.; Maximov, O.C.; Craciun, L.; Ficai, D.; Ficai, A.; Andronescu, E. Bioactive Glass-An Extensive Study of the Preparation and Coating Methods. *Coatings* **2021**, *11*, 1386. [CrossRef]
113. Liang, J.; Lu, X.; Zheng, X.; Li, Y.R.; Geng, X.; Sun, K.; Cai, H.; Jia, Q.; Jiang, H.B.; Liu, K. Modification of titanium orthopedic implants with bioactive glass: A systematic review of in vivo and in vitro studies. *Front. Bioeng. Biotechnol.* **2023**, *11*, 1269223. [CrossRef] [PubMed]
114. Hadem, H.; Mitra, A.; Kumar Ojha, A.; Rajasekaran, R.; Satpathy, B.; Das, D.; Mukherjee, S.; Dhara, S.; Das, S.; Das, K. Electrophoretic Deposition of 58S Bioactive Glass-Polymer Composite Coatings on 316L Stainless Steel: An Optimization for Corrosion, Bioactivity, and Cytocompatibility. *ACS Appl. Bio Mater.* **2024**, *7*, 2966–2981. [CrossRef] [PubMed]
115. Joughehdoust, S.; Manafi, S. Synthesis and In Vitro Investigation of Sol-Gel Derived Bioglass-58S Nanopowders. *Mater. Sci.* **2012**, *30*, 45–52. [CrossRef]
116. Cañaveral, S.; Morales, D.; Vargas, A.F. Synthesis and characterization of a 58S bioglass modified with manganese by a sol-gel route. *Mater. Lett.* **2019**, *255*, 126575. [CrossRef]
117. Sepulveda, P.; Jones, J.R.; Hench, L.L. Characterization of Melt-Derived 45S5 and sol-gel-derived 58S Bioactive Glasses. J. Biomed. *Mater. Res.* **2001**, *58*, 734–740. [CrossRef]
118. Silver, I.A.; Deas, J.; Erecińska, M. Interactions of bioactive glasses with osteoblasts in vitro: Effects of 45S5 Bioglass®, and 58S and 77S bioactive glasses on metabolism, intracellular ion concentrations and cell viability. *Biomaterials* **2001**, *22*, 175–185. [CrossRef]
119. Bosetti, M.; Cannas, M. The effect of bioactive glasses on bone marrow stromal cells differentiation. *Biomaterials* **2005**, *26*, 3873–3879. [CrossRef]
120. Hamadouche, M.; Meunier, A.; Greenspan, D.C.; Blanchat, C.; Zhong, J.P.; La Torre, G.P.; Sedel, L. Long-term in vivo bioactivity and degradability of bulk sol-gel bioactive glasses. *J. Biomed. Mater. Res.* **2001**, *54*, 560–566. [CrossRef]
121. Arcos, D.; Greenspan, D.C.; Vallet-Regí, M. A new quantitative method to evaluate the in vitro bioactivity of melt and sol-gel-derived silicate glasses. *J. Biomed. Mater. Res.* **2003**, *65A*, 344–351. [CrossRef] [PubMed]
122. Eriksson, E.; Björkenheim, R.; Strömberg, G.; Ainola, M.; Uppstu, P.; Aalto-Setälä, L.; Leino, V.M.; Hupa, L.; Pajarinen, J.; Lindfors, N.C. S53P4 bioactive glass scaffolds induce BMP expression and integrative bone formation in a critical-sized diaphysis defect treated with a single-stage d induce d membrane technique. *Acta Biomater.* **2021**, *126*, 463–476. [CrossRef]
123. Lindfors, N.C.; Hyvönen, P.; Nyyssönen, M.; Kirjavainen, M.; Kankare, J.; Gullichsen, E.; Salo, J. Bioactive glass S53P4 as bone graft substitute in treatment of osteomyelitis. *Bone* **2010**, *47*, 212–218. [CrossRef] [PubMed]
124. van Gestel, N.A.P.; Geurts, J.; Hulsen, D.J.W.; van Rietbergen, B.; Hofmann, S.; Arts, J.J. Clinical Applications of S53P4 Bioactive Glass in Bone Healing and Osteomyelitic Treatment: A Literature Review. *BioMed Res. Int.* **2015**, *2015*, 684826. [CrossRef]
125. Kolana, K.C.R.; Leu, M.C.; Hilmas, G.E.; Velez, M. Effect of material, process parameters, and simulated body fluids on mechanical properties of 13-93 bioactive glass porous constructs made by selective laser sintering. *J. Mech. Behav. Biomed. Mater.* **2012**, *13*, 14–24. [CrossRef] [PubMed]
126. Hoppe, A.; Sarker, B.; Detsch, R.; Hild, N.; Mohn, D.; Stark, W.J.; Boccaccini, A.R. In vitro reactivity of Sr-containing bioactive glass (type 1393) nanoparticles. *J. Non-Cryst. Solids* **2014**, *387*, 41–46. [CrossRef]
127. Beltrán, A.M.; Trueba, P.; Borie, F.; Alcudia, A.; Begines, B.; Rodriguez-Ortiz, J.A.; Torres, Y. Bioactive Bilayer Glass Coating on Porous Titanium Substrates with Enhanced Biofunctional and Tribomechanical Behavior. *Coatings* **2022**, *12*, 245. [CrossRef]

128. Rahaman, M.N.; Day, D.E.; Bal, B.S.; Fu, Q.; Jung, S.B.; Bonewald, L.F.; Tomsia, A.P. Bioactive glass in tissue engineering. *Acta Biomater.* **2011**, *7*, 2355–2373. [CrossRef]
129. Balasubramanian, P.; Büttner, T.; Miguez Pacheco, V.; Boccaccini, A.R. Boron-containing bioactive glasses in bone and soft tissue engineering. *J. Eur. Ceram. Soc.* **2018**, *38*, 855–869. [CrossRef]
130. Ege, D.; Zheng, K.; Boccaccini, A.R. Borate Bioactive Glasses (BBG): Bone Regeneration, Wound Healing Applications, and Future Directions. *ACS Appl. Bio Mater.* **2022**, *5*, 3608–3622. [CrossRef]
131. Midha, S.; Kim, T.B.; van den Bergh, W.; Lee, P.D.; Jones, J.R.; Mitchell, C.A. Preconditioned 70S30C bioactive glass foams promote osteogenesis in vivo. *Acta Biomaterialia* **2013**, *9*, 9169–9182. [CrossRef]
132. Yan, X.; Huang, X.; Yua, C.; Deng, H.; Wang, Y.; Zhang, Z.; Qiao, S.; Lu, G.; Zhao, D. The in-vitro bioactivity of mesoporous bioactive glasses. *Biomaterials* **2006**, *27*, 3396–3403. [CrossRef]
133. Thi Hoa, B.; Trong Hoa, H.T.; Anh Tien, N.; Huu Duy Khang, N.; Guseva, E.V.; Tuan, T.H.; Xuan Vuong, B. Green synthesis of bioactive glass 70SiO$_2$-30CaO by hydrothermal method. *Mater. Lett.* **2020**, *274*, 128032. [CrossRef]
134. Saravanapavan, P.; Jones, J.R.; Pryce, R.S.; Hench, L.L. Bioactivity of gel-glass powders in the CaO-SiO$_2$ system: A comparison with ternary (CaO-P$_2$O$_5$-SiO$_2$) and quaternary glasses (SiO$_2$-CaO-P$_2$O$_5$-Na$_2$O). *J. Biomed. Mater. Res.* **2003**, *66A*, 110–119. [CrossRef]
135. Hench, L.L.; West, J.K. The Sol-Gel Process. *Chem. Rev.* **1990**, *90*, 33–72. [CrossRef]
136. Li, R.; Clark, A.E.; Hench, L.L. An investigation of bioactive glass powders by sol-gel processing. *J. Appl. Biomater.* **1991**, *2*, 231–239. [CrossRef]
137. Arcos, D.; Vallet-Regí, M. Sol-gel silica-based biomaterials and bone tissue regeneration. *Acta Biomater.* **2010**, *6*, 2874–2888. [CrossRef]
138. Owens, G.J.; Singh, R.K.; Foroutan, F.; Alqaysi, M.; Han, C.M.; Mahapatra, C.; Kim, H.W.; Knowles, J.C. Sol-gel based materials for biomedical applications. *Prog. Mater. Sci.* **2016**, *77*, 1–79. [CrossRef]
139. Song, X.; Segura-Egea, J.J.; Díaz-Cuenca, A. Sol-Gel Technologies to Obtain Advanced Bioceramics for Dental Therapeutics. *Molecules* **2023**, *28*, 6967. [CrossRef]
140. Sakka, S. Birth of the sol-gel method: Early history. *J. Sol.-Gel. Sci. Technol.* **2022**, *102*, 478–481. [CrossRef]
141. Baino, F.; Fiume, E.; Miola, M.; Verné, E. Bioactive sol-gel glasses: Processing, properties, and applications. *Int. J. Appl. Ceram. Technol.* **2018**, *15*, 841–860. [CrossRef]
142. Wen, J.; Wilkes, G.L. Organic/Inorganic Hybrid Network Materials by the Sol-Gel Approach. *Chem. Mater.* **1996**, *8*, 1667–1681. [CrossRef]
143. Miguez-Pacheco, V.; Hench, L.L.; Boccaccini, A.R. Bioactive glasses beyond bone and teeth: Emerging applications in contact with soft tissues. *Acta Biomater.* **2015**, *13*, 1–15. [CrossRef]
144. Mazzoni, E.; Iaquinta, M.R.; Lanzillotti, C.; Mazziotta, C.; Maritati, M.; Montesi, M.; Sprio, S.; Tampieri, A.; Tognon, M.; Martini, F. Bioactive Materials for Soft Tissue Repair. *Front. Bioeng. Biotechnol.* **2021**, *9*, 613787. [CrossRef]
145. Hench, L.L. Bioceramics: From Concept to Clinic. *J. Am. Ceram. Soc.* **1991**, *74*, 1487–1510. [CrossRef]
146. Hench, L.L. Bioceramics. *J. Am. Ceram. Soc.* **1998**, *81*, 1705–1728. [CrossRef]
147. Jones, J.R. Review of bioactive glass: From Hench to hybrids. *Acta Biomater.* **2013**, *9*, 4457–4486. [CrossRef]
148. Hench, L.L. Bioglass: 10 milestones from concept to commerce. *J. Non-Cryst. Solids* **2016**, *432*, 2–8. [CrossRef]
149. Hench, L.L. The future of bioactive ceramics. *J. Mater. Sci. Mater. Med.* **2015**, *26*, 86. [CrossRef]
150. Vallet-Regí, M. Ceramics for medical applications. *J. Chem. Soc. Dalton Trans.* **2001**, *2*, 97–108. [CrossRef]
151. Vallet-Regí, M.; Ragel, C.V.; Salinas, A.J. Glasses with Medical Applications. *Eur. J. Inorg. Chem.* **2003**, *2003*, 1029–1042. [CrossRef]
152. Fiume, E.; Barberi, J.; Verné, E.; Baino, F. Bioactive Glasses: From Parent 45S5 Composition to Scaffold-Assisted Tissue-Healing Therapies. *J. Funct. Biomater.* **2018**, *9*, 24. [CrossRef] [PubMed]
153. Skallevold, H.E.; Rokaya, D.; Khurshid, Z.; Zafar, M.S. Bioactive Glass Applications in Dentistry. *Int. J. Mol. Sci.* **2019**, *20*, 5960. [CrossRef] [PubMed]
154. Khalid, M.D.; Khurshid, Z.; Zafar, M.S.; Farooq, I.; Khan, R.S.; Najmi, A. Bioactive Glasses and their Applications in Dentistry. *J. Pak. Dent. Assoc.* **2017**, *26*, 32–38. [CrossRef]
155. Oliver, J.N.; Su, Y.; Lu, X.; Kuo, P.H.; Du, J.; Zhu, D. Bioactive glass coatings on metallic implants for biomedical applications. *Bioact. Mater.* **2019**, *4*, 261–270. [CrossRef] [PubMed]
156. Aparajita Dash, P.; Mohanty, S.; Kumar Nayak, S. A review on bioactive glass, its modifications and applications in healthcare sectors. *J. Non-Cryst. Solids* **2023**, *614*, 122404. [CrossRef]
157. Brink, M.; Turunen, T.; Happonen, R.P.; Yli-Urpo, A. Compositional dependence of bioactivity of glasses in the system Na$_2$O-K$_2$O-MgO-CaO-B$_2$O$_3$-P$_2$O$_5$-SiO$_2$. *J. Biomed. Mater. Res.* **1997**, *37*, 114–121. [CrossRef]
158. Cannio, M.; Bellucci, D.; Roether, J.A.; Boccaccini, D.N.; Cannillo, V. Bioactive Glass Applications: A Literature Review of Human Clinical Trials. *Materials* **2021**, *14*, 5440. [CrossRef]
159. Al-Noaman, A.; Rawlinson, S.C.F.; Hill, R.G. The role of MgO on thermal properties, structure and bioactivity of bioactive glass coating for dental implants. *J. Non-Cryst. Solids* **2012**, *358*, 3019–3027. [CrossRef]
160. Jafari, N.; Habashi, M.S.; Hashemi, A.; Shirazi, R.; Tanideh, N.; Tamadon, A. Application of bioactive glasses in various dental fields. *Biomater. Res.* **2022**, *26*, 31. [CrossRef]

161. Shearer, A.; Montazerian, M.; Sly, J.J.; Hill, R.G.; Mauro, J.C. Trends and perspectives on the commercialization of bioactive glasses. *Acta Biomater.* **2023**, *160*, 14–31. [CrossRef]
162. Drevet, R.; Zhukova, Y.; Dubinskiy, S.; Kazakbiev, A.; Naumenko, V.; Abakumov, M.; Fauré, J.; Benhayoune, H.; Prokoshkin, S. Electrodeposition of cobalt-substituted calcium phosphate coatings on Ti22Nb6Zr alloy for bone implant applications. *J. Alloys Compd.* **2019**, *793*, 576–582. [CrossRef]
163. Wu, C.; Zhou, Y.; Fan, W.; Han, P.; Chang, J.; Yuen, J.; Zhang, M.; Xiao, Y. Hypoxia-mimicking mesoporous bioactive glass scaffolds with controllable cobalt ion release for bone tissue engineering. *Biomaterials* **2012**, *33*, 2076–2085. [CrossRef] [PubMed]
164. Al-Rashidy, Z.M.; Farag, M.M.; Abdel Ghany, N.A.; Ibrahim, A.M.; Abdel-Fattah, W.I. Orthopaedic bioactive glass/chitosan composites coated 316L stainless steel by green electrophoretic co-deposition. *Surf. Coat. Technol.* **2018**, *334*, 479–490. [CrossRef]
165. Ning, J.; Yao, A.; Wang, D.; Huang, W.; Fu, H.; Liu, X.; Jiang, X.; Zhang, X. Synthesis and in vitro bioactivity of a borate-based bioglass. *Mater. Lett.* **2007**, *61*, 5223–5226. [CrossRef]
166. Al-Rashidy, Z.M.; Farag, M.M.; Abdel Ghany, N.A.; Ibrahim, A.M.; Abdel-Fattah, W.I. Aqueous electrophoretic deposition and corrosion protection of borate glass coatings on 316 L stainless steel for hard tissue fixation. *Surf. Int.* **2017**, *7*, 125–133. [CrossRef]
167. Peddi, L.; Brow, R.K.; Brown, R.F. Bioactive borate glass coatings for titanium alloys. *J. Mater. Sci. Mater. Med.* **2008**, *19*, 3145–3152. [CrossRef]
168. Peitl, O.; Dutra Zanotto, E.; Hench, L.L. Highly bioactive P_2O_5-Na_2O-CaO-SiO_2 glass-ceramics. *J. Non-Cryst. Solids* **2001**, *292*, 115–126. [CrossRef]
169. Selvaraj, V.; Sekaran, S.; Dhanasekaran, A.; Warrier, S. Type 1 collagen: Synthesis, structure and key functions in bone mineralization. *Differentiation* **2024**, *136*, 100757. [CrossRef] [PubMed]
170. Hench, L.L.; Roki, N.; Fenn, M.B. Bioactive glasses: Importance of structure and properties in bone regeneration. *J. Mol. Struct.* **2014**, *1073*, 24–30. [CrossRef]
171. Besra, L.; Liu, M. A review on fundamentals and applications of electrophoretic deposition (EPD). *Progr. Mater. Sci.* **2007**, *52*, 1–61. [CrossRef]
172. Corni, I.; Ryan, M.P.; Boccaccini, A.R. Electrophoretic deposition: From traditional ceramics to nanotechnology. *J. Eur. Ceram. Soc.* **2008**, *28*, 1353–1367. [CrossRef]
173. Munyensanga, P.; Bricha, M.; El Mabrouk, K. Recent Developments of Bioactive Glass Electrophoretically Coated Cobalt-Chromium Metallic Implants. *Johns. Matthey Technol. Rev.* **2024**, *68*, 161–180. [CrossRef]
174. Boccaccini, A.R.; Keim, S.; Ma, R.; Li, Y.; Zhitomirsky, I. Electrophoretic deposition of Biomaterials. *J. R. Soc. Interface* **2010**, *7*, S581–S613. [CrossRef]
175. Azzouz, I.; Khlifi, K.; Faure, J.; Dhiflaoui, H.; Ben Cheikh Larbi, A.; Benhayoune, H. Mechanical behavior and corrosion resistance of sol-gel derived 45S5 bioactive glass coating on Ti6Al4V synthesized by electrophoretic deposition. *J. Mech. Behav. Biomed. Mater.* **2022**, *134*, 105352. [CrossRef] [PubMed]
176. Dusoulier, L. Elaboration de Dépôts d'$YBa_2Cu_3O_{7-x}$ par Électrophorèse et Projection Plasma, Chapitre 1, La Technique de Dépôt par Électrophorèse. Ph.D. Thesis, Liège University, Liège, Belgium, 2007.
177. Yang, J.; Wang, E.G. Reaction of water on silica surfaces. *Curr. Opin. Solid State Mater. Sci.* **2006**, *10*, 33–39. [CrossRef]
178. Lowe, B.M.; Skylaris, C.K.; Green, N.G. Acid-base dissociation mechanisms and energetics at the silica-water interface: An activation less process. *J. Colloid Interface Sci.* **2015**, *451*, 231–244. [CrossRef]
179. Image by Mjones1984. Available online: https://commons.wikimedia.org/wiki/File:Diagram_of_zeta_potential_and_slipping_planeV2.svg (accessed on 1 July 2024).
180. Behrens, S.H.; Grier, D.G. The Charge of Glass and Silica Surface. *J. Chem. Phys.* **2001**, *115*, 6716–6721. [CrossRef]
181. Van der Biest, O.; Vandeperre, L.J. Electrophoretic Deposition of Materials. *Annu. Rev. Mater. Sci.* **1999**, *29*, 327–352. [CrossRef]
182. Ferrari, B.; Moreno, R. EPD kinetics: A review. *J. Eur. Ceram. Soc.* **2010**, *30*, 1069–1078. [CrossRef]
183. Sujith, S.V.; Kim, H.; Lee, J. A Review on Thermophysical Property Assessment of Metal Oxide-Based Nanofluids: Industrial Perspectives. *Metals* **2022**, *12*, 165. [CrossRef]
184. O'Brien, R.W.; White, L.R. Electrophoretic Mobility of a Spherical Colloidal Particle. *J. Chem. Soc. Faraday Trans. 2* **1978**, *74*, 1607–1626. [CrossRef]
185. Zhitomirsky, I. Cathodic electrodeposition of ceramic and organoceramic materials. Fundamental aspects. *Adv. Colloid Interface Sci.* **2002**, *97*, 279–317. [CrossRef] [PubMed]
186. Sarkar, P.; Nicholson, P.S. Electrophoretic Deposition (EPD): Mechanisms, Kinetics, and Application to Ceramics. *J. Am. Ceram. Soc.* **1996**, *79*, 1987–2002. [CrossRef]
187. Zhang, Z.; Huang, Y.; Jiang, Z. Electrophoretic Deposition Forming of Sic-TZP Composites in a Nonaqueous Sol Media. *J. Am. Ceram. Soc.* **1994**, *77*, 1946–1949. [CrossRef]
188. Kargozar, S.; Baino, F.; Hamzehlou, S.; Hill, R.G.; Mozafari, M. Bioactive Glasses: Sprouting Angiogenesis in Tissue Engineering. *Trends Biotechnol.* **2018**, *36*, 430–444. [CrossRef]
189. Rabiee, S.M.; Nazparvar, N.; Azizian, M.; Vashaee, D.; Tayebi, L. Effect of ion substitution on properties of bioactive glasses: A review. *Ceram. Int.* **2015**, *41*, 7241–7251. [CrossRef]
190. Baino, F.; Hamzehlou, S.; Kargozar, S. Bioactive Glasses: Where Are We and Where Are We Going? *J. Funct. Biomater.* **2018**, *9*, 25. [CrossRef]

191. Vafa, E.; Tayebi, L.; Abbasi, M.; Azizli, M.J.; Bazargan-Lari, R.; Talaiekhozani, A.; Zareshahrabadi, Z.; Vaez, A.; Amani, A.M.; Kamyab, H.; et al. A better roadmap for designing novel bioactive glasses: Effective approaches for the development of innovative revolutionary bioglasses for future biomedical applications. *Environ. Sci. Pollut. Res.* **2023**, *30*, 116960–116983. [CrossRef]

192. Pantulap, U.; Arango-Ospina, M.; Boccaccini, A.R. Bioactive glasses incorporating less-common ions to improve biological and physical properties. *J. Mater. Sci. Mater. Med.* **2022**, *33*, 3. [CrossRef]

193. Zhu, H.; Hu, C.; Zhang, F.; Feng, X.; Li, J.; Liu, T.; Chen, J.; Zhang, J. Preparation and antibacterial property of silver-containing mesoporous 58S bioactive glass. *Mater. Sci. Eng. C* **2014**, *42*, 22–30. [CrossRef] [PubMed]

194. Yu, O.Y.; Ge, K.X.; Lung, C.Y.K.; Chu, C.H. Developing a novel glass ionomer cement with enhanced mechanical and chemical properties. *Dent. Mater.* **2024**, *40*, e1–e13. [CrossRef] [PubMed]

195. Vernè, E.; Di Nunzio, S.; Bosetti, M.; Appendino, P.; Vitale Brovarone, C.; Maina, G.; Cannas, M. Surface characterization of silver-doped bioactive glass. *Biomaterials* **2005**, *26*, 5111–5119. [CrossRef]

196. Shendage, S.S.; Gaikwad, K.; Kachare, K.; Kashte, S.; Chang, J.Y.; Ghule, A.V. In situ silver-doped antibacterial bioactive glass for bone regeneration application. *J. Mater. Sci.* **2024**, *59*, 10744–10762. [CrossRef]

197. Chen, X.; Karpukhina, N.; Brauer, D.S.; Hill, R.G. Novel Highly Degradable Chloride Containing Bioactive Glasses. *Biomed. Glasses* **2015**, *1*, 108–118. [CrossRef]

198. Chen, X.; Karpukhina, N.; Brauer, D.S.; Hill, R.G. High chloride content calcium silicate glasses. *Phys. Chem. Chem. Phys.* **2017**, *19*, 7078. [CrossRef]

199. Chen, X.; Chen, X.; Pedone, A.; Apperley, D.; Hill, R.G.; Karpukhina, N. New Insight into Mixing Fluoride and Chloride in Bioactive Silicate Glasses. *Sci. Rep.* **2018**, *8*, 1316. [CrossRef]

200. Brauer, D.S.; Karpukhina, N.; Law, R.V.; Hill, R.G. Structure of fluoride-containing bioactive glasses. *J. Mater. Chem.* **2009**, *19*, 5629–5636. [CrossRef]

201. Lusvardi, G.; Malavasi, G.; Menabue, L.; Aina, V.; Morterra, C. Fluoride-containing bioactive glasses: Surface reactivity in simulated body fluids solutions. *Acta Biomater.* **2009**, *5*, 3548–3562. [CrossRef]

202. Gentleman, E.; Stevens, M.M.; Hill, R.G.; Brauer, D.S. Surface properties and ion release from fluoride-containing bioactive glasses promote osteoblast differentiation and mineralization in vitro. *Acta Biomater.* **2013**, *9*, 5771–5779. [CrossRef] [PubMed]

203. Cannillo, V.; Sola, A. Potassium-based composition for a bioactive glass. *Ceram. Int.* **2009**, *35*, 3389–3393. [CrossRef]

204. Bellucci, D.; Sola, A.; Cannillo, V. Low Temperature Sintering of Innovative Bioactive Glasses. *J. Am. Ceram. Soc.* **2012**, *95*, 1313–1319. [CrossRef]

205. Bellucci, D.; Cannillo, V.; Sola, A. A new potassium-based bioactive glass: Sintering behaviour and possible applications for bioceramic scaffolds. *Ceram. Int.* **2011**, *37*, 145–157. [CrossRef]

206. de Magalhães Gomes, G.H.; Oliveira Guimarães, G.; Dorion Rodas, A.C.; Barbosa Milesi, M.T.; Nascimento Costa, F.; Rodrigues Pais Alves, M.F.; Santos, C.; Barboza Daguano, J.K.M. In vitro biodegradation, blood, and cytocompatibility studies of a bioactive lithium silicate glass-ceramic. *Mater. Chem. Phys.* **2024**, *314*, 128828. [CrossRef]

207. Khorami, M.; Hesaraki, S.; Behnamghader, A.; Nazarian, H.; Shahrabi, S. In vitro bioactivity and biocompatibility of lithium substituted 45S5 bioglass. *Mater. Sci. Eng. C* **2011**, *31*, 1584–1592. [CrossRef]

208. Moghanian, A.; Firoozi, S.; Tahriri, M. Synthesis and in vitro studies of sol-gel derived lithium substituted 58S bioactive glass. *Ceram. Int.* **2017**, *43*, 12835–12843. [CrossRef]

209. Miguez-Pacheco, V.; Büttner, T.; Maçon, A.L.B.; Jones, J.R.; Fey, T.; de Ligny, D.; Greil, P.; Chevalier, J.; Malchere, A.; Boccaccini, A.R. Development and characterization of lithium-releasing silicate bioactive glasses and their scaffolds for bone repair. *J. Non-Cryst. Solids* **2016**, *432*, 65–72. [CrossRef]

210. Pantulap, U.; Unalan, I.; Zheng, K.; Boccaccini, A.R. Hydroxycarbonate apatite formation, cytotoxicity, and antibacterial properties of rubidium-doped mesoporous bioactive glass nanoparticles. *J. Porous Mater.* **2024**, *31*, 685–696. [CrossRef]

211. Ouyang, S.; Zheng, K.; Huang, Q.; Liu, Y.; Boccaccini, A.R. Synthesis and characterization of rubidium-containing bioactive glass nanoparticles. *Mater. Lett.* **2020**, *273*, 127920. [CrossRef]

212. Majumdar, S.; Tiwari, A.; Mallick, D.; Patel, D.K.; Trigun, S.K.; Krishnamurthy, S. Oral Release Kinetics, Biodistribution, and Excretion of Dopants from Barium-Containing Bioactive Glass in Rats. *ACS Omega* **2024**, *9*, 7188–7205. [CrossRef]

213. Singh, J.; Kumar, V.; Vermani, Y.K.; Al-Buriahi, M.S.; Alzahrani, J.S.; Singh, T. Fabrication and characterization of barium based bioactive glasses in terms of physical, structural, mechanical and radiation shielding properties. *Ceram. Int.* **2021**, *47*, 21730–21743. [CrossRef]

214. Arepalli, S.K.; Tripathi, H.; Vyas, V.K.; Jain, S.; Suman, S.K.; Pyare, R.; Singh, S.P. Influence of barium substitution on bioactivity, thermal and physico-mechanical properties of bioactive glass. *Mater. Sci. Eng. C* **2015**, *49*, 549–559. [CrossRef]

215. Majumdar, S.; Hira, S.K.; Tripathi, H.; Kumar, A.S.; Manna, P.P.; Singh, S.P.; Krishnamurthy, S. Synthesis and characterization of barium-doped bioactive glass with potential anti-inflammatory activity. *Ceram. Int.* **2021**, *47*, 7143–7158. [CrossRef]

216. Barrioni, B.R.; Norris, E.; Jones, J.R.; Pereira, M.d.M. The influence of cobalt incorporation and cobalt precursor selection on the structure and bioactivity of sol-gel-derived bioactive glass. *J. Sol.-Gel. Sci. Technol.* **2018**, *88*, 309–321. [CrossRef]

217. Jiménez-Holguín, J.; Lozano, D.; Saiz-Pardo, M.; de Pablo, D.; Ortega, L.; Enciso, S.; Fernández-Tomé, B.; Díaz-Güemes, I.; Sánchez-Margallo, F.M.; Portolés, M.T.; et al. Osteogenic-angiogenic coupled response of cobalt-containing mesoporous bioactive glasses in vivo. *Acta Biomater.* **2024**, *176*, 445–457. [CrossRef] [PubMed]

218. Baino, F.; Montazerian, M.; Verné, E. Cobalt-Doped Bioactive Glasses for Biomedical Applications: A Review. *Materials* **2023**, *16*, 4994. [CrossRef] [PubMed]
219. Azari, Z.; Kermani, F.; Mollazadeh, S.; Alipour, F.; Sadeghi-Avalshahr, A.; Ranjbar-Mohammadi, M.; Jalali Kondori, B.; Mollaei, Z.; Atefe Hosseini, S.; Nazarnezhad, S.; et al. Fabrication and characterization of cobalt- and copper-doped mesoporous borate bioactive glasses for potential applications in tissue engineering. *Ceram. Int.* **2023**, *49*, 38773–38788. [CrossRef]
220. Kargozar, S.; Mozafari, M.; Ghodrat, S.; Fiume, E.; Baino, F. Copper-containing bioactive glasses and glass-ceramics: From tissue regeneration to cancer therapeutic strategies. *Mater. Sci. Eng. C* **2021**, *121*, 111741. [CrossRef] [PubMed]
221. Miola, M.; Verné, E. Bioactive and Antibacterial Glass Powders Doped with Copper by Ion-Exchange in Aqueous Solutions. *Materials* **2016**, *9*, 405. [CrossRef]
222. Vecchio, G.; Darcos, V.; Le Grill, S.; Brouillet, F.; Coppel, Y.; Duttine, M.; Pugliara, A.; Combes, C.; Soulié, J. Spray-dried ternary bioactive glass microspheres: Direct and indirect structural effects of copper-doping on acellular degradation behavior. *Acta Biomater.* **2024**, *181*, 453–468. [CrossRef]
223. Bari, A.; Bloise, N.; Fiorilli, S.; Novajra, G.; Vallet-Regí, M.; Bruni, G.; Torres-Pardo, A.; González-Calbet, J.M.; Visai, L.; Vitale-Brovarone, C. Copper-containing mesoporous bioactive glass nanoparticles as multifunctional agent for bone regeneration. *Acta Biomater.* **2017**, *55*, 493–504. [CrossRef]
224. Tiama, T.M.; Ibrahim, M.A.; Sharaf, M.H.; Mabied, A.F. Effect of germanium oxide on the structural aspects and bioactivity of bioactive silicate glass. *Sci. Rep.* **2023**, *13*, 9582. [CrossRef]
225. Mokhtari, S.; Krull, E.A.; Sanders, L.M.; Coughlan, A.; Mellott, N.P.; Gong, Y.; Borges, R.; Wren, A.W. Investigating the effect of germanium on the structure of SiO_2-ZnO-CaO-SrO-P_2O_5 glasses and the subsequent influence on glass polyalkenoate cement formation, solubility and bioactivity. *Mater. Sci. Eng. C* **2019**, *103*, 109843. [CrossRef] [PubMed]
226. Silva Pereira, A.F.; da Silva Neto, O.C.; Gomes Dias, T.; Silva Reis, A.; Pedrochi, F.; Steimacher, A.; Barboza, M.J. The role of MgO on physical and bioactive properties of borophosphate glasses for biomedical applications. *Ceram. Int.* **2024**, *50*, 17532–17543. [CrossRef]
227. Watts, S.J.; Hill, R.G.; O'Donnell, M.D.; Law, R.V. Influence of magnesia on the structure and properties of bioactive glasses. *J. Non-Cryst. Solids* **2010**, *356*, 517–524. [CrossRef]
228. Saboori, A.; Rabiee, M.; Moztarzadeh, F.; Sheikhi, M.; Tahriri, M.; Karimi, K. Synthesis, characterization and in vitro bioactivity of sol-gel-derived SiO_2-CaO-P_2O_5-MgO bioglass. *Mater. Sci. Eng. C* **2009**, *29*, 335–340. [CrossRef]
229. Vallet-Regí, M.; Salinas, A.J.; Román, J.; Gil, M. Effect of magnesium content on the in vitro bioactivity of CaO-MgO-SiO_2-P_2O_5 sol-gel glasses. *J. Mater. Chem.* **1999**, *9*, 515–518. [CrossRef]
230. Abati, M.; Contreras Jaimes, A.T.; Rigamonti, L.; Carrozza, D.; Lusvardi, G.; Brauer, D.S.; Malavasi, G. Assessing Mn as an antioxidant agent in bioactive glasses by quantification of catalase and superoxide dismutase enzymatic mimetic activities. *Ceram. Int.* **2024**, *50*, 2574–2587. [CrossRef]
231. Curcio, M.; Bochicchio, B.; Pepe, A.; Laezza, A.; De Stefanis, A.; Rau, J.V.; Teghil, R.; De Bonis, A. Mn-Doped Glass-Ceramic Bioactive (Mn-BG) Thin Film to Selectively Enhance the Bioactivity of Electrospun Fibrous Polymeric Scaffolds. *Coatings* **2022**, *12*, 1427. [CrossRef]
232. Miola, M.; Vitale Brovarone, C.; Maina, G.; Rossi, F.; Bergandi, L.; Ghigo, D.; Saracino, S.; Maggiora, M.; Canuto, R.A.; Muzio, G.; et al. In vitro study of manganese-doped bioactive glasses for bone regeneration. *Mater. Sci. Eng. C* **2014**, *38*, 107–118. [CrossRef]
233. Nawaz, Q.; Atiq Ur Rehman, M.; Burkovski, A.; Schmidt, J.; Beltrán, A.M.; Shahid, A.; Alber, N.K.; Peukert, W.; Boccaccini, A.R. Synthesis and characterization of manganese containing mesoporous bioactive glass nanoparticles for biomedical applications. *J. Mater. Sci. Mater. Med.* **2018**, *29*, 64. [CrossRef]
234. Boukhris, I.; Alalawi, A.; Al-Buriahi, M.S.; Kebaili, I.; Sayyed, M.I. Radiation attenuation properties of bioactive glasses doped with NiO. *Ceram. Int.* **2020**, *46*, 19880–19889. [CrossRef]
235. Vyas, V.K.; Kumar, A.S.; Singh, S.P.; Pyare, R. Effect of nickel oxide substitution on bioactivity and mechanical properties of bioactive glass. *Bull. Mater. Sci.* **2016**, *39*, 1355–1361. [CrossRef]
236. Vyas, V.K.; Kumar, A.S.; Singh, S.P.; Pyare, R. Destructive and non-destructive behavior of nickel oxide doped bioactive glass and glass-ceramic. *J. Aust. Ceram. Soc.* **2017**, *53*, 939–951. [CrossRef]
237. Silva, A.V.; Gomes, D.d.S.; Victor, R.d.S.; Santana, L.N.d.L.; Neves, G.A.; Menezes, R.R. Influence of Strontium on the Biological Behavior of Bioactive Glasses for Bone Regeneration. *Materials* **2023**, *16*, 7654. [CrossRef]
238. Rohilla, R.; Dahiya, M.S.; Agarwal, A.; Khasa, S. Influence of Sr^{2+} ions on structural, optical and bioactive behaviour of phosphoborate glass system. *J. Mol. Struct.* **2023**, *1291*, 136095. [CrossRef]
239. Ryu, J.H.; Mangal, U.; Lee, M.J.; Seo, J.Y.; Jeong, I.J.; Park, J.Y.; Na, J.Y.; Lee, K.J.; Yu, H.S.; Cha, J.K.; et al. Effect of strontium substitution on functional activity of phosphate-based glass. *Biomater. Sci.* **2023**, *11*, 6299. [CrossRef]
240. Bahati, D.; Bricha, M.; El Mabrouk, K. Synthesis, characterization, and in vitro apatite formation of strontium-doped sol-gel-derived bioactive glass nanoparticles for bone regeneration applications. *Ceram. Int.* **2023**, *49*, 23020–23034. [CrossRef]
241. Gentleman, E.; Fredholm, Y.C.; Jell, G.; Lotfibakhshaiesh, N.; O'Donnell, M.D.; Hill, R.G.; Stevens, M.M. The effects of strontium-substituted bioactive glasses on osteoblasts and osteoclasts in vitro. *Biomaterials* **2010**, *31*, 3949–3956. [CrossRef]
242. Aina, V.; Perardi, A.; Bergandi, L.; Malavasi, G.; Menabue, L.; Morterra, C.; Ghigo, D. Cytotoxicity of zinc-containing bioactive glasses in contact with human osteoblasts. *Chem. Biol. Interact.* **2007**, *167*, 207–218. [CrossRef]

243. Paramita, P.; Ramachandran, M.; Narashiman, S.; Nagarajan, S.; Kumar Sukumar, D.; Chung, T.W.; Ambigapathi, M. Sol–gel based synthesis and biological properties of zinc integrated nano bioglass ceramics for bone tissue regeneration. *J. Mater. Sci. Mater. Med.* **2021**, *32*, 5. [CrossRef]

244. Oki, A.; Parveen, B.; Hossain, S.; Adeniji, S.; Donahue, H. Preparation and in vitro bioactivity of zinc containing sol-gel-derived bioglass materials. *J. Biomed. Mater. Res. A* **2004**, *69*, 216–221. [CrossRef] [PubMed]

245. Atkinson, I.; Anghel, E.M.; Predoana, L.; Mocioiu, O.C.; Jecu, L.; Raut, I.; Munteanu, C.; Culita, D.; Zaharescu, M. Influence of ZnO addition on the structural, in vitro behavior and antimicrobial activity of sol-gel derived CaO-P_2O_5-SiO_2 bioactive glasses. *Ceram. Int.* **2016**, *42*, 3033–3045. [CrossRef]

246. Clavijo-Mejía, G.A.; Michálek, M.; Youssef, L.; Kaňková, H.; Galusek, D.; Boccaccini, A.R. Bioactivity of radiopaque 45S5 bioactive glass with progressive additions of Bi2O3: A dissolution study under static conditions. *Ceram. Int.* **2024**, *50*, 27216–27226. [CrossRef]

247. Wang, L.; Long, N.J.; Li, L.; Lu, Y.; Li, M.; Cao, J.; Zhang, Y.; Zhang, Q.; Xu, S.; Yang, Z.; et al. Multi-functional bismuth-doped bioglasses: Combining bioactivity and photothermal response for bone tumor treatment and tissue repair. *Light Sci. Appl.* **2018**, *7*, 1. [CrossRef]

248. Du, J.; Ding, H.; Fu, S.; Li, D.; Yu, B. Bismuth-coated 80S15C bioactive glass scaffolds for photothermal antitumor therapy and bone regeneration. *Front. Bioeng. Biotechnol.* **2023**, *10*, 1098923. [CrossRef]

249. Meng, X.; Wang, W.D.; Li, S.R.; Sun, Z.J.; Zhang, L. Harnessing cerium-based biomaterials for the treatment of bone diseases. *Acta Biomater.* **2024**, *183*, 30–49. [CrossRef] [PubMed]

250. Chromčíková, M.; Svoboda, R.; Hruška, B.; Pecušová, B.; Nowicka, A. Thermo-kinetic and structural characterization of Ce-doped glasses based on Bioglass 45S5. *Mater. Chem. Phys.* **2023**, *304*, 127833. [CrossRef]

251. Leonelli, C.; Lusvardi, G.; Malavasi, G.; Menabue, L.; Tonelli, M. Synthesis and characterization of cerium-doped glasses and in vitro evaluation of bioactivity. *J. Non-Cryst. Solids* **2003**, *316*, 198–216. [CrossRef]

252. Nicolini, V.; Malavasi, G.; Menabue, L.; Lusvardi, G.; Benedetti, F.; Valeri, S.; Luches, P. Cerium-doped bioactive 45S5 glasses: Spectroscopic, redox, bioactivity and biocatalytic properties. *J. Mater. Sci.* **2017**, *52*, 8845–8857. [CrossRef]

253. Zhang, W.; Zhang, X.; Zhou, Y.; Zhang, Y. Luminescence biomonitoring and antibacterial properties of Er^{3+}-doped SiO_2-CaO-P_2O_5 bioactive glass microspheres. *Ceram. Int.* **2023**, *49*, 22924–22931. [CrossRef]

254. Zhang, Y.; Zhang, W.; Zhang, X.; Zhou, Y. Erbium-ytterbium containing upconversion mesoporous bioactive glass microspheres for tissue engineering: Luminescence monitoring of biomineralization and drug release. *Acta Biomater.* **2023**, *168*, 628–636. [CrossRef] [PubMed]

255. Yoo, K.H.; Son, S.A.; Park, J.K.; Yoon, S.Y. Influence of glass composition on the network structure and mineralization of europium containing mesoporous bioactive glass nanoparticles. *Mater. Chem. Phys.* **2024**, *317*, 129179. [CrossRef]

256. Baranowska, A.; Leśniak, M.; Kochanowicz, M.; Żmojda, J.; Miluski, P.; Dorosz, D. Crystallization Kinetics and Structural Properties of the 45S5 Bioactive Glass and Glass-Ceramic Fiber Doped with Eu^{3+}. *Materials* **2020**, *13*, 1281. [CrossRef] [PubMed]

257. Wu, C.; Xia, L.; Han, P.; Mao, L.; Wang, J.; Zhai, D.; Fang, B.; Chang, J.; Xiao, Y. Europium-Containing Mesoporous Bioactive Glass Scaffolds for Stimulating in Vitro and in Vivo Osteogenesis. *ACS Appl. Mater. Interfaces* **2016**, *8*, 11342–11354. [CrossRef]

258. Zhou, T.; Xu, Z.; Sun, H.; Beltrán, A.M.; Nawaz, Q.; Sui, B.; Boccaccini, A.R.; Zheng, K. Unlocking the potential of iron-containing mesoporous bioactive glasses: Orchestrating osteogenic differentiation in bone marrow mesenchymal stem cells and osteoblasts. *Colloids Surf. A Physicochem. Eng. Asp.* **2024**, *694*, 134188. [CrossRef]

259. Nitu; Fopase, R.; Mohan Pandey, L.; Seal, P.; Prasad Borah, J.; Srinivasan, A. Assessment of sol-gel derived iron oxide substituted 45S5 bioglass-ceramics for biomedical applications. *J. Mater. Chem. B* **2023**, *11*, 7502. [CrossRef]

260. Baino, F.; Fiume, E.; Miola, M.; Leone, F.; Onida, B.; Verné, E. Fe-doped bioactive glass-derived scaffolds produced by sol-gel foaming. *Mater. Lett.* **2019**, *235*, 207–211. [CrossRef]

261. Abou Neel, E.A.; Ahmed, I.; Blaker, J.J.; Bismarck, A.; Boccaccini, A.R.; Lewis, M.P.; Nazhat, S.N.; Knowles, J.C. Effect of iron on the surface, degradation and ion release properties of phosphate-based glass fibres. *Acta Biomater.* **2005**, *1*, 553–563. [CrossRef] [PubMed]

262. Kurtuldu, F.; Mutlu, N.; Friedrich, R.P.; Beltrán, A.M.; Liverani, L.; Detsch, R.; Alexiou, C.; Galusek, D.; Boccaccini, A.R. Gallium-containing mesoporous nanoparticles influence in-vitro osteogenic and osteoclastic activity. *Biomater. Adv.* **2024**, *162*, 213922. [CrossRef] [PubMed]

263. Maha Lakshmi, A.; Prasad, A.; Murimadugula, S.; Venkateswara Rao, P.; Madaboosi, N.; Özcan, M.; Kumari, K.; Syam Prasad, P. Efficacy of Ga^{3+} ions on structural, biological and antimicrobial activity of mesoporous lithium silicate bioactive glasses for tissue engineering. *Microporous Mesoporous Mater.* **2024**, *373*, 113132. [CrossRef]

264. Pourshahrestani, S.; Zeimaran, E.; Kadri, N.A.; Gargiulo, N.; Samuel, S.; Vasudevaraj Naveen, S.; Kamarul, T.; Towler, M.R. Gallium-containing mesoporous bioactive glass with potent hemostatic activity and antibacterial efficacy. *J. Mater. Chem. B* **2016**, *4*, 71. [CrossRef]

265. Franchini, M.; Lusvardi, G.; Malavasi, G.; Menabue, L. Gallium-containing phospho-silicate glasses: Synthesis and in vitro bioactivity. *Mater. Sci. Eng. C* **2012**, *32*, 1401–1406. [CrossRef]

266. Piagentini Delpino, G.; Borges, R.; Zambanini, T.; Sousa Joca, J.F.; Gaubeur, I.; Santos de Souza, A.C.; Marchi, J. Sol-gel-derived 58S bioactive glass containing holmium aiming brachytherapy applications: A dissolution, bioactivity, and cytotoxicity study. *Mater. Sci. Eng. C* **2021**, *119*, 111595. [CrossRef] [PubMed]

267. Baino, F.; Marchi, J.; Borges, R.; Verné, E. Holmium-doped 58S glass-derived foam-like scaffolds. *Mater. Lett.* **2023**, *341*, 134256. [CrossRef]

268. Ashemary, A.Z.; Haidary, S.M.; Muhammed, Y.; Çardakli, İ.S. Preparation and characterization of mesoporous Lanthanum-doped bioactive glass nanoparticles. *Digest J. Nanomater. Biostruct.* **2023**, *18*, 681–688. [CrossRef]

269. Jodati, H.; Güner, B.; Evis, Z.; Keskin, D.; Tezcaner, A. Synthesis and characterization of magnesium-lanthanum dual doped bioactive glasses. *Ceram. Int.* **2020**, *46*, 10503–10511. [CrossRef]

270. Ben-Arfa, B.A.E.; Miranda Salvado, I.M.; Ferreira, J.M.F.; Pullar, R.C. The effects of Cu^{2+} and La^{3+} doping on the sintering ability of sol-gel derived high silica bioglasses. *Ceram. Int.* **2019**, *45*, 10269–10278. [CrossRef]

271. Orgaz, F.; Dzika, A.; Szycht, O.; Amat, D.; Barba, F.; Becerra, J.; Santos-Ruiz, L. Surface nitridation improves bone cell response to melt-derived bioactive silicate/borosilicate glass composite scaffolds. *Acta Biomater.* **2016**, *29*, 424–434. [CrossRef]

272. Bachar, A.; Mercier, C.; Tricoteaux, A.; Leriche, A.; Follet, C.; Hampshire, S. Bioactive oxynitride glasses: Synthesis, structure and properties. *J. Eur. Ceram. Soc.* **2016**, *36*, 2869–2881. [CrossRef]

273. Kushan Akin, S.R.; Dolekcekic, E.; Webster, T.J. Effect of nitrogen on the antibacterial behavior of oxynitride glasses. *Ceram. Int.* **2021**, *47*, 18213–18217. [CrossRef]

274. Ershad, M.; Vyas, V.K.; Prasad, S.; Ali, A.; Pyare, R. Effect of Sm2O3 substitution on mechanical and biological properties of 45S5 bioactive glass. *J. Aust. Ceram. Soc.* **2018**, *54*, 621–630. [CrossRef]

275. Deliormanli, A.M.; ALMisned, G.; Tekin, H.O. Synthesis and characterization of Nb^{5+} and Sm^{3+}-doped 13-93 bioactive glass particles with improved photon transmission properties for advanced biomedical and dental applications. *Ceram. Int.* **2024**, *50*, 31211–31224. [CrossRef]

276. Wang, X.; Zhang, Y.; Lin, C.; Zhong, W. Sol-gel derived terbium-containing mesoporous bioactive glasses nanospheres: In vitro hydroxyapatite formation and drug delivery. *Colloids Surf. B* **2017**, *160*, 406–415. [CrossRef]

277. Deliormanli, A.M.; Issa, S.A.M.; Al-Buriahi, M.S.; Rahman, B.; Zakaly, H.M.H.; Tekin, H.O. Erbium (III)- and Terbium (III)-containing silicate-based bioactive glass powders: Physical, structural and nuclear radiation shielding characteristics. *Appl. Phys. A Mater. Sci. Process.* **2021**, *127*, 463. [CrossRef]

278. Hadush Tesfay, A.; Chou, Y.J.; Tan, C.Y.; Bakare, F.F.; Tsou, N.T.; Huang, E.W.; Shih, S.J. Control of Dopant Distribution in Yttrium-Doped Bioactive Glass for Selective Internal Radiotherapy Applications Using Spray Pyrolysis. *Materials* **2019**, *12*, 986. [CrossRef]

279. Cacaina, D.; Ylänen, H.; Simon, S.; Hupa, M. The behaviour of selected yttrium containing bioactive glass microspheres in simulated body environments. *J. Mater. Sci. Mater. Med.* **2008**, *19*, 1225–1233. [CrossRef]

280. Christie, J.K.; Tilocca, A. Integrating biological activity into radioisotope vectors: Molecular dynamics models of yttrium-doped bioactive glasses. *J. Mater. Chem.* **2012**, *22*, 12023–12031. [CrossRef]

281. Zambanini, T.; Borges, R.; Faria, P.C.; Delpino, G.P.; Pereira, I.S.; Marques, M.M.; Marchi, J. Dissolution, bioactivity behavior, and cytotoxicity of rare earth containing bioactive glasses (RE = Gd, Yb). *Int. J. Appl. Ceram. Technol.* **2019**, *16*, 2028–2039. [CrossRef]

282. Chen, D.; Liang, Z.; Su, Z.; Huang, J.; Pi, Y.; Ouyang, Y.; Luo, T.; Guo, L. Selenium-Doped Mesoporous Bioactive Glass Regulates Macrophage Metabolism and Polarization by Scavenging ROS and Promotes Bone Regeneration In Vivo. *ACS Appl. Mater. Interfaces* **2023**, *15*, 34378–34396. [CrossRef]

283. Hu, M.; Fang, J.; Zhang, Y.; Wang, X.; Zhong, W.; Zhou, Z. Design and evaluation a kind of functional biomaterial for bone tissue engineering: Selenium/mesoporous bioactive glass nanospheres. *J. Colloid Interface Sci.* **2020**, *579*, 654–666. [CrossRef]

284. Karakuzu-İkizler, B.; Terzioğlu, P.; Oduncu-Tekerek, B.S.; Yücel, S. Effect of selenium incorporation on the structure and in vitro bioactivity of 45S5 bioglass. *J. Aust. Ceram. Soc.* **2020**, *56*, 697–709. [CrossRef]

285. Zhang, Y.; Hu, M.; Zhang, W.; Zhang, X. Construction of tellurium-doped mesoporous bioactive glass nanoparticles for bone cancer therapy by promoting ROS-mediated apoptosis and antibacterial activity. *J. Colloid Interface Sci.* **2022**, *610*, 719–730. [CrossRef]

286. Miola, M.; Massera, J.; Cochis, A.; Kumar, A.; Rimondini, L.; Vernè, E. Tellurium: A new active element for innovative multifunctional bioactive glasses. *Mater. Sci. Eng. C* **2021**, *123*, 111957. [CrossRef]

287. Moghanian, A.; Zohourfazeli, M.; Mahdi Tajer, M.H. The effect of zirconium content on in vitro bioactivity, biological behavior and antibacterial activity of sol-gel derived 58S bioactive glass. *J. Non-Cryst. Solids* **2020**, *546*, 120262. [CrossRef]

288. Hammami, I.; Gavinho, S.R.; Pádua, A.S.; Sá-Nogueira, I.; Silva, J.C.; Borges, J.P.; Valente, M.A.; Graça, M.P.F. Bioactive Glass Modified with Zirconium Incorporation for Dental Implant Applications: Fabrication, Structural, Electrical, and Biological Analysis. *Int. J. Mol. Sci.* **2023**, *24*, 10571. [CrossRef]

289. Zhu, Y.; Zhang, Y.; Wu, C.; Fang, Y.; Yang, J.; Wang, S. The effect of zirconium incorporation on the physiochemical and biological properties of mesoporous bioactive glasses scaffolds. *Microporous Mesoporous Mater.* **2011**, *143*, 311–319. [CrossRef]

290. de Souza, L.P.L.; Lopes, J.H.; Ferreira, F.V.; Martin, R.A.; Bertran, C.A.; Camilli, J.A. Evaluation of effectiveness of 45S5 bioglass doped with niobium for repairing critical-sized bone defect in in vitro and in vivo models. *J. Biomed. Mater. Res.* **2020**, *108A*, 446–457. [CrossRef]

291. de Souza Balbinot, G.; Branco Leitune, V.C.; da Cunha Bahlis, E.A.; Ponzoni, D.; Visioli, F.; Mezzomo Collares, F. Niobium-containing bioactive glasses modulate alkaline phosphatase activity during bone repair. *J. Biomed. Mater. Res.* **2023**, *111*, 1224–1231. [CrossRef]

292. Hammami, I.; Gavinho, S.R.; Pádua, A.S.; Lança, M.d.C.; Borges, J.P.; Silva, J.C.; Sá-Nogueira, I.; Jakka, S.K.; Graça, M.P.F. Extensive Investigation on the Effect of Niobium Insertion on the Physical and Biological Properties of 45S5 Bioactive Glass for Dental Implant. *Int. J. Mol. Sci.* **2023**, *24*, 5244. [CrossRef] [PubMed]

293. Lopes, J.H.; Souza, L.P.; Domingues, J.A.; Ferreira, F.V.; de Alencar Hausen, M.; Camilli, J.A.; Martin, R.A.; de Rezende Duek, E.A.; Mazali, I.O.; Bertran, C.A. In vitro and in vivo osteogenic potential of niobium-doped 45S5 bioactive glass: A comparative study. *J. Biomed. Mater. Res.* **2020**, *108B*, 1372–1387. [CrossRef] [PubMed]

294. Patel, S.; Azeem, P.A.; Manavathi, B.; Adhikari, A.; Padala, C. In-vitro biomineralization, mechanical properties and drug release efficacy of tantalum containing borophosphate bioactive glasses. *J. Drug Deliv. Sci. Technol.* **2023**, *84*, 104436. [CrossRef]

295. Mendonca, A.; Rahman, M.S.; Alhalawani, A.; Rodriguez, O.; Gallant, R.C.; Ni, H.; Clarkin, O.M.; Towler, M.R. The effect of tantalum incorporation on the physical and chemical properties of ternary silicon-calcium-phosphorous mesoporous bioactive glasses. *J. Biomed. Mater. Res. B Part B* **2019**, *107B*, 2229–2237. [CrossRef] [PubMed]

296. Alhalawani, A.M.F.; Towler, M.R. A novel tantalum-containing bioglass. Part I. Structure and solubility. *Mater. Sci. Eng. C* **2017**, *72*, 202–211. [CrossRef]

297. Alhalawani, A.M.F.; Mehrvar, C.; Stone, W.; Waldman, S.D.; Towler, M.R. A novel tantalum-containing bioglass. Part II. Development of a bioadhesive for sternal fixation and repair. *Mater. Sci. Eng. C* **2017**, *71*, 401–411. [CrossRef] [PubMed]

298. Cong, D.; Zhang, Z.; Xu, M.; Wang, J.; Pu, X.; Huang, Z.; Liao, X.; Yin, G. Vanadium-Doped Mesoporous Bioactive Glass Promotes Osteogenic Differentiation of rBMSCs via the WNT/β-Catenin Signaling Pathway. *ACS Appl. Bio Mater.* **2023**, *6*, 3863–3874. [CrossRef] [PubMed]

299. Li, J.; Li, J.; Wei, Y.; Xu, N.; Li, J.; Pu, X.; Wang, J.; Huang, Z.; Liao, X.; Yin, G. Ion release behavior of vanadium-doped mesoporous bioactive glass particles and the effect of the released ions on osteogenic differentiation of BMSCs via the FAK/MAPK signaling pathway. *J. Mater. Chem. B* **2021**, *9*, 7848. [CrossRef]

300. Li, J.; Li, X.; Li, J.; Pu, X.; Wang, J.; Huang, Z.; Yin, G. Effects of incorporated vanadium and its chemical states on morphology and mesostructure of mesoporous bioactive glass particles. *Microporous Mesoporous Mater.* **2021**, *319*, 111061. [CrossRef]

301. Ponta, O.; Ciceo-Lucacel, R.; Vulpoi, A.; Radu, T.; Simon, S. Molybdenum effect on the structure of SiO_2-CaO-P_2O_5 bioactive xerogels and on their interface processes with simulated biofluids. *J. Biomed. Mater. Res. Part A* **2014**, *102A*, 3177–3185. [CrossRef]

302. Ciceo Lucacel, R.; Ponta, O.; Licarete, E.; Radu, T.; Simon, V. Synthesis, structure, bioactivity and biocompatibility of melt-derived P_2O_5-CaO-B_2O_3-K_2O-MoO_3 glasses. *J. Non-Cryst. Solids* **2016**, *439*, 67–73. [CrossRef]

303. Rho, J.Y.; Kuhn-Spearing, L.; Zioupos, P. Mechanical properties and the hierarchical structure of bone. *Med. Eng. Phys.* **1998**, *20*, 92–102. [CrossRef]

304. Dhiflaoui, H.; Ben Salem, S.; Salah, M.; Dabaki, Y.; Chayoukhi, S.; Gassoumi, B.; Hajjaji, A.; Ben Cheikh Larbi, A.; Amlouk, M.; Benhayoune, H. Influence of TiO_2 on the Microstructure, Mechanical Properties and Corrosion Resistance of Hydroxyapatite HaP + TiO_2 Nanocomposites Deposited Using Spray Pyrolysis. *Coatings* **2023**, *13*, 1283. [CrossRef]

305. Elalmis, Y.B.; Ikizler, B.K.; Depren, S.K.; Yucel, S.; Aydin, I. Investigation of alumina doped 45S5 glass as a bioactive filler for experimental dental composites. *Int. J. Appl. Glass Sci.* **2021**, *12*, 313–327. [CrossRef]

306. Pawlik, J.; Widziołek, M.; Cholewa-Kowalska, K.; Łączka, M.; Osyczka, A.M. New sol-gel bioactive glass and titania composites with enhanced physico-chemical and biological properties. *J. Biomed. Mater. Res. Part A* **2014**, *102A*, 2383–2394. [CrossRef]

307. Radice, S.; Kern, P.; Bürki, G.; Michler, J.; Textor, M. Electrophoretic deposition of zirconia-Bioglass® composite coatings for biomedical implants. *J. Biomed. Mater. Res.* **2007**, *82A*, 436–444. [CrossRef]

308. Laybidi, F.H.; Bahrami, A. Antibacterial properties of ZnO-containing bioactive glass coatings for biomedical applications. *Mater. Lett.* **2024**, *365*, 136433. [CrossRef]

309. Marin, E.; Adachi, T.; Boschetto, F.; Zanocco, M.; Rondinella, A.; Zhu, W.; Bock, R.; McEntirec, B.; Bal, S.B.; Pezzotti, G. Biological response of human osteosarcoma cells to Si_3N_4-doped Bioglasses. *Mater. Des.* **2018**, *159*, 79–89. [CrossRef]

310. Farnoush, H.; Muhaffel, F.; Cimenoglu, H. Fabrication and characterization of nano-HA-45S5 bioglass composite coatings on calcium-phosphate containing micro-arc oxidized CP-Ti substrates. *Appl. Surf. Sci.* **2015**, *324*, 765–774. [CrossRef]

311. Soares, V.O.; Daguano, J.K.M.B.; Lombello, C.B.; Bianchin, O.S.; Gonçalves, L.M.G.; Zanotto, E.D. New sintered wollastonite glass-ceramic for biomedical applications. *Ceram. Int.* **2018**, *44*, 20019–20027. [CrossRef]

312. Hsu, P.Y.; Kuo, H.C.; Tuan, W.H.; Shih, S.J.; Naito, M.; Lai, P.L. Manipulation of the degradation behavior of calcium sulfate by the addition of bioglass. *Prog. Biomater.* **2019**, *8*, 115–125. [CrossRef] [PubMed]

313. He, F.; Yang, F.; Zhu, J.; Peng, Y.; Tian, X.; Chen, X. Fabrication of a Novel Calcium Carbonate Composite Ceramic as Bone Substitute. *J. Am. Ceram. Soc.* **2015**, *98*, 223–228. [CrossRef]

314. Batool, S.A.; Wadood, A.; Hussain, S.W.; Yasir, M.; Ur Rehman, M.A. A Brief Insight to the Electrophoretic Deposition of PEEK-, Chitosan-, Gelatin-, and Zein-Based Composite Coatings for Biomedical Applications: Recent Developments and Challenges. *Surfaces* **2021**, *4*, 205–239. [CrossRef]

315. Ur Rehman, M.A.; Bastan, F.E.; Nawaz, A.; Nawaz, Q.; Wadood, A. Electrophoretic deposition of PEEK/bioactive glass composite coatings on stainless steel for orthopedic applications: An optimization for in vitro bioactivity and adhesion strength. *Int. J. Adv. Manuf. Technol.* **2020**, *108*, 1849–1862. [CrossRef]

316. Manzur, J.; Akhtar, M.; Aizaz, A.; Ahmad, K.; Yasir, M.; Minhas, B.Z.; Avcu, E.; Ur Rehman, M.A. Electrophoretic Deposition, Microstructure, and Selected Properties of Poly(lactic-co-glycolic) Acid-Based Antibacterial Coatings on Mg Substrate. *ACS Omega* **2023**, *8*, 18074–18089. [CrossRef]

317. Sadeghinia, Z.; Emadi, R.; Shamoradi, F. A study of the electrophoretic deposition of polycaprolactone-chitosan-bioglass nanocomposite coating on stainless steel (316L) substrates. *J. Bioact. Compat. Pol.* **2022**, *37*, 53–71. [CrossRef]
318. Chen, Q.; Cabanas-Polo, S.; Goudouri, O.M.; Boccaccini, A.R. Electrophoretic co-deposition of polyvinyl alcohol (PVA) reinforced alginate-Bioglass® composite coating on stainless steel: Mechanical properties and in-vitro bioactivity assessment. *Mater. Sci. Eng. C* **2014**, *40*, 55–64. [CrossRef]
319. Rouein, Z.; Jafari, H.; Pishbin, F.; Mohammadi, R.; Simchi, A. Biodegradation behavior of polymethyl methacrylate-bioactive glass 45S5 composite coated magnesium in simulated body fluid. *Trans. Nonferrous Met. Soc. China* **2022**, *32*, 2216–2228. [CrossRef]
320. Clifford, A.; Luo, D.; Zhitomirsky, I. Colloidal strategies for electrophoretic deposition of organic-inorganic composites for biomedical applications. *Colloids Surf. A Physicochem. Eng. Asp.* **2017**, *516*, 219–225. [CrossRef]
321. Deen, I.; Singh Selopal, G.; Wang, Z.M.; Rosei, F. Electrophoretic deposition of collagen/chitosan films with copper-doped phosphate glasses for orthopaedic implants. *J. Colloid Interface Sci.* **2022**, *607*, 869–880. [CrossRef]
322. Sikkema, R.; Keohan, B.; Zhitomirsky, I. Hyaluronic-Acid-Based Organic-Inorganic Composites for Biomedical Applications. *Materials* **2021**, *14*, 4982. [CrossRef]
323. Chen, Q.; Cordero-Arias, L.; Roether, J.A.; Cabanas-Polo, S.; Virtanen, S.; Boccaccini, A.R. Alginate/Bioglass® composite coatings on stainless steel deposited by direct current and alternating current electrophoretic deposition. *Surf. Coat. Technol.* **2012**, *233*, 49–56. [CrossRef]
324. Cordero-Arias, L.; Boccaccini, A.R. Electrophoretic deposition of chondroitin sulfate-chitosan/bioactive glass composite coatings with multilayer design. *Surf. Coat. Technol.* **2017**, *315*, 417–425. [CrossRef]
325. Ramos Rivera, L.; Cochis, A.; Biser, S.; Canciani, E.; Ferraris, S.; Rimondini, L.; Boccaccini, A.R. Antibacterial, pro-angiogenic and pro-osteointegrative zein-bioactive glass/copper based coatings for implantable stainless steel aimed at bone healing. *Bioact. Mater.* **2021**, *6*, 1479–1490. [CrossRef]
326. Avcu, E.; Baştan, F.E.; Abdullah, H.Z.; Ur Rehman, M.A.; Avcu, Y.Y.; Boccaccini, A.R. Electrophoretic deposition of chitosan-based composite coatings for biomedical applications: A review. *Progr. Mater. Sci.* **2019**, *103*, 69–108. [CrossRef]
327. Smart, S.K.; Cassady, A.I.; Lu, G.Q.; Martin, D.J. The biocompatibility of carbon nanotubes. *Carbon* **2006**, *44*, 1034–1047. [CrossRef]
328. Meng, D.; Narayan Rath, S.; Mordan, N.; Salih, V.; Kneser, U.; Boccaccini, A.R. In vitro evaluation of 45S5 Bioglass®-derived glass-ceramic scaffolds coated with carbon nanotubes. *J. Biomed. Mater. Res. Part A* **2011**, *99A*, 435–444. [CrossRef]
329. Meng, D.; Ioannou, J.; Boccaccini, A.R. Bioglass®-based scaffolds with carbon nanotube coating for bone tissue engineering. *J. Mater. Sci. Mater. Med.* **2009**, *20*, 2139–2144. [CrossRef]
330. Charlotte Schausten, M.; Meng, D.; Telle, R.; Boccaccini, A.R. Electrophoretic deposition of carbon nanotubes and bioactive glass particles for bioactive composite coatings. *Ceram. Int.* **2010**, *36*, 307–312. [CrossRef]
331. Sun, F.; Sask, K.N.; Brash, J.L.; Zhitomirsky, I. Surface modifications of Nitinol for biomedical applications. *Colloids Surf. B Biointerfaces* **2008**, *67*, 132–139. [CrossRef]
332. Pishbin, F.; Mouriño, V.; Flor, S.; Kreppel, S.; Salih, V.; Ryan, M.P.; Boccaccini, A.R. Electrophoretic Deposition of Gentamicin-Loaded Bioactive Glass/Chitosan Composite Coatings for Orthopaedic Implants. *ACS Appl. Mater. Interfaces* **2014**, *6*, 8796–8806. [CrossRef]
333. Khanmohammadi, S.; Aghajani, H.; Farrokhi-Rad, M. Vancomycin loaded-mesoporous bioglass/hydroxyapatite/chitosan coatings by electrophoretic deposition. *Ceram. Int.* **2022**, *48*, 20176–20186. [CrossRef]
334. Zhao, P.; Liu, H.; Deng, H.; Xiao, L.; Qin, C.; Du, Y.; Shi, X. A study of chitosan hydrogel with embedded mesoporous silica nanoparticles loaded by ibuprofen as a dual stimuli-responsive drug release system for surface coating of titanium implants. *Colloids Surf. B Biointerfaces* **2014**, *123*, 657–663. [CrossRef]
335. Akhtar, M.A.; Mariotti, C.E.; Conti, B.; Boccaccini, A.R. Electrophoretic deposition of ferulic acid loaded bioactive glass/chitosan as antibacterial and bioactive composite coatings. *Surf. Coat. Technol.* **2021**, *405*, 126657. [CrossRef]
336. Radda'a, N.S.; Goldmann, W.H.; Detsch, R.; Roether, J.A.; Cordero-Arias, L.; Virtanen, S.; Moskalewicz, T.; Boccaccini, A.R. Electrophoretic deposition of tetracycline hydrochloride loaded halloysite nanotubes chitosan/bioactive glass composite coatings for orthopedic implants. *Surf. Coat. Technol.* **2017**, *327*, 146–157. [CrossRef]
337. Moses, J.C.; Mandal, B.B. Mesoporous Silk-Bioactive Glass Nanocomposites as Drug Eluting Multifunctional Conformal Coatings for Improving Osseointegration and Bactericidal Properties of Metal Implants. *ACS Appl. Mater. Interfaces* **2022**, *14*, 14961–14980. [CrossRef]
338. Patel, K.D.; El-Fiqi, A.; Lee, H.Y.; Singh, R.K.; Kim, D.A.; Lee, H.H.; Kim, H.W. Chitosan-nanobioactive glass electrophoretic coatings with bone regenerative and drug delivering potential. *J. Mater. Chem.* **2012**, *22*, 24945. [CrossRef]
339. Ur Rehman, M.A.; Bastan, F.E.; Nawaz, Q.; Goldmann, W.H.; Maqbool, M.; Virtanen, S.; Boccaccini, A.R. Electrophoretic deposition of lawsone loaded bioactive glass (BG)/chitosan composite on polyetheretherketone (PEEK)/BG layers as antibacterial and bioactive coating. *J. Biomed. Mater. Res. Part A* **2018**, *106A*, 3111–3122. [CrossRef]

Review

Plasma-Sprayed Osseoconductive Hydroxylapatite Coatings for Endoprosthetic Hip Implants: Phase Composition, Microstructure, Properties, and Biomedical Functions

Robert B. Heimann

Am Stadtpark 2A, D-02826 Görlitz, Germany; robert.heimann@ocean-gate.de

Abstract: This contribution attempts to provide a state-of-the-art account of the physicochemical and biomedical properties of the plasma-sprayed hydroxylapatite (HAp) coatings that are routinely applied to the surfaces of metallic endoprosthetic and dental root implants designed to replace or restore the lost functions of diseased or damaged tissues of the human body. Even though the residence time of powder particles of HAp in the plasma jet is extremely short, the high temperature applied induces compositional and structural changes in the precursor HAp that severely affect its chemical and physical properties and in turn its biomedical performance. These changes are based on the incongruent melting behavior of HAp and can be traced, among many other analytical techniques, by high resolution synchrotron X-ray diffraction, vibrational (Raman) spectroscopy, and nuclear magnetic resonance (NMR) spectroscopy. In vivo reactions of the plasma-sprayed coatings to extracellular fluid (ECF) can be assessed and predicted by in vitro testing using simulated body fluids (SBFs) as proxy agents. Ways to safeguard the appropriate biological performance of HAp coatings in long-term service by controlling their phase content, porosity, surface roughness, residual stress distribution, and adhesion to the implant surface are being discussed.

Keywords: plasma spraying; hydroxylapatite; implants; foreign body reaction; bond coats; Raman spectroscopy; nuclear magnetic resonance spectroscopy

Citation: Heimann, R.B. Plasma-Sprayed Osseoconductive Hydroxylapatite Coatings for Endoprosthetic Hip Implants: Phase Composition, Microstructure, Properties, and Biomedical Functions. *Coatings* **2024**, *14*, 787. https://doi.org/10.3390/coatings14070787

Academic Editor: Alina Vladescu

Received: 18 May 2024
Revised: 19 June 2024
Accepted: 20 June 2024
Published: 24 June 2024

1. Introduction

There is a growing demand for safe, high performance, and increasingly sophisticated medical implants, because our society is moving toward an increasingly aging population. Presently, much research is focused on advanced medical implants that are not only biocompatible but have modified surfaces able to direct specific immunomodulation at the cellular level. An important surface modification tool is the deposition by plasma spraying of hydroxylapatite (HAp) coatings, which promote improved implant performance and longevity, effectively counteract foreign body reaction by avoiding the formation of an acellular fibrous capsule, and provide by their osseoconductive nature enhanced bone cell adhesion, migration, spreading, and proliferation.

Total hip (THA) and knee arthroplasty (TKA) are among the most frequently performed and in their long-term outcome highly successful and effective surgical interventions worldwide. The need for endoprosthetic hip and knee implants is deeply rooted in the kinematics of human locomotion. The forces acting on the knee and hip joints can be expressed by multiples of body mass. They are approximately one times the body mass while resting (~1 kN), up to three times during walking, up to five times during running, and up to eight times when jumping. The risk of serious joint damage increases dramatically beyond this threshold. Intensifying the socio-economic impact of this physiologically given situation is the fact that people generally live longer and gain weight by overeating, excessive consumption of sugar, fat, and alcohol, and a sedentary lifestyle. Eventually, the cartilage tissue lining the articulating parts of the joints wears away. This leads to increased friction during movement, and in turn to inflammation, pain and, finally, immobilization.

Common root causes for the need of hip joint replacement are osteoarthritis, rheumatoid arthritis, osteonecrosis, injuries or broken bones from trauma or disease, and degeneration due to wear and tear. In all these cases, a THA (total hip arthroplasty) is the only reasonable option to restore mobility and hence, to promote a rewarding life without pain. This type of operation has a long pedigree, starting with the pioneering work by Themistocles Gluck (1853–1942) and continuing with Robert Judet (1909–1980) and Austin T. Moore (1899–1962) and then on to Sir John Charnley (1911–1982), who designed the hip joint replacement system on which all subsequent solutions are based.

In 2022, some 3.1 million THA and TKA procedures were reported in the USA alone, among them 37% primary THA and 54% primary TKA, the remainder being revision operations [1]. In 2021, ceramic-on-polyethylene (CoP) bearing couples were, at about 63%, the dominant choice for THA. Cementless fixation of the femoral stem constituted about 95% of all THA, in contrast to only 19% of all TKA.

In Germany, about 178,000 primary THA and 137,000 primary TKA were registered in 2022. Fifty-eight percent of all these THAs were performed with a ceramic-on-polyethylene (CoP) sliding couple [2]. Most other remaining implants used a metal-on-polyethylene (MoP) combination. Some 10% of all hip operation surgeries were revisions, the causes of which were, in descending order, implant loosening (23%), periprosthetic infection (16%), post-operative periprosthetic fracture (16%), dislocation (14%), wear (6%), and others (24%). Purely mechanical implant failure was remarkably low; it was reported in only 2% of all THA revision operations, attesting to the high reliability and solidly entrenched quality standards of modern biomedical devices and operation techniques. Indeed, the revision probability after 5 years was found to be only around 4%. In a global context, in Switzerland, Germany, and Austria, the number of THAs performed were a high 3230/1 million population (ppm), 3010 ppm, and 2870 ppm, respectively, compared to only 400 ppm in Chile, 180 in Costa Rica, and a paltry 70 ppm in Mexico. In 2021, the THA average of all OECD countries was 1720 ppm [2].

The high incidence of THA and TKA operations in the developed and increasingly, in the developing world, is reflected in the global hip replacement implant market valued at USD 7.26 billion in 2022, expected to grow at a CAGR of 5.2% to USD 11.5 billion in 2031 [3]. Hence, hip and knee arthroplasties constitute a meaningful part of the global GNP. This is also mirrored by the growing amount of research and development effort designed to improve existing and discover novel solutions to the problem of the growing demand for endoprosthetic devices. Several aspects of this research activity will be discussed in this contribution. It is this continuously accelerating research effort that makes the present review needed, timely, and meaningful.

2. Why Hydroxylapatite Coatings?

Figure 1A shows a high-end THA implant that consists of a titanium alloy stem with an alumina femoral ball attached. The stem will be anchored in the femur, and an acetabular cup with an alumina inset anchored in the hip bone. During the absence of synovial fluid that lubricates the joints, this type of ceramic-on-ceramic (CoC) bearing couple is highly effective and reliable, even though it may produce a disturbing squeaking noise during walking and a high risk of fracture [4]. In addition, its comparatively high cost puts a severe strain on the healthcare systems of many countries. Hence, for economic reason, the standard endoprosthesis does not use the bearing couple option shown in Figure 1A but that of a ceramic (CoP) or metal (MoP) femoral head articulating against a highly cross-linked polyethylene-lined acetabular cup. Figure 1B shows the cross-sectional image of the implant-bone interface, demonstrating that a plasma-sprayed hydroxylapatite (HAp) layer provides a tight and continuous connection between an implant and bone [5].

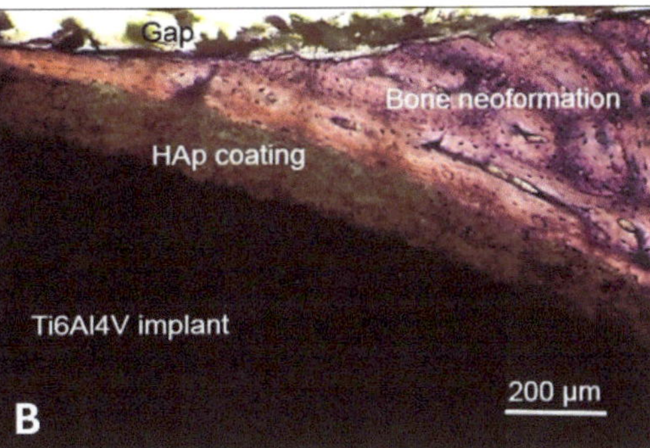

Figure 1. (**A**) State-of-the-art THA implant consisting of a titanium alloy femoral stem to which a spherical ball made from high performance alumina (α-Al$_2$O$_3$) is attached (left) as well as an acetabular cup fashioned from titanium with an alumina insert (right). A coating of plasma-sprayed HAp has been deposited onto the femoral stem and the acetabular cup casing. Image courtesy CeramTec AG, Plochingen, Germany. (**B**) In this cross-section, a HAp coating air plasma-sprayed onto a titanium alloy rod and implanted into the proximal femoral medulla of an adult sheep provides a tight and continuous interface between implant and cortical bone matter. Sample stained with toluidine blue [5]. © With permission by Springer Science and Business Media.

As shown in Figure 1A, modern endoprosthetic implants are frequently coated with a thin layer of hydroxylapatite (HAp) that assists in tight and continuous osseointegration (Figure 1B). At present, a cementless implant coated with HAp is regarded as the 'gold standard' in orthopedics and dentistry. The reason for this is grounded in the chemical and structural similarity of synthetic HAp with biological, i.e., bone-like apatite (bone mineral), making HAp a highly biocompatible material. Nevertheless, although synthetic HAp noticeably differs in its chemical make-up, stoichiometry, crystallinity, defect density, presence of hydroxyl ions, and type and degree of ion substitution from bone mineral, in biomedical applications it spans a wide range from hip and dental root implant coatings and porous bone-growth supporting 3D-printed scaffolds to bone cavity filling material and drug and gene delivery vehicles.

Any foreign material or device incorporated into the body triggers a host/recipient response that manifests itself in the five time-dependent foreign body reaction (FBR) phases, i.e., (i) blood protein adsorption completed a few seconds after contact, (ii) neutrophil recruitment and associated acute inflammation within minutes, (iii) monocyte recruitment and their differentiation to macrophages, associated with chronic inflammation within hours, (iv) foreign body multinucleated giant cell formation after some days, and (v) fibrotic encapsulation within weeks. Fibrotic encapsulation is a protective measure of the body, but unfortunately, it prevents solid and lasting osseointegration of the implant (Figure 2A).

The fibrous encapsulation of the implanted device can affect the implant's function and frequently leads to failure. For example, interface loosening of a press-fitted implant may be caused by mechanically induced periprosthetic osteolysis, in particular when associated with micro-movements of the patient during the healing phase, which limits implant longevity. Such early loosening is best defined as prosthetic migration. Initiation of loosening during or shortly after surgery is promoted by poor implant interlock, poor bone quality, and osteolytic resorption of a necrotic bone bed.

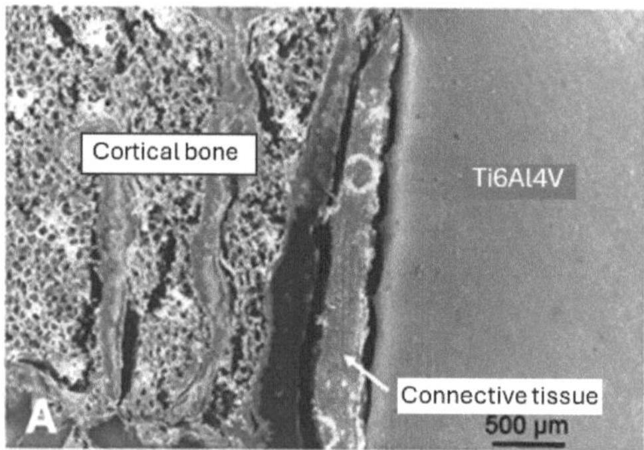

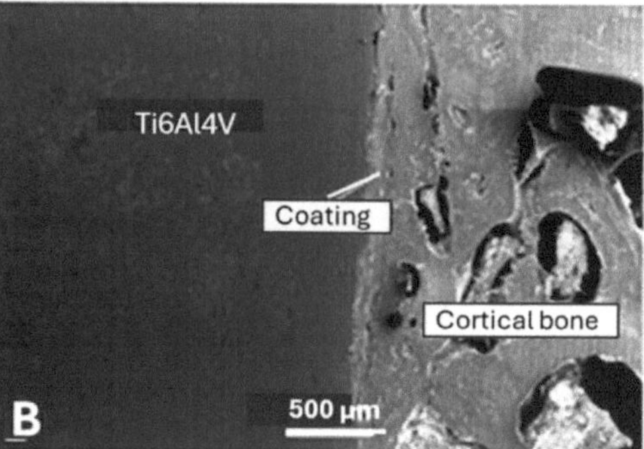

Figure 2. (**A**) An acellular connective tissue capsule formed around an implant is the result of a typical foreign body reaction (FBR). It has been formed around a titanium alloy cube of $5 \times 5 \times 5$ mm^3 size that was implanted into the femoral condyle of an adult dog for 6 months [6]. (**B**) The cross-sectional image shows, in the presence of a thin plasma-sprayed HAp coating, absence of a connective tissue capsule and thus, tight and continuous osseointegration. The titanium alloy cube was implanted into the lateral femoral condyle of a dog for 6 months [6]. © With permission by Wiley-VCH.

The quest for reducing the FBR is an effective way to promote better implant performance and longevity, including enhanced adhesion and strengthening of the implant-bone interface. Among many possible ways to combat FBR [7], coating the implant stem and the casing of the acetabular cup of a hip joint implant with a thin layer of osseoconductive HAp (Figures 1A and 2B) is an economically viable and biomedically reliable route toward alleviating the FBR and thus preventing the formation of a fibrous connective tissue capsule. Recast in simple terms, an HAp coating acts as an agent of biological subterfuge, triggering the body into accepting the implant as part of its own metabolic environment, thus avoiding the negative consequences of FBR.

In addition to preventing the formation of a fibrous acellular connective tissue capsule surrounding the implant, other advantages of a plasma-sprayed HAp coating include:

- Enhanced bone apposition rate by osseoinduction, owing to preferential adsorption of bone growth-supporting factors such as BMPs as well as NCPs such as osteocalcin, osteopontin, sialylated glycoproteins, proteoglycans, and several other hormones, chemokines and cytokines;
- Enhanced bone-bonding ability that provides a strong and continuous interface between bone tissue and implant and thus enables the transmission of compressive as well as (limited) tensile and also some shear forces;
- Variable HAp coating thicknesses between 50 and 250 µm can be selected, dependent on medical requirements;
- The option to apply some novel deposition techniques including SPS and SPPS allows the deposition of coatings with thicknesses $\ll 50$ µm;
- Acceleration of the healing process when compared to implants without an osseoconductive coating;
- Supporting attachment of the epithelium in the case of transmucosal dental implants;
- Reducing the health risk of potentially toxic heavy metal ions released from the surface of the metallic implant into the periprosthetic tissue and thus minimizing a possible cytotoxic response, and
- Support available by quality control and standards according to ASTM F1185-03 (2014), ASTM F1044-05 (2017), ASTM F1160 (2014), ISO 13179-2: 2018, and others.

However, despite the fact that modern endoprosthetic hip joint implants have reached a high degree of mechanical resilience, chemical inertness, biocompatibility, and overall performance reliability as is manifest in a high Weibull modulus, safeguarding their long-term successful behavior in the aggressive environment of the human body remains a major challenge. Stability problems may be associated with so-called aseptic loosening long after surgery, which to some degree can be related to the large gradient of the modulus of the metallic implant (around 110 GPa for Ti6Al4V) and cortical bone (10–15 GPa). Mechanical loads will then be transmitted through the high modulus metal implant rather than through bone matter, causing the latter to atrophy by forced bone mass loss, because bony health depends strongly on continuous tensile loading and bone remodeling along Wolff's law. This mechanism is known by the term of 'stress shielding'. Modern ß-type Ti alloys with elastic moduli closer to that of bone, such as Ti29Nb13Ta4.6Zr (TNTZ), which has a modulus of < 60 GPa, are thought to reduce the risk of stress shielding.

In parallel, the very process of plasma spraying triggers large tensile quenching and thermal stresses at the interface and throughout the bulk coating that in subsequent medical service may result in chipping and even delamination of the osseoconductive HAp coating layer. The reason underlying residual stresses are manifold, but all of them are related to differences in the coefficients of thermal expansion between the metallic substrate and the ceramic coating (see Section 5.3.8).

A possible way to avoid this is to apply bond coats that act as intermediates, binding both the implant and HAp coatings tightly to cortical bone matter [8]. In addition, ceramic bond coats are thought to function as thermal barriers, able to reduce the steep gradient of the coefficient of thermal expansion between metallic implant and ceramic coating. They decrease the rate of heat transfer and thus the degree of thermal decomposition of HAp during plasma spraying, which helps to reduce the occurrence of amorphous calcium phosphate (ACP) formed by a quenching contact at the immediate coating–substrate interface. In addition, a bond coat prevents direct contact between the metallic implant surface and the HAp layer, reducing the risk of thermal decomposition of the HAp, which is known to be catalyzed by metal ions.

However, despite their clear advantages, in the clinical praxis no bond coat of any nature has been applied to endoprosthetic implants yet, even though the ISO 13779-2:2018 norm refers to dual coatings (with the caveat that the testing methods recommended for single HAp coatings cannot be applied to dual layer coatings). Consequently, to date, the biomedical potential of bond coats, although promising, has only been experimentally tested and assessed in animal models [5,6].

Justification for applying biocompatible bond coats can be found in the fact that owing to the chemical and structural similarity of HAp to the inorganic component of living bone (bone mineral), the bonding between HAp and bone matter is much tighter than that between HAp and the metallic implant. A suitable bond coat is designed to overcome this impediment by providing a strong gap bridging potential, conducive to forming three undisturbed and continuous interfaces: bone-HAp, HAp-bond coat, and bond coat-implant metal. Such bond coats are only at the beginning of their complex and involved development cycle. Hence, reliable data on their long-term mechanical, chemical, and biological stability, their compatibility with implants of titanium alloy, CoCrMo alloy, and austenitic stainless steel, and their effect on bone cell adhesion, spreading, and proliferation are still absent and thus need to be collected and assessed by future research. Preliminary animal studies using sheep and dog models [5,6] have shown that titania-based bond coats may be well suited to withstand the high local tensile and shear stresses occurring during micromovement of the animal patient during the first stages of the healing process after implantation.

3. A Short History of Calcium Orthophosphate Research

The close association of calcium phosphates with living matter such as bone has been known for a rather long time. Calcium orthophosphates including hydroxylapatite, $Ca_{10}(PO_4)_6(OH)_2$, have been known to be associated with organic tissue, more specifically with bone, for at least 250 years [9]. Early evidence points to 1769, when the Swedish chemists Johan Gottlieb Gahn (1745–1818) and Carl Wilhelm Scheele (1742–1786) found tricalcium orthophosphate, $Ca_3(PO_4)_2$ to be the product of burning bone.

Subsequently, the roles that calcium orthophosphates play in bone, teeth, and pathologic urinary and renal calculi were in part already known by the early 1800s. Some calcium phosphate phases involved in biomineralization, such as amorphous calcium phosphate, octacalcium phosphate, and dicalcium phosphate dihydrate (brushite) [10] were already known or suggested by the end of the nineteenth century.

However, only the application of X-rays to crystal structure analysis over one hundred years ago [11] allowed a paradigmatic shift from a descriptive to a predictive acquisition of information on the structural chemistry of biological calcium phosphates [12–16]. Willem F. de Jong [17] identified the crystal structure of bone calcium phosphate as comparable to geological apatite. Hendricks et al. [12] concluded (erroneously) that the inorganic constituent of bone contains carbonate apatite, $Ca_{10}[CO_3(PO_4)_6]\cdot H_2O$, which they thought to be structurally akin to fluorapatite. The same authors mentioned oxyapatite, $Ca_{10}O(PO_4)_6$, as an anhydrous product of heating HAp up to 900°C. This was disputed by Bredig and coworkers [16], who instead proposed the existence of only partly dehydroxylated oxy-hydroxylapatite (OHAp), $Ca_{10}(PO_4)_6X_{2m}O_n$ ($X = OH^-$; $m + n = 1$). In their opinion, an empty X site would destabilize the structure, a contention that was confirmed much later by computer modeling applied to investigate the dehydration of HAp to oxyapatite (OAp) and the defect chemistry of calcium-deficient HAp [18]. However, despite these thermodynamically based constraints, there is evidence that under certain experimental conditions, OAp can be stabilized [19–22] (see Section 5.3.5).

A thorough account of the composition and crystal structure of Ca-deficient HAp with the approximate formula of $Ca_{10-x}(HPO_4)_x(PO_4)_{6-x}$ (OH, O, Cl, F, CO_3, $\square$)$_{2-x}\cdot nH_2O$; $0 < x < 1$; $n = 0$–2.5 [23,24] was given in a review by Rey et al. [25]. The role water molecules play in the development and structure of bone mineral was emphasized by Pasteris [26]. More details on the structure and biological function of biological apatite can be extracted from Ref. [8].

4. Hierarchical Structure of Bone

The realization that calcium orthophosphates are crucial constituents of bone matter has led eventually to their application in several medical disciplines. A major segment of the utilization of the calcium orthophosphate HAp is in the challenging fields of bone reconstruction, bone scaffolds, coatings for hip, knee, and dental implants, and vertebrae replacement [27].

Bone is a biological composite structure that during its growth is composed initially of soft, unmineralized tissue (osteoid) with embedded osteoblasts and osteocytes. These living cells secrete collagen I as well as non-collagenous bone matrix proteins, forming a complex network that will eventually be mineralized, whereby hydroxylapatite platelets crystallize in an oriented fashion in the gaps provided by the tropocollagen fibrils (Figure 3A).

Hence, bone can be regarded as a biocomposite consisting of Ca-deficient defect biological HAp nanocrystals of about $30 \times 50 \times 2$ nm^3 size, orientationally intergrown with triple helical strands of matrix proteins such as collagen I (ossein). Consequently, a spatially hierarchical organization exists that forms the basic structural units of bone [28,29]. Figure 3A shows schematically the hierarchical architecture of bone microfibrils [30]. Figure 3B shows a molecular mechanics model of the oriented intergrowth of hydroxylapatite nanocrystals with tropocollagen fibrils at different degrees of mineralization (0 to 40%) [31].

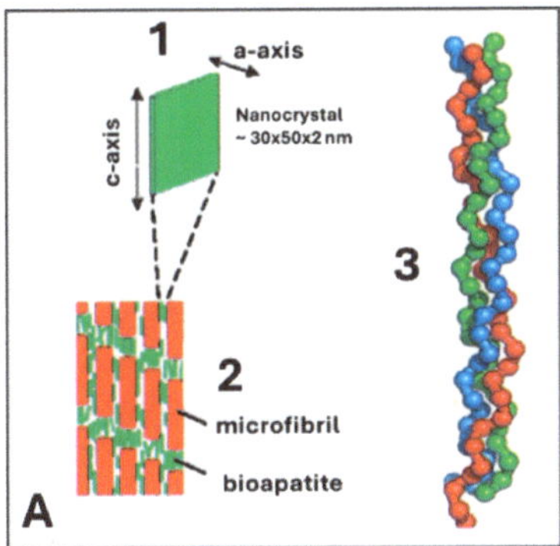
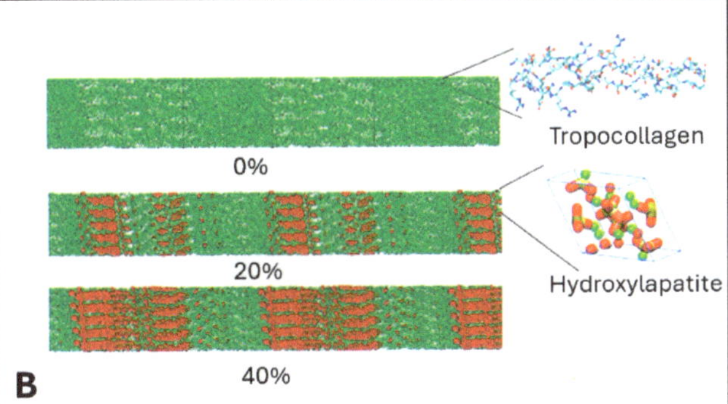

Figure 3. (**A**) Hierarchical nano-architecture of cortical bone, showing an individual crystalline platelet of HAp (1), an array of five collagen microfibrils with oriented intergrown HAp nanocrystals, which constitutes the smallest unit of the bone microstructure (2), and a triple helical strand of tropocollagen consisting of three polypeptide α-chains (3). Image modified after Ref. [30]. © Permission granted under Creative Commons Attribution 4.0 International License. (**B**) Collagen microfibril model with different degrees of mineralization. The HAp crystals are arranged such that their c-axes align with the fibril axis [31]. © Permission granted under Creative Commons Attribution-NC ShareAlike 3.0 International License.

5. Osseoconductive Hydroxylapatite Coatings

Given the important role that HAp coatings applied to the surface of an endoprosthetic implant play, it is not surprising that synthetic HAp, with its pronounced biocompatibility, was first suggested as a useful material to be incorporated in the human body [32] 50 years ago. Soon after, HAp was deposited as a bioactive, i.e., osseoconductive coating on dental implants, followed by coating the metallic parts of hip implants [33].

5.1. Osseoconduction, Osseoinduction, and Osseointegration

Osseoconduction refers to the ability of a biocompatible material including hydroxylapatite to assist the development and growth of osteoblasts, blood capillaries, and perivascular tissue into the gap between the existing cortical bone bed and the implant that was created during the operation [27]. The process results in a bony surface with various channels, pores, and a matrix that provides a scaffold for the newly formed bone matter.

The term osseoinduction refers to the recruitment of immature cells and their stimulation to develop into preosteoblasts [34]. HAp-based bioactive implant coatings assist normal cell differentiation in surrounding tissues by creating a fertile environment for enhanced cell adhesion. Cytoskeletal microfilaments such as actin, myosin, actinin, and tropomyosin control cell shape and migration. They are coupled through specialized cell membrane proteins (integrins) to extracellular adhesion molecules (fibronectin, laminin, vitronectin, thrombospondin). An interfacial layer of HAp will adsorb these integrins in a favorable conformation and promote the formation of focal adhesion centers. Particular growth factors (cytokines) may also be recruited and adsorbed at specific HAp surface sites, further promoting osseoinduction. Cytokines include transforming growth factor-β, insulin-like growth factor-1, tumor necrosis factor-α, and recombinant human bone morphogenetic proteins (rhBMPs). They stimulate osseogenesis by supporting the transformation of undifferentiated mesenchymal precursor cells to osteoprogenitor cells preceding endochondral ossification.

Finally, osseointegration describes the process by which interfacial bone is formed between an implant and living bone as a defense reaction akin to FBR. Osseointegration is the requirement for the desired long-term stability of an endoprosthesis. The process is characterized by the firm anchoring of a surgical implant by the growth of bone around it without fibrous tissue formation at the interface [34]. When an implant is integrated into the cortical bone bed, no progressive relative movement occurs between the implant and the bone with which it is in direct contact.

5.2. Deposition Techniques

A plethora of techniques are available and used to apply HAp coatings. These can be divided into non-thermal and thermal methods. Non-thermal methods carried out at or near ambient temperature include biomimetic deposition, electrochemical (ECD) and electrophoretic deposition (EPD), sol-gel deposition by dip and spin coating, hydrothermal deposition, and radiofrequency magnetron sputtering (r.f.MS). Thermal methods include atmospheric (air) plasma spraying (APS), micro plasma spraying (MPS), low-energy plasma spraying (LEPS), low pressure (vacuum) plasma spraying (LPPS/VPS), suspension plasma spraying (SPS), solution precursor plasma spraying (SPPS), plasma electrolytic oxidation (PEO), pulsed laser deposition (PLD), and other less frequently utilized techniques such as cold gas dynamic spraying (CGDS), flame spraying (FS), high velocity suspension flame spraying (HVSFS), high velocity oxyfuel spraying (HVOF), and chemical vapor deposition (CVD) [8]. The properties and application of hydroxylapatite coatings deposited by suspension plasma spraying will be briefly discussed in Section 5.3.7.

The preferred deposition method of biomedical HAp was and still is atmospheric plasma spraying [27,35–37] (Figure 4). This technique offers a fast, well-controlled, and mature way to coat metal implant surfaces with a thin, continuous layer of HAp, even though thermal dehydroxylation and decomposition, line-of-sight limitations, and porosity, residual stresses, and cohesion and adhesion issues are affecting the process and thus require close control of the intrinsic plasma spraying parameters. A comparison of the properties of HAp coatings deposited by different thermal and non-thermal methods, as well as their benefits and limitations, can be obtained from the comprehensive monograph by Heimann and Lehmann [8].

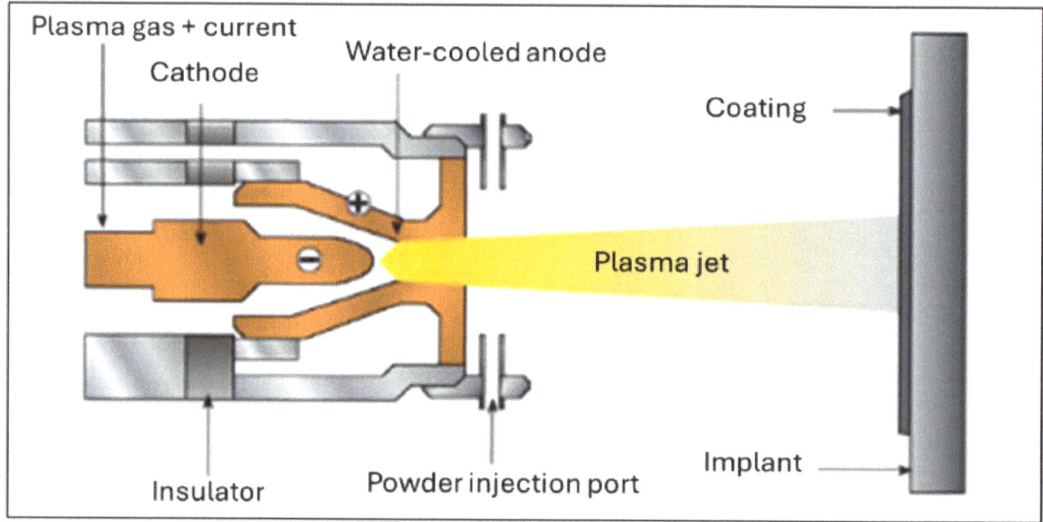

Figure 4. Schematic cross-section of an atmospheric plasmatron. © With permission of Walter de Gruyter GmbH, Berlin, Germany.

Plasma spraying is a useful tool to coat the surface of any reasonably thermally stable metallic, ceramic, or even polymeric material with a thin layer of a second material possessing a well-defined congruent melting point. Since the hot molten droplets will be rapidly quenched when arriving at the relatively cool implant surface, they form easily soluble amorphous phases by a mechanism akin to Ostwald's rule of steps. Accordingly, during spraying of HAp, ACP of variable composition forms as will be discussed below. Despite the apparent simplicity of the process, in practice several limitations exist; these limitations render many issues of adhesion, porosity, thickness, and thermal alteration of coatings highly challenging. In addition, the ubiquitous presence of residual stresses and line-of-sight restrictions require close attention.

The physics of plasma spraying involves the action of a plasma that is generated by ionization in an electric potential field of a suitable gas, preferentially argon or nitrogen. A plasma is considered the 'fourth state of matter' and is composed of positively charged ions, electrons, neutral gas atoms, and photons that make the plasma luminous. Charged ions and electrons moving within the plasma column generate a magnetic field $\vec{B}$ that is perpendicular to the direction of the electric field with a current $\vec{j}$. The vector cross-product $[\vec{j} \times \vec{B}]$ constitutes the magnetohydrodynamic Lorentz force with a vector that is mutually perpendicular to the vectors $\vec{j}$ and $\vec{B}$. As a result, an inward moving force that constricts the plasma column by the so-called magnetic pinch is generated. In addition, there exists a thermal pinch originating from the decreased conductivity of the plasma gas at the water-cooled wall of the anode (Figure 4). This effect results in an increase of the current density at the center of the plasma column. Consequently, the charged plasma concentrates at the central axis of the plasmatron, thereby further constricting the plasma column. This effects a drastic rise of the pressure in the plasma core, causing it to be blown out of the plasmatron as a plasma jet at a supersonic velocity [36,38].

5.3. Property Requirements and Performance Profile of Hydroxylapatite Coatings

HAp suffers dehydroxylation and subsequent decomposition in the plasma jet, which reaches temperatures up to 15,000 K and beyond. Partially dehydroxylated calcium orthophosphate phases such as OHAp and/or OAp are restored to fully stoichiometric HAp by contact with extracellular fluid (ECF) in vivo. Nevertheless, despite its apparent drawbacks, to this date atmospheric plasma spraying is the only biomedical coating deposition method certified and approved by the US Food and Drug Administration (FDA) [39] and many other national standardization authorities.

Figure 5A shows the typical surface of a plasma-sprayed HAp coating. One of the most crucial properties of HAp coatings is their biostability in vivo, which increases with increasing (i) phase purity, (ii) crystallinity, (iii) coatings density, and (iv) coating thickness. As discussed below, requirements (i) and (ii) are poorly met by plasma spray technology owing to the incongruent melting behavior of HAp, and requirement (iii) must be relaxed to guarantee appropriate biological performance because a certain degree of porosity must be obtained for optimum cell ingrowth. Finally, requirement (iv) calls for a coating thickness of at least 50 μm. Although thicker coatings are more stable against in vivo resorption, they bear the risk of cracking, spalling, and delamination owing to increased residual coating stresses (Figure 5B; see also Section 5.3.8).

Coating integrity, cohesive and adhesion strengths, and surface roughness can be estimated by checking the flow characteristics of the molten particles using a simple empirical 'wipe' test. A flat surface is quickly brought into the way of a molten particle trajectory with the intention to capture only a few particles that after solidification can be investigated with light optical or electron microscopy. Figure 6 shows typical examples of HAp particle splats deposited under low pressure conditions [36]. The plasma enthalpy increases from Figure 6A–D. Figure 6A shows that the enthalpy supplied to the particle is insufficient to achieve melting. Figure 6B shows a splat pattern of a particle the outer rim of which has been melted, but its core has remained highly viscous, as revealed by its porous sponge-like appearance. Finally, Figure 6C shows a well-molten splat with only a few micropores, whereas Figure 6D depicts an 'exploded' splat typical of a severely overheated particle. By comparing Figure 6A,D, this technique reveals how sensitive the parameter selection is: varying the stand-off distance, i.e., the distance between the plasmatron exit nozzle and the target surface by only a few millimeters changes the flow characteristics dramatically.

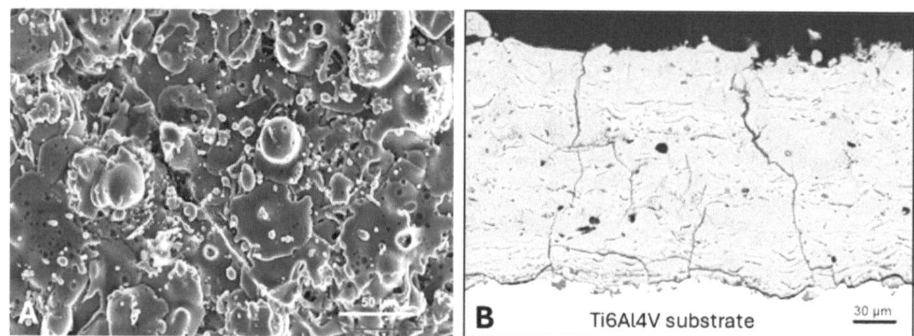

Figure 5. (**A**) Typical surface morphology of a plasma-sprayed HAp coating with well-spread 'pancake-like' particle splats. Some are only loosely adhered. Incompletely melted spherical particles are also visible. (**B**) Cross-section of a thick plasma-sprayed HAp coating with all-through microcracks generated by tensile quenching stress. Coating thickness ~200 µm; average surface roughness 10 ± 0.5 µm; porosity 5 ± 1 vol%; tensile adhesion strength 50 ± 9 MPa [40,41]. © With permission by John Wiley & Sons.

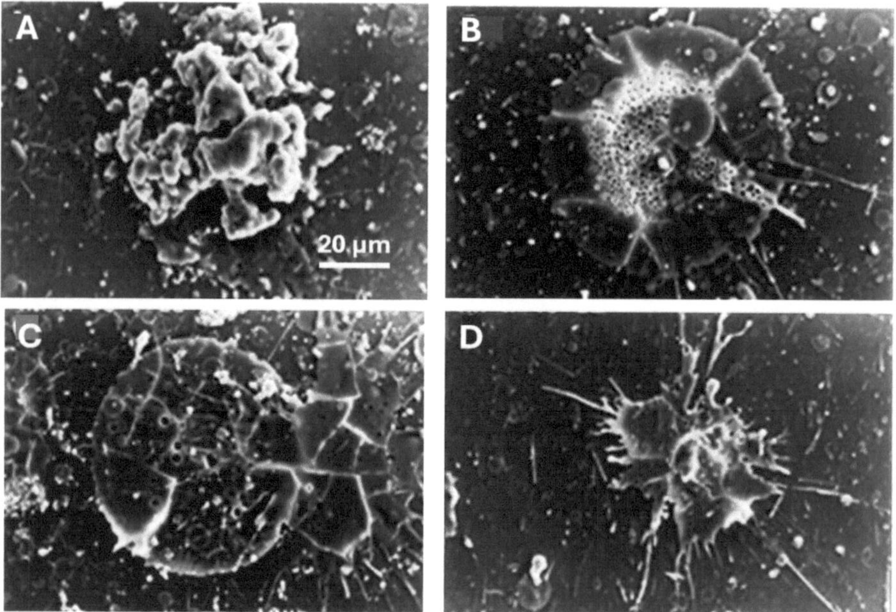

Figure 6. HAp particle splats produced by a 'wipe' test applied to determine the optimum settings of plasma power and stand-off distance. (**A**) Plasma power 45 kW, stand-off distance 26 mm; (**B**) Plasma power 30 kW, stand-off distance 24 mm; (**C**) Plasma power 30 kW, stand-off distance 22 mm; (**D**) Plasma power 45 kW, stand-off distance 22 mm [36]. © With permission of Wiley-VCH.

5.3.1. Incongruent Melting and Thermal Decomposition of HAp: Phase Composition

The temperature of the plasma jet of commercial APS equipment exceeds 15,000 K [36]. This extremely high temperature causes severe structural changes of HAp, even though the residence time of the powder particles in the plasma jet is very short, in the range of hundreds of microseconds to a few milliseconds. Since HAp melts incongruently at 1570 °C (Figure 7A), it decomposes to tricalcium orthophosphate (α'-TCP, $Ca_3(PO_4)_2$), tetracalcium orthophosphate (TTCP, $Ca_4O(PO_4)_2$) and finally CaO during the four consecutive steps indicated in Table 1.

Table 1. Thermal decomposition sequence of hydroxylapatite [41].

Step 1:	$Ca_{10}(PO_4)_6(OH)_2$	$\rightarrow$	$Ca_{10}(PO_4)_6(OH)_{2-x}O_x\square_x + xH_2O$	Oxyhydroxylapatite (OHAp)
Step 2:	$Ca_{10}(PO_4)_6(OH)_{2-x}O_x\square_x$	$\rightarrow$	$Ca_{10}(PO_4)_6O_x\square_x + (1-x)H_2O$	Oxyapatite (OAp)
Step 3:	$Ca_{10}(PO_4)_6O_x\square_x$	$\rightarrow$	$2\,Ca_3(PO_4)_2 + Ca_4O(PO_4)_2$	TCP + TTCP ($C_3P + C_4P$)
Step 4a:	$Ca_3(PO_4)_2$	$\rightarrow$	$3\,CaO + P_2O_5$	Stepwise decomposition of TCP
Step 4b:	$Ca_4O(PO_4)_2$	$\rightarrow$	$4\,CaO + P_2O_5$	and TTCP

Based on this decomposition sequence, a simple shell model of the inflight evolution of calcium phosphate phases was adopted by Graßmann and Heimann [41] and Dyshlovenko et al. [42] (Figure 7B). Owing to the low thermal conductivity of HAp of ~1 W/mK, the inner core of a particle is still below the incongruent melting point and only dehydroxylation steps 1 and 2 come to bear, resulting in OHAp and OAp, short-range ordered (SRO) structures with lattice vacancies $\square$. In contrast, the outer shells consist of a mixture of liquid TCP + TTCP (step 3) and liquid + solid CaO (step 4). According to Dyshlovenko et al. [42], there may exist a thin shell of solid TCP + TTCP. On impact with the cool surface of an implant to be coated, the outer liquid parts of the incoming particle splash across the surface and solidify preferentially as ACP.

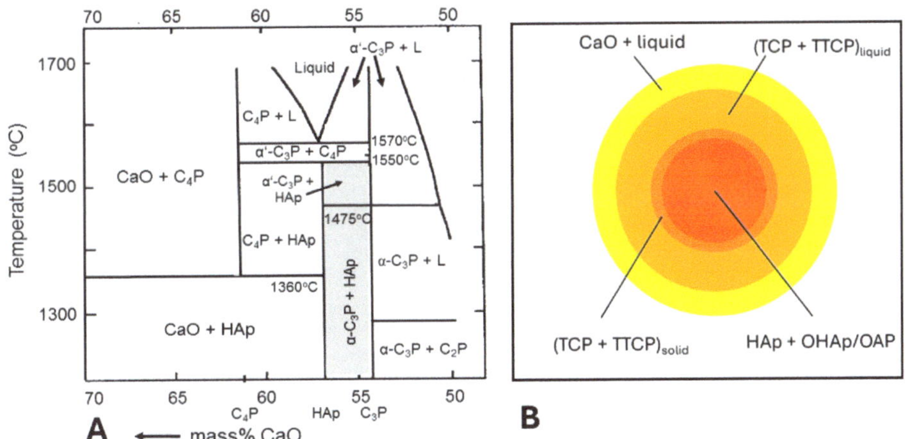

Figure 7. (**A**) Partial quasi-ternary CaO-P_2O_5-(H_2O) phase diagram at a water partial pressure of 65.5 kPa (after Riboud [43]). The area of interest is shaded, showing the transformation of α-C_3P to α'-C_3P above 1475 °C and the incongruent melting of HAp and decomposition beyond 1570 °C to form C_3P (α-TCP) and C_4P (TTCP). © With permission of Wiley-VCH. (**B**) Model of the phase composition of a spherical HAp particle thermally decomposing during flight in the high temperature plasma jet at a low water partial pressure of 1.3 kPa [41,42].

5.3.2. Degree of Crystallinity

As far as coating crystallinity is concerned, there are several aspects to consider. Fully crystalline and well-ordered HAp exhibits low solubility that does not lend itself easily to osseoconductivity as it causes the coating to behave largely like a bioinert material. Hence, fully crystalline, stoichiometric HAp inhibits cell proliferation as evidenced by decreased alkaline phosphatase (ALP) activity [44]. The enzyme ALP is highly expressed in the cells of mineralized tissue and plays a critical function in the formation of hard tissue. Also, reduced osteocalcin secretion has been observed in the presence of fully crystallized and well-ordered HAp [45]. Osteocalcin has routinely been used as a serum marker of osteoblastic bone formation and is believed to act in the bone matrix to regulate mineralization.

Synchrotron radiation X-ray diffraction of coating cross-sections reveals that the crystallinity of the coating decreases exponentially with depth, from about 93% at the surface to about 50% at the substrate-coating interface (Figure 8). In parallel, the amount of more or less pristine HAp decreases linearly from some 80 mass% (the remaining phases were 15 mass% TTCP, 4 mass% ß-TCP, and 1.5 mass% CaO) at the free, i.e., outermost coating surface to less than 15 mass% immediately at the metal interface [46]. Additional studies of the depth-resolved cross-sectional phase composition of plasma-sprayed HAp support this finding [47].

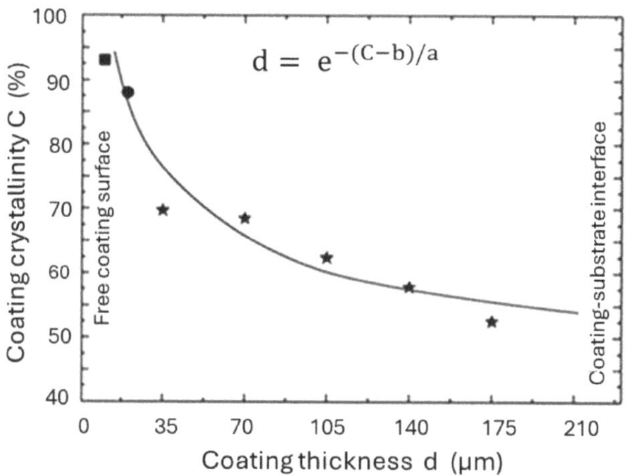

Figure 8. With increasing thickness d, a plasma-sprayed HAp coating reveals an exponential decrease of its crystallinity C. The coefficients a and b of the exponential decay equation were calculated to be 128.3 and 14.3, respectively. Data were obtained by conventional X-ray diffraction at 8 keV (square) and synchrotron radiation X-ray diffraction at 11 keV (dot) and 100 keV (stars) [46]. © With permission by Elsevier.

The amorphous calcium phosphate (ACP) (Figure 9A,B) that accumulates at the immediate interface of coating and substrate as the first product of solidification during plasma spraying provides a low-energy path of coating delamination [48] as it dissolves preferentially during in vivo exposure to extracellular fluid (ECF). A certain degree of resorbability/solubility is required to obtain a sufficiently high concentration of Ca^{2+} and PO_4^{3-} ions conducive to precipitating secondary HAp micro- to nanocrystals for sustained biomineralization and thus, osseointegration. Consequently, careful engineering of the plasma spray parameters is mandatory. To control coating crystallinity, the application of a thin bond coat separating the metallic substrate and the HAp coating has been suggested [49]. The advantages of titania, zirconia or zirconium titanate bond coats include [50]:

- Increase of adhesion to both metal and HAp. For example, a titania bond coat is thought to act as an extension of the native oxide layer on metallic titanium that may interact with HAp to form a thin reaction layer of perovskitic calcium titanate.

- Reduction of the thermal decomposition of HAp by inhibiting the heat flow using a thin titania bond coat film with low thermal conductivity (~1 W/mK) as opposed to a Ti6Al4V substrate (~7 W/mK).
- Reduction of the formation of amorphous phase that forms by a quenching contact immediately at the metal interface. An increase in crystallinity is caused by the thermal barrier function of a bond coat that prolongs solidification time and thus allows the ACP to nucleate apatite and crystallize. Experimental NMR results [50] show that as-sprayed coatings without a bond coat contain only 46 mass% well-ordered HAp at the free coating surface as contrasted with 62 mass% in coatings with a titania bond coat. During incubation for 12 weeks in r-SBF [51], these values increase by dissolving TCP, TTCP, CaO, and ACP phases to 74 mass% and 92 mass%, respectively.
- Reduction of residual coating stresses by reducing the gradient of the coefficient of thermal expansion between the metal substrate and the ceramic overlayer.

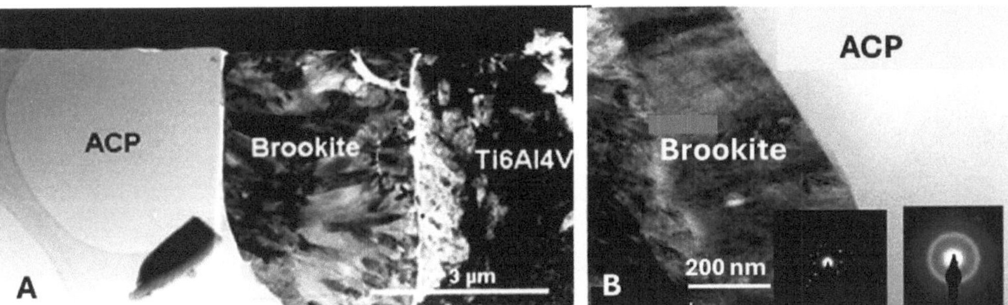

Figure 9. (**A**) A scanning transmission electron microscopy (STEM) image of the cross-section of the interface between a Ti6Al4V substrate (right) and a titanium dioxide (brookite) bond coat (center) and an ACP layer (left). The cross-section was obtained by focused ion beam (FIB) cutting using Ga ions. (**B**) Enlarged interface between columnar brookite crystals and ACP at high magnification STEM. The insets display the electron diffraction pattern of both phases [37]. © With permission by Elsevier.

Figure 9A shows a STEM image of a cross-section of a plasma-sprayed titanium dioxide bond coat that separates the titanium alloy from the ACP layer. As ascertained by Figure 9B, the bond coat was found to consist of the metastable orthorhombic brookite polymorph of titania with low surface energy instead of the expected thermodynamically stable rutile or Magnéli-type phases Ti_nO_{2n-1}. It was likely formed as the result of quenching by a mechanism governed by Oswald's rule of steps. The columnar brookite crystals interdigitate with both the ACP and the Ti6Al4V substrate and thus may provide some resistance against the shearing forces expected in vivo.

5.3.3. Assessment of Structural Order in Hydroxylapatite Coatings: Raman and NMR Studies

Application of a high temperature introduces severe structural disorder in as-received crystalline HAp. This has noticeable repercussions on its mechanical, chemical, and biological behavior. Detailed information on the structural order of calcium phosphate phases can be obtained by Raman and NMR spectroscopies [50,52–54]. Figure 10A shows the full range of the laser-Raman spectrum of an atmospheric plasma-sprayed HAp coating with the four principal vibration modes of the PO_4^{3-} tetrahedron indicated. Figure 10B shows the Gaussian–Lorentzian deconvolved principal Raman mode ν_1. The weak shoulders at 949 cm^{-1} and 971 cm^{-1} may be assigned to ACP and ß-TCP, respectively [55]. The splitting of the triply degenerate ν_3 and ν_4 modes shown in Figure 10A is presumably caused by an increasingly perturbed PO_4^{3-} environment owing to increasing dehydroxylation and to plasma spray-induced disorder of the hydroxyl ions located in the [00.1] lattice direction of HAp, respectively [52].

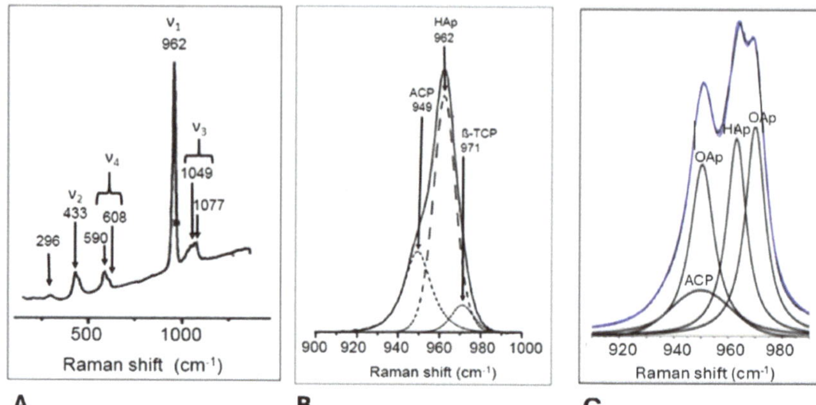

Figure 10. Laser-Raman spectra (λ = 514.6 nm; 5 mW) of plasma-sprayed HAp coatings [55]. (**A**) Overview of the four principal vibrational modes of the PO_4^{3-} tetrahedron of a high-energy plasma-sprayed HAp coating. Assignment of the vibration bands are as follows: ν_1, symmetric stretching ymode of the P-O bond; ν_2, doubly degenerate O-P-O bending mode; ν_3, triply degenerate asymmetric P-O stretching mode; and ν_4, triply degenerate O-P-O bending mode. (**B**) Gaussian–Lorentzian deconvolution of the ν_1 vibrational mode into the contribution of well-ordered HAp [56] and those of the thermal decomposition product ß-TCP and ACP [49,56]. © With permission by Wiley-VCH. (**C**) Gaussian–Lorentzian deconvolved laser-Raman spectrum of a HAp coating plasma-sprayed under low-energy conditions (LEPS), showing distinct and intense Raman bands of OAp [53]. © With permission by Elsevier.

Figure 10C reveals that during low-energy plasma spraying (LEPS), distinct high intensity bands of OAp are still present [53], in contrast to high-energy plasma spraying, during which OAp decomposes to form ß-TCP and TTCP, consistent with step 3 of the decomposition sequence (Table 1). Consequently, when it is desired to maintain oxyapatite in the coating, the electrical energy input into the plasma spray process and the attendant attainment of high temperatures must be drastically reduced.

More detailed data on the type and degree of structural disorder that HAp suffers during plasma spraying has been obtained by high resolution X-ray diffraction [57] and solid-state nuclear magnetic resonance (NMR) spectroscopy [49,50,58,59]. Determining the position as well as the shift of ^{1}H-magic angle spinning (MAS) and ^{31}P-MAS NMR bands provide important clues to determine the environment of PO_4^{3-} tetrahedra and thus, allows identifying dehydroxylation phases such as OHAp and OAp as well as decomposition phases such as TCP, TTCP, and CaO as well as ACP.

Figure 11 shows the ^{1}H-MAS (A) and the ^{31}P-MAS (B) NMR spectra of a plasma-sprayed HAp coating [50,58,60]. The insets refer to completely ordered, stoichiometric, and highly crystalline HAp. In Figure 11A, the high intensity band L at -0.1 ± 0.1 ppm can be assigned to the proton band position of well-ordered crystalline and stoichiometric HAp, and the isotropically shifted weak band L* at -1.3 ± 0.3 ppm may be assigned to protons present in OHAp. The broad M band at 5.2 ± 0.2 ppm indicates isolated pairs of strongly coupled protons in the channels parallel to the c-axis in HAp [58]. Band G at ~1.3 ppm may belong to free water molecules attached to the surface of HAp particles [59]. The ^{31}P-MAS NMR spectrum shown in Figure 11B is more complex. The principal band A at 2.3 ± 0.1 ppm is associated with highly crystalline hydroxylapatite (see inset) whereas the other bands of the Gaussian–Lorentzian deconvolved NMR spectrum represent dehydroxylation (B,C) and decomposition (D) phases. Band B at 1.5 ± 0.2 ppm signals a strongly distorted PO_4^{3-} environment without OH^- neighbors as suggested for OAp, and band C at 3.0 ± 0.2 ppm has been assigned to distorted PO_4^{3-} tetrahedra associated with single or paired OH^- ions as in OHAp [58]. Finally, band D at 5.0 ± 0.2 ppm may represent very strongly distorted PO_4^{3-} tetrahedra lacking OH^- ions as present in TCP, TTCP, and OAp.

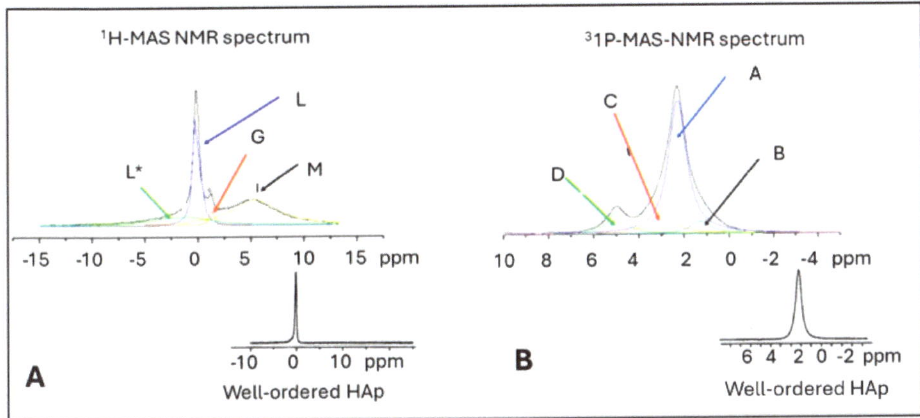

Figure 11. ^{1}H-MAS (**A**) and the ^{31}P-MAS (**B**) NMR spectra of a plasma-sprayed HAp coating. For assignment of spectral bands see text. © Image courtesy Dr. Thi Hong Van Tran [60].

Supporting 2D-double quantum ^{1}H/^{31}P cross-polarization (CP) heteronuclear correlation (HETCOR) NMR spectroscopy (Figure 12) further suggests that the D-band may indeed represent OAp, the chemical shift of which is identical to that of TCP and TTCP, thus confirming the thermal decomposition sequence HAp → OHAp → OAp → TTCP/TCP shown in Table 1.

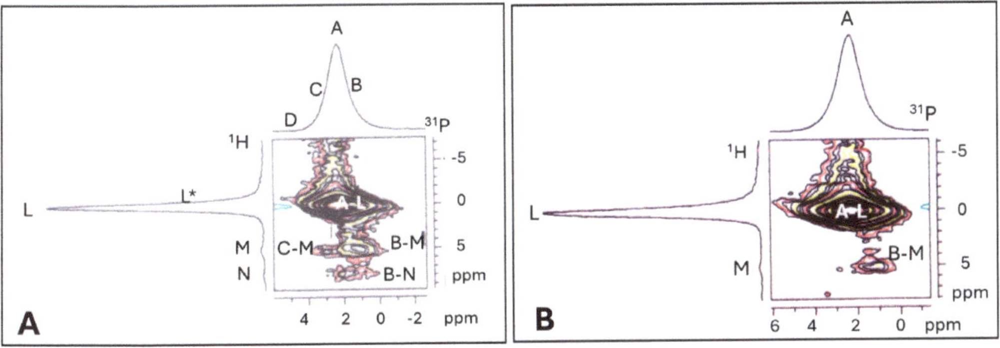

Figure 12. 2D-^{1}H/^{31}P-CP-HETCOR NMR spectra of a plasma-sprayed HAp coatings. (**A**) As plasma-sprayed (**B**) Incubated in r-SBF [50] for 8 weeks. © Image courtesy Dr. Thi Hong Van Tran [60].

Figure 12A shows the main A-L correlation band of crystalline, stoichiometric HAp together with the weak satellite correlation bands C-M and B-M that can be associated with a partially dehydroxylated apatite structure, i.e., OHAp sensu lato and Ca-deficient ACP, respectively [49]. The correlation bands B-L and C-L are swamped by the intense A-L band but can be visualized by Gaussian–Lorentzian deconvolution of the A-L band (not shown here but contained in Ref. [60]). The weak and broad band N in the individual proton spectrum at ~7.5 ppm was assigned to isolated OH-groups in the c-channel of HAp [61]. Figure 12B shows the same coating incubated in r-SBF [51] for 8 weeks at 37 ± 0.5 °C. The satellite correlation bands C-M and B-N have disappeared and B-M has weakened, indicating that the SRO phases were dissolved. In parallel, the intensity of the A-L band of more or less pure, stoichiometric hydroxylapatite has increased, reflecting its relative increase from about 67 mass% in the as-sprayed coating to 85 mass% after 8 weeks of incubation. In parallel, the amount of TTCP decreased from about 26 mass% to 10 mass%, whereas that of ß-TCR remained nearly constant at 5 mass%. Minor amounts of CaO (around 1.5 mass%) have disappeared completely after a few days of incubation [60].

In addition, quantitative ^{31}P magic angle spinning (MAS) solid-state NMR allows distinguishing between PO_4^{3-} groups of apatitic calcium phosphates and HPO_4^{2-} groups of non-apatitic calcium phosphates (Figure 13). Non-apatitic calcium phosphate is thought to be a measure of the maturity of bone; with time, the non-apatitic precursor transforms to true apatite [62].

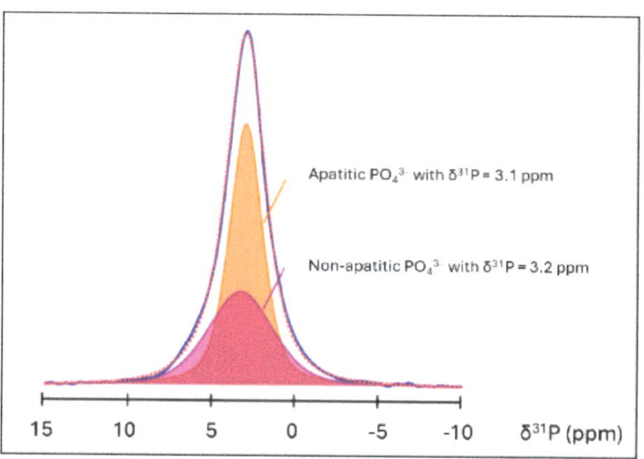

Figure 13. Solid-state ^{31}P NMR-MAS spectrum of bone of an immature sheep (blue envelope) and its deconvolution into a well-ordered and stoichiometric apatitic core (orange) and a HPO_4^{3-}-containing amorphous non-apatitic calcium phosphate surface layer (purple) [63]. © Permission granted under Creative Commons Attribution 4.0 International License.

For a long time, biological apatite has been described as Ca- and OH-deficient carbonated hydroxyapatite (CHA) in which a fraction of the PO_4^{3-} lattice sites is occupied by HPO_4^{2-} ions, resulting in the tentative formula $Ca_{10-x}(HPO_4)_x(PO_4)_{6-x}$ (OH, O, Cl, F, CO_3, □)$_{2-x} \cdot nH_2O$; $0 < x < 1$; $n = 0$–2.5 (see above). However, more recent solid-state NMR studies have revealed that the surface of mature bone mineral particles does not consist of well-ordered HAp at all but of hydrated ACP [63], a contention that mirrors an earlier suggestion by Jäger et al. [59]. They proposed that HAp nanoparticles comprise a stoichiometric and highly crystalline core that is covered by an extremely thin (~1 nm) layer of a disordered calcium phosphate phase having a Ca/P-ratio of ~1.5.

5.3.4. Crystallographic Structure of Hydroxylapatite

Figure 14A shows a ball-and-spoke model of the crystallographic structure of HAp. Hydroxylapatite $Ca_{10}(PO_4)_6(OH)_2$ is a member of a large group of chemically different but structurally identical compounds with space group $P6_3/m$ (176), and with the general formula $M_{10}(ZO_4)_6X_2$ (M = Ca, Pb, Cd, Zn, Sr, La, Ce, K, Na; Z = P, V, As, Cr, Si, C, Al, S; X = OH, Cl, F, CO_3, H_2O, □). The crystallographic structure of HAp comprises Ca $^{[9]}$ polyhedra sharing faces and thus forming chains running parallel to the crystallographic c-axis [00.1], that is, a 6_3 screw axis. These chains form a hexagonal net and share edges and corners with PO_4 tetrahedra. OH^- ions are situated in wide hexagonal channels parallel to the crystallographic c-axis [00.1].

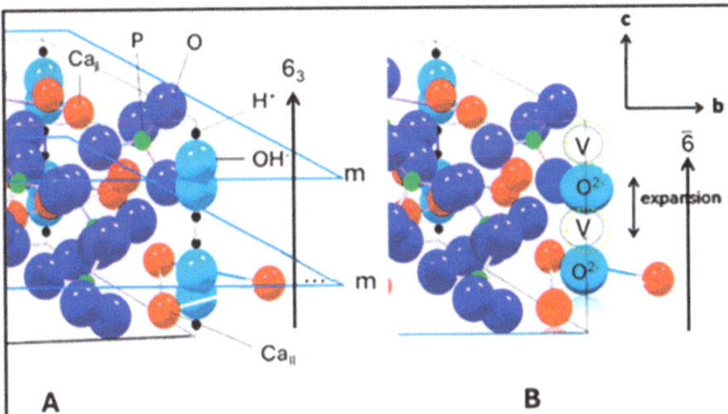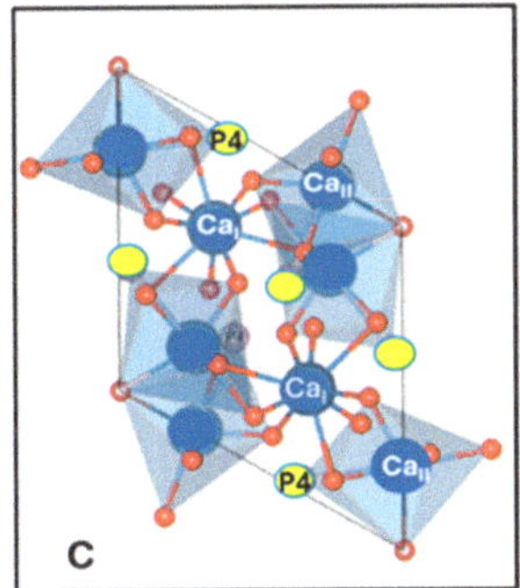

Figure 14. (**A**) Ball-and-spoke model of the crystallographic unit cell of HAp with a hexagonal space group P6$_3$/m. Two mirror planes m are located at z = $\frac{1}{4}$ and $\frac{3}{4}$, perpendicular to the 6$_3$ screw axis [64]. (**B**) Hypothetical structure of OAp (space group P$\bar{6}$) with a chain of O^{2-} ions separated by vacancies V parallel to [00.1] [19,65] © With permission by Wiley-VCH. (**C**) The Ca atoms (blue gray) are arranged in two crystallographic positions: Ca$_I$ coordinated by nine oxygen atoms (red) at z = 0 and $\frac{1}{2}$ along the threefold a$_i$ axes, and Ca$_{II}$ atoms irregularly coordinated with six oxygen atoms and five orthophosphate groups at z = $\frac{1}{4}$ and $\frac{3}{4}$. PO$_4{}^3$—tetraeder positions (P4) are shown in yellow. © Permission granted under Creative Commons Attribution 4.0 International License.

Caused by the presence of open channels, HAp is able to incorporate many other large ions that fill either cationic Ca^{2+} positions or anionic OH$^-$ and PO$_4{}^{3-}$ positions. This happens without large distortions of the crystal lattice, thus maintaining the P6$_3$/m space group symmetry of pure stoichiometric HAp. In biological apatite, Ca^{2+} is partially replaced by metabolically important cations such as Na$^+$, Mg^{2+}, Sr^{2+}, K$^+$ and some trace elements such as Pb^{2+}, Ba^{2+}, Zn^{2+}, and Fe^{2+}. Substitution of PO$_4{}^{3-}$ anions with CO$_3{}^{2-}$, SiO$_4{}^{4-}$, and SO$_4{}^{2-}$ and OH$^-$ with Cl$^-$, F$^-$, and CO$_3{}^{2-}$ enables several biochemical pathways leading to the formation and transformation of bony matter [27,66,67]. This compositional variability of biological apatite is the root cause of its high biocompatibility and osseoconductivity. In particular, the OH$^-$ positions can be occupied by mobile O^{2-} ions and lattice vacancies and thus assist in the kinetics of the dehydroxylation of HAp during plasma spraying and biomineralization in vivo.

5.3.5. Oxyapatite: Fact or Fiction?

Oxyapatite (OAp), Ca$_{10}$O(PO$_4$)$_6$ is considered the product of the complete dehydroxylation of HAp [12], but it converts back to stoichiometric HAp in the presence of water either during cooling of the as-sprayed coating in moist air or by an in vivo reaction with extracellular fluid (ECF). As discussed in Section 3 above, there is evidence against the existence of OAp as a thermodynamically stable phase.

Investigation into the structure of OAp has suggested that a linear chain of O^{2-} ions and associated vacancies parallel to the c$_0$-axis may exist [19], as shown in Figure 14B. More recent calculations involving a density-functional theory supported by a local-density approximation (DFT-LDA) and first-principles pseudo-potentials [20] hinted at a hexagonal 'c-empty' structure Ca$_{10}$(PO$_4$)$_6\square_2$ having a stable total-energy minimum. During thermal dehydroxylation, the mirror planes m (Figure 14A) of the parent HAp are lost, thereby transforming the symmetry of the screw axis 6$_3$ to that of the polar axis $\bar{6}$.

Confirming the existence of OAp by classic X-ray powder diffraction is challenging or even impossible because the length of the c$_0$-axis of OAp is only slightly larger than that of HAp [68]. This accounts for only a small shift of the (00.2) interplanar spacing in the direction of lower diffraction angles. This means that very accurate measurements are called for when using X-ray diffraction using synchrotron radiation. Indeed, high resolution synchrotron X-ray diffraction (Figure 15) reveals that the average c$_0$ distance of OAp is (at 0.6900 nm) about 0.3% longer than the c_0 lattice distance of stoichiometric HAp (0.6880 nm), confirming the postulated expansion of the unit cell (Figure 14B) that is arguably related to the larger Shannon radius of the O^{2-} ion of 135 pm, in contrast to that of the OH^- ion (118 pm) [65].

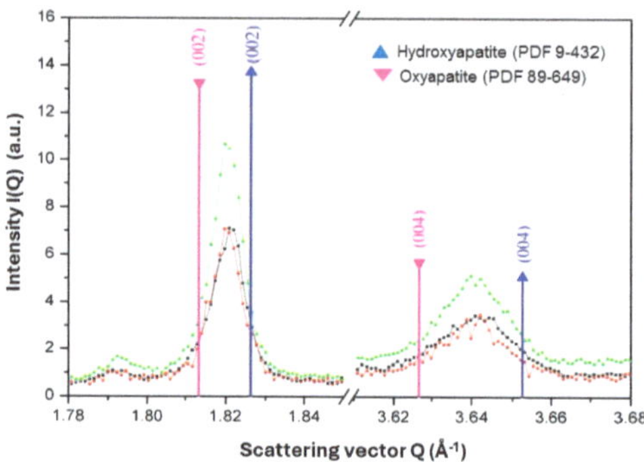

Figure 15. Scattering vectors $Q = 4\pi\cdot\sin\Theta/\lambda$ of synchrotron X-ray diffraction profiles of the (00.2) and (00.4) interplanar spacings of three different plasma-sprayed HAp coatings [22]. © With permission by Wiley-VCH.

5.3.6. Transformation of Amorphous Calcium Phosphate (ACP)

Of the sixteen available positions for OH^- ions in the unit cell of HAp, only 50% are statistically occupied, leaving 8 vacancies along [00.1]. This leads to direction-dependent differences in the mobility of OH^- ions as well as the associated Ca^{2+} ions, relevant for the transformation of ACP to crystalline HAp, either in vitro in contact with simulated body fluid (SBF) (Figure 16) or in vivo by extracellular fluid (ECF) contact. The extent of electrical conductivity [69] as well the kinetics of the stepwise dehydroxylation forming OAp [22] is also dependent on the mobility of OH^- ions.

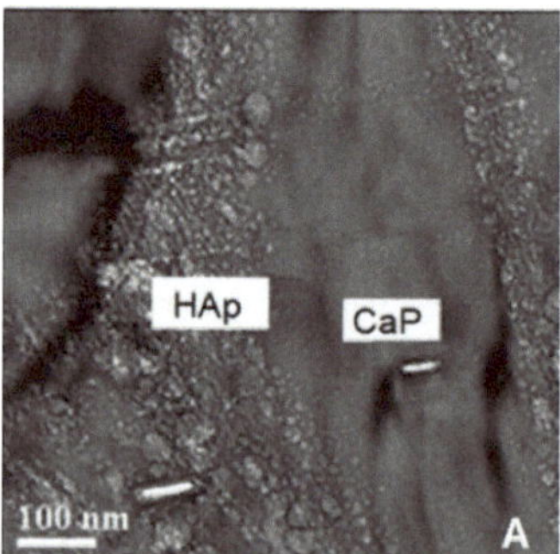

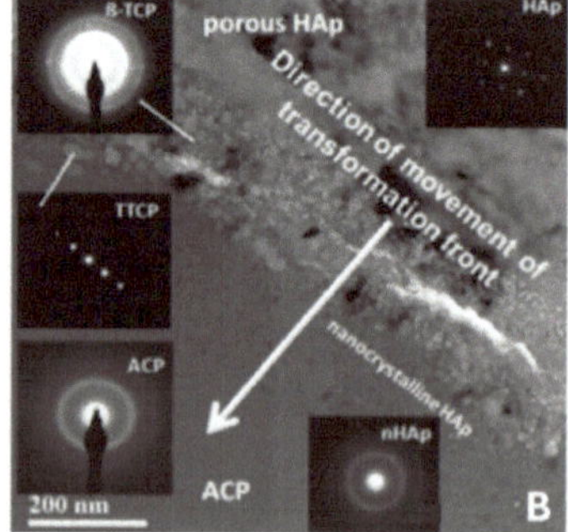

Figure 16. (**A**) Porous crystalline HAp with Ca/P = 1.65 formed by transformation of amorphous calcium phosphate (CaP; ACP) with Ca/P = 1.36 in contact with r-SBF for 24 weeks [70]. © With permission by Elsevier. (**B**) STEM image of the transformation of ACP to crystalline calcium phosphate phases in contact with r-SFB for 1 week. The insets show the electron diffraction pattern of ACP with a single diffuse ring at d = 0.809 nm corresponding to {10.0} of HAp (bottom left), well-crystallized HAp (upper right), nanocrystalline HAp with two diffuse rings at d = 0.288 assigned to {21.0/21.1} and 0.251 nm assigned to {30.1} (bottom right), as well as ß-TCP and TTCP (left) [49]. © With permission by Wiley-VCH.

Figure 16A shows the formation of porous crystalline HAp from ACP during in vitro contact with r-SBF [70]. Crystallization likely occurs by a fluid flow of SBF along cracks and fissures within the coating to form porous HAp with Ca/P = 1.65 as well as dense, needle-like crystalline Ca-depleted calcium orthophosphate CaP with Ca/P~1.36. Such needles with a comparable composition thought to be akin to bone-like apatite can nucleate from ACP and are implicated in mediating osseointegration [71].

The STEM image of Figure 16B demonstrates how during incubation of the coating in r-SBF the transformation front sweeping across the coating layer changes ACP to crystalline phases. Porous but well-crystallized HAp has been formed at the trailing edge of the transformation front (upper right corner), whereas nanocrystalline HAp appears at the leading edge. In addition, within the transformed section of ACP, ß-TCP and TTCP crystallites were detected as decomposition phases of the original HAp [49].

5.3.7. Coating Porosity, Surface Roughness, and Surface Nanotopography

Close control of the coating's porosity and surface roughness are required when attempting to enhance the mechanical and biomedical performance of medical implants. Optimum coating porosity, pore size distribution, and fractal surface roughness are vital preconditions for the uninhibited ingrowth of bone cells [72]. However, the risk of bonding degradation involving mechanisms such as cracking, spalling, delamination, or dissolution during in vivo contact with aggressive ECF decreases with increasing coating density [73]. These two antagonistic requirements have to be carefully assessed, balanced, and controlled [41]. This is particularly important when considering the risk of a release of coating particles into the body environment. HAp particles are known to be distributed by the lymphatic system throughout the body, leading to inflammatory FBR events characterized by the formation of giant cells and phagocytes as described above in Section 2 [74]. Thus, balancing the two antagonistic porosity requirements is challenging and needs target-oriented design and control of the appropriate intrinsic plasma spray parameters by statistical design of experiments (SDE) and statistical process control (SPC) methodology. Parameters influencing the coating porosity include powder particle size and the degree of particle melting. Particle melting degree and kinetics have been found to be complex functions of plasma gas composition, plasma enthalpy, powder injection geometry and position, and spray distance [37]. This type of porosity control is difficult because atmospheric plasma spraying frequently yields dense coatings that cannot satisfy the biomedical requirements that stipulate pore sizes of at least 75 μm [75]. Figure 17 shows the effect of the degree of particle melting, expressed as the fraction of molten particles formed, on crystallinity, porosity, and adhesive bond strength of plasma-sprayed HAp [76]. As evident, an increasing fraction of molten HAp particles causes the porosity and the degree of crystallinity of the coating to decrease, but the adhesive bond strength to increase. Hence, careful adjustment of the critical intrinsic plasma spray parameters is required to guarantee optimum coating performance.

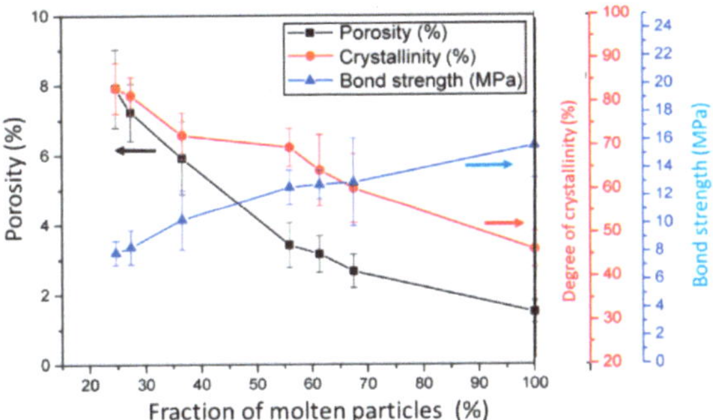

Figure 17. Dependence of porosity (**left**), degree of crystallinity and bond strength (**right**) of plasma-sprayed hydroxylapatite coatings on the degree of particle melting [76]. © With permission by Elsevier.

Dense, i.e., pore-free hydroxylapatite coatings deposited by APS inhibit the attachment, migration, spreading, and proliferation of bone cells. Hence, research is underway to increase the coating porosity to conform to biomedical requirements. Suspension plasma spraying (SPS) provides coatings with drastically enhanced porosity [27,77] that allows the uninhibited ingrowth of osteoblasts. Consequently, HAp coatings deposited by SPS may impart improved biocompatibility, provided sufficient adhesion can be engineered. One way to increase porosity consists of adding pore-forming agents such as ammonium carbonate to the suspension [78].

An important prerequisite for optimum cell adhesion and proliferation is an appropriate surface nanotopography. Assessing the nature of micro- and nano-roughened plasma-sprayed surfaces may be done by involving fractality theory [79,80]. Fractality considerations were used by Gentile et al. [81] in experiments studying cell proliferation on silicon proxy surfaces that were electrochemically etched to obtain varying degrees of roughness while maintaining comparable surface energies. The surface profiles were found to be self-affine fractals. Their average roughness Ra increased with increasing etching time from ~2 nm to 100 nm, showing fractal dimensions ranging from $D = 2$ indicating a nominal flat surface, to $D = 2.6$. Etched silicon surfaces with Ra between 10 and 45 nm revealed a near Brownian surface topography with D~2.5. Such moderately rough surfaces with large fractal dimensions were found to support efficient cell proliferation. Gittens et al. [82] investigated the influence of surface topography on the osseointegration of spinal implants and concluded that apart from the implant design, the experience of the physician, and the age and bone status of the patients, the final success of a spinal implant's integration is to a high degree dependent on the surface characteristics of the implant, i.e., surface roughness, surface chemistry, and surface energy. This has been echoed by a recent review of the role of implant surface modification in osseointegration [83]. The nanotextured architecture of implant surfaces with asperities < 100 nm affects surface roughness, surface area, and surface energy. Nanotextured implant surfaces promote osteoblast contact signaling, adhesion, and proliferation and have shown enhanced cell spreading and filipodia extension [84]. Means to engineer implant surfaces toward suitable nanotopography include mechanical roughening by grit blasting, electron beam lithography, nanoimprinting, chemical etching, reactive ion etching, laser scribing, selective laser sintering, and electrospinning [85].

5.3.8. Residual Coating Stresses

Plasma spraying is a rapid solidification process during which the molten or semi-molten particles strike the substrate surface with supersonic velocity, which may lead to reverberating shock waves that provide additional heat to the deposit and slow down its solidification by adding a thermal pressure component [36,86]. Determination of the direction and the extent of residual coating stresses can be experimentally assessed by X-ray diffraction ($\sin^2\Psi$-technique), Almen-type curvature measurements, the hole-drilling strain gauge method, or photoluminescence and Raman piezospectroscopies [8].

Control of residual coating stresses is mandatory to obtain HAp deposits that adhere well to the implant substrate and thus guarantee reasonable resistance to chipping, spalling, and complete delamination. The large temperature gradient experienced during the plasma spray process generates residual stresses in the deposited coating [36]. There are two main types of stress, thermal and quenching stress. Combined with the complicated solidification process within the coating, these stresses are the two main contributors to the overall residual stress [27].

Thermal stress, σ_c can be expressed by the Dietzel equation

$$\sigma_c = \{E_c(\alpha_c - \alpha_s)\frac{\Delta T}{1 - v_c} + [\frac{1 - v_s}{E_s}](\frac{d_c}{d_s})\}, \tag{1}$$

where E is Young's modulus of elasticity, α is the thermal expansion coefficient, ΔT is the temperature difference between coating and substrate, v is the Poisson's number, and d is the thickness. The subscripts c and s refer to the coating and the substrate, respectively.

At given values of E and ν, σ_c increases with increasing coating thickness d_c. Hence, the risk of deleterious coating destruction is higher in thick coatings compared to thin ones. Depending on the sign of $(\alpha_c - \alpha_s)$, thermal stresses can be tensile or compressive.

The origin of quenching stress lies in the effect of molten particles impacting the cool substrate, whereby their contraction during solidification is restricted by clamping adherence to the roughened substrate surface. This leads to tensile stress in the coating [87–91], frequently resulting in cracking when the cohesive coating strength can be overcome (Figure 5B). The first coating layer adjacent to the interface, consisting of ACP [70], will critically control both the magnitude and the sign of the residual stress. In addition, the ACP layer is known to provide a low-energy fracture path and may eventually cause coating delamination when tensile or shear stresses occur during movement in the post-operative phase. The transformation of ACP (Figure 16) to crystalline calcium phosphate phases during incubation in SBF [70,92] and presumably during in vivo contact with ECF will result in stress relaxation [93]. This reduces the risk of coating failure by delamination as shown in Figure 18A. The figure shows the interfacial coating strain $\varepsilon = (d - d_0)/d_0 \cdot 10^{-3}$ of as-sprayed (dots) and incubated (triangles) HAp coatings deposited by APS on Ti6Al4V substrates as a function of $\sin^2\psi$, where ψ is the tilt angle toward the X-ray beam [94,95].

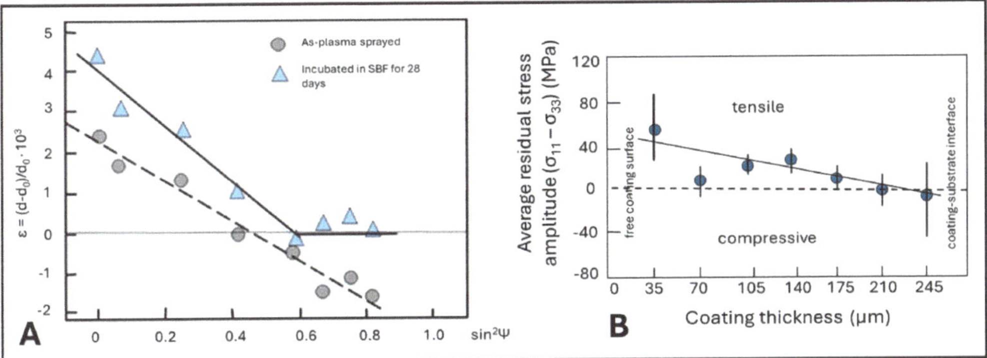

Figure 18. (**A**) Strain ε of an as plasma sprayed HAp coating (dots) and coatings incubated for 28 days in r-SBF [51] (triangles) [94,95]. (**B**) Residual stress amplitude $(\sigma_{11} - \sigma_{33})$ of a plasma-sprayed HAp coating showing that close to the coating–substrate interface the residual tensile stress amplitude changes sign and becomes weakly compressive [46]. © With permission by Elsevier.

When the coating is in tension, $\varepsilon \propto d - d_0$ increases. When in compression it decreases. As shown in Figure 18A, the plasma-sprayed HAp coatings measured by the $\sin^2\Psi$ method is at a rather strong tensile stress created by the thermal mismatch during the cooling of the deposited layer to room temperature. This tensile stress will relax during incubation in SBF, when the high levels of ACP thought to be a main contributor to the residual stress transform to different calcium phosphate phases (see Figure 16B), most notably Ca-depleted defect HAp. Hence, the layer of bone-like secondary apatite deposited at the outermost rim of the samples will be decoupled from the declining stress field, and thus reveal close to zero stress (triangles in Figure 18A). At the free coating surface, the tilt angle Ψ is 0° ($\sin^2\Psi = 0$), i.e., the X-ray beam is tangential to the surface in grazing incidence. With increasing tilt angle Ψ, deeper areas of the coating are being probed, until at 90° ($\sin^2\Psi = 1$) the coating–substrate interface is being reached by the probing X-ray beam, and the character of the stress changes from tensile to slightly compressive.

Residual stress analyses using synchrotron radiation (11 and 100 keV) X-ray diffraction allow more detailed insight into the mechanisms of stress development and relaxation [46]. Although the principal Cauchy stress tensor components σ_{11} and σ_{33} were found both to be tensile close to the coating–substrate interface, they relax to zero within the first 80 μm of the coating. With further accumulating coating thickness, the component σ_{11} slightly increases to become tensile again, reaching +20 MPa at the free coating surface. However, the tensile stress tensor component σ_{33} decreases monotonously from the coating–substrate interface to become compressive with -30 MPa at the free coating surface. This interplay of the two principal tensor components causes the average residual stress amplitude ($\sigma_{11} - \sigma_{33}$) to become quasi-linear as shown in Figure 18B.

5.3.9. Adhesion of Plasma-Sprayed Hydroxylapatite Coatings

The adhesion strength of a plasma-sprayed HAp coating to the metallic implant surface determines, to a large extent, its mechanical performance. In clinical setting, the degree of adhesion between the HAp coating and bone has been assessed by investigating retrieved orthopedic implants [96,97]. These studies have confirmed that the clinical success of coated implants depends not only on sufficiently strong adhesion of the coating to the implant surface but is also influenced by many other non-material variables. Important factors prominently include the experience and skill of the surgeon, proper placement of the implant at the correct angle, the strengths, health, and quantity of the cortical bone bed, and the age and physical condition of the implant recipient.

The adhesion strength of plasma-sprayed HAp coatings to the metallic surface of an implant is rather low, in contrast to a desired high value beyond 35 MPa [75]. That, however, was lowered to at least 15 MPa as stipulated by the recent ISO 13779-2: 2018 norm [98]. For a long time, it was assumed that the adhesion of plasma-sprayed coatings is basically caused and controlled by strictly mechanical clamping of the solidified particle splats to the roughened substrate surface. However, in a modern view, more complex processes such as chemisorption and epitaxial/topotaxial orientational registry are seen as alternative mechanisms supporting coating adhesion [99,100]. In the case of HAp coatings, there are suggestions that extremely thin reaction layers of $CaTi_2O_5$ or $CaTiO_3$ may mediate adhesion [101,102]. The effect of calcium titanate surfaces on HAp nucleation has been explained by an epitaxial relationship of the (022) lattice plane of $CaTiO_3$ and the (00.1) lattice plane of HAp [103]. However, the study of adhesion-mediating reaction layers by TEM even at very high magnification is counteracted by their intrinsic tenuity. This is a consequence of the very short diffusion path lengths of Ca^{2+} and Ti^{4+} ions, rendering any potential reaction layer exceedingly thin.

To obtain high adhesion strengths beyond the application of bond coats (see Section 2), the enthalpy contained in the plasma jet may be enhanced, thereby increasing the degree of melting and superheating, respectively, of the HA particles. However, high plasma enthalpy inevitably leads to increased thermal decomposition of HAp, causing a decrease of its resorption resistance and thus the in vivo longevity of the coatings. Consequently, controlling the heat transfer from the hot core of the plasma jet to the center of the powder particles is mandatory to optimize the coating properties, including adhesion.

5.3.10. Other Implant Surface Functionalization Strategies

Beyond the deposition of hydroxylapatite as and osseoconductive coating on endoprosthetic implants, modification of implant surfaces by different means has been explored [104]. Since most biomaterials do not possess all of the ideal properties and desirable functions needed to fulfill their anticipated biological role completely, additional surface functionalization by physical, chemical, biological, and radiative processes may be required [27].

Chemical functionalization includes hydroxylation of titanium surfaces by NaOH treatment [105] and covalent attachment of proteins to implant surfaces [106]. Vertically oriented TiO_2 nanotubes enhance osteoblast differentiation and raise osteocalcin expression and integrin/focal contact [107]. Anti-infection coatings include silver-doped hydroxylapatite coatings [108], chitosan coatings [109] as well as gentamicin- and vancomycine-loaded PMMA and PLGA polymers [110]. Novel silicon nitride coatings that provide strong antiviral properties based on the formation of reactive nitrogen species by a surficial hydrolysis reaction are being developed. These nitrogen species create osmotic stress in the cytoplasmic space of viruses that leads to lysis, effectively inactivating the virus [111,112].

Biological functionalization routes comprise, among many others, collagen I [113] and arginylglycylaspartic acid (RGD) coatings [114]. The latter provides the peptide motif responsible for cell adhesion to the extracellular matrix (ECM). In addition, adsorption and/or incorporation of extracellular non-collagenous proteins (NCPs) such as bone morphogenetic proteins (BMPs) and other growth factors in HAp and bioglass coatings elicit specific biological responses that enhance the osseoinductive process and thus, the osseointegration ability of implants [115].

6. Concluding Remarks

At present, total hip arthroplasty (THA) is among the most frequently performed and in the long-term outcome most successful and effective surgeries worldwide. Applying atmospheric plasma-sprayed (APS) osseoconductive HAp coatings to the metallic stem and the casing of the acetabular cup of hip endoprostheses assists the ingrowth of bone cells (osteocytes). The HAp coating prevents an acellular connective tissue capsule forming around the implant as a response to the introduced foreign body. The HAp layer will support bonding osteogenesis that through 'bony in-growth' allows the transmission of the tensile and shear forces acting on the hip joint during locomotion. Clinical evidence has overwhelmingly confirmed that a long-term stable osseoconductive HAp coating will elicit a specific biological response at the interface of the implant material by controlling its surface chemistry through adsorption of bone growth-mediating factors. These factors include platelet-derived growth factor (PDGF) AA, insulin-like growth factors (IGFs) I and II, cytokines such as interleukin-6, colony-stimulating factors, and tumor necrosis factor-α, as well as non-collagenous matrix proteins such as osteocalcin, osteonectin, osteopontin, sialylated glycoproteins, and proteoglycans. Their action will result in the eventual establishment of a strong and lasting bond between living tissue and biomaterial by osseointegration.

However, there is a growing need to address several shortcomings, despite the fact that the deposition of HAp coatings by APS is a mature and well research-supported process. Exposure of HAp to the extreme temperature of the plasma jet during plasma spraying leads to dehydroxylation, forming OHAp and/or OAp as well as to partial or even complete thermal decomposition to TCP and TTCP as well as ACP, owing to incongruent melting of HAp. The large temperature gradient between the cool substrate and the superheated molten particle droplets and the solidification kinetics after deposition generate residual coating stresses that are the root cause of the formation of coating cracks and may lead to post-implantation delamination in vivo. Finally, line-of-sight limitation during the plasma spray deposition process prevents complex-shaped implant structures from being effectively coated by HAp.

During the past five decades, countless attempts have been made to optimize essential properties of osseoconductive HAp coatings. Unraveling the complexity of the interactions among numerous plasma spray parameters influencing key coating properties has produced an incalculable mass of information. These studies include research into phase composition, crystallinity, porosity, thickness, micro- and nano-roughness, adhesion and cohesion, and the nature of residual coating stresses on plasma-sprayed HAp coatings. Controlling phase composition at values stipulated by international norms [27,98] and implementing novel deposition techniques such as SPS and SPPS to optimize coating porosity are high up on the agenda of current research worldwide. Important areas of research and development are related to the improvement of adhesion of HAp coatings to titanium alloy implant surfaces [116] and the modification and control of surface nanotopography of implants [84,85,117].

In addition, research is progressing beyond HAp coatings, aiming at designing and developing novel intelligent coatings responsive to changes in pH, temperature, and piezoelectric and magnetic stimuli, thereby providing osseoimmunomodulation and angiogenesis that in turn promote osseogenesis and reduce inflammatory responses via foreign body reaction (FBR) [118].

Funding: This research received no external funding.

Conflicts of Interest: The author declares no conflict of interest.

Acronyms

ACP, amorphous calcium phosphate; ALP, alkaline phosphatase; APS, atmospheric (air) plasma spraying; ASTM, American Society for Testing and Materials; BMP, bone morphogenetic protein; CAGR, compound annual growth rate; CGDS, cold gas dynamic spraying; CoC, ceramic-on-ceramic; CoP, ceramic-on-polymer; CP, cross polarization; CVD, chemical vapor deposition; DFT-LDA, density functional theory-local density approximation; ECD, electrochemical deposition; ECF, extracellular fluid; ECM, extracellular matrix; EPD, electrophoretic deposition; FBR, foreign body reaction; FDA, Food and Drug Administration; FIB, focused ion beam; FS, flame spraying; GNP, gross national product; GPA, gigapascal; HAp, hydroxylapatite; HETCOR, heteronuclear correlation; HVOF, high velocity oxyfuel spraying; HVSFS, high velocity suspension flame spraying; IGF, insulin-like growth factor; ISO, International Organization for Standardization; LEPS, low-energy plasma spraying; LPPS, low pressure plasma spraying; MAS, magic angle spinning; MoP, metal-on-polymer; MPa, megapascal; MPS, micro plasma spraying; NCP, non-collagenous protein; NMR, nuclear magnetic resonance; OAp, oxyapatite; OECD, Organization for Economic Co-operation and Development; OHAp, oxyhydroxylapatite; PDGF, platelet-derived growth factor; PEO, plasma electrolytic oxidation; PLD, pulsed laser deposition; PLGA, poly(lactide-*co*-glycolic acid); PMMA, poly(methylmethacrylate); r.f.MS, radiofrequency magnetron sputtering; SBF, simulated body fluid; SDE, statistical design of experiments; SRO, short range order; SPC, statistical process control; SPS, suspension plasma spraying; SPPS, solution precursor plasma spraying; STEM, scanning transmission electron microscopy; TCP, tricalcium phosphate; TEM, transmission electron microscopy; THA, total hip arthroplasty; TKA, total knee arthroplasty; TNTZ, titanium-niobium-tantalum-zirconium alloy; TTCP, tetracalcium phosphate; USD, US dollar; VPS, vacuum plasma spraying.

References

1. Hegde, V.; Stambough, J.B.; Levine, B.R.; Springer, B.D. Highlights of the American Joint Replacement Registry Annual Report. *Arthroplast. Today* **2023**, *21*, 101137. [CrossRef]
2. Grimberg, A.; Lützner, J.; Melsheimer, O.; Morlock, M.; Steinbrück, A. *EPDR Annual Report 2023*; EPDR Deutsche Endoprothesenregister gGmbH: Berlin, Germany, 2023. Available online: https://www.eprd.de/en/about-the-eprd/news/article/annual-report-2023 (accessed on 16 March 2024).
3. Global Hip Replacement Implants Market Report and Forecast 2023–2031. Available online: https://www.researchandmarkets.com/reports/5797997 (accessed on 16 March 2024).
4. Merfort, R.; Maffulli, N.; Hofmann, U.K.; Hildebrand, F.; Simeone, F.; Eschweiler, J.; Migliorini, F. Head, acetabular liner composition, and rate of revision and wear in total hip arthroplasty: A Bayesian network meta-analysis. *Sci. Rep.* **2023**, *13*, 2032. [CrossRef]
5. Heimann, R.B.; Schürmann, N.; Müller, R.T. In vitro and in vivo performance of Ti6Al4V implants with plasma-sprayed osteoconductive hydroxylapatite-bioinert titania bond coat 'duplex' systems: An experimental study in sheep. *J. Mater. Sci. Mater. Med.* **2004**, *15*, 1045–1052. [CrossRef] [PubMed]
6. Itiravivong, P.; Promasa, A.; Laiprasert, T.; Techapongworachai, T.; Kuptnitazsaikul, S.; Thanakit, V.; Heimann, R.B. Comparison of tissue reaction and osteointegration of metal implants between hydroxyapatite/Ti alloy coat: An animal experimental study. *J. Medical Assoc. Thail.* **2003**, *86*, S422–S430.
7. Zhang, D.H.; Chen, Q.; Shi, C.; Chen, M.; Ma, K.; Wan, J.; Liu, R. Dealing with the foreign body response to implanted biomaterials: Strategies and applications of new materials. *Adv. Funct. Mater.* **2021**, *31*, 2007226. [CrossRef]
8. Heimann, R.B.; Lehmann, H.D. *Bioceramic Coatings for Medical Implants. Trends and Techniques*; Wiley-VCH: Weinheim, Germany, 2015; p. 467. ISBN 978-3-527-334743-9.
9. Brande, W.T.; Taylor, A.S. *Chemistry*; Blanchard and Lea: Philadelphia, PA, USA, 1863; p. 696.
10. Warington, R., Jr. Researches on the phosphates of calcium, and upon the solubility of tricalcic phosphate. *J. Chem. Soc.* **1866**, *19*, 296–318. [CrossRef]
11. Bragg, W.H. Application of the ionisation spectrometer to the determination of the structure of minute crystals. *Proc. Phys. Soc.* **1921**, *33*, 222–224. [CrossRef]

12. Hendricks, S.B.; Hill, W.A.; Jakobs, K.D.; Jefferson, M.E. Structural characteristics of apatite-like substances and composition of phosphate rock and bone as determined from microscopical and X-ray examinations. *Ind. Eng. Chem.* **1931**, *23*, 1413–1418. [CrossRef]
13. Roseberry, H.H.; Hastings, A.B.; Morse, J.K. X-ray analysis of bone and teeth. *J. Biol. Chem.* **1931**, *90*, 395–407. [CrossRef]
14. Trömel, G. Untersuchungen über die Bildung eines halogenfreien Apatits aus basischen Calciumphosphaten. *Z. Phys. Chem. A* **1932**, *158*, 422–432. [CrossRef]
15. Bredig, M.A. Zur Apatitstruktur der anorganischen Knochen- und Zahnsubstanz. *Hoppe Seyler's Z. Physiol. Chem.* **1933**, *216*, 239–243. [CrossRef]
16. Bredig, M.A.; Franck, H.H.; Füldner, H. Beiträge zur Kenntnis der Kalk-Phosphorsäure-Verbindungen II. *Z. Elektrochem.* **1933**, *39*, 959–969. [CrossRef]
17. De Jong, W.F. La substance minerale dans les os. *Recl. Trav. Chim. Pays. Bas. Belg.* **1926**, *45*, 445–448. [CrossRef]
18. De Leeuw, N.H.; Bowe, J.R.; Rabone, J.A.L. A computational investigation of stoichiometric and calcium-deficient oxy- and hydroxyapatites. *Faraday Discuss.* **2007**, *134*, 195–214. [CrossRef] [PubMed]
19. Alberius Henning, P.; Landa-Canovas, A.; Larsson, A.K.; Lidin, S. The structure of oxyapatite solved by HREM. *Acta Cryst.* **1999**, *B55*, 170–176. [CrossRef] [PubMed]
20. Calderin, L.; Stott, M.J.; Rubio, A. Electronic and crystallographic structure of apatites. *Phys. Rev.* **2003**, *B67*, 134106–134112. [CrossRef]
21. Gross, K.A.; Pluduma, L. Putting oxyhydroxyapatite into perspective. A pathway to oxyapatite and its application. In *Calcium Phosphate. Structure, Synthesis, Properties, and Applications*; Heimann, R.B., Ed.; Biochemistry Research Trends Series; Nova Science Publication: New York, NY, USA, 2012; pp. 95–120. ISBN 978-1-62257-299-1.
22. Heimann, R.B. Characterisation of as-sprayed and incubated hydroxyapatite coatings with high resolution techniques. *Mater. Wiss. Werkstofftechn.* **2009**, *40*, 21–30.
23. Hattori, T.; Iwadate, Y. Hydrothermal preparation of calcium hydroxylapatite powders. *J. Am. Ceram. Soc.* **1990**, *73*, 1803–1807. [CrossRef]
24. Liu, C.; Huang, A.; Shen, W.; Cui, J. Kinetics of hydroxyapatite precipitation at pH 10 and 11. *Biomaterials* **2001**, *23*, 301–306. [CrossRef]
25. Rey, C.; Combes, C.; Drouet, C.; Glimcher, M.J. Bone mineral: Update on chemical composition and structure. *Osteopor. Int.* **2009**, *20*, 1013–1021. [CrossRef]
26. Pasteris, J.D. Structurally incorporated water in bone apatite: A cautionary tale. In *Calcium Phosphate. Structure, Synthesis, Properties, and Applications*; Heimann, R.B., Ed.; Biochemistry Research Trends Series; Nova Science Publication: New York, NY, USA, 2012; pp. 63–94. ISBN 978-1-62257-299-1.
27. Heimann, R.B. *Materials for Medical Application*; Walter de Gruyter GmbH: Berlin, Germany, 2020; p. 615. ISBN 978-3-11-061919-5.
28. Glimcher, M.J. A basic architectural principle in the organisation of mineralized tissue. *Clin. Orthop.* **1968**, *61*, 16–36. [CrossRef] [PubMed]
29. Florencio-Silva, R.; Rodrigues da Silva Sasso, G.; Sasso-Cerri, E.; Simões, M.J.; Cerri, P.S. Biology of bone tissue: Structure, function, and factors that influence bone cells. *Biomed. Res. Int.* **2015**, *2015*, 421746. [CrossRef] [PubMed]
30. Allo, B.A.; Costa, D.O.; Dixon, S.J.; Mequanint, K.; Rizkalla, A.S. Bioactive and biodegradable nanocomposites and hybrid biomaterials for bone regeneration. *J. Funct. Mater.* **2012**, *5*, 432–463. [CrossRef] [PubMed]
31. Nair, A.K.; Gautieri, A.; Chang, S.W.; Buehler, M.J. Molecular mechanics of mineralized collagen fibrils in bone. *Nature Comm.* **2013**, *4*, 1724. [CrossRef]
32. Jarcho, M.; Bolen, C.M.; Thomas, M.B.; Bobick, J.; Kay, J.F.; Doremus, R.H. Hydroxylapatite synthesis and characterization in dense polycrystalline form. *J. Mater. Sci.* **1976**, *11*, 2027–2035. [CrossRef]
33. Ducheyne, P.; Hench, L.L.; Kagan, A.; Martens, M.; Mulier, J.C.; Burssens, A. The effect of hydroxyapatite impregnation on bonding of porous coated implants. *J. Biomed. Mater. Res.* **1980**, *14*, 225–237. [CrossRef] [PubMed]
34. Albrektsson, T.; Johansson, C. Osteoinduction, osseoconduction and osseointegration. *Eur. Spine J.* **2001**, *10* (Suppl. S2), S96–S101.
35. León, B.; Jansen, J.A. *Thin Calcium Phosphate Coatings for Medical Implants*; Springer: New York, NY, USA, 2009; p. 328, ISBN 978-0-387-77718-4.
36. Heimann, R.B. *Plasma Spray Coating. Principles and Applications*, 2nd ed.; Wiley-VCH: Weinheim, Germany, 2008; ISBN 978-3-527-32050-9.
37. Heimann, R.B. Hydroxylapatite coatings: Applied mineralogy research in the bioceramics field. In *Highlights in Applied Mineralogy*; Heuss-Aßbichler, S., Amthauer, G., John, M., Eds.; Walter de Gruyter GmbH: Berlin, Germany, 2018; pp. 301–316, ISBN 978-3-11-049122-7.
38. Heimann, R.B. The nature of plasma spraying. *Coatings* **2023**, *13*, 622. [CrossRef]
39. Sun, L. Thermal spray coatings on orthopedic devices: When and how the FDA reviews your coatings. *J. Thermal Spray Technol.* **2018**, *27*, 1280–1290. [CrossRef]
40. Graßmann, O.; Heimann, R.B. Compositional and microstructural changes of engineered plasma-sprayed hydroxyapatite coatings on Ti6Al4V substrates during incubation in protein-free simulated body fluid. *J. Biomed. Mater. Res.* **2000**, *53*, 685–693. [CrossRef]
41. Heimann, R.B.; Graßmann, O.; Zumbrink, T.; Jennissen, H.P. Biomimetic processes during in vitro leaching of plasma-sprayed hydroxyapatite coatings for endoprosthetic application. *Mater. Wiss. Werkstofftechn.* **2001**, *32*, 913–921. [CrossRef]

42. Dyshlovenko, S.; Pateyron, B.; Pawlowski, L.; Murano, D. Numerical simulation of hydroxyapatite powder behaviour in plasma jet. *Surf. Coat. Technol.* **2004**, *179*, 110–117. [CrossRef]

43. Riboud, P.V. Composition et stabilité des phases a structure d'apatite dans le systeme CaO-P$_2$O$_5$-oxide de Fer-H$_2$O a haute temperature. *Ann. Chim.* **1973**, *8*, 381–390.

44. Frayssinet, P.; Tourenne, F.; Roquet, N.; Conte, P.; Delga, C.; Bonel, G. Comparative biological properties of HA plasma-sprayed coatings having different crystallinities. *J. Mater. Sci. Mater. Med.* **1994**, *5*, 11–17. [CrossRef]

45. De Santis, D.; Guerriero, C.; Nocini, P.F.; Ungersbock, A.; Richards, G.; Gotte, P.; Armato, U. Adult human bone cells from jaw bones cultured on plasma-sprayed or polished surfaces of titanium or hydroxyapatite discs. *J. Mater. Sci. Mater. Med.* **1996**, *7*, 21–28. [CrossRef]

46. Ntsoane, T.P.; Topić, M.; Härting, M.; Heimann, R.B.; Theron, C. Spatial and depth-resolved studies of air plasma-sprayed hydroxyapatite coatings by means of diffraction techniques: Part 1. *Surf. Coat. Technol.* **2016**, *294*, 153–163. [CrossRef]

47. Hesse, C.; Hengst, M.; Kleeberg, R.; Götze, J. Influence of experimental parameters on spatial phase distribution in as-sprayed and incubated hydroxyapatite coatings. *J. Mater. Sci. Mater. Med.* **2008**, *19*, 3225–3241. [CrossRef]

48. Park, E.; Condrate, R.A.; Lee, D.H. Infrared spectral investigation of plasma spray coated hydroxyapatite. *Mater. Lett.* **1998**, *36*, 38–43. [CrossRef]

49. Heimann, R.B. Novel approaches towards design and biofunctionality of plasma-sprayed osteoconductive calcium phosphate coatings for biomedical implants: The concept of bond coats. In *Trends in Biomaterials Research*; Pannone, P.J., Ed.; Nova Science Publishers Inc.: New York, NY, USA, 2007; pp. 1–80. ISBN 978-1-60021-361-8.

50. Heimann, R.B.; Tran, H.V.; Hartmann, P. Laser-Raman and nuclear magnetic resonance (NMR) studies on plasma-sprayed hydroxyapatite coatings: Influence of bioinert bond coats on phase composition and resorption kinetics in simulated body fluid. *Mater. Wiss. Werkstofftechn.* **2003**, *34*, 1163–1169. [CrossRef]

51. Kim, H.M.; Miyazaki, M.; Kokubo, T.; Nakamura, T. Revised simulated body fluid. *Key Eng. Mater.* **2001**, *192/195*, 47–50. [CrossRef]

52. Park, E.; Condrate, R.A.; Lee, D.H.; Kociba, K.; Gallagher, P.K. Characterization of hydroxyapatite before and after plasma spraying. *J. Mater. Sci. Mater. Med.* **2002**, *13*, 211–218. [CrossRef] [PubMed]

53. Demnati, I.; Parco, M.; Grossin, D.; Fagoaga, I.; Drouet, C.; Barykin, G.; Combes, C.; Braceras, I.; Goncalves, S.; Rey, C. Hydroxyapatite coating on titanium by a low energy plasma spraying mini-gun. *Surf. Coat. Technol.* **2012**, *206*, 2346–2353. [CrossRef]

54. Demnati, I.; Grossin, D.; Combes, C.; Rey, C. Plasma-sprayed apatite coatings: Review of physical-chemical aspects and their biological consequences. *J. Med. Biol. Eng.* **2014**, *34*, 1–7. [CrossRef]

55. Heimann, R.B.; Vu, T.A. Low-pressure plasma-sprayed (LPPS) bioceramic coatings with improved adhesion strength and resorption resistance. *J. Thermal Spray Technol.* **1997**, *6*, 145–149. [CrossRef]

56. Cusco, R.; Guitian, F.; de Aza, S.; Artus, L. Differentiation between hydroxyapatite and ß-tricalcium phosphate by means of μ-Raman spectroscopy. *J. Eur. Ceram. Soc.* **1998**, *18*, 1301–1305. [CrossRef]

57. Shamray, V.F.; Sirotinkin, V.P.; Kalita, V.I.; Komlev, V.S.; Barinov, S.M.; Fedotov, A.Y.; Gordeev, A.S. Studies of the crystal structure of hydroxyapatite in plasma coating. *Surf. Coat. Technol.* **2019**, *372*, 201–208. [CrossRef]

58. Hartmann, P.; Jäger, C.; Barth, S.; Vogel, J.; Meyer, K. Solid state NMR, X-ray diffraction, and infrared characterization of local structure in heat-treated oxyhydroxylapatite microcrystals: An analogy of the thermal deposition of hydroxyapatite during plasma-spray procedure. *J. Solid State Chem.* **2001**, *160*, 460–468. [CrossRef]

59. Jäger, C.; Welzel, T.; Meyer-Zaika, W.; Epple, M. A solid-state NMR investigation of the structure of nanocrystalline hydroxyapatite. *Magn. Reason. Chem.* **2006**, *44*, 573–580. [CrossRef] [PubMed]

60. Tran, T.H.V. Investigation into the Thermal Dehydroxylation and Decomposition of Hydroxylapatite during Atmospheric Plasma Spraying: NMR and Raman Spectroscopic Study of as-Sprayed Coatings and Coatings Incubated in Simulated Body Fluid. Ph.D. Thesis, Department of Mineralogy, Technische Universität Bergakademie, Freiberg, Germany, 2004.

61. Hartmann, P.; Barth, S.; Vogel, J.; Jäger, C. Investigation of structural changes in plasma-sprayed hydroxy/apatite coatings. In *Applied Mineralogy in Research, Economy, Technology, Ecology and Culture*; Rammlmair, D., Mederer, J., Oberthür, T., Heimann, R.B., Pentinghaus, H., Eds.; A.A. Balkema: Rotterdam, The Netherlands, 2000; Volume 1, pp. 147–150. ISBN 90-5809-164-3.

62. LeGeros, R.Z. Formation and transformation of calcium phosphates: Relevance to vascular calcification. *Z. Kardiol.* **2001**, *90* (Suppl. S3), 116–124. [CrossRef]

63. von Euw, S.; Wang, Y.; Laurent, G.; Drouet, C.; Babonneau, F.; Nassif, N.; Azaïs, T. Bone mineral: New insights into its chemical composition. *Sci. Rep.* **2019**, *9*, 8456. [CrossRef]

64. Posner, A.S.; Perloff, A.; Diorio, A.F. Refinement of the hydroxyapatite structure. *Acta Cryst.* **1958**, *11*, 308–309. [CrossRef]

65. Heimann, R.B. Plasma-sprayed hydroxylapatite coatings as biocompatible intermediaries between inorganic implant surfaces and living tissue. *J. Thermal Spray Technol.* **2018**, *27*, 1212–1237. [CrossRef]

66. Laskus, A.; Kolmas, J. Ionic substitution in non-apatitic calcium phosphates. *Int. J. Mol. Sci.* **2017**, *18*, 2542. [CrossRef] [PubMed]

67. Graziani, G.; Boi, M.; Bianchi, M. A review on ionic substitution in hydroxyapatite thin films. *Coatings* **2018**, *8*, 269. [CrossRef]

68. Trombe, J.C.; Montel, C. Sur la préparation del'oxyapatite phospho-calcique. *C. R. Acad. Sci. Paris* **1971**, *273*, 462–465.

69. Takahashi, T.; Tanase, S.; Yamamoto, O. Electrical conductivity of some hydroxyapatites. *Electrochim. Acta* **1978**, *23*, 369–373. [CrossRef]

70. Heimann, R.B.; Wirth, R. Formation and transformation of amorphous calcium phosphates on titanium alloy surfaces during atmospheric plasma spraying and their subsequent in vitro performance. *Biomaterials* **2006**, *27*, 823–831. [CrossRef]

71. Weng, J.; Liu, Q.; Wolke, C.D.; Zhang, D.; de Groot, K. The role of amorphous phase in nucleating bone-like apatite on plasma-sprayed hydroxyapatite coating in simulated body fluid. *J. Mater. Sci. Letters* **1997**, *16*, 335–337. [CrossRef]

72. Cook, S.D.; Thomas, K.A.; Kay, J.F.; Jarcho, M. Hydroxyapatite coated titanium for orthopaedic implant applications. *Clin. Orthop.* **1988**, *232*, 225–243. [CrossRef]

73. Yang, C.Y.; Wang, B.C.; Chang, E.; Wu, J.D. Bond degradation at the plasma-sprayed HA coating/Ti-6Al-4V alloy interface: An in vitro study. *J. Mater. Sci. Mater. Med.* **1995**, *6*, 258–265. [CrossRef]

74. Lemons, J.E. Biodegradation and wear of total joint replacements. In *Bone Implant Interface*; Cameron, H.U., Ed.; Mosby: Baltimore, Germany, 1994; pp. 307–317.

75. Wintermantel, E.; Ha, S.W. Biokompatible Werkstoffe und Bauweisen. In *Implantate für Medizin und Umwelt*; Springer: Berlin/Heidelberg, Germany, 1996; p. 225.

76. Li, H.; Ma, Y.L.; Zhao, Z.C.; Tian, Y.L. Fatigue behavior of plasma sprayed structural-grade hydroxyapatite coating under simulated body fluid. *Surf. Coat. Technol.* **2019**, *368*, 110–118. [CrossRef]

77. Gross, K.A.; Saber-Samandari, S. Revealing mechanical properties of a suspension plasma sprayed hydroxyapatite coating with nanoindentation. *Surf. Coat. Technol.* **2009**, *203*, 2995–2999. [CrossRef]

78. Wu, F.; Huang, Y.; Song, L.; Liu, X.; Xiao, Y.; Feng, J.; Chen, J. Method for Preparing Porous Hydroxyapatite Coatings by Suspension Plasma Spraying. U.S. Patent 8877283 B2, 4 November 2014.

79. Reisel, G.; Heimann, R.B. Correlation between surface roughness of plasma-sprayed chromium oxide coatings and powder grain size distribution: A fractal approach. *Surf. Coat. Technol.* **2004**, *185*, 215–221. [CrossRef]

80. Heimann, R.B. On the self-affine fractal geometry of plasma-sprayed surfaces. *J. Thermal Spray Technol.* **2011**, *20*, 898–908. [CrossRef]

81. Gentile, F.; Tirinato, I.; Battista, E.; Causa, F.; Liberale, C.; di Fabrizio, E.M.; Decuzzi, P. Cells preferentially grow on rough substrates. *Biomaterials* **2010**, *31*, 7205–7212. [CrossRef] [PubMed]

82. Gittens, R.A.; Olivares-Navarrete, R.; Schwartz, Z.; Boyan, B.D. Implant osseointegration and the role of microroughness and nanostructures: Lessons for spine implants. *Acta Biomater.* **2014**, *10*, 3363–3371. [CrossRef]

83. Liu, Y.; Rath, B.; Tingart, M.; Eschweiler, J. Role of implant modification in osseointegration: A systematic review. *J. Biomed. Mater. Res.* **2019**, *108*, 470–484. [CrossRef]

84. Dalby, M.J.; McCloy, D.; Robertson, M.; Wilkinson, C.D.; Oreffo, R.O. Osteoprogenitor response to defined topographies with nanoscale depths. *Biomaterials* **2006**, *27*, 1306–1315. [CrossRef]

85. Ross, A.M.; Jiang, Z.; Bastmeyer, M.; Lahann, J. Physical aspects of cell culture substrates: Topography, roughness, and elasticity. *Small* **2012**, *8*, 336–355. [CrossRef] [PubMed]

86. Heimann, R.B.; Kleiman, J.I. Shock-induced growth of superhard materials. In *Crystals, Growth, Properties, and Applications*; Freyhardt, H.C., Ed.; Springer: Berlin/Heidelberg, Germany, 1988; Volume 11, pp. 1–73. ISBN 3-540-18602-6.

87. Gill, S.C.; Clyne, T.W. Stress distribution and material response in thermal spraying of metallic and ceramic deposits. *Metall. Trans.* **1990**, *B21*, 377–385. [CrossRef]

88. Kuroda, S.; Clyne, T.W. The quenching stress in thermally sprayed coatings. *Thin. Solid Film.* **1991**, *200*, 49–66. [CrossRef]

89. Kuroda, S.; Dendo, T.; Kitahara, S. Quenching stress in plasma sprayed coatings and its correlation with the deposit microstructure. *J. Thermal Spray Technol.* **1995**, *4*, 75–84. [CrossRef]

90. Matejicek, J.; Sampath, S. Intrinsic residual stresses in single splats produced by thermal spray processes. *Acta Mater.* **2001**, *49*, 1993–1999. [CrossRef]

91. Matejicek, J.; Sampath, S. In situ measurement of residual stresses and elastic moduli in thermal sprayed coatings. Part 1: Apparatus and analysis. *Acta Mater.* **2003**, *51*, 863–872. [CrossRef]

92. Stammeier, J.A.; Purgstaller, B.; Hippler, D.; Mavromatis, V.; Dietzel, M. In-situ Raman spectroscopy of amorphous calcium phosphate to crystalline hydroxyapatite transformation. *MethodX* **2018**, *5*, 1241–1250. [CrossRef] [PubMed]

93. Topíc, M.; Ntsoane, T.; Hüttel, T.; Heimann, R.B. Microstructural characterisation and stress determination in as-plasma sprayed and incubated bioconductive hydroxyapatite coatings. *Suf. Coat. Technol.* **2006**, *201*, 3633–3641. [CrossRef]

94. Heimann, R.B.; Ntsoane, T.; Pineda-Vargas, C.A.; Przybylowicz, W.J.; Topíc, M. Biomimetic formation of hydroxyapatite investigated by analytical techniques with high resolution. *J. Mater. Sci. Mater. Med.* **2008**, *19*, 3295–3302. [CrossRef] [PubMed]

95. Heimann, R.B.; Graßmann, O.; Hempel, M.; Bucher, R.; Härting, M. Phase content, resorption resistance and residual stresses of bioceramic coatings. In *Applied Mineralogy in Research, Economy, Technology, Ecology and Culture*; Rammlmair, D., Mederer, J., Oberthür, T., Heimann, R.B., Pentinghaus, H., Eds.; A.A. Balkema: Rotterdam, The Netherlands, 2000; Volume 1, pp. 155–158. ISBN 90-5809-164-3.

96. Geesink, R.G. Hydroxyapatite-coated total hip prostheses two-year clinical and roentgenographic results of 100 cases. *Clin. Orthop. Relat. Res.* **1990**, *261*, 39–58. [CrossRef]

97. *ISO 12891-2*; Retrieval and Analysis of Surgical Implants. Part 2: Analysis of Retrieved Surgical Implants. International Organization for Standardization: Geneva, Switzerland, 2020.

98. *ISO 13779-2*; Implants for Surgery-Hydroxyapatite. Part 2: Coatings of Hydroxyapatite. International Organization for Standardization: Geneva, Switzerland, 2018.

99. Heimann, R.B. Plasma-sprayed hydroxyapatite-based coatings: Chemical, mechanical, microstructural, and biomedical properties. *J. Thermal Spray Technol.* **2016**, *25*, 827–850. [CrossRef]

100. Lacombe, R. *Adhesion Measurement Methods: Theory and Practice*; CRC Taylor & Francis: Boca Raton, FL, USA, 2006.

101. Filiaggi, M.J.; Coombs, N.A.; Pilliar, R.M. Characterization of the interface in the plasma-sprayed HApcCoating/Ti-6Al-4V implant system. *J. Biomed. Mater. Res.* **1991**, *25*, 1211–1229. [CrossRef]

102. Webster, T.J.; Ergun, C.; Doremus, R.H.; Lanford, W.A. Increased osteoblast adhesion on titanium-coated hydroxylapatite that forms CaTiO$_3$. *J. Biomed. Mater. Res. A* **2003**, *67*, 975–980. [CrossRef] [PubMed]

103. Wei, D.; Zhou, Y.; Jia, D.; Wang, Y. Structure of calcium titanate/titania bioceramic composite coatings on titanium alloy and apatite deposition on their surfaces in a simulated body fluid. *Surf. Coat. Technol.* **2007**, *201*, 8715–8722. [CrossRef]

104. Zhang, B.G.X.; Myers, D.E.; Wallace, G.G.; Brandt, M.; Choong, P.F. Bioactive coatings for orthopaedic implants—recent trends in development of implant coatings. *Int. J. Mol. Sci.* **2014**, *15*, 11878–11921. [CrossRef]

105. Fujibayashi, S.; Neo, M.; Kim, H.M.; Kokubo, T.; Nakamura, T. Osteoinduction of porous bioactive titanium metal. *Biomaterials* **2004**, *25*, 443–450. [CrossRef] [PubMed]

106. Stewart, C.; Akhavan, B.; Wise, S.G.; Bilek, M.M.M. A review of biomimetic surface functionalization for bone-integrating orthopedic implants: Mechanisms, current approaches, and further directions. *Progr. Mater. Sci.* **2019**, *106*, 100588. [CrossRef]

107. Oh, S.; Brammer, K.S.; Li, Y.S.; Teng, D.; Engler, A.J.; Chien, S.; Jin, S. Stem cell fate dictated solely by altered nanotube dimension. *Proc. Natl. Acad. Sci. USA* **2009**, *106*, 2130–2135. [CrossRef]

108. Shimazaki, T.; Miyamoto, H.; Ando, Y.; Noda, I.; Yonekura, Y.; Kawano, S.; Miyazaki, M.; Mawatari, M.; Hotokebuchi, T. In vivo antibacterial and silver-releasing properties of novel thermal sprayed silver-containing hydroxyapatite coating. *J. Biomed. Mater. Res. B Appl. Biomater.* **2010**, *92*, 386–389. [CrossRef]

109. Rabea, E.I.; Badawy, M.E.; Stevens, C.V.; Smagghe, G.; Steurbaut, W. Chitosan as antimicrobial agent: Applications and mode of action. *Biomacromolecules* **2003**, *4*, 1457–1465. [CrossRef] [PubMed]

110. Neut, D.; van de Belt, H.; Stokroos, I.; van Horn, J.R.; van der Mei, H.C.; Busscher, H.J. Biomaterial-associated infection of gentamicin-loaded PMMA beads in orthopaedic revision surgery. *J. Antimicrob. Chemother.* **2001**, *47*, 885–891. [CrossRef]

111. Pezzotti, G.; Boschetto, F.; Ohgitani, E.; Fujita, Y.; Shin-Ya, M.; Adachi, T.; Yamamoto, T.; Kanamura, N.; Marin, E.; Zhu, W.; et al. Mechanism of instantaneous inactivation of SARS-CoV-2 by silicon nitride bioceramics. *Mater. Today Bio.* **2021**, *12*, 100144. [CrossRef]

112. Heimann, R.B. Silicon nitride ceramics: Structure, synthesis, properties, and biomedical applications. *Materials* **2023**, *16*, 5142. [CrossRef] [PubMed]

113. Rammelt, S.; Schulze, E.; Bernhardt, R.; Hanisch, U.; Scharnweber, D.; Worch, H.; Zwipp, H.; Biewener, A. Coating of titanium implants with type-I collagen. *J. Orthop. Res.* **2004**, *22*, 1025–1034. [CrossRef] [PubMed]

114. Bitschnau, A.; Alt, V.; Bohner, F.; Heerich, K.E.; Margesin, E.; Hartmann, S.; Sewing, A.; Meyer, C.; Wenisch, S.; Schnettler, R. Comparison of new bone formation, implant integration, and biocompatibility between RGD-hydroxyapatite and pure hydroxyapatite coating for cementless joint prostheses—An experimental study in rabbits. *J. Biomed. Mater. Res. B Appl. Biomater.* **2009**, *88*, 66–74. [CrossRef] [PubMed]

115. Lobel, K.D.; Hench, L.L. In vitro adsorption and activity of enzymes on reaction layers of bioactive glass substrates. *J. Biomed. Mater. Res.* **1998**, *39*, 575–579. [CrossRef]

116. Mohseni, E.; Zalnezhad, E.; Bushroa, A.R. Comparative investigation on the adhesion of hydroxyapatite coating on Ti-6Al-4V implant: A review paper. *Int. J. Adhes.* **2014**, *48*, 238–257. [CrossRef]

117. Harazawa, K.; Cousins, B.; Roach, P.; Fernandez, A. Modification of the surface nanotopography of implant devices: A translational perspective. *Mater. Today Bio.* **2021**, *12*, 100152. [CrossRef]

118. Joshi, M.U.; Kulkarni, S.P.; Choppadandi, M.; Keerthana, M.; Kapusetti, G. Current state of art smart coatings for orthopedic implants: A comprehensive review. *Smart Mater. Med.* **2023**, *4*, 661–679. [CrossRef]

Review

Calcium Orthophosphate (CaPO$_4$)-Based Bioceramics: Preparation, Properties, and Applications

Sergey V. Dorozhkin

Kudrinskaja sq. 1-155, 123242 Moscow, Russia; sedorozhkin@yandex.ru

Abstract: Various types of materials have been traditionally used to restore damaged bones. In the late 1960s, a strong interest was raised in studying ceramics as potential bone grafts due to their biomechanical properties. A short time later, such synthetic biomaterials were called bioceramics. Bioceramics can be prepared from diverse inorganic substances, but this review is limited to calcium orthophosphate (CaPO$_4$)-based formulations only, due to its chemical similarity to mammalian bones and teeth. During the past 50 years, there have been a number of important achievements in this field. Namely, after the initial development of bioceramics that was just tolerated in the physiological environment, an emphasis was shifted towards the formulations able to form direct chemical bonds with the adjacent bones. Afterwards, by the structural and compositional controls, it became possible to choose whether the CaPO$_4$-based implants would remain biologically stable once incorporated into the skeletal structure or whether they would be resorbed over time. At the turn of the millennium, a new concept of regenerative bioceramics was developed, and such formulations became an integrated part of the tissue engineering approach. Now, CaPO$_4$-based scaffolds are designed to induce bone formation and vascularization. These scaffolds are usually porous and harbor various biomolecules and/or cells. Therefore, current biomedical applications of CaPO$_4$-based bioceramics include artificial bone grafts, bone augmentations, maxillofacial reconstruction, spinal fusion, and periodontal disease repairs, as well as bone fillers after tumor surgery. Prospective future applications comprise drug delivery and tissue engineering purposes because CaPO$_4$ appear to be promising carriers of growth factors, bioactive peptides, and various types of cells.

Keywords: calcium orthophosphates; hydroxyapatite; tricalcium phosphate; bioceramics; biomaterials; grafts; biomedical applications; tissue engineering

Citation: Dorozhkin, S.V. Calcium Orthophosphate (CaPO$_4$)-Based Bioceramics: Preparation, Properties, and Applications. *Coatings* **2022**, *12*, 1380. https://doi.org/10.3390/coatings12101380

Academic Editors: Anton Ficai and James Kit-Hon Tsoi

Received: 1 August 2022
Accepted: 14 September 2022
Published: 21 September 2022

Publisher's Note: MDPI stays neutral with regard to jurisdictional claims in published maps and institutional affiliations.

1. Introduction

One of the most exciting and rewarding areas of the engineering discipline involves development of various devices for healthcare. Some of them are implantable. Examples comprise sutures, catheters, heart valves, pacemakers, breast implants, fracture fixation plates, nails and screws in orthopedics, various filling formulations, orthodontic wires, total joint replacement prostheses, etc. However, in order to be accepted by the living body without any unwanted side effects, all implantable items must be prepared from a special class of tolerable materials, called biomedical materials or biomaterials, in short. The physical character of the majority of the available biomaterials is solids [1,2].

From the material point of view, all types of solids are divided into four major groups: metals, polymers, ceramics, and various blends thereof, called composites. Similarly, all types of solid biomaterials are also divided into the same groups: biometals, biopolymers, bioceramics, and biocomposites. All of them play very important roles in both replacement and regeneration of various human tissues; however, setting biometals, biopolymers, and biocomposites aside, this review is focused on bioceramics only. In general, bioceramics comprise various polycrystalline materials, amorphous materials (glasses), and blends thereof (glass-ceramics). Nevertheless, the chemical elements used to manufacture bioceramics form just a small set of the periodic table; namely, bioceramics might be prepared from alumina, zirconia, magnesia, carbon, silica-contained, and calcium-contained compounds, as well as from a limited number of other compounds. All these compounds might be manufactured in both dense and porous forms in bulk, as well as in the forms of crystals, powders, particles, granules, scaffolds, and/or coatings [1–3].

As seen from the above, the entire subject of bioceramics is still rather broad. To specify it further, let me limit myself by a description of calcium orthophosphate (abbreviated as $CaPO_4$)-based formulations only. If compared with other types of bioceramics (such as alumina, zirconia, calcium silicates, calcium sulfate, etc.), the main feature and superiority of $CaPO_4$ is based on their chemical similarity to the composition of calcified tissues of mammals (bones, teeth, and deer antlers) and the need for versatile and risk-free bone substitute biomaterials immediately available without the constraint of bone grafts. One of the major properties of most types of $CaPO_4$ is their osteoconductivity, an ability to favor bone healing and to bind firmly to bone tissues. In addition, some types of $CaPO_4$ have been shown to be able to initiate bone formation de novo in nonosseous sites [1–3]. Therefore, $CaPO_4$ bioceramics are widely used in a number of different applications throughout the body, covering all areas of the skeleton. The examples include healing of bone defects, fracture treatment, total joint replacement, bone augmentation, orthopedics, cranio-maxillofacial reconstruction, spinal surgery, otolaryngology, ophthalmology, and percutaneous devices [1–3], as well as dental fillings and periodontal treatments [4]. Furthermore, they are also used in nonosseous applications, such as ocular implants, allowing eye movements. Depending upon the required properties, different types of $CaPO_4$ might be used. For example, Figure 1 displays some randomly chosen samples of the commercially available $CaPO_4$ bioceramics for bone graft applications. One should note that the global bone grafts and substitutes market was valued at USD 2.65 billion in 2020 and is projected to reach USD 3.36 billion by 2028, registering a cumulative annual growth rate of ~4.3% from 2021 to 2028 [5]. This clearly demonstrates the biomedical perspectives of $CaPO_4$-based bioceramics.

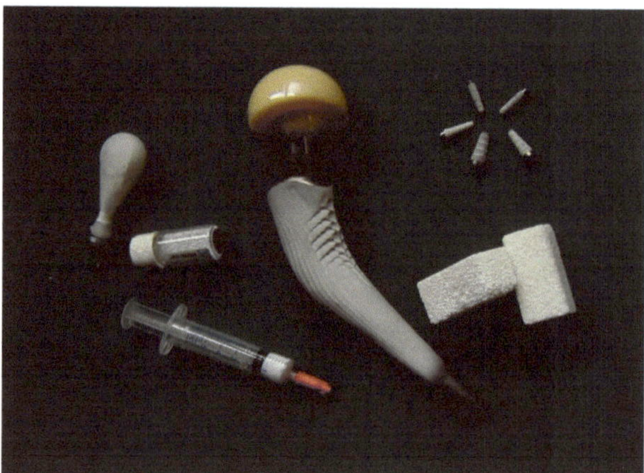

Figure 1. Several examples of the commercial $CaPO_4$-based bioceramics.

A list of the available $CaPO_4$, including their standard abbreviations and major properties, is summarized in Table 1 [3,6]. To narrow the subject further, with a few important exceptions, bioceramics prepared from undoped and unsubstituted $CaPO_4$ will be considered and discussed only. Due to this reason, $CaPO_4$-based bioceramics prepared from biological resources, such as bones, teeth, corals, antlers, etc. [7–14], including food [15] and animal wastes [16], as well as various types of ion-substituted $CaPO_4$ [17–41], including rhenanite $NaCaPO_4$ and chlorapatite $Ca_{10}(PO_4)_6Cl_2$, are not considered. The readers interested in both topics are advised to study the original publications.

Table 1. Existing calcium orthophosphates and their major properties [3,6].

Ca/P Molar Ratio	Compounds and Their Typical Abbreviations	Chemical Formula	Solubility at 25 °C, -log(K_s)	Solubility at 25 °C, g/L	pH Stability Range in Aqueous Solutions at 25 °C
0.5	Monocalcium phosphate monohydrate (MCPM)	$Ca(H_2PO_4)_2 \cdot H_2O$	1.14	~18	0.0–2.0
0.5	Monocalcium phosphate anhydrous (MCPA or MCP)	$Ca(H_2PO_4)_2$	1.14	~17	[c]
1.0	Dicalcium phosphate dihydrate (DCPD), mineral brushite	$CaHPO_4 \cdot 2H_2O$	6.59	~0.088	2.0–6.0
1.0	Dicalcium phosphate anhydrous (DCPA or DCP), mineral monetite	$CaHPO_4$	6.90	~0.048	[c]
1.33	Octacalcium phosphate (OCP)	$Ca_8(HPO_4)_2(PO_4)_4 \cdot 5H_2O$	96.6	~0.0081	5.5–7.0
1.5	α-Tricalcium phosphate (α-TCP)	$\alpha\text{-}Ca_3(PO_4)_2$	25.5	~0.0025	[a]
1.5	β-Tricalcium phosphate (β-TCP)	$\beta\text{-}Ca_3(PO_4)_2$	28.9	~0.0005	[a]
1.2–2.2	Amorphous calcium phosphates (ACP)	$Ca_xH_y(PO_4)_z \cdot nH_2O$, $n = 3\text{–}4.5$; 15%–20% H_2O	[b]	[b]	~5–12 [d]
1.5–1.67	Calcium-deficient hydroxyapatite (CDHA or Ca-def HA) [e]	$Ca_{10-x}(HPO_4)_x(PO_4)_{6-x}(OH)_{2-x}$ $(0 < x < 1)$	~85	~0.0094	6.5–9.5
1.67	Hydroxyapatite (HA, HAp, or OHAp)	$Ca_{10}(PO_4)_6(OH)_2$	116.8	~0.0003	9.5–12
1.67	Fluorapatite (FA or FAp)	$Ca_{10}(PO_4)_6F_2$	120.0	~0.0002	7–12
1.67	Oxyapatite (OA, OAp, or OXA) [f], mineral voelckerite	$Ca_{10}(PO_4)_6O$	~69	~0.087	[a]
2.0	Tetracalcium phosphate (TTCP or TetCP), mineral hilgenstockite	$Ca_4(PO_4)_2O$	38–44	~0.0007	[a]

[a] These compounds cannot be precipitated from aqueous solutions. [b] Cannot be measured precisely. However, the following values were found: 25.7 ± 0.1 (pH = 7.40), 29.9 ± 0.1 (pH = 6.00), 32.7 ± 0.1 (pH = 5.28). The comparative extent of dissolution in acidic buffer is ACP >> α-TCP >> β-TCP > CDHA >> HA > FA. [c] Stable at temperatures above 100 °C. [d] Always metastable. [e] Occasionally, it is called "precipitated HA (PHA)". [f] Existence of OA remains questionable.

2. General Knowledge and Definitions

A number of definitions have been developed for the term "biomaterials". For example, by the end of the 20th century, the consensus developed by the experts was the following: biomaterials were defined as synthetic or natural materials to be used to replace parts of a living system or to function in intimate contact with living tissues [42]. In September 2009, a more advanced definition was introduced: "A biomaterial is a substance that has been engineered to take a form which, alone or as part of a complex system, is used to direct, by control of interactions with components of living systems, the course of any therapeutic or diagnostic procedure, in human or veterinary medicine" [43]. Further, in 2018, the term biomaterial was redefined as "a material designed to take a form that can direct, through interactions with living systems, the course of any therapeutic or diagnostic procedure" [44]. According to the Williams, "The two critical parts of this definition relate to the objectives of the systems in which a biomaterial is used and the fact that the material has to interact with living systems, in most cases parts of the human body, in order for these objectives to be realized. This definition, and indeed, the whole concept of biomaterials science, applies equally to situations involving implantable devices, artificial organs, tissue engineering templates, nonviral gene vectors, drug delivery systems and contrast agents" [45]. The definition alterations were accompanied by a shift in both the conceptual ideas and the expectations of biological performance, which mutually changed in time. However, one should stress that any artificial materials that are in contact with skin, such as hearing aids and wearable artificial limbs, are not included in the definition of biomaterials since the skin acts as a protective barrier between the body and the external world [1,2].

In general, the biomaterials discipline is founded in the knowledge of the synergistic interaction of material science, biology, chemistry, medicine, and mechanical science and it requires the input of comprehension from all these areas so that potential implants perform adequately in a living body and interrupt normal body functions as little as possible [46]. As biomaterials deal with all aspects of the material synthesis and processing, the knowledge in chemistry, material science, and engineering appears to be essential. On the other hand, since clinical implantology is the main purpose of biomaterials, biomedical sciences become the key part of the research. These include cell and molecular biology, histology, anatomy, and physiology. The final aim is to achieve the correct biological interaction of the artificial grafts with living tissues of a host. Thus, to achieve the goals, several stages have to be performed, such as material synthesis, design, and manufacturing of prostheses, followed by various types of tests. Furthermore, before clinical applications, any potential biomaterial must also pass all regulatory requirements [47].

The major difference between biomaterials and other classes of materials lays in their ability to remain in a biological environment while neither damaging the surroundings nor being damaged in that process. Therefore, biomaterials must be distinguished from *biological materials* because the former are the materials that are accepted by living tissues and, therefore, they might be used for tissue replacements, while the latter are just the materials being produced by various biological systems (wood, cotton, bones, chitin, etc.) [48]. Furthermore, there are *biomimetic materials*, which are not made by living organisms but have composition, structure, and properties similar to those of biological materials. Concerning the subject of the current review, *bioceramics* (or biomedical ceramics) are defined as biomaterials having a ceramic origin. Now, it is important to define the meaning of ceramics. According to Wikipedia, the free encyclopedia: "The word *ceramic* comes from the Greek word κεραμικός (*keramikos*), "of pottery" or "for pottery", from κέραμος (*keramos*), "potter's clay, tile, pottery". The earliest known mention of the root "ceram-" is the Mycenaean Greek *ke-ra-me-we*, "workers of ceramics", written in Linear B syllabic script. The word "ceramic" may be used as an adjective to describe a material, product or process, or it may be used as a noun, either singular, or, more commonly, as the plural noun "ceramics". A ceramic material is an inorganic, nonmetallic, often crystalline oxide, nitride or carbide material. Some elements, such as carbon or silicon, may be considered ceramics. Ceramic materials are brittle, hard, strong in compression, weak in shearing and tension. They withstand chemical erosion that occurs in other materials subjected to acidic or caustic environments. Ceramics generally can withstand very high temperatures, such as temperatures that range from 1000 to 1600 °C (1800 to 3000 °F). Glass is often not considered a ceramic because of its amorphous (noncrystalline) character. However, glassmaking involves several steps of the ceramic process and its mechanical properties are similar to ceramic materials" [49]. Similar to any other type of biomaterials, bioceramics can have structural functions as joint or tissue replacements, and be used as coatings to improve the biocompatibility, as well as function as resorbable lattices, providing temporary structures and frameworks that are dissolved and/or replaced as the body rebuilds the damaged tissues [50–53]. Some types of bioceramics feature a drug-delivery capability [54–57].

In medicine, bioceramics are needed to alleviate pain and restore functions of diseased or damaged calcified tissues (bones and teeth) of the body. A great challenge facing its medical application is, first, to replace and, second, to regenerate old and deteriorating bones with a biomaterial that can be replaced by a new mature bone without transient loss of a mechanical support [1,2]. The excellent performance of the specially designed bioceramics that have survived these clinical conditions represents one of the most remarkable accomplishments of research, development, production, and quality assurance before the end of the past century [50]. Regarding $CaPO_4$ bioceramics, a surface bioactivity appears to be its major feature. It contributes to a bone bonding ability and enhances new bone formation [58].

3. Bioceramics of $CaPO_4$

3.1. History

The detailed history of HA and other types of $CaPO_4$, including the subject of $CaPO_4$ bioceramics, as well as description of their past biomedical applications, might be found elsewhere [59,60], where the interested readers are referred. One should just note that the earliest book devoted to $CaPO_4$ bioceramics was published in 1983 [61].

3.2. Chemical Composition and Preparation

Currently, $CaPO_4$ bioceramics can be prepared from various sources [7–16]. Nevertheless, up to now, all attempts to synthesize bone replacement materials for clinical applications featuring the physiological tolerance, biocompatibility, and a long-term stability have had only relative success; this clearly demonstrates both the superiority and a complexity of the natural structures [62].

In general, a characterization of $CaPO_4$ bioceramics should be performed from various viewpoints such as the chemical composition (including stoichiometry and purity), homogeneity, phase distribution, morphology, grain sizes and shape, grain boundaries, crystallite size, crystallinity, pores, cracks, surface roughness, etc. From the chemical point of view, the vast majority of $CaPO_4$ bioceramics are based on HA [63–67], both types of TCP [68–78], and various multiphasic formulations thereof [79]. Biphasic formulations (commonly abbreviated as BCP–biphasic calcium phosphate) are the simplest among the latter ones. They include β-TCP + HA [80–88], α-TCP + HA [89–91], and biphasic TCP (commonly abbreviated as BTCP), consisting of α-TCP and β-TCP [92–97]. In addition, triphasic formulations (HA + α-TCP + β-TCP) have been prepared as well [98–101]. Further details on this topic can be found in a special review [79]. Leaving aside a big subject of DCPD-forming self-setting formulations [102,103], one should note that just a few publications on bioceramics prepared from other types of $CaPO_4$ are available [104–112].

The preparation techniques of various types of $CaPO_4$ have been extensively reviewed in the literature [6,113–117], where the interested readers are referred. Briefly, when compared to both α- and β-TCP, HA is a more stable phase under the physiological conditions, as it has a lower solubility (Table 1) and, thus, slower resorption kinetics [118–120]. Therefore, the BCP concept is determined by the optimum balance of a more stable phase of HA and a more soluble TCP. Due to a higher biodegradability of the α- or β-TCP component, the reactivity of BCP increases with the TCP/HA ratio increasing. Thus, in vivo bioresorbability of BCP can be controlled through the phase composition [81]. Similar conclusions are also valid for the biphasic TCP (in which α-TCP is a more soluble phase), as well as for both triphasic (HA, α-TCP, and β-TCP) and yet more complex formulations [79].

As implants made of sintered HA are found in bone defects for many years after implantation (Figure 2, bottom), bioceramics made of more soluble types of $CaPO_4$ [68–112,121,122] are preferable for the biomedical purposes (Figure 2, top). Furthermore, the experimental results showed that BCP had a higher ability to adsorb fibrinogen, insulin, or type I collagen than HA [123]. Thus, according to both observed and measured bone formation parameters, $CaPO_4$ bioceramics have been ranked as follows: low sintering temperature BCP (rough and smooth) $\approx$ medium sintering temperature BCP $\approx$ TCP > calcined low sintering temperature HA > non-calcined low sintering temperature HA > high sintering temperature BCP (rough and smooth) > high sintering temperature HA [124]. This sequence was developed in the year 2000 and, thus, neither multiphase formulations nor other $CaPO_4$ are included.

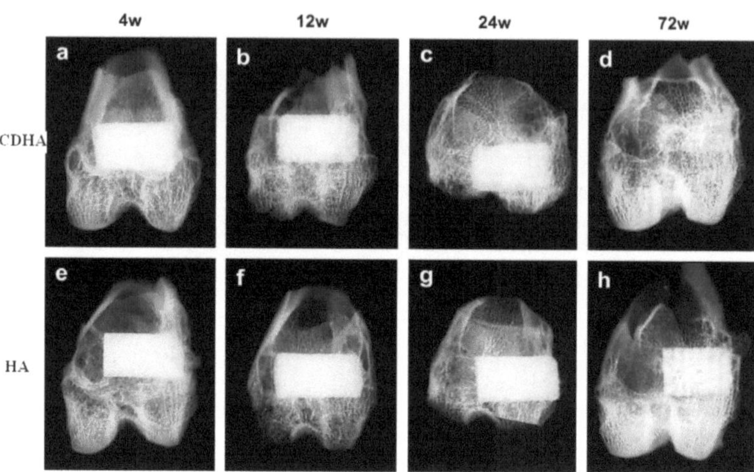

Figure 2. Soft X-ray photographs of the operated portion of the rabbit femur. Four weeks (**a**), 12 weeks (**b**), 24 weeks (**c**), and 72 weeks (**d**) after implantation of CDHA; 4 weeks (**e**), 12 weeks (**f**), 24 weeks (**g**), and 72 weeks (**h**) after implantation of sintered HA. Reprinted from Ref. [121] with permission.

3.3. Forming and Shaping

In order to fabricate $CaPO_4$ bioceramics in progressively complex shapes, scientists are investigating the use of both old and new manufacturing techniques. These techniques range from an adaptation of the age-old pottery techniques to the newest manufacturing methods for high-temperature ceramic parts for airplane engines; namely, reverse engineering [125,126] and rapid prototyping [127–129] technologies have revolutionized a generation of physical models, allowing the engineers to efficiently and accurately produce physical models and customized implants with high levels of geometric intricacy. Combined with computer-aided design and manufacturing (CAD/CAM), complex physical objects of the anatomical structure can be fabricated in a variety of shapes and sizes. In a typical application, an image of a bone defect in a patient can be taken and used to develop a three-dimensional (3D) CAD computer model [130–134]. Then, a computer can reduce the model to slices or layers. Afterwards, 3D objects and coatings are constructed layer-by-layer using rapid prototyping techniques. The examples comprise fused deposition modeling [135,136], selective laser sintering [137–142], laser cladding [143–146], 3D printing and/or plotting [73,147–153], robocasting [154–156], solid freeform fabrication [157–162], stereolithography [163–166], and direct light processing [167]. More advanced techniques, such as 4D [168,169] and 5D [170] printing techniques, have been introduced as well. Three-dimensional printing of the $CaPO_4$-based self-setting formulations is known as well [151]. Additional details of these techniques are available in the literature [171–174].

In the specific case of ceramic scaffolds, a sintering step is usually applied after printing the green bodies (see *Section 3.4. Sintering and Firing* below). Furthermore, a thermal printing process of melted $CaPO_4$ was proposed [175], while, in some cases, laser processing might be applied as well [176,177]. A schematic of the 3D-printing technique as well as some 3D-printed items are shown in Figure 3 [56]. A custom-made implant of actual dimensions would reduce the time it takes to perform the medical implantation procedure and subsequently lower the risk to the patient. Another advantage of a prefabricated, exact-fitting implant is that it can be used more effectively and applied directly to the damaged site rather than a replacement, which is formulated during surgery from a paste or granular material [158,177–179].

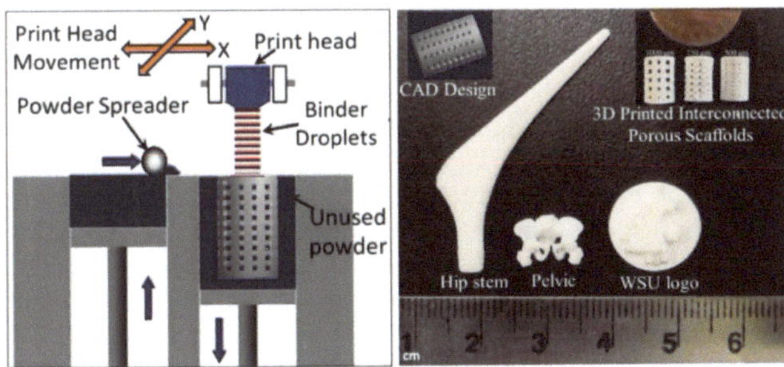

Figure 3. A schematic of 3D printing and some 3D-printed parts (fabricated at Washington State University) showing the versatility of 3D-printing technology for ceramic scaffolds fabrication with complex architectural features. Reprinted from Ref. [56] with permission.

In addition to the aforementioned modern techniques, classical forming and shaping approaches are still widely used. The selection of the desired technique depends greatly on the ultimate application of the bioceramic device, e.g., whether it is for a hard-tissue replacement or an integration of the device within the surrounding tissues. In general, three types of processing technologies might be used: (1) employment of a lubricant and a liquid binder with ceramic powders for shaping and subsequent firing; (2) application of self-setting and self-hardening properties of water-wet molded powders; (3) materials are melted to form a liquid and are shaped during cooling and solidification [180–182]. Since $CaPO_4$ are either thermally unstable (MCPM, MCPA, DCPA, DCPD, OCP, ACP, CDHA) or have a melting point at temperatures exceeding ~1400 °C with a partial decomposition (α-TCP, β-TCP, HA, FA, TTCP), only the first and the second consolidation approaches are used to prepare bulk bioceramics and scaffolds. The methods include uniaxial compaction [154,183,184], isostatic pressing (cold or hot) [87,185–191], granulation [192–198], loose packing [199], slip casting [75,200–205], gel casting [163,206–211], pressure mold forming [212–214], injection molding [215–218], polymer replication [219–226], ultrasonic machining [227], extrusion [228–234], and slurry dipping and spraying [235]. In addition, to form ceramic sheets from slurries, tape casting [207,236–240], doctor blade [241], and colander methods can be employed [180–182]. In addition, flexible, ultrathin (of 1 to several microns thick), freestanding HA sheets were produced by a pulsed laser deposition technique, followed by thin film isolation technology [242]. Various combinations of several techniques are also possible [77,207,243–245]. Furthermore, some of those processes might be performed under the electromagnetic field, which helps crystal aligning [201,204,246–249]. Finally, the prepared $CaPO_4$ bioceramics might be subjected to additional treatments (e.g., chemical, thermal, and/or hydrothermal ones) to convert one type of $CaPO_4$ into another one [226].

To prepare bulk bioceramics, powders are usually pressed damp in metal dies or dry in lubricated dies at pressures high enough to form sufficiently strong structures to hold together until they are sintered [250]. An organic binder, such as polyvinyl alcohol, helps to bind the powder particles altogether. Afterwards, the binder is removed by heating in air to oxidize the organic phases to carbon dioxide and water. Since many binders contain water, drying at ~100 °C is a critical step in preparing damp-formed pieces for firing. Too much or too little water in the compacts can lead to blowing apart the ware on heating or crumbling, respectively [180–182,186]. Furthermore, removal of water during drying often results in subsequent shrinkage of the product. In addition, due to local variations in water content, warping and even cracks may be developed during drying. Dry pressing and hydrostatic molding can minimize these problems [182]. Finally, the manufactured green samples are sintered.

It is important to note that forming and shaping of any ceramic products require a proper selection of the raw materials in terms of particle sizes and size distribution; namely, tough and strong bioceramics consist of pure, fine, and homogeneous microstructures. To attain this, pure powders with small average size and high surface area must be used as the starting sources. However, for maximum packing and least shrinkage after firing, mixing of ~70% coarse and ~30% fine powders have been suggested [182]. Mixing is usually carried out in a ball mill for uniformity of properties and reaction during subsequent firing. Mechanical die forming or sometimes extrusion through a die orifice can be used to produce a fixed cross-section.

Finally, to produce the accurate shaping, necessary for the fine design of bioceramics, machine finishing might be essential [132,180,251,252]. Unfortunately, cutting tools developed for metals are usually useless for bioceramics due to their fragility; therefore, grinding and polishing appear to be the most convenient finishing techniques [132,180]. In addition, the surface of $CaPO_4$ bioceramics might be modified by various supplementary treatments [253,254], and $CaPO_4$ bioceramics might be subjected to post-processing actions, such as immersing into special solutions [255].

3.4. Sintering and Firing

After being formed and shaped, the $CaPO_4$ bioceramics are commonly sintered. A sintering (or firing) procedure is a thermal process in which loosely bound particles are converted into a consistent solid mass under the influence of heat and/or pressure without melting the particles. This process is of great importance to manufacture bulk bioceramics with the required mechanical properties. Usually, this technique is carried out according to controlled temperature programs of electric furnaces in adjusted ambience of air with necessary additional gasses; however, always at temperatures below the melting points of the materials. The firing step can include temporary holds at intermediate temperatures to burn out organic binders [180–182]. The heating rate, sintering temperature, and holding time depend on the starting materials. For example, in the case of HA, these values are in the ranges of 0.5–3 °C/min, 1000–1250 °C, and 2–5 h, respectively [256]. In the majority of cases, sintering allows a structure to retain its shape. However, this process might be accompanied by a considerable degree of shrinkage [257–259], which must be accommodated in the fabrication process. For instance, in the case of FA sintering, a linear shrinkage was found to occur at ~715 °C and the material reached its final density at ~890 °C. Above this value, grain growth became important and induced an intra-granular porosity, which was responsible for density decrease. At ~1180 °C, a liquid phase was formed due to formation of a binary eutectic between FA and fluorite contained in the powder as impurity. This liquid phase further promoted the coarsening process and induced formation of large pores at high temperatures [260].

In general, sintering occurs only when the driving force is sufficiently high, while the latter relates to the decrease in surface and interfacial energies of the system by matter (molecules, atoms, or ions) transport, which can proceed by solid, liquid, or gaseous phase diffusion. Namely, when solids are heated to high temperatures, their constituents are driven to move to fill up pores and open channels between the grains of powders, as well as to compensate for the surface energy differences among their convex and concave surfaces (matter moves from convex to concave). At the initial stages, bottlenecks are formed and grow among the particles (Figure 4). Existing vacancies tend to flow away from the surfaces of sharply curved necks; this is an equivalent of a material flow towards the necks, which grow as the voids shrink. Small contact areas among the particles expand and, at the same time, a density of the compact increases and the total void volume decreases. As the pores and open channels are closed during a heat treatment, the particles become tightly bonded together, and density, strength, and fatigue resistance of the sintered object improve greatly. Grain boundary diffusion was identified as the dominant mechanism for densification [261]. Furthermore, strong chemical bonds are formed among the particles, and loosely compacted green bodies are hardened to denser materials [180–182]. Further knowledge on the ceramic sintering process can be found elsewhere [262].

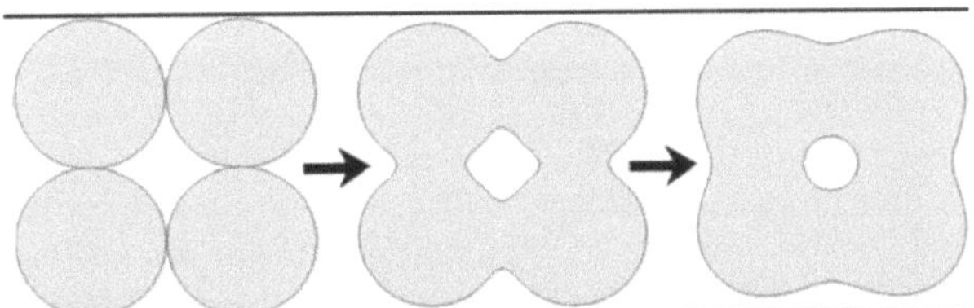

Figure 4. A schematic diagram representing the changes occurring with spherical particles under sintering. Shrinkage is noticeable.

In the case of $CaPO_4$, the earliest paper on their sintering was published in 1971 [263]. Since then, numerous papers on this subject have been published, and several specific processes have been found to occur during $CaPO_4$ sintering. Firstly, moisture, carbonates and all other volatile chemicals remaining from the synthesis stage, such as ammonia, nitrates, and any organic compounds, are removed as gaseous products. Secondly, unless powders are sintered, the removal of these gases facilitates production of denser ceramics with subsequent shrinkage of the samples (Figure 5). Thirdly, all chemical changes are accompanied by a concurrent increase in crystal size and a decrease in the specific surface area. Fourthly, a chemical decomposition of all acidic orthophosphates and their transformation into other phosphates (e.g., $2HPO_4{}^{2-} \rightarrow P_2O_7{}^{4-} + H_2O$) takes place. In addition, sintering causes toughening [66], densification [67,264], partial dehydroxylation (in the case of HA) [67], a partial evaporation and condensation of phosphates [265], and grain growth [261,266], as well as a mechanical strength increasing [267–269]. The latter events are due to presence of air and other gases filling gaps among the particles of unsintered powders. At sintering, the gases move towards the outside of powders, and green bodies shrink owing to decrease of distances among the particles. For example, sintering of biologically formed apatites was investigated [270,271] and the obtained products were characterized [272,273]. In all cases, the numerical value of the Ca/P ratio in sintered apatites of biological origin was higher than that of the stoichiometric HA. One should mention that in the vast majority of cases, $CaPO_4$ with Ca/P ratio < 1.5 (Table 1) are not sintered, since these compounds are thermally unstable, while sintering of nonstoichiometric $CaPO_4$ (CDHA and ACP) always leads to their transformation into various types of biphasic, triphasic, and multiphase formulations [79].

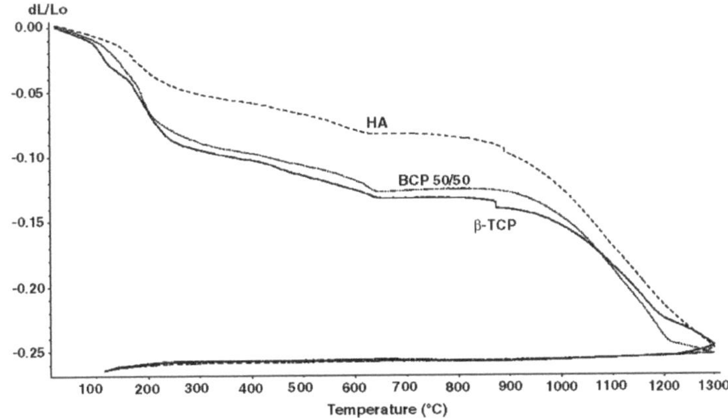

Figure 5. Linear shrinkage of the compacted ACP powders that were converted into β-TCP, BCP (50% HA + 50% β-TCP), and HA upon heating. According to the authors: "At 1300 °C, the shrinkage reached a maximum of approximately ~25, ~30 and ~35% for the compacted ACP powders that converted into HA, BCP 50/50 and β-TCP, respectively" [258]. Reprinted from Ref. [258] with permission.

An extensive study on the effects of sintering temperature and time on the properties of HA bioceramics revealed a correlation between these parameters and density, porosity, grain size, chemical composition, and strength of the scaffolds [274]. Namely, sintering below ~1000 °C was found to result in initial particle coalescence, with little or no densification and a significant loss of the surface area and porosity. The degree of densification appeared to depend on the sintering temperature, whereas the degree of ionic diffusion was governed by the period of sintering [274]. To enhance sinterability of $CaPO_4$, a variety of sintering additives might be added [275–278].

Solid-state pressureless sintering is the simplest procedure. For example, HA bioceramics can be pressurelessly sintered up to the theoretical density at 1000–1200 °C. Processing at even higher temperatures usually lead to exaggerated grain growth and decomposition because HA becomes unstable at temperatures exceeding ~1300 °C [6,113–117,279–281]. The decomposition temperature of HA bioceramics is a function of the partial pressure of water vapor. Moreover, processing under vacuum leads to an earlier decomposition of HA, while processing under high partial pressure of water prevents the decomposition. On the other hand, the presence of water in the sintering atmosphere was reported to inhibit densification of HA and accelerate grain growth [282]. Unexpectedly, an application of a magnetic field during sintering was found to influence the growth of HA grains [266]. A definite correlation between hardness, density, and a grain size in sintered HA bioceramics was found; despite exhibiting high bulk density, hardness started to decrease at a certain critical grain size limit [283–285].

Since grain growth occurs mainly during the final stage of sintering, to avoid this, a new method called "two-step sintering" (TSS) was proposed [286]. The method consists of suppressing grain boundary migration responsible for grain growth, while keeping grain boundary diffusion that promotes densification. The TSS approach was successfully applied to $CaPO_4$ bioceramics [77,86,287–290]. For example, HA compacts prepared from nanodimensional powders were two-step sintered. The average grain size of near full dense (>98%) HA bioceramics made via conventional sintering was found to be ~1.7 μm, while that for TSS HA bioceramics was ~190 nm (i.e., ~9 times less) with simultaneous increasing of the fracture toughness of samples from 0.98 ± 0.12 to 1.92 ± 0.20 MPa m$^{1/2}$. In addition, due to the lower second-step sintering temperature, no HA phase decomposition was detected in the TSS method [287].

Hot pressing [285,291–297], hot isostatic pressing [87,185,190,191], or hot pressing with post-sintering [298,299], as well as "cold sintering" (which is very similar to hot pressing) [300] processes make it possible to decrease the temperature of the densification process, diminish the grain size, and achieve higher densities. This leads to finer microstructures, higher thermal stability, and subsequently better mechanical properties of $CaPO_4$ bioceramics. In addition, microwave [301–306], spark plasma [69,104,307–315], flash [316,317], and ultrafast high-temperature [318] sintering techniques are alternative methods to the conventional sintering, hot pressing, and hot isostatic pressing. Both alternative methods were found to be time- and energy-efficient densification techniques. Further developments are still possible. For example, a hydrothermal hot pressing method was developed to fabricate OCP [105], CDHA [319], HA/β-TCP [294], and HA [295–298,320] bioceramics with neither thermal dehydration nor thermal decomposition. Further details on the sintering and firing processes of $CaPO_4$ bioceramics are available in the literature [115,321,322].

To conclude this section, one should note that the sintering stage is not always necessary. For example, $CaPO_4$-based bulk bioceramics with the reasonable mechanical properties might be prepared by means of self-setting (self-hardening) formulations (see *Section 6.1. Self-setting (Self-hardening) Formulations* below). Furthermore, the reader's attention is directed to an excellent review on various ceramic manufacturing techniques [323], in which various ceramic processing techniques are well described.

4. The Major Properties

4.1. Mechanical Properties

The modern generation of biomedical materials should stimulate the body's own self-repairing abilities [324]. Therefore, during healing, a mature bone should replace the modern grafts and this process must occur without transient loss of the mechanical support. Unluckily for material scientists, a human body provides one of the most inhospitable environments for the implanted biomaterials. It is warm, wet, and both chemically and biologically active. For example, a diversity of body fluids in various tissues might have a solution pH varying from 1 to 9. In addition, a body is capable of generating quite massive force concentrations, and the variance in such characteristics among individuals might be enormous. Typically, bones are subjected to ~4 MPa loads, whereas tendons and ligaments experience peak stresses in the range of 40–80 MPa. The hip joints are subjected to an average load of up to three times the body weight (3000 N), and peak loads experienced during jumping can be as high as 10 times the body weight. These stresses are repetitive and fluctuating depending on the nature of the activities, which can include standing, sitting, jogging, stretching, and climbing. Therefore, all types of implants must sustain attacks of a great variety of aggressive conditions [325]. Regrettably, there is presently no artificial material fulfilling all these requirements.

Now it is important to mention that the mechanical behavior of any ceramics is rather specific; namely, ceramics is brittle, which is attributed to high-strength ionic bonds. Thus, it is not possible for plastic deformation to happen prior to failure, as a slip cannot occur. Therefore, ceramics fail in a dramatic manner. Namely, if a crack is initiated, its progress will not be hindered by the deformation of material ahead of the crack, as would be the case in a ductile material (e.g., a metal). In ceramics, the crack will continue to propagate, rapidly resulting in a catastrophic breakdown. In addition, the mechanical data typically have a considerable amount of scatter [181]. Alas, all of these are applicable to $CaPO_4$ bioceramics.

For dense bioceramics, the strength is a function of the grain sizes. Namely, finer-grain-size bioceramics have smaller flaws at the grain boundaries and thus are stronger than ones with larger grain sizes. Thus, in general, the strength for ceramics is proportional to the inverse square root of the grain sizes [326]. In addition, the mechanical properties decrease significantly with increasing content of an amorphous phase, microporosity, and grain sizes, while a high crystallinity, a low porosity, and small grain sizes tend to give a higher stiffness, a higher compressive and tensile strength, and a greater fracture toughness. Furthermore, ceramics strength appears to be very sensitive to slow crack growth [327]. Accordingly, from the mechanical point of view, $CaPO_4$ bioceramics appear to be brittle polycrystalline materials for which the mechanical properties are governed by crystallinity, grain size, grain boundaries, porosity, and composition [328]. Thus, it possesses poor mechanical properties (for instance, a low impact and fracture resistances) that do not allow $CaPO_4$ bioceramics to be used in load-bearing areas, such as artificial teeth or bones [50–53]. For example, fracture toughness (this is a property that describes the ability of a material containing a crack to resist fracture and is one of the most important properties of any material for virtually all design applications) of HA bioceramics does not exceed the value of ~1.2 MPa·m$^{1/2}$ [329] (human bone: 2–12 MPa·m$^{1/2}$). It decreases exponentially with increasing porosity [330]. Generally, fracture toughness increases with grain size decreasing. However, in some materials, especially noncubic ceramics, fracture toughness reaches the maximum and rapidly drops with decreasing grain size. For example, a fracture toughness of pure hot-pressed HA with grain sizes between 0.2–1.2 μm was investigated. The authors found two distinct trends, where fracture toughness decreased with increasing grain size above ~0.4 μm and subsequently decreased with decreasing grain size. The maximum fracture toughness measured was 1.20 ± 0.05 MPa·m$^{1/2}$ at ~0.4 μm [291]. Fracture energy of HA bioceramics is in the range of 2.3–20 J/m^2, while the Weibull modulus (a measure of the spread or scatter in fracture strength) is low (~5–12) in wet environments, which means that HA behaves as a typical brittle ceramics and indicates a low reliability of HA implants [331]. Porosity has a great influence on the Weibull modulus [332,333]. In addition, the reliability of HA bioceramics was found to depend on deformation mode (bending or compression), along with pore size and pore size distribution: reliability was higher for smaller average pore sizes in bending but lower for smaller pore sizes in compression [334]. Interestingly, three peaks of internal friction were found at temperatures of about –40, 80, and 130 °C for HA but no internal friction peaks were obtained for FA in the measured temperature range; this effect was attributed to the differences of F^- and OH^- positions in FA and HA, respectively [335]. Differences in internal friction values were also found between HA and TCP [336].

Bending, compressive, and tensile strengths of dense HA bioceramics are in the ranges of 38–250, 120–900, and 38–300 MPa, respectively. Similar values for porous HA bioceramics are substantially lower: 2–11, 2–100, and ~3 MPa, respectively [331]. These wide variations in the properties are due to both structural variations (e.g., an influence of remaining microporosity, grain sizes, presence of impurities, etc.) and manufacturing processes, and they are also caused by a statistical nature of the strength distribution. Strength was found to increase with Ca/P ratio increasing, reaching the maximum value around Ca/P ~1.67 (stoichiometric HA) and decreasing suddenly when Ca/P > 1.67 [331]. Furthermore, strength decreases almost exponentially with increasing porosity [337,338]. However, by changing the pore geometry, it is possible to influence the strength of porous bioceramics. It is also worth mentioning that porous $CaPO_4$ bioceramics are considerably less fatigue-resistant than dense ones (in materials science, fatigue is the progressive and localized structural damage that occurs when a material is subjected to cyclic loading). Both grain sizes and porosity are reported to influence the fracture path, which itself has little effect on the fracture toughness of $CaPO_4$ bioceramics [328,339]. However, no obvious decrease in mechanical properties was found after $CaPO_4$ bioceramics had been aged in the various solutions during the different periods of time [340].

Young's (or elastic) modulus of dense HA bioceramics is in the range of 3–120 GPa [341,342], which is more or less similar to those of the most resistant components of the natural calcified tissues (dental enamel: ~74 GPa, dentine: ~21 GPa, compact bone: ~18–22 GPa). This value depends on porosity [343,344]. Nevertheless, dense bulk compacts of HA have mechanical resistances of the order of 100 MPa versus ~00 MPa of human bones, drastically diminishing their resistances in the case of porous bulk compacts [345]. Young's modulus measured in bending is between 44 and 88 GPa. To investigate the subject in more detail, various types of modeling and calculations are increasingly used [346–350]. For example, the elastic properties of HA appeared to be significantly affected by the presence of vacancies, which softened HA via reducing its elastic modules [350]. In addition, a considerable anisotropy in the stress–strain behavior of the perfect HA crystals was found by ab initio calculations [347]. The crystals appeared to be brittle for tension along the z-axis with the maximum stress of ~9.6 GPa at 10% strain. Furthermore, the structural analysis of the HA crystal under various stages of tensile strain revealed that the deformation behavior manifested itself mainly in the rotation of PO_4 tetrahedrons with concomitant movements of both the columnar and axial Ca ions [347]. Data for single crystals are also available [351]. Vickers hardness (a measure of the resistance to permanent indentation) of dense HA bioceramics is within 3–7 GPa, while the Poisson's ratio (the ratio of the contraction or transverse strain to the extension or axial strain) for HA is about 0.27, which is close to that of bones (~0.3). At temperatures within 1000–1100 °C, dense HA bioceramics were found to exhibit superplasticity with a deformation mechanism based on grain boundary sliding [312,352,353]. Furthermore, both wear resistance and friction coefficient of dense HA bioceramics are comparable to those of dental enamel [331].

Due to a high brittleness (associated with a low crack resistance), the biomedical applications of $CaPO_4$ bioceramics are focused on production of non-load-bearing implants, such as pieces for middle ear surgery, filling of bone defects in oral or orthopedic surgery, and coating of dental implants and metallic prosthesis (see below) [62,354,355]. Therefore, methods are continuously sought to improve the reliability of $CaPO_4$ bioceramics. Namely, the mechanical properties of sintered bioceramics might be improved by changing the morphology of the initial $CaPO_4$ [356]. In addition, diverse reinforcements (ceramics, metals, or polymers) have been applied to manufacture various biocomposites and hybrid biomaterials [357], but that is another story. However, successful hybrid formulations consisting of $CaPO_4$ only [358–365] are within the scope of this review. Namely, bulk HA bioceramics might be reinforced by HA whiskers [359–363]. Furthermore, various biphasic apatite/TCP formulations were tested [358,364,365] and, for example, a superior superplasticity of HA/β-TCP biocomposites to HA bioceramics was detected [364].

Another method to improve the mechanical properties of $CaPO_4$ bioceramics is to cover the items by polymeric coatings [366–368] or infiltrate porous structures by polymers [369–371]; however, this is another topic. Other approaches are also possible [154]. Further details on the mechanical properties of $CaPO_4$ bioceramics are available elsewhere [330,331,372], where interested readers are referred.

4.2. Electric/Dielectric and Piezoelectric Properties

Recently, an interest in both electric/dielectric [301,373–385] and piezoelectric [386,387] properties of $CaPO_4$ bioceramics has been expressed. In addition, some types of $CaPO_4$ bioceramics (namely, HA) appear to be electrets [388,389]. An electret is a dielectric material that has a quasi-permanent electric charge or dipole polarization. An electret generates internal and external electric fields, and is the electrostatic equivalent of a permanent magnet [390]. For example, a surface ionic conductivity of both porous and dense HA bioceramics was examined for humidity sensor applications, since the room temperature conductivity was influenced by relative humidity [374]. Namely, the ionic conductivity of HA is a subject of research for its possible use as a gas sensor for alcohol [375], carbon dioxide [373,382], or carbon monoxide [378]. Electric measurements were also used as a characterization tool to study the evolution of microstructure in HA bioceramics [376]. More to the point, the dielectric properties of HA were examined to understand its decomposition to β-TCP [375]. In the case of CDHA, the electric properties, in terms of ionic conductivity, were found to increase after compression of the samples at 15 t/cm^2, which was attributed to establishment of some order within the apatitic network [377]. The conductivity mechanism of CDHA appeared to be multiple [380]. Furthermore, there are attempts to develop HA and/or CDHA electrets for biomedical utilization [379,388,389].

The electric properties of $CaPO_4$ bioceramics appear to influence their biomedical applications. For example, there is an interest in polarization of HA bioceramics to generate a surface charge by the applying a constant DC electric field of 0.5–10.0 kV/cm at elevated temperatures (300–1000 °C) to samples previously sintered at ~ 1000–1250 °C for ~2 h. This technique is called thermally stimulated polarization and its results indicated that the polarization effects were a consequence of electrical dipoles associated with the formation of defects inside crystal grains, such as thermally-induced OH^- vacancies, and of the space charge polarization that originated in the grain boundaries [391–393]. The presence of surface charges on HA was shown to have a significant effect on both in vitro and in vivo crystallization of biological apatite [394–400], as well as on an ability to adsorb various types of phosphate ions [393]. Furthermore, a growth of both biomimetic $CaPO_4$ and bones was found to be accelerated on negatively charged surfaces and decelerated at positively charged surfaces [398–407]. A similar effect was found for adsorption of bovine serum albumin [408]. In addition, the electric polarization of $CaPO_4$ was found to accelerate a cytoskeleton reorganization of osteoblast-like cells [409–412], extend bioactivity [413], enhance bone ingrowth through the pores of porous implants [414], and influence the cell activity [415,416]. The positive effect of electric polarization was found for carbonated apatite as well [417]. There is an interesting study on the interaction of a blood coagulation factor on electrically polarized HA surfaces [418]. Further details on the electric properties of $CaPO_4$-based bioceramics are available in the literature [301,383,384,389].

4.3. Possible Transparency

Single crystals of all types of $CaPO_4$ are optically transparent for the visible light. As bioceramics of $CaPO_4$ have a polycrystalline nature with a random orientation of big amounts of small crystals, it is opaque and of white color, unless colored dopants have been added. However, in some cases, a transparency is convenient to provide some essential advantages (e.g., to enable direct viewing of living cells, their attachment, spreading, proliferation, and osteogenic differentiation cascade in a transmitted light). Thus, transparent $CaPO_4$ bioceramics (Figure 6) [419] have been prepared and investigated [69,87,185,187,310–315,419–427]. They can exhibit an optical transmittance of ~66% at a wavelength of 645 nm [425]. The preparation techniques include a hot isostatic pressing [87,185,187,426], an ambient-pressure sintering [420], a gel casting coupled with a low-temperature sintering [421,424], and a pulse electric current sintering [422], as well as spark plasma [69,307–315] and flash [316,317] sintering techniques. Fully dense, transparent $CaPO_4$ bioceramics are obtained at temperatures above ~800 °C. Depending on the preparation technique, the transparent bioceramics have a uniform grain size ranging from ~67 nm [87] to ~250 μm [421] and are always pore-free. Furthermore, translucent $CaPO_4$ bioceramics are also known [87,263,428–430]. Concerning possible biomedical applications, the optically transparent in visible light $CaPO_4$ bioceramics can be useful for direct viewing of other objects, such as cells, in some specific experiments [423]. In addition, the transparency for laser light $CaPO_4$ bioceramics may appear to be convenient for minimal invasive surgery by allowing passing the laser beam through it to treat the injured tissues located underneath. However, due to a lack of both porosity and the necessity to have see-through implants inside the body, the transparent and translucent forms of $CaPO_4$ bioceramics will hardly be extensively used in medicine, except for the aforementioned cases and possible eye implants.

Figure 6. Transparent HA bioceramics prepared by spark plasma sintering at 900 °C from nanosized HA single crystals. Reprinted from Ref. [419] with permission.

4.4. Porosity

Porosity is defined as a percentage of voids in solids, and this morphological property is independent of the material. The surface area of porous bodies is much higher, which guarantees a good mechanical fixation in addition to providing sites on the surface that allow chemical bonding between the bioceramics and bones [431]. Furthermore, a porous material may have both closed (isolated) pores and open (interconnected) pores. The latter look similar to tunnels and are accessible by gases, liquids, and particulate suspensions [432]. The open-cell nature of porous materials (also known as reticulated materials) is a unique characteristic essential in many applications. In addition, pore dimensions are also important. Namely, the dimensions of open pores are directly related to bone formation, since such pores grant both the surface and space for cell adhesion and bone ingrowth [433–435]. On the other hand, pore interconnection provides the ways for cell distribution and migration, and it allows an efficient in vivo blood vessel formation suitable for sustaining bone tissue neo-formation and possibly remodeling [123,414,436–440]. Namely, porous CaPO$_4$ bioceramics are colonized easily by cells and bone tissues [436,439,441–446]. Therefore, interconnecting macroporosity (pore size > 100 μm) [84,431,436,447,448] is intentionally introduced in solid bioceramics (Figure 7). Calcining of natural bones and teeth appears to be the simplest way to prepare porous CaPO$_4$ bioceramics [7–14]. In addition, macroporosity might be formed artificially due to a release of various easily removable compounds and, for that reason, incorporation of pore-creating additives (porogens) is the most popular technique to create macroporosity. The porogens are crystals, particles, or fibers of either volatile (they evolve gases at elevated temperatures) or soluble substances. The popular examples comprise paraffin [449–451], naphthalene [328,452–454], sucrose [455,456], NaHCO$_3$ [457–459], NaCl [460,461], polymethylmethacrylate [74,462–464], hydrogen peroxide [465–468], cellulose [469], and its derivatives [64]. Several other compounds [338,470–477], including carbon nanotubes [478], might be used as porogens as well. The ideal porogen should be nontoxic and be removed at ambient temperature, thereby allowing the bioceramic/porogen mixture to be injected directly into a defect site and allowing the scaffold to fit the defect [479]. Sintering particles, preferably spheres of equal size, is a similar way to generate porous 3D bioceramics of CaPO$_4$; however, pores resulting from this method are often irregular in size and shape and not fully interconnected with one another. Schematic drawings of various types of the ceramic porosity are shown in Figure 8 [480]. One should note that 3D-printing techniques allow producing structures with tailored pore orientations by changing framework directions in a controlled, periodic pattern (Figure 9) [481].

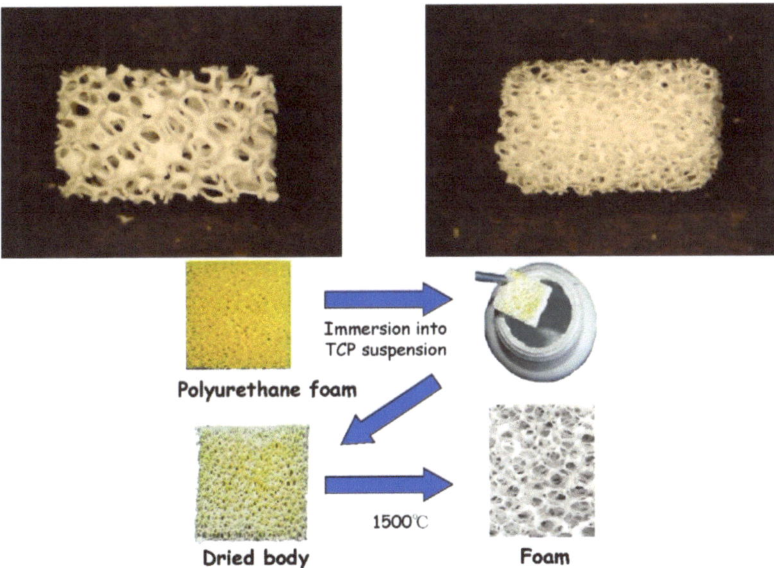

Figure 7. Photographs of commercially available porous CaPO$_4$ bioceramics with different porosity (**top**) and a method of their production (**bottom**). For photos, the horizontal field width is 20 mm.

Many other techniques, such as replication of polymer foams by impregnation [219–221, 224,482–486] (Figure 7), various types of casting [202,203,207,209,468,487–495], suspension foaming [101], surfactant washing [496], microemulsions [497,498], and ice templating [499–502], as well as many other approaches [68,71,74,75,140,503–528], have been applied to fabricate porous CaPO$_4$ bioceramics. Some of them are summarized in Table 2 [479]. In addition, both natural CaCO$_3$ porous materials, such as coral skeletons [529,530], shells [530,531], and even wood [532], as well as artificially prepared ones [533], can be converted into porous CaPO$_4$ under the hydrothermal conditions (250 °C, 24–48 h) with the microstructure undamaged. Porous HA bioceramics can also be obtained by hydrothermal hot pressing. This technique allows solidification of the HA powder at 100–300 °C (30 MPa, 2 h) [320]. In another approach, bi-continuous water-filled microemulsions are used as preorganized systems for the fabrication of needle-like frameworks of crystalline HA (2 °C, 3 weeks) [497,498]. In addition, porous $CaPO_4$ might be prepared by a combination of gel casting and foam burn out methods [243,245], as well as by hardening of the self-setting formulations [450,451,458–461,520]. Lithography was used to print a polymeric material, followed by packing with HA and sintering [507]. Hot pressing was applied as well [292,293]. More to the point, an HA suspension can be cast into a porous $CaCO_3$ skeleton, which is then dissolved, leaving a porous network [503]. A 3D periodic macroporous frame of HA was fabricated via a template-assisted colloidal processing technique [509,512]. In addition, porous HA bioceramics might be prepared by using different starting HA powders and sintering at various temperatures by a pressureless sintering [505]. Porous bioceramics with an improved strength might be fabricated from $CaPO_4$ fibers or whiskers. In general, fibrous porous materials are known to exhibit an improved strength due to fiber interlocking, crack deflection, and/or pullout [534]. Namely, porous bioceramics with well-controlled open pores were processed by sintering of fibrous HA particles [504]. In another approach, porosity was achieved by firing apatite-fiber compacts mixed with carbon beads and agar. By varying the compaction pressure, firing temperature and carbon/HA ratio, the total porosity was controlled in the ranges from ~40% to ~85% [64]. Finally, a superporous (~85% porosity) HA bioceramic was developed as well [515,517,518]. Additional information on the processing routes to produce porous ceramics can be found in the literature [535].

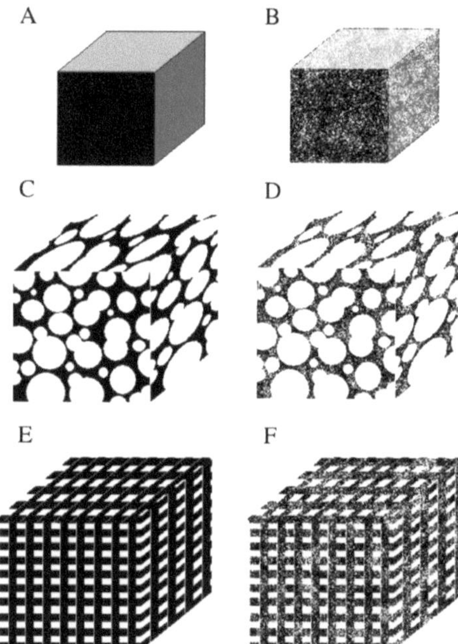

Figure 8. Schematic drawings of various types of the ceramic porosity: (**A**)—nonporous, (**B**)—microporous, (**C**)—macroporous (spherical), (**D**)—macroporous (spherical) + micropores, (**E**)—macroporous (3D-printing), (**F**)— macroporous (3D-printing) + micropores. Reprinted from Ref. [480] with permission.

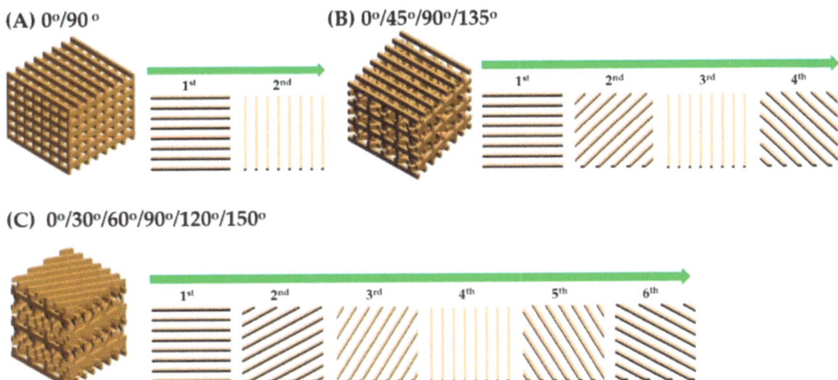

Figure 9. A schematic diagram showing the porous structures of various CaPO4 scaffolds with different pore orientations: (**A**) $0°/90°$, (**B**) $0°/45°/90°/135°$, and (**C**) $0°/30°/60°/90°/120°/150°$, with their top-down view of the repeating CaPO4 frameworks. Reprinted from Ref. [481] with permission.

Bioceramic microporosity (pore size < 10 μm), which is defined by its capacity to be impregnated by biological fluids [536], results from the sintering process, while the pore dimensions mainly depend on the material composition, thermal cycle, and sintering time. The microporosity provides both a greater surface area for protein adsorption and increased ionic solubility. For example, embedded osteocytes distributed throughout microporous rods might form a mechanosensory network, which would not be possible in scaffolds without microporosity [537,538]. $CaPO_4$ bioceramics with nanodimensional (<100 nm) pores might be fabricated as well [539–543]. It is important to stress that differences in porogens usually influence the bioceramics' macroporosity, while differences in sintering temperatures and conditions affect the percentage of microporosity. Usually, the higher the sintering temperature, the lower both the microporosity content and the specific surface area of bioceramics. Namely, HA bioceramics sintered at ~1200 °C show significantly less microporosity and a dramatic change in crystal sizes, if compared with those sintered at ~1050 °C (Figure 10) [544]. Furthermore, the average shape of pores was found to transform from strongly oblate to round at higher sintering temperatures [545]. The total porosity (macroporosity + microporosity) of $CaPO_4$ bioceramics was reported to be ~70% [546] or even ~85% [515,517,518] of the entire volume. In the case of coralline HA or bovine-derived apatites, the porosity of the original biologic material (coral or bovine bone) is usually preserved during processing [547]. To finalize the production topic, creation of the desired porosity in $CaPO_4$ bioceramics is a rather complicated engineering task and interested readers are referred to the additional publications on the subject [338,435,519,548–553].

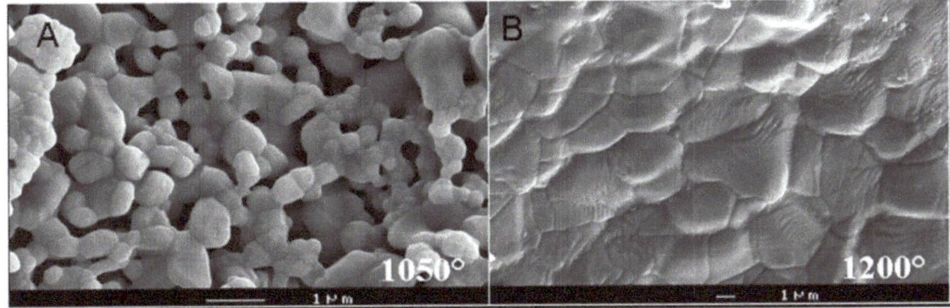

Figure 10. SEM pictures of HA bioceramics sintered at (**A**) 1050 °C and (**B**) 1200 °C. Note the presence of microporosity in **A** and not in **B**. Reprinted from Ref. [544] with permission.

Table 2. The procedures used to manufacture porous CaPO$_4$ scaffolds for tissue engineering [479].

Year	Location	Process	Apatite from:	Sintering	Compressive Strength	Pore Size	Porosity	Method of Porosity Control
2006	Deville et al., Berkeley, CA	HA + ammonium methacrylate in polytetrafluoroethylene mold, freeze dried and sintered.	HA #30	Yes: 1300 °C	16 MPa 65 MPa 145 MPa	Open unidi-rectional 50–150 μm.	>60% 56% 47%	Porosity control: slurry conc. Structure controlled by physics of ice front formation.
2006	Saiz et al., Berkeley, CA	Polymer foams coated, compressed after infiltration, then calcined.	HA powder	Yes: 700–1300 °C	–	100–200 μm.	–	Porosity control: extent of compression, HA loading.
2006	Murugan et al., Singapore + USA	Bovine bone cleaned, calcined.	bovine bone	Yes: 500 °C	–	Retention of nanosized pores.	–	Porosity control: native porosity of bovine bone.
2006	Xu et al., Gaithersburg, MD	Directly injectable CaPO$_4$ cement, self-hardens, mannitol as porogen.	nanocrystalline HA	No	2.2–4.2 MPa (flexural)	0%–50% macrop-orous.	65%–82%	Porosity control: mannitol mass fraction in mixture.
2004	Landi et al., Italy + Indonesia	Sponge impregnation, isotactic pressing, sintering of HA in simulated body fluid.	CaO + H$_3$PO$_4$	Yes: 1250 °C for 1 h	23 ± 3.8 MPa	Closed 6%, open 60%.	66%	Porosity control: possibly by controlling HA particle size. Not suggested by authors.
2003	Charriere et al., EPFL, Switzerland	Thermoplastic negative porosity by Inkjet printing, slip casting process for HA.	DCPA + calcite	No: 90 °C for 1 day.	12.5 ± 4.6 MPa	–	44%	Porosity control: negative printing.
2003	Almirall et al., Barcelona, Spain	α-TCP foamed with hydrogen peroxide at different conc., liq. Ratios, poured in polytetrafluoroethylene molds.	A-TCP + (10% and 20% H$_2$O$_2$)	No: 60 °C for 2 h.	1.41 ± 0.27 MPa 2.69 ± 0.91 MPa	35.7% macro. 29.7% micro. 26.8% macro. 33.8% micro.	65.5% 60.7%	Porosity control: different concentration, α-TCP particle sizes.
2003	Ramay et al., Seattle, WA	Slurries of HA prepared: gel-casting + polymer sponge technique, sintered.	HA powder	Yes: 600 °C for 1 h 1350 °C for 2·h.	0.5–5 MPa	200–400 μm.	70%–77%	Porosity control: replicate of polymer sponge template.

Table 2. *Cont.*

Year	Location	Process	Apatite from:	Sintering	Compressive Strength	Pore Size	Porosity	Method of Porosity Control
2003	Miao et al., Singapore	TTCP to CaPO$_4$ cement. Slurry cast on polymer foam, sintered.	TTCP	Yes: 1200 °C for 2 h.	–	1 mm macro. 5 µm micro.	~70%	Porosity control: Recoating time, polyurethane foam.
2003	Uemura et al., China + Japan	Slurry of HA with polyoxyethylene lauryl ether (cross-linked) and sintered.	HA powders	Yes: 1200 °C for 3 h.	2.25 MPa (0 wk) 4.92 MPa (12 wk) 11.2 MPa (24 wx)	500 µm. 200 µm interconnects.	~77%	Porosity control: polymer interconnects cross-linking.
2003	Ma et al., Singapore + USA	Electrophoretic deposition of HA, sintering.	HA powders	Yes: 1200 °C for 2 h.	860 MPa	0.5 µm. 130 µm.	~20%	Porosity control: electrophoresis field.
2002	Barralet et al., Birmingham, London, UK	CaPO$_4$ cement + sodium phosphate ice, evaporated.	CaCO$_3$ + DCPD	1st step: 1400 °C for 1 day.	0.6 ± 0.27 MPa	2 µm.	62% ± 9%	Porosity control: porogen shape.

Regarding the biomedical importance of porosity, studies revealed that increasing of both the specific surface area and pore volume of bioceramics might greatly accelerate the in vivo process of apatite deposition and, therefore, enhance the bone-forming bioactivity. More importantly, a precise control over the porosity, pore dimensions, and internal pore architecture of bioceramics on different length scales is essential for understanding the structure–bioactivity relationship and the rational design of better bone-forming biomaterials [551,554,555]. Namely, in antibiotic charging experiments, $CaPO_4$ bioceramics with nanodimensional (<100 nm) pores showed a much higher charging capacity (1621 μg/g) than those of commercially available $CaPO_4$ (100 μg/g), which did not contain nanodimensional porosity [549]. In other experiments, porous blocks of HA were found to be viable carriers with sustained release profiles for drugs [556] and antibiotics over 12 days [557] and 12 weeks [558], respectively. Unfortunately, porosity significantly decreases the strength of implants [334,339,372]. Thus, porous $CaPO_4$ implants cannot be loaded and are used to fill only small bone defects; however, their strength increases gradually when bones ingrow into the porous network of $CaPO_4$ implants [119,559–562]. For example, bending strengths of 4–60 MPa for porous HA implants filled with 50%–60% of cortical bone were reported [559], while in another study an ingrown bone increased strength of porous HA bioceramics by a factor of three to four [561].

Unfortunately, the biomedical effects of bioceramics' porosity are not straightforward. For example, the in vivo response of $CaPO_4$ to different porosity was investigated, and a hardly any effect of macropore dimensions (~150, ~260, ~510, and ~1220 μm) was observed [563]. In another study, a greater differentiation of mesenchymal stem cells was observed when cultured on ~200 μm pore size HA scaffolds when compared to those on ~500 μm pore size HA [564]. The latter finding was attributed to the fact that a higher pore volume in ~500 μm macropore scaffolds might contribute to a lack of cell confluency, leading to the cells proliferating before beginning differentiation. In addition, the authors hypothesized that bioceramics having less than the optimal pore dimensions induced quiescence in differentiated osteoblasts due to reduced cell confluency [564]. In still another study, the use of BCP (HA/TCP = 65/35 wt.%) scaffolds with cubic pores of ~500 μm resulted in the highest bone formation compared with the scaffolds with lower (~100 μm) or higher (~1000 μm) pore sizes [565]. Furthermore, $CaPO_4$ bioceramics with greater strut porosity appeared to be more osteoinductive [566]. As early as 1979, Holmes suggested that the optimal pore range was 200–400 μm with the average human osteon size of ~223 μm [567]. In 1997, Tsurga and coworkers implied that the optimal pore size of bioceramics that supported ectopic bone formation was 300–400 μm [568]. Thus, there is no need to create $CaPO_4$ bioceramics with very big pores; however, the pores must be interconnected [437,447,448,569]. Interconnectivity governs a depth of cells or tissue penetration into the porous bioceramics, and it allows development of blood vessels required for new bone nourishing and wastes removal [570,571]. Nevertheless, the total porosity of implanted bioceramics appears to be important. For example, 60% porous β-TCP granules achieved a higher bone fusion rate than 75% porous β-TCP granules in lumbar posterolateral fusion [537].

More details on the importance of $CaPO_4$ bioceramics porosity on bone regeneration are available in a topical review [572].

5. Biological Properties and In Vivo Behavior

The most important differences between bioactive bioceramics and all other implanted materials comprise inclusion in the metabolic processes of the organism, adaptation of either surface or the entire material to the biomedium, integration of a bioactive implant with bone tissues at the molecular level, or the complete replacement of a resorbable bioceramics by healthy bone tissues. All of the enumerated processes are related to the effect of an organism on the implant. Nevertheless, another aspect of implantation is also important—the effect of the implant on the organism. For example, use of bone implants from corpses or animals, even after they have been treated in various ways, provokes a substantially negative immune reaction in the organism, which substantially limits the application of such implants. In this connection, it is useful to dwell on the biological properties of bioceramic implants, particularly those of $CaPO_4$, which in the course of time may be resorbed completely [573].

5.1. Interactions with Surrounding Tissues and the Host Responses

All interactions between implants and the surrounding tissues are dynamic processes. Water, dissolved ions, various biomolecules, and cells surround the implant surface within the initial few seconds after the implantation. It is accepted that no foreign material placed inside a living body is completely compatible. The only substances that conform completely are those manufactured by the body itself (autogenous), while any other substance, which is recognized as foreign, initiates some types of reactions (a host-tissue response). The reactions occurring at the biomaterial/tissue interfaces lead to time-dependent changes in the surface characteristics of both the implanted biomaterials and the surrounding tissues [58,574].

In order to develop new biomaterials, it is necessary to understand the in vivo host responses. Similar to any other species, biomaterials and bioceramics react chemically with their environment and, ideally, they should neither induce any changes nor provoke undesired reactions in the neighboring or distant tissues. In general, living organisms can treat artificial implants as biotoxic (or bioincompatible [53]), bioinert (or biostable [47]), biotolerant (or biocompatible [53]; however, this term appears to be questionable [575]), and bioactive and bioresorbable materials [1–3,42,43,50–53,573,574,576]. Biotoxic (e.g., alloys containing cadmium, vanadium, lead, and other toxic elements) materials release to the body substances in toxic concentrations and/or trigger the formation of antigens that may cause immune reactions ranging from simple allergies to inflammation to septic rejection with the associated severe health consequences. They cause atrophy, pathological change, or rejection of living tissue near the material as a result of chemical, galvanic, or other processes. Bioinert (this term should be used with care, since it is clear that any material introduced into the physiological environment will induce a response; however, for the purposes of biomedical implants, the term can be defined as a minimal level of response from the host tissue), such as zirconia, alumina, carbon, and titanium, as well as biotolerant (e.g., polymethylmethacrylate, titanium, and Co–Cr alloy), materials do not release any toxic constituents but also do not show positive interaction with living tissue. They evoke a physiological response to form a fibrous capsule, thus isolating the material from the body. In such cases, thickness of the layer of fibrous tissue separating the material from other tissues of an organism can serve as a measure of bioinertness. Generally, both bioactivity and bioresorbability phenomena are fine examples of chemical reactivity, and $CaPO_4$ (both nonsubstituted and ion-substituted ones) fall into these two categories of bioceramics [1–3,42,43,50–53,573,574,576]. A bioactive material will dissolve slightly but promote formation of a surface layer of biological apatite before interfacing directly with the tissue at the atomic level, that results in formation of direct chemical bonds to bones. Such implants provide a good stabilization for materials that are subject to mechanical loading. A bioresorbable material will dissolve over time (regardless of the mechanism leading to the material removal) and allow a newly formed tissue to grow into any surface irregularities, but may not necessarily interface directly with the material. Consequently, the functions of bioresorbable materials are to participate in dynamic processes of formation and reabsorption occurring in bone tissues; thus, bioresorbable materials are used as scaffolds or filling spacers, allowing the tissues their infiltration and substitution [180,576–579].

It is important to stress that a distinction between the bioactive and bioresorbable bioceramics might be associated with structural factors only. Namely, bioceramics made from nonporous, dense, and highly crystalline HA behave as bioinert (but a bioactive) materials and are retained in an organism for at least 5–7 years without noticeable changes (Figure 2 bottom), while highly porous bioceramics of the same composition can be resorbed approximately within a year. Furthermore, submicron-sized HA powders are biodegraded even faster than the highly porous HA scaffolds. Other examples of bioresorbable materials comprise porous bioceramic scaffolds made of biphasic, triphasic, or multiphasic $CaPO_4$ formulations [79] or bone grafts (dense or porous) made of CDHA [121], TCP [74,580,581], and/or ACP [470,582]. One must note that at the beginning of the 2000s, the concepts of bioactive and bioresorbable materials were converged and bioactive materials were made bioresorbable, while bioresorbable ones were made bioactive [583].

Although in certain in vivo experiments inflammatory reactions were observed after implantation or injection of $CaPO_4$ [584–593], the general conclusion on using $CaPO_4$ with Ca/P ionic ratio within 1.0–1.7 is that all types of implants (bioceramics of various porosities and structures, powders, or granules) are not only nontoxic but also induce neither inflammatory nor foreign-body reactions [108,571,594]. The biological response to implanted $CaPO_4$ follows a similar cascade to that observed in fracture healing. This cascade includes a hematoma formation, inflammation, neovascularization, osteoclastic resorption, and a new bone formation. An intermediate layer of fibrous tissue between the implants and bones has been never detected. Furthermore, $CaPO_4$ implants display the ability to directly bond to bones [1–3,42,43,50–53,573,574,576]. For further details, interested readers are referred to a good review on cellular perspectives of bioceramic scaffolds for bone tissue engineering [479].

One should note that the aforementioned rare cases of the inflammatory reactions to $CaPO_4$ bioceramics were often caused by "other" reasons. For example, a high rate of wound inflammation occurred when highly porous HA was used. In that particular case, the inflammation was explained by sharp implant edges, which irritated surrounding soft tissues [585]. To avoid this, only rounded material should be used for implantation (Figure 11) [595]. Another reason for inflammation produced by porous HA could be due to micro movements of the implants, leading to simultaneous disruption of a large number of microvessels, which grow into the pores of the bioceramics. This would immediately produce an inflammatory reaction. Additionally, problems could arise in clinical tests connected with migration of granules used for alveolar ridge augmentation, because it might be difficult to achieve a mechanical stability of implants at the implantation sites [585]. In addition, presence of calcium pyrophosphate impurity might be the reason for inflammation [588]. Additional details on inflammatory cell responses to $CaPO_4$ can be found in a special review on this topic [589].

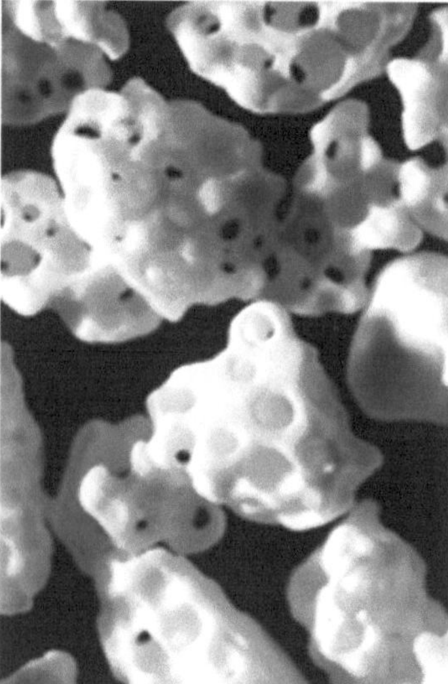

Figure 11. Rounded β-TCP granules 2.6–4.8 mm in size, providing no sharp edges for combination with bone cement. Reprinted from Ref. [595] with permission.

5.2. Osteoinduction

Until recently, it was generally considered that, alone, no type of synthetic bioceramic possessed either osteogenic (osteogenesis is the process of laying down new bone material by osteoblasts [596]) or osteoinductive (the property of the material to induce bone formation de novo or ectopically (i.e., in non-bone-forming sites) [596]) properties, and they demonstrated a minimal immediate structural support. However, a number of reports have already shown the osteoinductive properties of certain types of $CaPO_4$ bioceramics [544,566,597–605], and the amount of such publications is rapidly increasing. For example, bone formation was found to occur in dog muscle inside porous $CaPO_4$ with surface microporosity, while bone was not observed on the surface of dense bioceramics [601]. Furthermore, implantation of porous β-TCP bioceramics appeared to induce bone formation in soft tissues of dogs, while no bone formation was detected in any α-TCP implants [598]. More to the point, titanium implants coated with a microporous layer of OCP were found to induce ectopic bone formation in goat muscles, while a smooth layer of carbonated apatite on the same implants was not able to induce bone formation there [599,600]. In another study, β-TCP powder, biphasic (HA + β-TCP) powder, and intact biphasic (HA + β-TCP) rods were implanted into leg muscles of mice and dorsal muscles of rabbits [606]. One month and three months after implantation, samples were harvested for biological and histological analysis. New bone tissues were observed in 10 of 10 samples for β-TCP powder, 3 of 10 samples for biphasic powder, and 9 of 10 samples for intact biphasic rods at the third month in mice, but not in rabbits. The authors concluded that the chemical composition was the prerequisite in osteoinduction, while porosity contributed to more bone formation [606]. Therefore, researchers had already discovered the methods to prepare osteoinductive $CaPO_4$ bioceramics.

Unfortunately, the underlying mechanism(s) leading to bone induction by synthetic materials remains largely unknown. Nevertheless, besides the specific genetic factors [604] and chosen animals [606], the dissolution/precipitation behavior of $CaPO_4$ [607], their particle size [608,609], microporosity [572,603,610–614], physicochemical properties [601,603], composition [606], the specific surface area [614], and nanostructure [605], as well as the surface topography and geometry [602,615–621] have been pointed out as the relevant parameters [617]. A positive effect of increased microporosity on the ectopic bone formation could be both direct and indirect. Firstly, an increased microporosity is directly related to the changes in surface topography, i.e., it increases surface roughness, which affects the cellular differentiation [619]. Secondly, an increased microporosity indirectly means a larger surface that is exposed to the body fluids, leading to elevated dissolution/precipitation phenomena, as compared to non-microporous surfaces. In addition, other hypotheses are also available, namely, Reddi explained the apparent osteoinductive properties as an ability of particular bioceramics to concentrate bone growth factors, which are circulating in biological fluids, and those growth factors induce bone formation [615]. Other researchers proposed a similar hypothesis, that the intrinsic osteoinduction by $CaPO_4$ bioceramics is a result of adsorption of osteoinductive substances on their surface [602]. Moreover, Ripamonti [616] and Kuboki et al. [617] independently postulated that the geometry of $CaPO_4$ bioceramics is a critical parameter in bone induction. Specifically, bone induction by $CaPO_4$ was never observed on flat bioceramic surfaces. All osteoinductive cases were observed on either porous structures or structures that contained well-defined concavities. Furthermore, bone formation was never observed on the peripheries of porous implants and was always found inside the pores or concavities, aligning the surface [180]. Some researchers speculated that a low oxygen tension in the central region of implants might provoke a dedifferentiation of pericytes from blood microvessels into osteoblasts [622]. Finally, yet importantly, both nanostructured rough surfaces and a surface charge on implants were found to cause an asymmetrical division of the stem cells into osteoblasts, which is important for osteoinduction [613]. Additional details on this topic are available in the literature [623].

Nevertheless, to finalize this topic, it is worth citing a conclusion made by Boyan and Schwartz [624]: "Synthetic materials are presently used routinely as osteoconductive bone graft substitutes, but before purely synthetic materials can be used to treat bone defects in humans where an osteoinductive agent is required, a more complete appreciation of the biology of bone regeneration is needed. An understanding is needed of how synthetic materials modulate the migration, attachment, proliferation and differentiation of mesenchymal stem cells, how cells on the surface of a material affect other progenitor cells in the peri-implant tissue, how vascular progenitors can be recruited and a neovasculature maintained, and how remodeling of newly formed bone can be controlled." (p. 9).

5.3. Biodegradation

Shortly after implantation, a healing process is initiated by compositional changes of the surrounding bio-fluids and adsorption of biomolecules. Following this, various types of cells reach the $CaPO_4$ surface, and the adsorbed layer dictates the ways the cells respond. Further, a biodegradation (which can be envisioned as an in vivo process by which an implanted material breaks down into either simpler components or components of the smaller dimensions) of the implanted $CaPO_4$ bioceramics begins. This process can occur by three possible ways: (1) physical: due to abrasion, fracture and/or disintegration; (2) chemical: due to physicochemical dissolution of the implanted phases of $CaPO_4$ with a possibility of phase transformations into other phases of $CaPO_4$, as well as their precipitation; and (3) biological: due to cellular activity (so called, bioresorption). In biological systems, all these processes take place simultaneously and/or in competition with each other. For example, authors of interesting in vivo studies on a rat calvarial repair model showed that HA bioceramics degraded first, followed by diffusion of the degraded product, which was reconstructed to form new HA to repair the bone defect [625].

Since the existing $CaPO_4$ are differentiated by Ca/P ratio, basicity/acidity, and solubility (Table 1), in the first instance, their degradation kinetics and mechanisms depend on the chosen type of $CaPO_4$ [626,627]. Given the fact that dissolution is a physical chemistry process, it is controlled by some factors, such as $CaPO_4$ solubility, surface area to volume ratio, local acidity, fluid convection, and temperature. For HA and FA, the dissolution mechanism in acids has been described by a sequence of four successive chemical equations, in which several other $CaPO_4$, such as TCP, DCPD/DCPA and MCPM/MCPA, appear as virtual intermediate phases [628,629].

With a few exceptions, dissolution rates of $CaPO_4$ are inversely proportional to the Ca/P ratio (except for TTCP), phase purity, and crystalline size, and they are also directly related to both the porosity and the surface area. In addition, phase transformations might occur with DCPA, DCPD, OCP, α-TCP, β-TCP, and ACP because they are unstable in aqueous environments under the physiological conditions [630]. Bioresorption is a biological process mediated by cells (mainly osteoclasts and, to a lesser extent, macrophages) [631,632]. In vitro, this process may be followed up by various techniques, such as a spherical instrumented indentation [633]. Bioresorption depends on the response of cells to their environment. Osteoclasts attach firmly to the implant and dissolve $CaPO_4$ by secreting an enzyme carbonic anhydrase or any other acid, leading to a local pH drop to ~4–5 [634]. Formation of multiple spine-like crystals at the exposed areas of β-TCP was discovered [635]. Furthermore, nanodimensional particles of $CaPO_4$ can also be phagocytosed by cells, i.e., they are incorporated into cytoplasm and thereafter dissolved by acid attack and/or enzymatic processes [636]. A study is available [637] in which a comparison was made between the solubility and osteoclastic resorbability of three types of $CaPO_4$ (DCPA, ACP, and HA) + β-calcium pyrophosphate (β-CPP) powders having the monodisperse particle size distributions. The authors discovered that with the exception of β-CPP, the difference in solubility among different calcium phosphates became neither mitigated nor reversed but augmented in the resorptive osteoclastic milieu. Namely, DCPA (the phase with the highest solubility) was resorbed more intensely than any other calcium phosphate, whereas HA (the phase with the lowest solubility) was resorbed the least. B-CPP became retained inside the cells for the longest period of time, indicating hindered digestion of only this particular type of calcium phosphate. Genesis of osteoclasts was found to be mildly hindered in the presence of HA, ACP, and DCPA, but not in the presence of β-CPP. HA appeared to be the most viable compound with respect to the mitochondrial succinic dehydrogenase activity. The authors concluded that chemistry did have a direct effect on biology, while biology neither overrode nor reversed the chemical propensities of calcium phosphates with which it interacted, but rather augmented and took a direct advantage of them [637]. Similar conclusions on both the resorbability and dissolution behavior of OCP, β-TCP, and HA [630], as well as β-TCP, BCP (HA + β-TCP), and HA [638], were made by other researchers. In addition, in vivo biodegradation of MCPA was found to be faster than that of bovine HA [639]. Thus, one can conclude that in vivo biodegradation kinetics of $CaPO_4$ seem to correlate well with their solubility. Nevertheless, one must keep in mind that this is a very complicated combination of various nonequilibrium processes, occurring simultaneously and/or in competition with each other [640].

Strictly speaking, the processes that happen in vitro do not necessarily represent the ones occurring in vivo and vice versa; nevertheless, in vitro experiments are widely performed. Usually, an in vitro biodegradation of $CaPO_4$ bioceramics is simulated by suspending the material in a slightly acidic (pH~4) buffer and monitoring the release of major ions with time [627,641–644]. An acidic buffer, to some extent, mimics the acidic environment during osteoclastic activity. The authors of one study reviewed the available literature on acellular in vitro resorption of $CaPO_4$ bioceramics and found the following [645]: "The materials were certainly processed under different conditions, but this dispersion of data is also due to the large variety of tests performed. In fact, each work differs from the others in the type of sample (i.e., composition, shape, porosity, dimension.), of immersion condition (i.e., kind of solution, quantity, stirring, refresh.) and of performed analysis (i.e., microstructural, physicochemical, mechanical) and testing conditions. However, all these aspects can affect the final results." (p. 912). Further, the authors of that paper performed in vitro resorption of DCPD and β-TCP samples in TRIS and PBS solutions for different times with or without refresh of the medium and demonstrated the importance of choosing the appropriate immersion conditions according to the phenomenon being investigated (i.e., $CaPO_4$ dissolution, precipitation of new phases, etc.) [645].

For example, in vivo behavior of porous β-TCP bioceramics prepared from rod-shaped particles and those prepared from non-rod-shaped particles in the rabbit femur was compared. Although the porosities of both types of β-TCP bioceramics were almost the same, a more active osteogenesis was preserved in the region where rod-shaped bioceramics were implanted [646]. Furthermore, the dimensions of both the particles [608] and the surface microstructure [607] were found to influence the osteoinductive potential of $CaPO_4$ bioceramics. These results implied that the microstructure affected the activity of bone cells and subsequent bone replacement.

In addition, a quantitative and fast method was developed to measure the chemical changes occurring within the pores of β-TCP granules incubated in a simulated body fluid [647]. A factorial design of experiments revealed that the particle size, specific surface area, microporosity, and purity of the β-TCP granules influenced the chemical composition of the solution. Large pH, calcium, and phosphate concentration changes were observed inside the granules and lasted for several days. The kinetics and magnitude of these changes (up to 2 pH units) largely depended on the processing and properties of the granules. Small particles, low sintering temperature, high microporosity, and the presence of HA impurity magnified the intensity and duration of pH, calcium, and phosphate variations [647].

Regarding in vivo studies, terbium (Tb)-doped uniform nanodimensional CDHA crystals were implanted into bone tissue and compared with those of native bone apatite. The comparisons demonstrated an occurrence of compositional and structural alterations of the implanted CDHA crystals and their gradual degradation during bone reconstruction. They also revealed notable differences between implanted Tb-doped CDHA and bone apatite crystals in dimensions, distribution pattern, and state of existence in bone tissue.

The authors concluded that although synthetic nanodimensional CDHA crystals could osteointegrate with bone tissue, they still seemed to be treated as foreign material and thus were gradually degraded [648].

The experimental results demonstrated that both the dissolution kinetics and in vivo biodegradation of biologically relevant $CaPO_4$ proceed in the following decreasing order: β-TCP > bovine bone apatite (unsintered) > bovine bone apatite (sintered) > coralline HA > HA. In the case of biphasic (HA + TCP), triphasic, and multiphasic $CaPO_4$ formulations, the biodegradation kinetics depend on the HA/TCP ratio: the higher the ratio, the lower the degradation rate. Similarly, the in vivo degradation rate of biphasic TCP (α-TCP + β-TCP) bioceramics appeared to be lower than that of α-TCP and higher than that of β-TCP bioceramics, respectively [93]. Furthermore, incorporation of doping ions can either increase (e.g., CO_3^{2-}, Mg^{2+}, or Sr^{2+}) or decrease (e.g., F^-) the solubility (therefore, biodegradability) of CDHA and HA. Contrarily to apatites, solubility of β-TCP is decreased by incorporation of either Mg^{2+} or Zn^{2+} ions [544]. Here, one should remember that ion-substituted $CaPO_4$ are not considered in this review; the interested readers are advised to read the original publications [17–41].

5.4. Bioactivity

Generally, bioactive materials interact with surrounding bone, resulting in formation of a chemical bond to this tissue (bone bonding). The bioactivity phenomenon is determined by both chemical factors, such as crystal phases and molecular structures of a biomaterial, and physical factors, such as surface roughness and porosity. Currently, it is agreed that the newly formed bone bonds directly to biomaterials through a carbonated CDHA layer precipitating at the bone/biomaterial interface. Strangely enough, a careful search of the literature resulted in just a few publications [544,623,649–651] where the bioactivity mechanism of $CaPO_4$ was briefly described. For example, the chemical changes occurring after exposure of a synthetic HA bioceramic to both in vivo (implantation in human) and in vitro (cell culture) conditions were studied. A small amount of HA was phagocytozed but the major remaining part behaved as a secondary nucleator, as evidenced by the appearance of a newly formed mineral [649]. In vivo, cellular activity (e.g., of macrophages or osteoclasts; however, this may depend on the cellular origin [652]) associated with an acidic environment was found to result in partial dissolution of $CaPO_4$, causing liberation of calcium and orthophosphate ions to the microenvironment. The liberated ions increased the local supersaturation degree of the surrounding biologic fluids, causing precipitation of nanosized crystals of biological apatite with simultaneous incorporating of various ions present in the fluids. Infrared spectroscopic analyses demonstrated that these nanodimensional crystals were intimately associated with bioorganic components (probably proteins), which might also have originated from the biologic fluids, such as serum [544]. However, in 2019, the concept of a local supersaturation degree that caused $CaPO_4$ precipitation was criticized: on the contrary, intrinsic osteoinduction was proposed to be the result of calcium and/or phosphate depletion (blood supply must be insufficient to maintain the physiological calcium and/or phosphate ion concentrations) [623].

Therefore, one should consider the bioactivity mechanism of other biomaterials, particularly of bioactive glasses—the concept introduced by Prof. Larry L. Hench [50,51]. The bonding mechanism of bioactive glasses to living tissues involves a sequence of 11 successive reaction steps (Figure 12), some of which comprise $CaPO_4$. The initial five steps occurring on the surface of bioactive glasses are "chemistry" only, while the remaining six steps belong to "biology" because the latter include colonization by osteoblasts, followed by proliferation and differentiation of the cells to form a new bone that has a mechanically strong bond to the implant surface. Therefore, in the case of bioactive glasses, the border between "dead" and "alive" is postulated between stages 5 and 6. According to Hench, all bioactive materials "form a bone-like apatite layer on their surfaces in the living body and bond to bone through this apatite layer. The formation of bone-like apatite on artificial material is induced by functional groups, such as Si–OH (in the case of biological glasses),

Ti–OH, Zr–OH, Nb–OH, Ta–OH, –COOH and –H$_2$PO$_4$ (in the case of other materials). These groups have specific structures revealing negatively charge and induce apatite formation via formations of an amorphous calcium compound, e.g., calcium silicate, calcium titanate and ACP" [50,51].

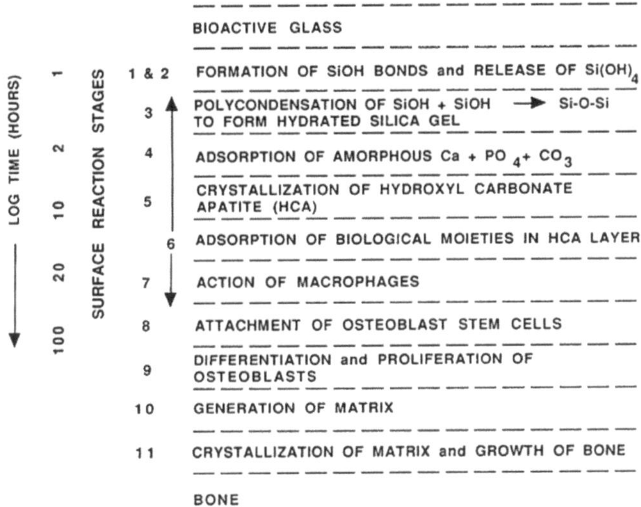

Figure 12. A sequence of interfacial reactions involved in forming a bond between tissue and bioactive ceramics. Reprinted from Refs. [50,51] with permission.

In addition, one should mention another set of 11 successive reaction steps for bonding mechanism of unspecified bioceramics, developed by Prof. Paul Ducheyne (Figure 13) [58]. One can see that the Ducheyne's model is rather similar to that proposed by Hench; however, there are noticeable differences between them. For example, Ducheyne mentions ion exchange and structural rearrangement at the bioceramic/tissue interface (stage 3), as well as on interdiffusion from the surface boundary layer into bioceramics (stage 4) and deposition with integration into the bioceramics (stage 7), which are absent in the Hench's model. On the other hand, Hench describes six biological stages (stages 6–11), while Ducheyne describes only four (stages 8–11). Both models were developed more than two decades ago and, to the best of my knowledge, remain unchanged since then. Presumably, both approaches have pro et contra of their own and, obviously, should be updated and/or revised. Furthermore, in the literature there are at least two other descriptions of the biological and cellular events occurring at the bone/implant interface [653,654]. Unfortunately, both of them comprise fewer stages. In 2010, one more hypothesis was proposed (Figure 14). For the first time, it describes reasonable surface transformations, happening with CaPO$_4$ bioceramics (in that case, HA) shortly after the implantation [651]. However, one must note that the schemes displayed in Figures 12–14 do not represent the real mechanisms, but are only descriptions of the observable events occurring at the CaPO$_4$ interface after implantation. Furthermore, many events occur simultaneously; therefore, none of the schemes should be considered in terms of the strict time sequences.

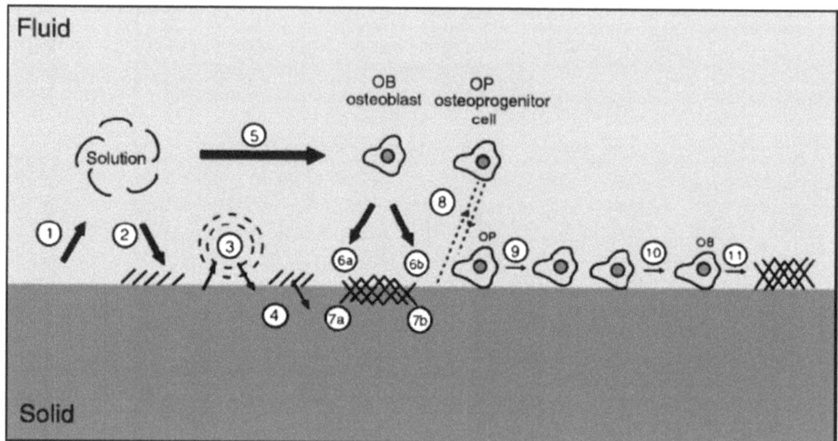

Figure 13. A schematic diagram representing the events which take place at the interface between bioceramics and the surrounding biological environment: (1) dissolution of bioceramics; (2) precipitation from solution onto bioceramics; (3) ion exchange and structural rearrangement at the bioceramic/tissue interface; (4) interdiffusion from the surface boundary layer into the bioceramics; (5) solution-mediated effects on cellular activity; (6) deposition of either the mineral phase (a) or the organic phase (b) without integration into the bioceramic surface; (7) deposition with integration into the bioceramics; (8) chemotaxis to the bioceramic surface; (9) cell attachment and proliferation; (10) cell differentiation; (11) extracellular matrix formation. All phenomena, collectively, lead to the gradual incorporation of a bioceramic implant into developing bone tissue. Reprinted from Ref. [58] with permission.

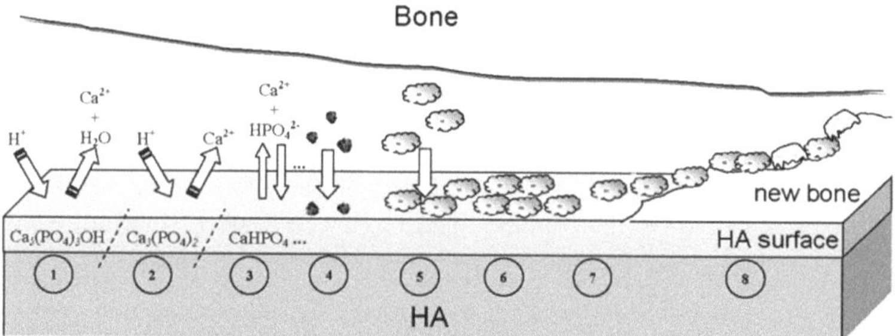

Figure 14. A schematic diagram representing the phenomena that occur on the HA surface after implantation: (1) beginning of the implant procedure, where a solubilization of the HA surface starts; (2) continuation of the solubilization of the HA surface; (3) the equilibrium between the physiological solutions and the modified surface of HA has been achieved (changes in the surface composition of HA do not mean that a new phase of DCPA or DCPD forms on the surface); (4) adsorption of proteins and/or other bioorganic compounds; (5) cell adhesion; (6) cell proliferation; (7) beginning of a new bone formation; (8) new bone has been formed. Reprinted from Ref. [651] with permission.

An important study on formation of $CaPO_4$ precipitates on various types of bioceramic surfaces in both simulated body fluid and rabbit muscle sites was performed [655]. The bioceramics were sintered porous solids, including bioglass, glass-ceramics, α-TCP, β-TCP, and HA. The ability to induce $CaPO_4$ precipitation was compared among these types of bioceramics. The following conclusions were made: (1) OCP formation ubiquitously occurred on all types of bioceramic surfaces both in vitro and in vivo, except on β-TCP. (2) Apatite formation did not occur on every type of bioceramic surface; it was less likely to occur on the surfaces of HA and α-TCP. (3) Precipitation of $CaPO_4$ on the bioceramic surfaces was more difficult in vivo than in vitro. (4) Differences in $CaPO_4$ precipitation among the bioceramic surfaces were less noticeable in vitro than that in vivo. (5) β-TCP bioceramics showed poor ability of $CaPO_4$ precipitation both in vitro and in vivo [655]. These findings clearly revealed that apatite formation in the physiological environments could not be confirmed as the common feature of bioceramics. Nevertheless, for want of anything better, currently, the bioactivity mechanism of $CaPO_4$ bioceramics should be described by a reasonable combination of Figures 12–14, e.g., by updating the Ducheyne's and Hench's models with the three initial stages taken from Figure 14. Additional details on this topic are available in the literature [656].

Interestingly, bioactivity of HA bioceramics might be enhanced by a high-energy ion irradiation [657]. The effect was attributed to formation of a unique 3D macroporous apatite layer of decreased crystallinity and crystal size on the irradiated surfaces. Obviously, to obtain further insights into the bioactivity phenomenon, the atomic and molecular processes occurring at the bioceramic surface in aqueous solutions and their effects on the relevant reaction pathways of cells and tissues must be elucidated in more details.

5.5. Cellular Response

Fixation of any implants in the body is a complex dynamic process that remodels the interface between the implants and living tissues at all dimensional levels, from the molecular up to the cell and tissue morphology level, and at all time scales, from the first second up to several years after implantation. Immediately following the implantation, a space filled with biological fluids appears next to the implant surface. With time, cells are adsorbed at the implant surface that will give rise to their proliferation and differentiation towards bone cells, followed by revascularization and eventual gap closing. Ideally, a strong bond is formed between the implants and surrounding tissues [53]. An interesting study on the interfacial interactions between calcined HA and substrates was performed [658], where the interested readers are referred for further details.

The aforementioned paragraph clearly demonstrates the importance of studies on cellular responses to $CaPO_4$ bioceramics. Such investigations have been performed extensively for several decades [589,659–671]. For example, bioceramic discs made of seven different types of $CaPO_4$ (TTCP, HA, carbonate apatite, β-TCP, α-TCP, OCP, and DCPD) were incubated in osteoclastic cell cultures for 2 days. In all cases, similar cell morphologies and good cell viability were observed; however, different levels of resorbability of various types of $CaPO_4$ were detected [661]. Similar results were found for fluoridated HA coatings [663]. Chemical composition of $CaPO_4$, which contributed to pH changes, and concentration of calcium ions in the medium were found to make up particularly significant factors for cellular responses; moreover, it was proved that the number of material types represented a further important aspect [670]. Experiments performed with human osteoblasts revealed that nanostructured bioceramics prepared from nanosized HA showed significant enhancement in mineralization compared to microstructured HA bioceramics [662]. In addition, the influence of lengths and surface areas of rod-shaped HA on cellular response were studied. Again, similar cell morphologies and good cell viability were observed; however, it was concluded that high surface area could increase cell–particle interaction [665]. Nevertheless, another study with cellular response to rod-shaped HA bioceramics revealed that some types of crystals might trigger a severe inflammatory response [666]. In addition, $CaPO_4$-based sealers appeared to show fewer cytotoxicity and inflammatory mediators compared with other sealers [664]. More examples are available in the literature [589,659–671].

Cellular biodegradation of $CaPO_4$ bioceramics is known to depend on its phases. For example, a higher solubility of β-TCP was shown to prevent L-929 fibroblast cell adhesion, thereby leading to damage and rupture of the cells [672]. A mouse ectopic model study indicated the maximal bone growth for the 80:20 β-TCP:HA biphasic formulations preloaded with human mesenchymal stem cells when compared to other $CaPO_4$ [673]. The effects of substrate microstructure and crystallinity have been corroborated with an in vivo rabbit femur model, where rod-like crystalline β-TCP was reported to enhance osteogenesis when compared to non-rod-like crystalline β-TCP [649]. Additionally, using a dog mandibular defect model, a higher bone formation on a scaffold surface coated by nanodimensional HA was observed when compared to that coated by a microdimensional HA [674]. Furthermore, studies revealed a stronger stress signaling response by osteoblast precursor cells in 3D scaffolds when compared to 2D surfaces [675].

Mesenchymal stem cells are one of the most attractive cellular lines for application as bone grafts [676,677]. Early investigations by Okumura et al. indicated an adhesion, proliferation, and differentiation, which ultimately became new bone and integrated with porous HA bioceramics [660]. Later, a sustained coculture of endothelial cells and osteoblasts on HA scaffolds for up to 6 weeks was demonstrated [678]. Furthermore, a release of factors by endothelial and osteoblast cells in coculture-supported proliferation and differentiation was suggested to ultimately result in microcapillary-like vessel formation and supported a neo-tissue growth within the scaffold [479]. More to the point, investigation of rat calvaria osteoblasts cultured on transparent HA bioceramics, as well as the analysis of osteogenic-induced human bone marrow stromal cells at different time points of culturing, indicated a good cytocompatibility of HA bioceramics and revealed favorable cell proliferation [424]. Positive results for other types of cells were obtained in other studies [187,423,443,444,679–681]. In addition, $CaPO_4$ are used for cell transfections [682].

Interestingly, HA scaffolds with marrow stromal cells in a perfused environment were reported to result in ~85% increase in mean core strength, a ~130% increase in failure energy, and a ~355% increase in post-failure strength. The increase in mineral quantity and promotion of the uniform mineral distribution in that study was suggested to be attributed to the perfusion effect [560]. Additionally, other investigators indicated mechanical properties increasing for other $CaPO_4$ scaffolds after induced osteogenesis [559,562].

To finalize this subsection, one should note the recent developments to influence the cellular response. First, to facilitate interactions with cells, the $CaPO_4$ surfaces could be functionalized [683–686]. Second, it appears that crystals of biological apatite of calcified tissues exhibit different orientations depending on the tissue; namely, in vertebrate bones and tooth enamel surfaces, the respective *a, b*-planes and *c*-planes of the apatite crystals are preferentially exposed. Therefore, ideally, this should be taken into account in artificial bone grafts. Recently, a novel process to fabricate dense HA bioceramics with highly preferred orientation to the *a,b*-plane was developed. The results revealed that increasing the *a,b*-plane orientation degree shifted the surface charge from negative to positive and decreased the surface wettability with simultaneous decreasing of cell attachment efficiency [687–689]. The latter finding resulted in further developments on preparation of oriented $CaPO_4$ compounds [690–692].

Finally, to conclude the entire *Biological Properties and in vivo Behavior* section, let me quote several sentences from Ref. [265]: "Variations in surface chemistry resulting from variable thermal processing conditions of otherwise identical samples might thus explain inconsistencies in biological behavior reported in the literature. For instance, β-TCP has been reported to be both bioactive [693] and non-bioactive [655], non-osteoinductive [694] and highly osteoinductive [695,696], highly resorbable [694,697] and poorly resorbable [697]. Authors have related this dichotomous behavior with the effect of sintering temperature on specific surface area, bulk composition, and scaffold or pore topography." (p. 6096). In addition, simple thermal treatment at 500 °C was found to reduce body reactions to irregular α- and β-TCP granules as foreign bodies, due to a partial evaporation of phosphate species during thermal treatment [698]. Thus, there are still many uncertainties in our understanding of the biological properties of $CaPO_4$ bioceramics.

6. Biomedical Applications

Since Levitt et al. described a method of preparing FA bioceramics and suggested their possible use in medical applications in 1969 [699], $CaPO_4$ bioceramics have been widely tested for clinical applications. Namely, over 400 forms, compositions, and trademarks

(Table 3) are currently either in use or under consideration in many areas of orthopedics and dentistry [700], with even more in development. In addition, various formulations containing demineralized bone matrix (commonly abbreviated as DBM) are produced for bone grafting. For example, bulk materials, available in dense and porous forms, are used for alveolar ridge augmentation, immediate tooth replacement, and maxillofacial reconstruction [4,701]. Other examples comprise burr-hole buttons [702,703], cosmetic (nonfunctional) eye replacements such as Bio-Eye® [704–709], increment of the hearing ossicles [710–712], and spine fusion [713–716], as well as repair of bone [118,717,718], craniofacial [719], and dental [720] defects. In order to permit growth of new bone into defects, a suitable bioresorbable material should fill these defects. Otherwise, ingrowth of fibrous tissue might prevent bone formation within the defects.

Table 3. Registered commercial trademarks (current and past) of CaPO$_4$-based bioceramics and biomaterials.

Calcium Orthophosphate	Trade Name and Producer (When Available)
CDHA	Calcibon (Zimmer Biomet, IN, USA)
	Cementek (Teknimed, France)
	CHT Ceramic Hydroxyapatite (Bio-Rad, CA, USA)
	nanoXIM (Fluidinova, Portugal)
	OsteoGen (Impladent, NY, USA)
	without trade name (Himed, NY, USA)
HA	Actifuse (ApaTech, UK)
	Alveograf (Cooke-Waite Laboratories, USA)
	Apaceram (HOYA Technosurgical, Japan)
	Apafill-G (Habana, Cuba)
	ApaPore (ApaTech, UK)
	BABI-HAP (Berkeley Advanced Biomaterials, CA, USA)
	Bio-Eye (Integrated Orbital Implants, CA, USA)
	BIOGAP (Connectbiopharm, Russia)
	BioGraft (IFGL BIO CERAMICS, India)
	Bioroc (Depuy Bioland, France)
	Blue Bone (Regener Biomateriais, Brazil)
	Boneceram (Sumitomo Osaka Cement, Japan)
	Bonefil (Pentax, Japan)
	BoneSource (Stryker Orthopaedics, NJ, USA)
	Bonetite (Pentax, Japan)
	Bonfil (Mitsubishi Materials, Japan)
	Bongros-HA (Daewoong Pharmaceutical, Korea)
	CAFOS DT (Chemische Fabrik Budenheim, Germany)
	Calcitite (Sulzer Calcitek, CA, USA)
	CAMCERAM HA (CAM Implants, Netherlands)
	CAPTAL (Plasma Biotal, UK)
	CELLYARD (HOYA Technosurgical, Japan)
	Cerapatite (Ceraver, France)
	Ceros HA (Mathys, Switzerland)

Table 3. *Cont.*

Calcium Orthophosphate	Trade Name and Producer (When Available)
	CHT Ceramic Hydroxyapatite (Bio-Rad, CA, USA)
	Durapatite (unknown producer)
	ENGIpore (JRI Orthopaedics, UK)
	G-Bone (Surgiwear, India)
	GranuMas (GranuLab, Malaysia)
	HA BIOCER (CHEMA – ELEKTROMET, Poland)
	HA^{nano} Surface (Promimic, Sweden)
	HAP-91 (JHS Biomateriais, Brazil)
	HAP-99 (Polystom, Russia)
	HAP–Bionnovation (Bionnovation, Brazil)
	IngeniOs HA (Zimmer Dental, CA, USA)
	Micro Crystalline Hydroxyapatite Complex (MCHC) (Clarion Pharmaceutical, India)
	nanoXIM (Fluidinova, Portugal)
	Neobone (Covalent Materials, Japan)
	Osbone (Curasan, Germany)
	OsproLife HA (Lincotek Medical, Italy)
	Ossein Hydroxyapatite (Clarion Pharmaceutical, India)
	OssaBase-HA (Lasak, Czech Republic)
	Ostegraf (Ceramed, CO, USA)
	Ostim (Heraeus Kulzer, Germany)
	Ovis Bone HA (DENTIS, Korea)
	Periograf (Cooke-Waite Laboratories, USA)
	PermaOS (Mathys, Switzerland)
	PRINT3D Hydroxyapatite (Prodways, France)
	Pro Osteon (Zimmer Biomet, IN, USA)
	PurAtite (PremierBiomaterials, Ireland)
	REGENOS (Kuraray, Japan)
	SHAp (SofSera, Japan)
	Synatite (SBM, France)
	Synthacer (KARL STORZ Recon, Germany)
	Theriridge (Therics, OH, USA)
	without trade name (Cam Bioceramics, Netherlands)
	without trade name (CaP Biomaterials, WI, USA)
	without trade name (DinganTec, China)
	without trade name (Ensail Beijing, China)
	without trade name (Himed, NY, USA)
	without trade name (MedicalGroup, France)
	without trade name (SANGI, Japan)
	without trade name (Shanghai Rebone Biomaterials, China)
	without trade name (SigmaGraft, CA, USA)

Table 3. *Cont.*

Calcium Orthophosphate	Trade Name and Producer (When Available)
	without trade name (SkySpring Nanomaterials, TX, USA)
	without trade name (SofSera, Japan)
	without trade name (Taihei Chemical Industrial, Japan)
	without trade name (Xpand Biotechnology, Netherlands)
Mg-HA	SINTlife (JRI Orthopaedics, UK)
HA powder suspended in water	Ostibone (FH Orthopedics, France)
	NANOSTIM (Medtronic Sofamor Danek, TN, USA)
	n-IBS (Bioceramed, Portugal)
	Skelifil (Osteotec, UK)
HA embedded or suspended in a gel	Bio-Gel HT hydroxyapatite (Bio-Rad, CA, USA)
	Coaptite (Boston Scientific, MA, USA)
	Facetem (Daewoong, Korea)
	NanoBone (Artoss, Germany)
	Nanogel (Teknimed, France)
	Radiesse (Merz Aesthetics, Germany)
	Renú Calcium Hydroxylapatite Implant (Cytophil, WI, USA)
HA/collagen, CDHA/collagen and/or carbonate apatite/collagen	AUGMATRIX (Wright Medical Technology, TN, USA)
	Bioimplant (Connectbiopharm, Russia)
	Bio-Oss Collagen (Geitslich, Switzerland)
	Boneject (Koken, Japan)
	COL.HAP-91 (JHS Biomateriais, Brazil)
	Collagraft (Zimmer and Collagen Corporation, USA)
	CollaOss (SK Bioland, Korea)
	CollapAn (Intermedapatite, Russia)
	COLLAPAT (Symatese, France)
	DualPor collagen (OssGen, Korea)
	G-Graft (Surgiwear, India)
	HAPCOL (Polystom, Russia)
	Healos (DePuy Spine, USA)
	LitAr (LitAr, Russia)
	Ossbone Collagen (SK Bioland, Korea)
	OssFill (Sewon Cellontech, Korea)
	OssiMend (Collagen Matrix, NJ, USA)
	Osteomatrix (Connectbiopharm, Russia)
	OsteoTape (Impladent, NY, USA)
	ReFit (HOYA Technosurgical, Japan
	RegenOss (JRI Orthopaedics, UK)
	RegenerOss Synthetic (Zimmer Dental, CA, USA)
	Straumann XenoFlex (Straumann, Switzerland)

Table 3. *Cont.*

Calcium Orthophosphate	Trade Name and Producer (When Available)
HA/sodium alginate	Bialgin (Biomed, Russia)
HA/poly-L-lactic acid	Biosteon (Biocomposites, UK)
	ReOss (ReOss, Germany)
	OSTEOTRANS MX (Teijin Medical Technologies, Japan)
	SuperFIXSORB30 (Takiron, Japan)
HA/polyethylene	HAPEX (Gyrus, TN, USA)
HA/CaSO$_4$	BioWrist Bone Void Filler (Skeletal Kinetics, CA, USA)
	Bond Apatite (Augma Biomaterials, NJ, USA)
	Hapset (LifeCore, MN, USA)
	PerOssal (aap Implantate, Germany)
HA/CaSO$_4$ powders suspended in a liquid	CERAMENT (BONESUPPORT, Sweden)
Coralline HA	Biocoral (Bio Coral Calcium Bone, France)
	BoneMedik-S (Meta Biomed, Korea)
	Interpore (Interpore, CA, USA)
	ProOsteon (Interpore, CA, USA)
Carbonate apatite	Cytrans (GC, Japan)
	Norian SRS (Norian, CA, USA)
Algae-derived HA	Algipore (AlgOss Biotechnologies, Austria)
	Algisorb (AlgOss Biotechnologies, Austria)
	FRIOS Algipore (DENTSPLY Implants, Sweden)
	SIC nature graft (AlgOss Biotechnologies, Austria)
HA/glass	Bonelike (unknwn producer)
Bovine bone (unsintered)	Unilab Surgibone (Unilab, NJ, USA)
Bovine bone (unsintered) + polymer	Alpha-Bio's Graft (Alpha-Bio Tec, Israel)
	C-Graft Putty (unknwn producer)
Bovine bone apatite (unsintered)	Apatos (OsteoBiol, Italy)
	Bio-Oss (Geistlich Biomaterials, Switzerland)
	Bonefill (Bionnovation, Brazil).
	CANCELLO-PURE (Wright Medical Technology, TN, USA)
	CenoBone (Tissue Regeneration Corporation, Iran)
	CopiOs Cancellous Particulate Xenograft (Zimmer, IN, USA)
	GenOs (OsteoBiol, Italy)
	InterOss (SigmaGraft, CA, USA)
	Laddec (Ost-Developpement, France)
	Lubboc (Ost-Developpement, France)
	MatrixCellect (Curasan, Germany)
	Mega-Oss Bovine (Megagen Implant, Korea)
	Orthoss (Geitslich, Switzerland)
	OssiGuide (Collagen Matrix, NJ, USA)
	Oxbone (Bioland biomateriaux, France)

Table 3. *Cont.*

Calcium Orthophosphate	Trade Name and Producer (When Available)
	Straumann XenoGraft (Straumann, Switzerland)
	Surgibone (Surgibon, Ecuador)
	Tutobone (Tutogen Medical, Germany)
	Tutofix (Tutogen Medical, Germany)
	Tutoplast (Tutogen Medical, Germany)
	without trade name (MedicalGroup, France)
Porcine bone apatite (unsintered)	A-OSS (Osstem Implant, Korea)
	GEM Bone Graft (Lynch Biologics, USA)
	Gen-Os (OsteoBiol, Italy)
	MatrixOss (Collagen Matrix, NJ, USA)
	OsteoBiol (OsteoBiol, Italy)
	Symbios Xenograft (DENTSPLY Implants, Sweden)
	THE Graft (Purgo Biologics, Korea)
Equine bone apatite (unsintered)	BIO-GEN (BioTECK, Italy)
	Sp-Block (OsteoBiol, Italy)
Bovine bone apatite (sintered)	4Bone XBM (MIS Implants, Israel)
	BonAP (unknown producer)
	Cerabone (aap Implantate, Germany and botiss, Germany)
	Endobon (Merck, Germany)
	GenoxInorgânico (Baumer, SP, Brazil)
	Iceberg oss (Global Medical Implants, Spain)
	Navigraft (Zimmer Dental, USA)
	Osteograf (Ceramed, CO, USA)
	OVIS XENO (DENTIS, Korea)
	PepGen P-15 (DENTSPLY Implants, Sweden)
	Pyrost (Osteo AG, Germany)
	Sinbone (Purzer Pharmaceutical, Taiwan)
	SynOss (Collagen Matrix, NJ, USA)
	Straumann cerabone (Straumann, Switzerland)
Human bone allograft	ALLOPURE (Wright Medical Technology, TN, USA)
	Allosorb (Curasan, Germany)
	CancellOss (Impladent, NY, USA)
	CurOss (Impladent, NY, USA)
	J Bone Block (Impladent, NY, USA)
	maxgraft (botiss, Germany)
	Mega-Oss (Megagen Implant, Korea)
	NonDemin (Impladent, NY, USA)
	Osnatal (aap Implantate, Germany)
	OsteoDemin (Impladent, NY, USA)
	OsteoWrap (Curasan, Germany)
	OVIS ALLO (DENTIS, Korea)

Table 3. *Cont.*

Calcium Orthophosphate	Trade Name and Producer (When Available)
	PentOS OI (Citagenix, QC, Canada)
	RAPTOS (Citagenix, QC, Canada)
	Straumann AlloGraft (Straumann, Switzerland)
	TenFUSE (Wright Medical Technology, TN, USA)
α-TCP	BioBase (Biovision, Germany)
	Tetrabone (unknown producer)
	without trade name (Cam Bioceramics, Netherlands)
	without trade name (DinganTec, China)
	without trade name (Ensail Beijing, China)
	without trade name (Himed, NY, USA)
	without trade name (InnoTERE, Germany)
	without trade name (PremierBiomaterials, Ireland)
	without trade name (Taihei Chemical Industrial, Japan)
β-TCP	AdboneTCP (Medbone Medical Devices, Portugal)
	AFFINOS (Kuraray, Japan)
	Allogran-R (Biocomposites, UK)
	Antartik TCP (MedicalBiomat, France)
	ArrowBone (Brain Base Corporation, Japan)
	AttraX scaffold (NuVasive, CA, USA)
	BABI-TCP (Berkeley Advanced Biomaterials, CA, USA)
	Betabase (Biovision, Germany)
	BioGraft (IFGL BIO CERAMICS, India)
	Bioresorb (Sybron Implant Solutions, Germany)
	Biosorb (SBM, France)
	Bi-Ostetic (Berkeley Advanced Biomaterials, CA, USA)
	Bonegraft (Bonegraft biomaterials, Turkey)
	BoneSigma TCP (SigmaGraft, CA, USA)
	C 13-09 (Chemische Fabrik Budenheim, Germany)
	Calc-i-oss classic (Degradable Solutions, Switzerland)
	Calciresorb (Ceraver, France)
	CAMCERAM TCP (CAM Implants, Netherlands)
	CAPTAL β-TCP (Plasma Biotal, UK)
	CELLPLEX (Wright Medical Technology, TN, USA)
	Cerasorb (Curasan, Germany)
	Ceros TCP (Mathys, Switzerland)
	ChronOS (Synthes, PA, USA)
	Cidemarec (KERAMAT, Spain)
	Conduit (DePuy Spine, USA)
	cyclOS (Mathys, Switzerland)
	ExcelOs (BioAlpha, Korea)
	GenerOs (Berkeley Advanced Biomaterials, CA, USA)

Table 3. *Cont.*

Calcium Orthophosphate	Trade Name and Producer (When Available)
	HT BIOCER (CHEMA – ELEKTROMET, Poland)
	Iceberg TCP (Global Medical Implants, Spain)
	IngeniOs β-TCP (Zimmer Dental, CA, USA)
	ISIOS+ (Kasios, France)
	JAX (Smith and Nephew Orthopaedics, USA)
	Keramedic (Keramat, Spain)
	KeraOs (Keramat, Spain)
	Mega-TCP (Megagen Implant, Korea)
	microTCP (Conmed, USA)
	nanoXIM (Fluidinova, Portugal)
	Orthograft (DePuy Spine, USA)
	Ossaplast (Ossacur, Germany)
	Osferion (Olympus Terumo Biomaterials, Japan)
	Osfill (Olympus Terumo Biomaterials, Japan)
	OsproLife β-TCP (Lincotek Medical, Italy)
	OsSatura TCP (Integra Orthobiologics, CA, USA)
	Ossoconduct (SteinerBio, NV, USA)
	Osteoblast (Galimplant, Spain)
	Osteocera (Hannox, Taiwan)
	Osteopore TCP (SpiteCraft, IL, USA)
	OSTEOwelt (Biolot Medical, Turkey)
	Periophil β-TCP (Cytophil, WI, USA)
	Platon Pearl Bone (Platon, Japan)
	PolyBone (Kyungwon Medical, Korea)
	PORESORB-TCP (Lasak, Czech Republic)
	Powerbone (Medical Expo Bonegraft Biomaterials, Spain)
	PRINT3D Tricalcium Phosphate (Prodways, France)
	Repros (JRI Orthopaedics, UK)
	R.T.R. (Septodont, PA, USA)
	SigmaOs TCP (SigmaGraft, CA, USA)
	Socket Graft (SteinerBio, NV, USA)
	Sorbone (Meta Biomed, Korea)
	SUPERPORE (HOYA Technosurgical, Japan)
	Suprabone TCP (BMT Group, Turkey)
	Syncera (Oscotec, Korea)
	SynthoGraft (Bicon, MA, USA)
	Synthos (unknown producer)
	Syntricer (KARL STORZ Recon, Germany)
	TCP (Kasios, France)
	Terufill (Olympus Terumo Biomaterials, Japan)
	TKF-95 (Polystom, Russia)

Table 3. *Cont.*

Calcium Orthophosphate	Trade Name and Producer (When Available)
	TriCaFor (BioNova, Russia)
	Triha+ (Teknimed, France)
	TriOSS (Bioceramed, Portugal)
	Vitomatrix (Orthovita, PA, USA)
	Vitoss (Orthovita, PA, USA)
	without trade name (CaP Biomaterials, WI, USA)
	without trade name (Cam Bioceramics, Netherlands)
	without trade name (DinganTec, China)
	without trade name (Ensail Beijing, China)
	without trade name (Himed, NY, USA)
	without trade name (Shanghai Bio-lu Biomaterials, China)
	without trade name (Shanghai Rebone Biomaterials, China)
	without trade name (SigmaGraft, CA, USA)
	without trade name (Taihei Chemical Industrial, Japan)
	without trade name (Xpand Biotechnology, Netherlands)
β-TCP/CaSO$_4$	Fortoss vital (Biocomposites, UK)
	Genex (Biocomposites, UK)
β-TCP/poly-lactic acid	Bilok (Biocomposites, UK)
	Duosorb (SBM, France)
	Matryx Interference Screws (Conmed, USA)
β-TCP/poly-lactic-*co*-glycolic acid	Evolvemer TCP30PLGA (Arctic Biomaterials, Finland)
β-TCP/polymer	AttraX putty (NuVasive, CA, USA)
	Therigraft (Therics, OH, USA)
β-TCP/bone marrow aspirate	Induce (Skeletal Kinetics, CA, USA)
β-TCP/collagen	Integra Mozaik (Integra Orthobiologics, CA, USA)
β-TCP/growth-factor	GEM 21S (Lynch Biologics, USA)
β-TCP/rhPDGF-BB solution	AUGMENT Bone Graft (Wright Medical Group, TN, USA)
	4Bone BCH (MIS Implants, Israel)
	adboneBCP (Medbone Medical Devices, Portugal)
	Antartik (MedicalBiomat, France)
	ARCA BONE (ARCA-MEDICA, Switzerland)
	Artosal (aap Implantate, Germany)
	BABI-HATCP (Berkeley Advanced Biomaterials, CA, USA)
	Bicera (Hannox, Taiwan)
	BCP BiCalPhos (Medtronic, MN, USA)
	BIO-C (Cowellmedi, Korea)
	BioActys (Graftys, France)
	BioGraft (IFGL BIO CERAMICS, India)
	Biosel (Depuy Bioland, France)
	BonaGraft (Biotech One, Taiwan)
	Boncel-Os (BioAlpha, Korea)

Table 3. *Cont.*

Calcium Orthophosphate	Trade Name and Producer (When Available)
	Bone Plus BCP (Megagen Implant, Korea)
	Bone Plus BCP Eagle Eye (Megagen Implant, Korea)
	BoneMedik-DM (Meta Biomed, Korea)
	BoneSave (Stryker Orthopaedics, NJ, USA)
	BoneSigma BCP (SigmaGraft, CA, USA)
	BONITmatrix (DOT, Germany)
	Calcicoat (Zimmer, IN, USA)
	Calciresorb (Ceraver, France)
	Calc-i-oss crystal (Degradable Solutions, Switzerland)
	CellCeram (Scaffdex, Finland)
	Ceraform (Teknimed, France)
	Ceratite (NGK Spark Plug, Japan)
	Cross.Bone (Biotech Dental, France)
	CuriOs (Progentix Orthobiology BV, Netherlands)
	DM-Bone (Meta Biomed, Korea)
	Eclipse (Citagenix, QC, Canada)
	Eurocer (FH Orthopedics, France)
	Frabone (Inobone, Korea)
	Genesis-BCP (DIO, Korea)
	GenPhos HA TCP (Baumer, Brazil)
	Graftys BCP (Graftys, France)
	Hatric (Arthrex, Naples, FL, USA)
	Hydroxyapol (Polystom, Russia)
	Kainos (Signus, Germany)
	MagnetOs (Kuros Biosciences, Switzerland)
	MasterGraft (Medtronic Sofamor Danek, TN, USA)
	Maxresorb (botiss, Germany)
	MBCP (Biomatlante, France)
	MimetikOss (Mimetis Biomaterials, Spain)
	Neobone (Bioceramed, Portugal)
	New Bone (GENOSS, Korea)
	NT-BCP (OssGen, Korea)
	NT-Ceram (Meta Biomed, Korea)
	OdonCer (Teknimed, France)
	OpteMX (Exactech, FL, USA)
	OrthoCer HA TCP (Baumer, Brazil)
	OsproLife HA-βTCP (Lincotek Medical, Italy)
	OsSatura BCP (Integra Orthobiologics, CA, USA)
	ossceram nano (bredent medical, Germany)
	OSSEOPLUS (JHS Biomateriais, Brazil)
	Osspol (Genewel, Korea)

Table 3. *Cont.*

Calcium Orthophosphate	Trade Name and Producer (When Available)
	OsteoFlux (VIVOS-Dental, Switzerland)
	Osteon (GENOSS, Korea)
	Osteosynt (Einco, Brazil)
	Ostilit (Stryker Orthopaedics, NJ, USA)
	Ovis Bone BCP (DENTIS, Korea)
	Periophil biphasic (Cytophil, WI, USA)
	Q-OSS+ (Osstem Implant, Korea)
	ReproBone (Ceramisys, UK)
	R.T.R.+ (Septodont, PA, USA)
	SBS (Expanscience, France)
	Scaffdex (Scaffdex Oy, Finland)
	SigmaOs BCP (SigmaGraft, CA, USA)
	SinboneHT (Purzer Pharmaceutical, Taiwan)
	SkeliGraft (Osteotec, UK)
	Straumann BoneCeramic (Straumann, Switzerland)
	SYMBIOS Biphasic Bone Graft Material (DENTSPLY Implants, Sweden)
	SynMax (BioHorizons, Spain)
	Synergy (unknown producer)
	TCH (Kasios, France)
	Topgen-S (Toplan, Korea)
	Tribone (Stryker, Europe)
	Triosite (Zimmer, IN, USA)
	without trade name (AlgOss Biotechnologies, Austria)
	without trade name (Cam Bioceramics, Netherlands)
	without trade name (CaP Biomaterials, WI, USA)
	without trade name (Himed, NY, USA)
	without trade name (MedicalGroup, France)
	without trade name (SigmaGraft, CA, USA)
	without trade name (Xpand Biotechnology, Netherlands)
BCP (HA + α-TCP)	Skelite (Millennium Biologix, ON, Canada)
	Allograft (Zimmer, IN, USA)
	collacone max (botiss, Germany)
	Collagraft (Zimmer, IN, USA)
	Cross.Bone Matrix (Biotech Dental, France)
BCP (HA + β-TCP)/collagen	Indost (Polystom, Russia)
	MasterGraft (Medtronic Sofamor Danek, TN, USA)
	MATRI BONE (Biom'Up, France)
	Osteon III collagen (GENOSS, Korea)
	SynergOss (Nobil Bio Ricerche, Italy)
	without trade name (MedicalGroup, France)

Table 3. *Cont.*

Calcium Orthophosphate	Trade Name and Producer (When Available)
BCP (HA + β-TCP)/hydrogel	4MATRIX+ (MIS Implants, Israel)
	Eclipse (Citagenix, QC, Canada)
BCP (HA + β-TCP)/polymer	In'Oss (Biomatlante, France)
	Hydros (Biomatlante, France)
	Osteocaf (Texas Innovative Medical Devices, TX, USA)
	Osteotwin (Biomatlante, France)
BCP (HA + TTCP)	OsproLife HA-TTCP (Lincotek Medical, Italy)
BCP (HA + β-TCP)/chitosan	k-IBS (Bioceramed, Portugal)
BCP (HA + β-TCP)/fibrin	TricOS (Baxter BioScience, France)
BCP (HA + β-TCP)/silicon	FlexHA (Xomed, FL, USA)
Bioglass + α-TCP + β-TCP + HA + polymers	OsteoFlo NanoPutty (SurGenTec, FL, USA)
FA	without trade name (CaP Biomaterials, WI, USA)
FA + BCP (HA + β-TCP)	FtAP (Polystom, Russia)
DCPA	without trade name (Himed, NY, USA)
	without trade name (Shanghai Rebone Biomaterials, China)
DCPA + $MgHPO_4 \cdot 3H_2O$ + SiO_2 + carboxymethyl cellulose	Novogro (OsteoNovus, OH, USA)
DCPD	without trade name (Himed, NY, USA)
DCPD/collagen	CopiOs Bone Void Filler (Zimmer, IN, USA)
DCPD + β-TCP/$CaSO_4$	PRO-DENSE (Wright Medical Group, TN, USA)
DCPD + β-TCP/$CaSO_4$ + collagen	PRO-STIM (Wright Medical Group, TN, USA)
ACP	CAPTAL ACP (Plasma Biotal, UK)
	without trade name (Himed, NY, USA)
OCP	Bontree (HudensBio, Korea)
	OctoFor (BioNova, Russia)
	without trade name (Himed, NY, USA)
OCP/fibrin	FibroFor (BioNova, Russia)
OCP/collagen	Bonarc (Toyobo, Japan)
TTCP	without trade name (Ensail Beijing, China)
	without trade name (Himed, NY, USA)
	without trade name (Shanghai Rebone Biomaterials, China)
	without trade name (Taihei Chemical Industrial, Japan)
Undisclosed $CaPO_4$	Arex Bone (Osteotec, UK)
	Inno-CaP (Cowellmedi, Korea)
Undisclosed $CaPO_4$ + biologics	i-FACTOR (Cerapedics, CO, USA)
MCPM	Phosfeed MCP (OCP group, Morocco)
MCPM + DCPD	Phosfeed MDCP (OCP group, Morocco)

In spite of the aforementioned serious mechanical limitations (see *Section 4.1. Mechanical Properties*), bioceramics of $CaPO_4$ are available in various physical forms: powders, particles, granules (or granulates), dense blocks, porous scaffolds, self-setting formulations, implant coatings, and composite components of different origin (natural, biological, or synthetic), often with specific shapes, such as implants, prostheses, or prosthetic devices. In addition, $CaPO_4$ are also applied as nonhardening injectable formulations [721–726] and pastes [726–730]. Generally, they consist of a mixture of $CaPO_4$ powders or granules and a "glue", which can be a highly viscous hydrogel. More to the point, custom-designed shapes such as wedges for tibial opening osteotomy, cones for spine and knee, and inserts for vertebral cage fusion are also available [546]. Various trademarks of the commercially available types of $CaPO_4$-based bioceramics and biomaterials are summarized in Table 3, while their surgical applications are schematically shown in Figure 15 [731]. A long list of both trademarks and producers clearly demonstrates that $CaPO_4$ bioceramics are easy to make and not very difficult to register for biomedical applications. There is an ISO standard for $CaPO_4$-based bone substitutes [732].

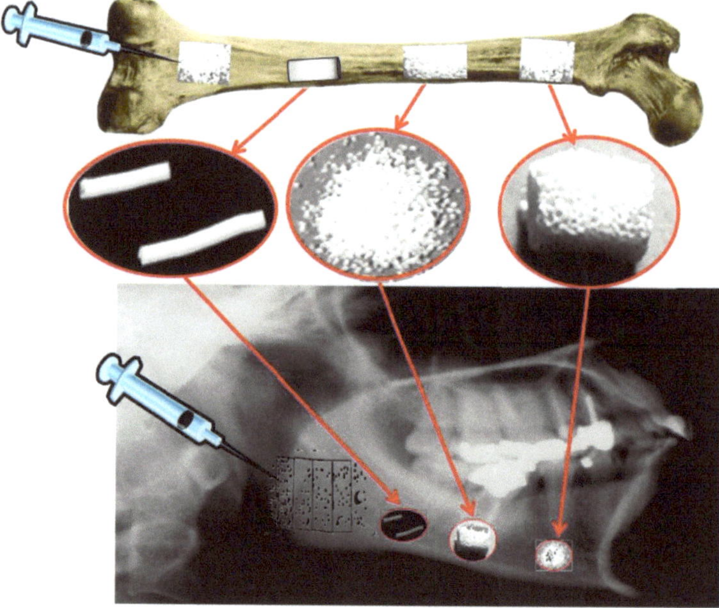

Figure 15. Different types of biomedical applications of $CaPO_4$ bioceramics. Reprinted from Ref. [731] with permission.

One should note that among the existing $CaPO_4$ (Table 1), only certain compounds are useful for biomedical applications, because those having a Ca/P ionic ratio less than 1 are not suitable for implantation due to their high solubility and acidity. Furthermore, due to its basicity, TTCP alone cannot be suitable either. Nevertheless, researchers try [141]. In addition, to simplify biomedical applications, these "of little use" $CaPO_4$ can be successfully combined with either other types of $CaPO_4$ or other chemicals.

6.1. Self-Setting (Self-Hardening) Formulations

The need for bioceramics for minimal invasive surgery has induced the concept of self-setting (or self-hardening) formulations consisting of $CaPO_4$ only to be applied as injectable and/or moldable bone substitutes [102,103,124,507,733]. After hardening, they form bulk $CaPO_4$ bioceramics. In addition, there are reinforced formulations that, in a certain sense, might be defined as $CaPO_4$ concretes [102]. Furthermore, self-setting formulations able to produce porous bulk $CaPO_4$ bioceramics are also available [450,451,458–461,507,520,733–736].

All types of the self-setting $CaPO_4$ formulations belong to low-temperature bioceramics. They are divided into two major groups. The first one is a dry mixture of two different types of $CaPO_4$ (a basic one and an acidic one), in which, after being wetted, the setting reaction occurs according to an acid–base reaction. The second group contains only one $CaPO_4$, such as ACP with Ca/P molar ratio within 1.50–1.67 or α-TCP: both of them form CDHA upon contact with an aqueous solution [102,124]. Chemically, setting (= hardening, curing) is due to the succession of dissolution and precipitation reactions. Mechanically, it results from crystal entanglement and intergrowth (Figure 16) [737]. By influencing dimensions of forming $CaPO_4$ crystals, it is possible to influence the mechanical properties of the hardened bulk bioceramics [738]. Sometimes, the self-set formulations are sintered to prepare high-temperature $CaPO_4$ bioceramics [739]. Despite a large number of initial compositions, all types of self-setting $CaPO_4$ formulations can form three products only: CDHA, DCPD, and, rarely, DCPA [102,103,124,507,733]. Special reviews on the topic are available in [102,103,739], where interested readers are referred for further details.

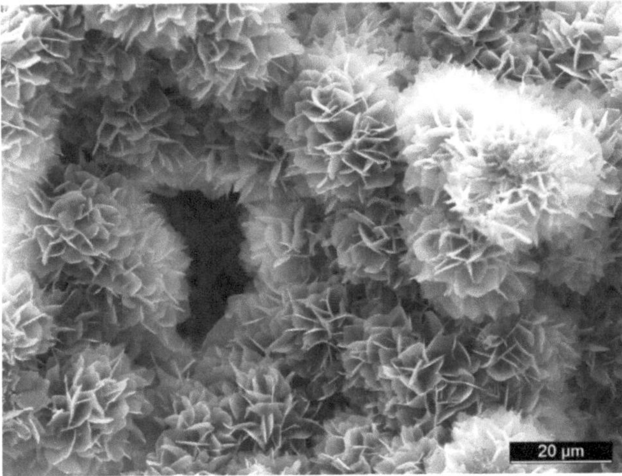

Figure 16. A typical microstructure of a $CaPO_4$ cement after hardening. The mechanical stability is provided by the physical entanglement of crystals. Reprinted from Ref. [737] with permission.

6.2. CaPO₄ Deposits (Coatings, Films, and Layers)

For many years, the clinical application of $CaPO_4$-based bioceramics has been largely limited to non-load-bearing parts of the skeleton due to their inferior mechanical properties. Therefore, materials with better mechanical properties appear to be necessary. For example, metallic implants are encountered in endoprostheses (total hip joint replacements) and artificial teeth sockets. As metals do not undergo bone bonding, i.e., they do not form a mechanically stable link between the implant and bone tissue, methods have been sought to improve contacts at the interface. One major method is to coat metals with $CaPO_4$, which enables bonding ability between the metal and the bone [180,190,397,740–742].

A number of factors influence the properties of $CaPO_4$ deposits (coatings, films, and layers). They include thickness (this will influence coating adhesion and fixation—the agreed optimum now seems to be within 50–100 µm), crystallinity (this affects the dissolution and biological behavior), phase and chemical purity, porosity, and adhesion. The coated implants combine the surface biocompatibility and bioactivity of $CaPO_4$ with the core strength of strong substrates (Figure 17). Moreover, $CaPO_4$ deposits decrease a release of potentially hazardous chemicals from the core implant and shield the substrate surface from environmental attack. In the case of porous implants, the $CaPO_4$-coated surface enhances bone ingrowth into the pores [331]. The clinical results for $CaPO_4$-deposited implants reveal that they have much longer lifetimes after implantation than uncoated devices and they are found to be particularly beneficial for younger patients. Further details on this topic are available in the special reviews [740–742].

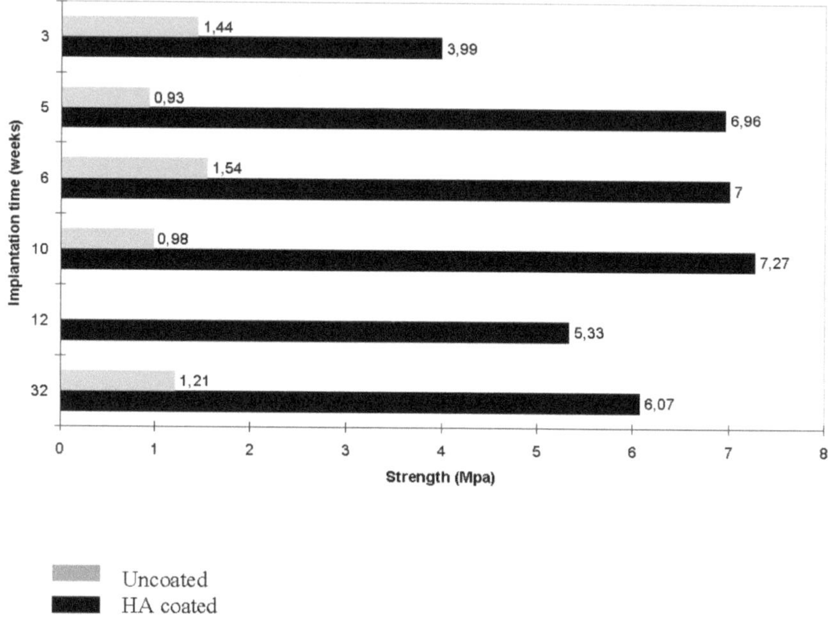

Figure 17. Shows how a plasma-sprayed HA coating on a porous titanium (dark bars), dependent on the implantation time, will improve the interfacial bond strength compared to uncoated porous titanium (light bars). Reprinted from Ref. [50] with permission.

6.3. Functionally Graded Bioceramics

In general, functionally gradient materials (FGMs) are defined as materials having either compositional or structural gradient from their surface to the interior. The idea of FGMs allows one device to possess two different properties. One of the most important combinations for the biomedical field is that of mechanical strength and biocompatibility. Namely, only surface properties govern a biocompatibility of the entire device. In contrast, the strongest material determines the mechanical strength of the entire device. Although this subject belongs to the previous section on coatings, films, and layers, in a certain sense, all types of implants covered by $CaPO_4$ might be also considered as FGMs.

Within the scope of this review, functionally graded bioceramics consisting of $CaPO_4$ are considered and discussed only. Such formulations have been developed [74,491,494, 550,743–753]. For example, dense sintered bodies with gradual compositional changes from α-TCP to HA were prepared by sintering diamond-coated HA compacts at 1280 °C under a reduced pressure, followed by heating under atmospheric conditions [743]. The content of α-TCP gradually decreased, while the content of HA increased with increasing depth from the surface. This functionally gradient bioceramic consisting of HA core and α-TCP surface showed potential value as a bone-substituting biomaterial [743]. Two types of functionally gradient FA/β-TCP biocomposites were prepared in another study [744]. As shown in Figure 18, one of the graded biocomposites was in the shape of a disk and contained four different layers of about 1 mm thick. The other graded biocomposite was also in the shape of a disk but contained two sets of the four layers, each layer being 0.5 mm thick controlled by using a certain amount of the mixed powders. The final FA/β-TCP graded structures were formed at 100 MPa and sintered at 1300 °C for 2 h [744]. The same approach was used in yet another study, but HA was used instead of FA and CDHA was used instead of β-TCP [752]. $CaPO_4$ coatings with graded crystallinity were prepared as well [748].

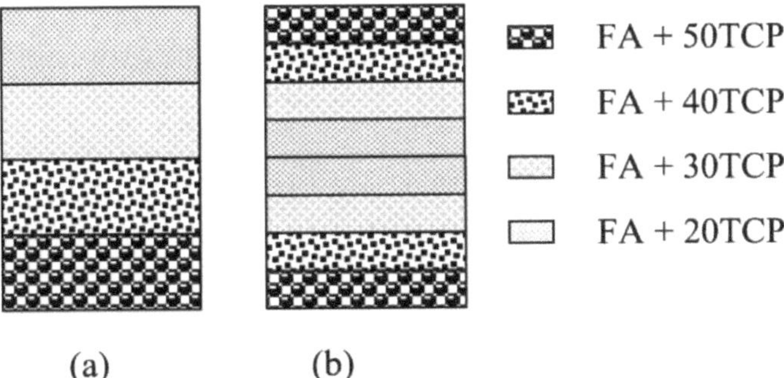

Figure 18. A schematic diagram showing the arrangement of the FA/β-TCP biocomposite layers. (**a**) A nonsymmetric functionally gradient material (FGM); (**b**) symmetric FGM. Reprinted from Ref. [744] with permission.

In addition, it is well known that a bone cross-section from cancellous to cortical bone is nonuniform in porosity and pore dimensions. Thus, in various attempts to mimic the porous structure of bones, $CaPO_4$ bioceramics with graded porosity have been fabricated [74,432,478,491,494,550,743–746]. For example, graded porous $CaPO_4$ bioceramics can be produced by means of tape casting and lamination (Figure 19, top). Other manufacturing techniques, such as a compression molding process (Figure 19, bottom) followed by impregnation and firing, are known as well [432]. In the first method, an HA slurry was mixed with a pore former. The mixed slurry was then cast into a tape. Using the same method, different tapes with different pore former sizes were prepared individually. The different tape layers were then laminated together. Firing was then performed to remove the pore formers and sinter the HA particle compacts, resulting in graded porous bioceramics [746]. This method was also used to prepare graded porous HA with a dense part (core or layer) in order to improve the mechanical strength, as dense ceramics are much stronger than porous ceramics. However, as in the pressure infiltration of mixed particles, this multiple tape casting also has the problem of poor connectivity of pores, although the pore size and the porosity are relatively easy to control. Furthermore, the lamination step also introduces additional discontinuity of the porosity on the interfaces between the stacked layers.

Since diverse biomedical applications require different configurations and shapes, the graded (or gradient) porous bioceramics can be grouped according to both the overall shape and the structural configuration [432]. The basic shapes include rectangular blocks and cylinders (or disks). For the cylindrical shape, there are configurations of dense core–porous layer, less porous core–more porous layer, dense layer–porous core, and less porous layer–more porous core. For the rectangular shape, in the gradient direction, i.e., the direction with varying porosity, pore size, or composition, there are configurations of porous top–dense bottom (same as porous bottom–dense top), porous top–dense center–porous bottom, dense top–porous center–dense bottom, etc. Concerning biomedical applications, a dense core–porous layer structure is suitable for implants of a high mechanical strength and with bone ingrowth for stabilization, whereas a less porous layer–more porous core configuration can be used for drug delivery systems. Furthermore, a porous top –dense bottom structure can be shaped into implants of articulate surfaces for wear resistance and with porous ends for bone ingrowth fixation, while a dense top–porous center–dense bottom arrangement mimics the structure of head skull. Further details on bioceramics with graded porosity can be found in the literature [432].

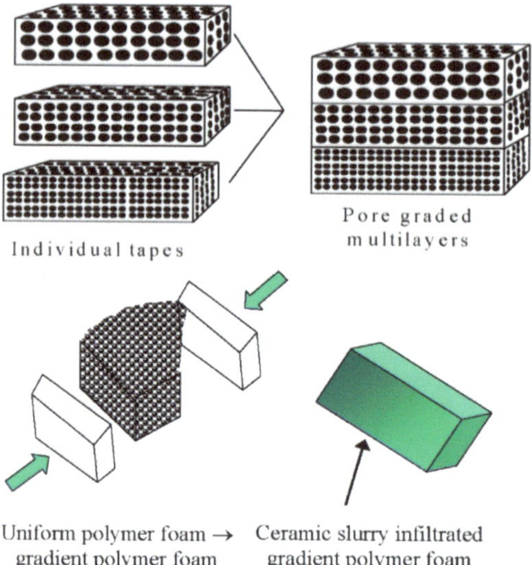

Figure 19. Schematic illustrations of fabrication of pore-graded bioceramics: top—lamination of individual tapes, manufactured by tape casting; bottom—a compression molding process. Reprinted from Ref. [432] with permission.

7. CaPO$_4$ Bioceramics in Tissue Engineering

7.1. Tissue Engineering

Tissue/organ repair has been the ultimate goal of surgery from ancient times to nowadays [56,57]. The repair has traditionally taken two major forms: tissue grafting followed by organ transplantation, and alloplastic or synthetic material replacement. Both approaches, however, have limitations. Grafting requires second surgical sites with associated morbidity and is restricted by limited amounts of material, especially for organ replacement. Synthetic materials often integrate poorly with host tissue and fail over time due to wear and fatigue or adverse body response [754]. In addition, all modern artificial orthopedic implants lack three of the most critical abilities of living tissues: (i) self-repairing; (ii) maintaining of blood supply; (iii) self-modifying their structure and properties in response to external aspects such as a mechanical load [755]. It is needless to mention that bones not only possess all of these properties but, in addition, they are self-generating, hierarchical, multifunctional, nonlinear, composite, and biodegradable; therefore, the ideal artificial bone grafts must possess similar properties [62].

The last decades have seen a surge in creative ideas and technologies developed to tackle the problem of repairing or replacing diseased and damaged tissues, leading to the emergence of a new field in healthcare technology now referred to as *tissue engineering*, which might be defined as "the creation of new tissue for the therapeutic reconstruction of the human body, by the deliberate and controlled stimulation of selected target cells through a systematic combination of molecular and mechanical signals" [756]. Briefly, this is an interdisciplinary field that exploits a combination of living cells, engineering materials, and suitable biochemical factors (Figure 20) in a variety of ways to improve, replace, restore, maintain, or enhance living tissues and whole organs [757–759]. However, since two of three major components (namely, cells and biochemical factors) of the tissue engineering subject appear to be far beyond the scope of this review, the topic of bone tissue engineering that aims to mimic the in vivo bone regeneration processes in a laboratory environment is narrowed down to the engineering materials prepared from CaPO$_4$ bioceramics only.

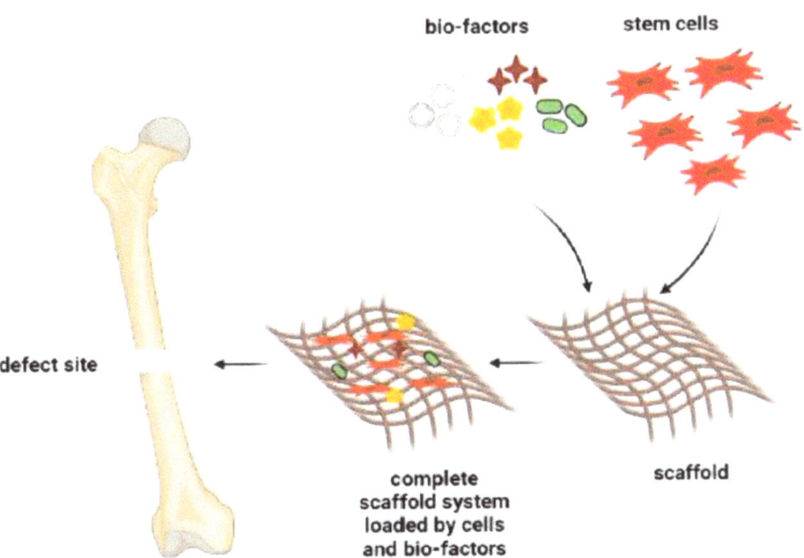

Figure 20. A tissue engineering approach for developing an advanced bone scaffold. Reprinted from Ref. [759] with permission.

Regeneration, rather than a repair, is the central goal of any tissue engineering strategy; therefore, it aims to create tissues and organs de novo [758]. This field of science started more than two decades ago [760,761], and the famous publication by Langer and Vacanti [762] has greatly contributed to the promotion of tissue engineering research worldwide. The field of tissue engineering, particularly when applied to bone substitutes where tissues often function in a mechanically demanding environment [763–765], requires a collaboration of excellence in cell and molecular biology, biochemistry, material sciences, bioengineering, and clinical research [766]. For success, it is necessary that researchers with expertise in one area have an appreciation of the knowledge and challenges of the other areas. However, since the technical, regulatory, and commercial challenges might be substantial, the introduction of new products is likely to be slow [758].

Nowadays, tissue engineering is at full research potential due to the following key advantages: (i) the solutions it provides are long-term, much safer than other options, and cost-effective as well; (ii) the need for donor tissue is minimal, which eliminates the immunosuppression problems; (iii) the presence of residual foreign material is eliminated as well [767,768].

7.2. Scaffolds and Their Properties

It would be very convenient for both patients and physicians if devastated tissues or organs of patients could be regenerated by simple cell injections to the target sites, but such cases are rare. The majority of large-sized tissues and organs with distinct 3D form require a support for their formation from cells. The support is called a scaffold, template, and/or artificial extracellular matrix [127,128,534,760,763–772]. The major function of scaffolds is similar to that of the natural extracellular matrix that assists proliferation, differentiation, and biosynthesis of cells. In addition, scaffolds placed at the regeneration sites will prevent disturbing cells from invasion into the sites of action [771,772]. The role of scaffolds was perfectly described by a Spanish classical guitarist Andrés Segovia (1893–1987): "When one puts up a building one makes an elaborate scaffold to get everything into its proper place. But when one takes the scaffold down, the building must stand by itself with no trace of the means by which it was erected. That is how a musician should work". However, for the future of tissue engineering, the term "template" might become more suitable because, according to David F. Williams, the term scaffold "conveys an old fashioned meaning of an inert external structure that is temporarily used to assist in the construction or repair of inanimate objects such as buildings, taking no part in the characteristics of the finished product." [773] (p. 1129).

Therefore, the idea behind tissue engineering is to create or engineer autografts by either expanding autologous cells in vitro guided by a scaffold or implanting an acellular template in vivo and allowing the patient's cells to repair the tissue guided by the scaffold. The first phase is the in vitro formation of a tissue constructed by placing the chosen cells and scaffolds in a metabolically and mechanically supportive environment with growth media (in a bioreactor), in which the cells proliferate and elaborate extracellular matrix. It is expected that cells infiltrate into the porous matrix and consequently proliferate and differentiate therein [774,775]. In the second phase, the construct is implanted in the appropriate anatomic location, where remodeling in vivo is intended to recapitulate the normal functional architecture of an organ or a tissue [776,777]. The key processes occurring during both in vitro and in vivo phases of the tissue formation and maturation are (1) cell proliferation, sorting, and differentiation, (2) extracellular matrix production and organization, (3) biodegradation of the scaffold, and (4) remodeling and potentially growth of the tissue [778].

To achieve the goal of tissue reconstruction, the scaffolds (templates) must meet a number of the specific requirements [127,128,769–773]. Five features of the scaffold's architecture appear to influence the biological response: (1) a macroscopic shape, (2) a porous network, (3) pore dimensions and geometry, (4) surface microtopography, and (5) micro-, submicro-, and nanoporosities. In addition, scaffolds should be biodegradable. Among them, a reasonable surface roughness is necessary to facilitate cell seeding and fixation [619,779–784]. A high porosity and the adequate pore dimensions (Tables 2 and 4) are very important to allow cell migration and vascularization, as well as a diffusion of nutrients [437,758]. A French architect, Robert le Ricolais (1894–1977), stated: "The art of structure is where to put the holes". Therefore, to enable proper tissue ingrowth, vascularization, and nutrient delivery, scaffolds should have a highly interconnected porous network, formed by a combination of macro- and micropores, in which more than ~60% of the pores should have a size ranging from ~150 to ~400 µm and at least ~20% should be smaller than ~20 µm [437,442,443,448,529,530,536,538,544,563–570,572,754,785–791]. Furthermore, a sufficient mechanical strength and stiffness are mandatory to oppose contraction forces and later for the remodeling of damaged tissues [792,793]. In addition, scaffolds must be manufactured from the materials with controlled biodegradability and/or bioresorbability, such as $CaPO_4$, so that a new bone will eventually replace the scaffold [763,786,794]. Furthermore, the degradation byproducts of scaffolds must be noncytotoxic. More to the point, the resorption rate has to coincide as much as possible with the rate of bone formation (i.e., between a few months and about 2 years) [795]. This means that while cells are fabricating their own natural matrix structure around themselves, the scaffold is able to provide a structural integrity within the body, and eventually it will break down, leaving the newly formed tissue that will take over the mechanical load. However, one should bear in mind that the scaffold's architecture changes with the degradation process, and the degradation byproducts affect the biological response. In addition, scaffolds should be easily fabricated into a variety of shapes and sizes [796] and be malleable to fit irregularly shaped defects, while the fabrication processes should be effortlessly scalable for mass production. In many cases, ease of processability, as well as easiness of conformation and injectability, which self-setting $CaPO_4$ formulations possess (see *Section 6.1. Self-setting (Self-hardening) Formulations*), can determine the choice of a certain biomaterial. Finally, sterilization with no loss of properties is a crucial step in scaffold production at both a laboratory and an industrial level [763–765]. Thus, each scaffold (template) should fulfill many functions before, during, and after implantation.

Table 4. A hierarchical pore size distribution that an ideal scaffold should exhibit [797].

Pore Sizes of a 3D Scaffold	A Biochemical Effect or Function
<1 μm	Interaction with proteins
	Responsible for bioactivity
1–20 μm	Type of cells attracted
	Cellular development
	Orientation and directionality of cellular ingrowth
100–1000 μm	Cellular growth
	Bone ingrowth
	Predominant function in the mechanical strength
>1000 μm	Implant functionality
	Implant shape
	Implant esthetics

Many fabrication techniques are available to produce porous $CaPO_4$ scaffolds (Table 2) with varying architectural features (see the aforementioned Section 3.3 and 4.4). In order to achieve the desired properties with the minimum expenses, the production process should be optimized [798]. The main goal is to develop a high potential synthetic bone substitute (so-called "smart scaffold") which will not only promote osteoconduction, i.e., bone growth on a surface, but also osteopromotion, i.e., the ability to enhance osteoinduction [799]. In the case of $CaPO_4$, a smart scaffold represents a biphasic (HA/β–TCP ratio of 20/80) formulation with a total porosity of ~73%, constituted of macropores (>100 μm), mesopores (10–100 μm), and a high content (~40%) of micropores (<10 μm) with the crystal dimensions within <0.5 to 1 μm and the specific surface area ~6m^2/g [800]. With the advent of $CaPO_4$ in tissue engineering, the search is on for the ultimate option consisting of a synthetic smart scaffold impregnated with cells and growth factors. Figure 21 schematically depicts a possible fabrication process of such item that, afterwards, will be implanted into a living organism to induce bone regeneration [47].

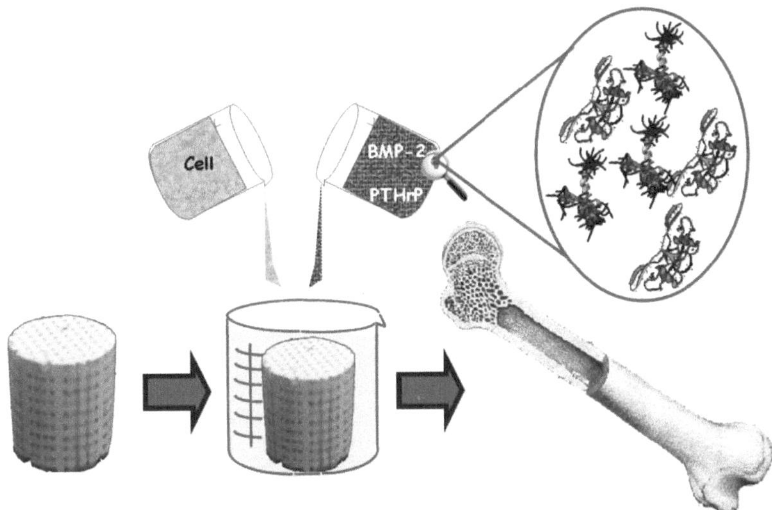

Figure 21. A schematic view of a third-generation biomaterial, in which porous $CaPO_4$ bioceramic acts as a scaffold or a template for cells, growth factors, etc. Reprinted from Ref. [47] with permission.

To finalize this topic, one should note the fundamental unfeasibility to create the so-called "ideal scaffold" for bone grafting. Since bones of the human skeleton have very different dimensions, shapes, and structures depending on their functions and locations, synthetic bone grafts of various sizes, shapes, porosity, mechanical strength, composition, and resorbability appear to be necessary. Therefore, HA bioceramics of 0 to 15% porosity are used as both ilium and intervertebral spacers, where a high strength is required, HA bioceramics of 30 to 40% porosity are useful as spinous process spacers for laminoplasty, where both bone formation and middle strength are necessary, while HA bioceramics of 40% to 60% porosity are useful for the calvarias plate, where a fast bone formation is needed (Figure 22) [518]. Furthermore, defining the optimum parameters for artificial scaffolds is, in fact, an attempt to find a reasonable compromise between various conflicting functional requirements. Namely, an increased mechanical strength of bone substitutes requires solid and dense structures, while colonization of their surfaces by cells requires interconnected porosity. Additional details and arguments on this subject are well described elsewhere [801], in which the authors concluded that "there is enough evidence to postulate that ideal scaffold architecture does not exist." (p. 478).

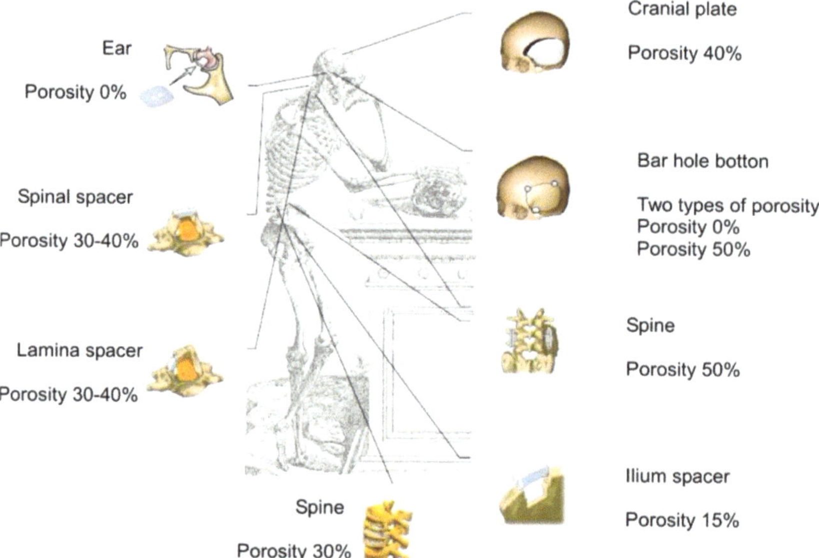

Figure 22. A schematic drawing presenting the potential usage of HA with various degrees of porosity. Reprinted from Ref. [518] with permission.

7.3. Bioceramic Scaffolds from CaPO₄

Philosophically, the increase in life expectancy requires biological solutions to all biomedical problems, including orthopedic ones, which were previously managed with mechanical solutions. Therefore, since the end of the 1990s, biomaterials research has focused on tissue regeneration instead of tissue replacement [802]. The alternatives include using hierarchical bioactive scaffolds to engineer in vitro living cellular constructs for transplantation or using bioresorbable bioactive particulates or porous networks to activate, in vivo, the mechanisms of tissue regeneration [803,804]. Thus, the aim of $CaPO_4$ is to prepare artificial porous bioceramic scaffolds able to provide the physical and chemical cues to guide cell seeding, differentiation, and assembly into 3D tissues of a newly formed bone. Particle sizes, shape, and surface roughness of the scaffolds are known to affect cellular adhesion, proliferation, and phenotype [619,779–784]. Additionally, the surface energy might play a role in attracting particular proteins to the bioceramic surface and, in turn, this will affect the cells' affinity to the material. More to the point, cells are exceedingly sensitive to the chemical composition, and their bone-forming functions can be dependent on grain morphology of the scaffolds. For example, osteoblast functions were found to increase on nanodimensional fibers when compared to nanodimensional spheres because the former more closely approximated the shape of biological apatite in bones [805]. In addition, a significantly higher osteoblast proliferation on HA bioceramics sintered at 1200 °C, compared to that on HA bioceramics sintered at 800 and 1000 °C, was reported [806]. Furthermore, since ions of calcium and orthophosphate are known to regulate bone metabolism, $CaPO_4$ appears to be among the few bone graft substitute materials that can be considered as a drug. A schematic drawing of the key scaffold properties affecting a cascade of biological processes occurring after $CaPO_4$ implantation is shown in Figure 23 [807].

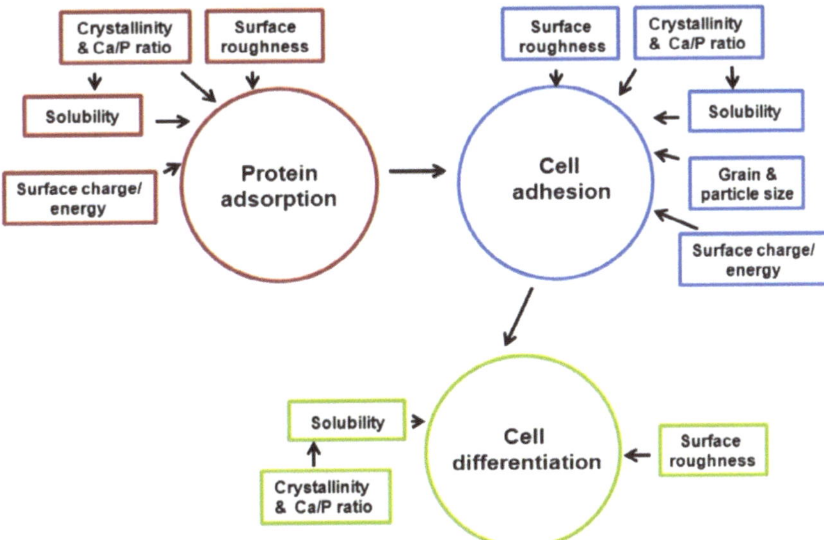

Figure 23. A schematic drawing of the key scaffold properties affecting a cascade of biological processes occurring after $CaPO_4$ implantation. Reprinted from Ref. [807] with permission.

Thus, to meet the tissue engineering requirements, much attention is devoted to further improvements of $CaPO_4$ bioceramics [808–810]. From the chemical point of view, the developments include synthesis of novel ion-substituted $CaPO_4$ [17–41]. For example, a recent systematic review and meta-analysis indicated a significant positive effect on new bone formation by supplementing $CaPO_4$-based bone substitutes with bioinorganics compared to those without dopants, especially when strontium, silicon, or magnesium were used [811]. A positive influence of $CaPO_4$ doping by carbonates was also noticed [812]. From the material point of view, the major research topics include nanodimensional and nanocrystalline structures [813–816], amorphous compounds [817,818], (bio)organic/$CaPO_4$ biocomposites and hybrid formulations [357,819,820], and biphasic, triphasic, and multiphasic formulations [79], as well as various types of structures, forms, and shapes. The latter comprise fibers, wires, whiskers, and filaments [135,231,821–834], macro-, micro-, and nanosized spheres, beads, and granules [833–846], tetrapods and pyramids [847], micro- and nanosized tubes [848–852], aerogels [853], "flowers" [854], porous 3D scaffolds [172] made of ACP [470,582,855], DCPD/DCPA [148,856–859], OCP [860,861], TCP [68,71,139,140,862–865], HA [147,447,448,487,513,514,798,852,866–870], TTCP [141] and biphasic formulations [245,474,491,531,836,843,865,871–877], structures with graded porosity [74,432,478,491,494,550,743–746], and hierarchically organized ones [877,878]. More advanced combined techniques are also possible. For example, to increase the degree of ossification and homogenization, the surface of porous $CaPO_4$ scaffolds might be modified by generation of $CaPO_4$ whiskers on their surface [879,880], followed by reinforcing with multiple layers of releasable nanodimensional CDHA particles [880]. Recently, research progress has been made in the deformable $CaPO_4$-based biomaterials with a high flexibility, softness, and/or elasticity based on nanodimensional ultralong HA wires [881,882]. Furthermore, an addition of defects through an intensive milling [883,884] or their removal by a thermal treatment [885] can be used to modify the chemical reactivity of $CaPO_4$. In addition, more attention should be paid to a crystallographically aligned $CaPO_4$ bioceramics [687–692,886].

In general, there are three principal therapeutic strategies for treating diseased or injured tissues in patients: (i) implantation of freshly isolated or cultured cells; (ii) implantation of tissues assembled in vitro from cells and scaffolds; (iii) in situ tissue regeneration. For cellular implantation, individual cells or small cellular aggregates from the patient or a donor are either injected into the damaged tissue directly or are combined with a degradable scaffold in vitro and then implanted. For tissue implantation, a complete 3D tissue is grown in vitro using patient or donor cells and a bioresorbable scaffold and then is implanted into the patients to replace diseased or damaged tissues. For in situ regeneration, a scaffold implanted directly into the injured tissue stimulates the body's own cells to promote local tissue repair [324,757]. In any case, simply trapping cells at the particular point on a surface is not enough: the cells must be encouraged to differentiate, which is impossible without the presence of suitable biochemical factors [887]. All previously mentioned points clearly indicate that, for the purposes of tissue engineering, $CaPO_4$ bioceramics play an auxiliary role; namely, they acts as a suitable material to manufacture the appropriate 3D templates, substrates, or scaffolds to be colonized by living cells before the successive implantation [799,800,888–890]. The in vitro evaluation of potential $CaPO_4$ scaffolds for tissue engineering is described elsewhere [891], while the data on the mechanical properties of $CaPO_4$ bioceramics for use in tissue engineering are also available [892–894]. The effect of an HA-based biomaterial on gene expression in osteoblast-like cells was reported as well [895]. To conclude this part, the excellent biocompatibility of $CaPO_4$ bioceramics, their possible osteoinductivity [544,566,597–605], and high affinity for drugs [54–57,896–900], proteins, and cells [897,901] make them very functional for the tissue engineering applications [902]. The feasible production of scaffolds with tailored structures and properties opens up a spectacular future for $CaPO_4$ bioceramics [895–903].

7.4. A Clinical Experience

To date, there are just a few publications on clinical application of cell-seeded $CaPO_4$ bioceramics for bone tissue engineering of humans. Namely, Quarto et al. [904] were the first to report a treatment of large (4–7 cm) bone defects of the tibia, ulna, and humerus in three patients aged from 16 to 41 years old, where the conventional surgical therapies had failed. The authors implanted a custom-made unresorbable porous HA scaffold seeded with in vitro expanded autologous bone marrow stromal cells. In all three patients, radiographs and computed tomographic scans revealed abundant callus formation along the implants and good integration at the interfaces with the host bones by the second month after surgery [904]. In the same year, Vacanti et al. [905] reported the case of a man who had a traumatic avulsion of the distal phalanx of a thumb. The phalanx was replaced with a specially treated natural coral (porous HA; 500-pore ProOsteon (see Table 3)) implant that was previously seeded with in vitro expanded autologous periosteal cells. The procedure resulted in the functional restoration of a stable and biomechanically sound thumb of normal length, without the pain and complications that are usually associated with harvesting a bone graft.

Morishita et al. [906] treated a defect resulting from surgery of benign bone tumors in three patients using HA scaffolds seeded with in vitro expanded autologous bone marrow stromal cells after osteogenic differentiation of the cells. Two bone defects in a tibia and one defect in a femur were treated. Although ectopic implants in nude mice were mentioned to show the osteogenicity of the cells, details such as the percentage of the implants containing bone and at what quantities were not reported. Furthermore, cell-seeded $CaPO_4$ scaffolds were found to be superior to autograft, allograft or cell-seeded allograft in terms of bone formation at ectopic implantation sites [907]. In addition, it has been hypothesized that dental follicle cells combined with β-TCP bioceramics might become a novel therapeutic strategy to restore periodontal defects [908]. In yet another study, the behavior of human periodontal ligament stem cells on an HA-coated genipin-chitosan scaffold in vitro was studied followed by evaluation on bone repair in vivo [909]. The study demonstrated the potential of this formulation for bone regeneration.

A research group from Holland evaluated vascularization in relation to bone formation potential of adipose-stem-cells-containing stromal vascular fractions of adipose tissues, seeded on two types of $CaPO_4$ carriers, within the human maxillary sinus floor elevation model, in a phase I study [910]. Autologous stromal vascular fractions were obtained from 10 patients and seeded on either β-TCP scaffolds with 60% porosity (n = 5) or BCP (HA + β-TCP) ones with 90% porosity (n = 5) and used for maxillary sinus floor elevations in a one-step surgical procedure. After 6 months, biopsies were obtained during dental implant placements and the quantifications of the number of blood vessels were performed using histomorphometric analysis and immunohistochemical stainings for blood vessel markers. Bone percentages seemed to correlate with blood vessel formation and were higher in the study versus control biopsies in the cranial area, in particular for β-TCP-treated patients. That study showed the safety, feasibility, and efficiency of using of adipose-stem-cells-seeded $CaPO_4$ scaffolds for human maxillary sinus floor elevations and indicated a proangiogenic effect of stromal vascular fraction [910]. A brief description of several other cases is available in the literature [911].

To finalize this section, one must mention that $CaPO_4$ bioceramics are also used in veterinary orthopedics for favoring animal bone healing in areas in which bony defects exist [912,913].

8. Non-Biomedical Applications of $CaPO_4$

Due to their strong adsorption ability, surface acidity or basicity, and ion exchange abilities, some types of $CaPO_4$ possess a catalytic activity [914–926]. As seen from the references, $CaPO_4$ are able to catalyze oxidation and reduction reactions, as well as formation of C–C bonds. Namely, the application in oxidation reactions mainly includes oxidation of alcohol and dehydrogenation of hydrocarbons, while the reduction reactions include hydrogenolysis and hydrogenation. The formation of C–C bonds mainly comprises Claisen–Schmidt and Knoevenagel condensation reactions, Michael addition reaction, as well as Friedel–Crafts, Heck, Diels–Alder, and aldol reactions [921].

In addition, due to the chemical similarity to the inorganic part of mammalian calcified tissues, $CaPO_4$ powders appear to be good solid carriers for chromatography of biological substances. Namely, high-value biological materials such as recombinant proteins, therapeutic antibodies, and nucleic acids are separated and purified [927–933]. Furthermore, some types of $CaPO_4$ are used as a component of various sensors [373,374,378,379,382,934–938]. Finally, $CaPO_4$ ceramics appear to be good adsorbents of fluorides [939]; however, since these subjects are almost irrelevant to bioceramics, they are not detailed further. Additional details and examples are available elsewhere [940].

9. Conclusions and Outlook

The available chronology of seeking suitable bioceramics for bone substitutes is as follows: since the 1950s, the first aim was to use bioinert bioceramics, which had no reaction with living tissues. They included inert and tolerant compounds, which were designed to withstand physiological stress without, however, stimulating any specific cellular responses. Later on, in the 1980s, the trend changed towards exactly the opposite: the idea was to implant bioceramics that reacted with the surrounding tissues by producing newly formed bone (a "responsive" bioceramic because it was able to elicit biological responses).

These two stages are referred to as the first and the second generations of bioceramics, respectively [941] and, currently, both of them are extensively commercialized. Thus, the majority of the marketable products listed in Table 3 belong to the first and the second generations of bone substitute biomaterials. However, the progress has continued and, in the current century, scientists are searching for the third generation of bioceramics [324], which will be able to "instruct" the physiological environment toward desired biological responses (i.e., bioceramics will be able to regenerate bone tissues by stimulating specific responses at the molecular level) [47,942]. Since each generation represents an evolution of the requirements and properties of the biomaterials involved, one should stress that these three generations should not be interpreted as the chronological, but instead as the conceptual ones. This means that at present, research and development is still devoted to biomaterials and bioceramics that, according to their properties, could be considered to be of the first or the second generations, because the second generation of bioceramics with added porosity is one of the initial approaches in developing the third generation of bioceramics [943]. Furthermore, there is another classification of the history of biomaterials introduced by Prof. James M. Anderson. According to Anderson, from 1950–1975 the researchers studied bioMATERIALS, from 1975–2000 they studied BIOMATERIALS, and since 2000 the time for BIOmaterials has been approaching [944]. Here, the capital letters emphasis the major direction of the research efforts in the complex subject of biomaterials. As bioceramics are biomaterials of the ceramic origin (see *Section 2. General Knowledge and Definitions*), Anderson's historical classification appears to be applicable to the bioceramics field as well.

The historical development of biomaterials informs that their widespread use experiences two major difficulties. The first is an incomplete understanding of the physical and chemical functioning of biomaterials and of the human response to these materials. Recent advances in material characterization and computer science, as well as in cell and molecular biology, are expected to play a significant role in the study of biomaterials. A second difficulty is that many biomaterials do not perform as desirably as we would like. This is not surprising, since many materials used in medicine were not designed for medical purposes. It needs to be mentioned here that biomaterials are expected to perform in our body's internal environment, which is very aggressive. For example, solution pH of body fluids in various tissues varies in the range from 1 to 9. During daily activities, bones are subjected to a stress of ~4 MPa, whereas the tendons and ligaments experience peak stresses in the range of 40–80 MPa. The mean load on a hip joint is up to three times the body weight (3000 N) and peak load during jumping can be as high as ~10 times the body weight. More importantly, these stresses are repetitive and fluctuating, depending on the activities, such as standing, sitting, jogging, stretching, and climbing. All of these require careful designing of biomaterials in terms of composition, shape, physical, and biocompatibility properties. Therefore, a significant challenge is the rational design of human biomaterials based on a systematic evaluation of desired biological, chemical, and engineering requirements [945].

Nevertheless, the field of biomaterials is in the midst of a revolutionary change in which the life sciences are becoming equal in importance to materials science and engineering as the foundation of the field. Simultaneously, advances in engineering (for example, nanotechnology) are greatly increasing the sophistication with which biomaterials are designed and have allowed fabrication of biomaterials with increasingly complex functions [797]. Specifically, during the last ~50 years, $CaPO_4$ bioceramics have become an integral and vital segment of our modern healthcare delivery system. In the modern fields of the third-generation bioceramics (Hench) or BIOceramics (Anderson), the full potential of $CaPO_4$ has only begun to be recognized. Namely, $CaPO_4$, which were intended as osteoconductive bioceramics in the past, represent materials to fabricate osteoinductive implants nowadays [544,566,597–605]. Some steps in this direction have been already made by fabricating scaffolds for bone tissue engineering through the design of controlled 3D-porous structures and increasing the biological activity through development of novel ion-substituted $CaPO_4$ bioceramics [546,946]. The future of biosynthetic bone implants will point to better mimicking the autologous bone grafts. Therefore, the composition, structure, and molecular surface chemistry of various types of $CaPO_4$ will be tailored to match the specific biological and metabolic requirements of tissues or disease states [947,948]. This new generation of $CaPO_4$ bioceramics should enhance the quality of life of millions of people as they grow older.

However, in spite of the great progress, there is still a great potential for major advances to be made in the field of $CaPO_4$ bioceramics. This includes requirements for [949]:

- Improvement of the mechanical performance of existing types of bioceramics.
- Enhanced bioactivity in terms of gene activation.
- Improvement in the performance of biomedical coatings in terms of their mechanical stability and ability to deliver biological agents.
- Development of smart biomaterials capable of combining sensing with bioactivity.
- Development of improved biomimetic composites.

Furthermore, there is still need for a better understanding of the biological systems. For example, the bonding mechanism between the bone mineral and collagen remains unclear. It is also unclear whether a rapid repair that is elicited by the new generation of bioceramics is a result of the enhancement of mineralization, per se, or whether there is a more complex signaling process involving proteins in collagen. If we were able to understand the fundamentals of bone response to specific ions and the signals they activate, then we would be able to design better bioceramics for the future [949].

To finalize this review, it is completely obvious that the present status of research and development in the field of $CaPO_4$ bioceramics is still at the starting point for the solution of new problems at the confluence of materials science, biology, and medicine, concerned with the restoration of damaged functions in the human organisms. A large increase in active elderly people has dramatically raised the need for load-bearing bone graft substitutes, for example, for bone reconstruction during revision arthroplasty or for the reinforcement of osteoporotic bones. Strategies applied in the last four decades towards this goal have failed, so new strategies, possibly based on self-assembling and/or nanofabrication, will have to be proposed and developed [950]. Angiogenesis (the process and stimulation of new blood vessel formation via sprouting from existing blood vessels) is also very important to the success of hard tissue regeneration, and $CaPO_4$ seem to be useful for this purpose [951]. In addition, some $CaPO_4$-containing formulations were tested with soft tissues [952,953], mainly for various types of soft-tissue augmentations [721,954–956] and eyeball replacements [704–709], as well as for cancer diagnostics and therapy [957–959], wound healing [960], as components of various cosmetic formulations [961], and vaccine adjuvants [962]. Recently, a concept of black bioceramics for Ca- and Mg-silicates, as well as for HA and TCP, was introduced [963]. The black bioceramics were prepared through a partial thermal reduction of traditional white ceramics ($CaSiO_3$, $MgSiO_3$, TCP, and HA) by magnesium. Due to the presence of oxygen vacancies and structural defects, the black bioceramics were found to possess a photothermal functionality while maintaining their initial high bioactivity and regenerative capacity. These black bioceramics showed excellent photothermal antitumor effects for both skin and bone tumors. At the same time, they significantly improved bioactivity for skin/bone tissue repair both in vitro and in vivo [963]. Bioceramics prepared from other types of calcium phosphates, such as calcium pyrophosphate, are worth investigating as well [964]. Additive manufacturing techniques will be further developed [965]. A bioinformatics approach to study the role of $CaPO_4$ properties in bone regeneration might be promising as well [966].

In future, it should be feasible to design a new generation of gene-activating $CaPO_4$-based scaffolds tailored for specific patients and disease states. In addition, further developments of 3D-printing technologies [967] will allow designing of personalized bone grafts, which will provide an accurate control of the geometry. The design of the implant shape, based on X-ray computed tomography data, will ensure the perfect fit between the graft and the anatomical defect [968]. Personalized implants will be produced with tailored characteristics better adapted to the patient-specific bone tissue regions/defects that need to be replaced/reinforced. To transfer these technologies to clinical practice, material science and tissue engineering need to be closely assisted by biomedical researchers in order to confer the safety risk assessment, as well as efficacy at high standards. In addition, the development of complex testing strategies will help to unveil the network of biological events elicited by $CaPO_4$-based bioceramics as bulks, coatings, or nanodimensional forms, which are essential to ensure a longer and safer implant life in orthopedic and dentistry applications [969]. Perhaps, sometime-bioactive stimuli will be used to activate genes in a preventative treatment to maintain the health of aging tissues. Currently, all the aforementioned seem hardly possible. However, we need to remember that only ~50 years ago, the concept of a material that would not be rejected by living tissues also seemed impossible [583].

Funding: This research received no external funding.

Institutional Review Board Statement: Not applicable.

Informed Consent Statement: Not applicable.

Data Availability Statement: Not applicable.

Conflicts of Interest: The author declares no conflict of interest.

References

1. Ratner, B.D.; Hoffman, A.S.; Schoen, F.J.; Lemons, J.E. (Eds.) *Biomaterials Science: An Introduction to Materials in Medicine*, 3rd ed.; Academic Press: Oxford, UK, 2013; p. 1573.
2. Ducheyne, P.; Healy, K.; Hutmacher, D.E.; Grainger, D.W.; Kirkpatrick, C.J. (Eds.) *Comprehensive Biomaterials II*, 2nd ed.; Seven-Volume Set; Elsevier: Amsterdam, The Netherlands, 2017; p. 4858.
3. Dorozhkin, S.V. *Calcium Orthophosphate-Based Bioceramics and Biocomposites*; Wiley-VCH: Weinheim, Germany, 2016; p. 405.
4. Dorozhkin, S.V. Dental applications of calcium orthophosphates ($CaPO_4$). *J. Dent. Res.* **2019**, *1*, 24–54.
5. Available online: https://www.alliedmarketresearch.com/bone-graft-substitutes-market (accessed on 24 June 2022).
6. Dorozhkin, S.V. *Calcium Orthophosphates: Applications in Nature, Biology, and Medicine*; Pan Stanford: Singapore, 2012; p. 850.
7. Balazsi, C.; Weber, F.; Kover, Z.; Horvath, E.; Nemeth, C. Preparation of calcium-phosphate bioceramics from natural resources. *J. Eur. Ceram. Soc.* **2007**, *27*, 1601–1606. [CrossRef]
8. Gergely, G.; Wéber, F.; Lukács, I.; Illés, L.; Tóth, A.L.; Horváth, Z.E.; Mihály, J.; Balázsi, C. Nano-hydroxyapatite preparation from biogenic raw materials. *Cent. Eur. J. Chem.* **2010**, *8*, 375–381. [CrossRef]
9. Mondal, S.; Mahata, S.; Kundu, S.; Mondal, B. Processing of natural resourced hydroxyapatite ceramics from fish scale. *Adv. Appl. Ceram.* **2010**, *109*, 234–239. [CrossRef]
10. Lim, K.T.; Suh, J.D.; Kim, J.; Choung, P.H.; Chung, J.H. Calcium phosphate bioceramics fabricated from extracted human teeth for tooth tissue engineering. *J. Biomed. Mater. Res. B Appl. Biomater.* **2011**, *99B*, 399–411. [CrossRef]
11. Seo, D.S.; Hwang, K.H.; Yoon, S.Y.; Lee, J.K. Fabrication of hydroxyapatite bioceramics from the recycling of pig bone. *J. Ceram. Proc. Res.* **2012**, *13*, 586–589.
12. Ho, W.F.; Hsu, H.C.; Hsu, S.K.; Hung, C.W.; Wu, S.C. Calcium phosphate bioceramics synthesized from eggshell powders through a solid state reaction. *Ceram. Int.* **2013**, *39*, 6467–6473. [CrossRef]
13. González-Rodríguez, L.; López-Álvarez, M.; Astray, S.; Solla, E.L.; Serra, J.; González, P. Hydroxyapatite scaffolds derived from deer antler: Structure dependence on processing temperature. *Mater. Charact.* **2019**, *155*, 109805. [CrossRef]
14. Arokiasamy, P.; Al Bakri Abdullah, M.M.; Abd Rahim, S.Z.; Luhar, S.; Sandu, A.V.; Jamil, N.H.; Nabiałek, M. Synthesis methods of hydroxyapatite from natural sources: A review. *Ceram. Int.* **2022**, *48*, 14959–14979. [CrossRef]
15. Grigoraviciute-Puroniene, I.; Zarkov, A.; Tsuru, K.; Ishikawa, K.; Kareiva, A. A novel synthetic approach for the calcium hydroxyapatite from the food products. *J. Sol-Gel Sci. Technol.* **2019**, *91*, 63–71. [CrossRef]
16. Tosun, G.U.; Sakhno, Y.; Jaisi, D.P. Synthesis of hydroxyapatite nanoparticles from phosphorus recovered from animal wastes. *ACS Sustain. Chem. Eng.* **2021**, *9*, 15117–15126. [CrossRef]
17. Ergun, C.; Webster, T.J.; Bizios, R.; Doremus, R.H. Hydroxylapatite with substituted magnesium, zinc, cadmium, and yttrium. I. Structure and microstructure. *J. Biomed. Mater. Res.* **2002**, *59*, 305–311. [CrossRef]
18. Webster, T.J.; Ergun, C.; Doremus, R.H.; Bizios, R. Hydroxylapatite with substituted magnesium, zinc, cadmium, and yttrium. II. Mechanisms of osteoblast adhesion. *J. Biomed. Mater. Res.* **2002**, *59*, 312–317. [CrossRef]
19. Kim, S.R.; Lee, J.H.; Kim, Y.T.; Riu, D.H.; Jung, S.J.; Lee, Y.J.; Chung, S.C.; Kim, Y.H. Synthesis of Si, Mg substituted hydroxyapatites and their sintering behaviors. *Biomaterials* **2003**, *24*, 1389–1398. [CrossRef]
20. Landi, E.; Celotti, G.; Logroscino, G.; Tampieri, A. Carbonated hydroxyapatite as bone substitute. *J. Eur. Ceram. Soc.* **2003**, *23*, 2931–2937. [CrossRef]

21. Vallet-Regí, M.; Arcos, D. Silicon substituted hydroxyapatites. A method to upgrade calcium phosphate based implants. *J. Mater. Chem.* **2005**, *15*, 1509–1516. [CrossRef]

22. Gbureck, U.; Thull, R.; Barralet, J.E. Alkali ion substituted calcium phosphate cement formation from mechanically activated reactants. *J. Mater. Sci. Mater. Med.* **2005**, *16*, 423–427. [CrossRef]

23. Gbureck, U.; Knappe, O.; Grover, L.M.; Barralet, J.E. Antimicrobial potency of alkali ion substituted calcium phosphate cements. *Biomaterials* **2005**, *26*, 6880–6886. [CrossRef]

24. Reid, J.W.; Tuck, L.; Sayer, M.; Fargo, K.; Hendry, J.A. Synthesis and characterization of single-phase silicon substituted α-tricalcium phosphate. *Biomaterials* **2006**, *27*, 2916–2925. [CrossRef]

25. Tas, A.C.; Bhaduri, S.B.; Jalota, S. Preparation of Zn-doped β-tricalcium phosphate (β-Ca$_3$(PO$_4$)$_2$) bioceramics. *Mater. Sci. Eng. C* **2007**, *27*, 394–401.

26. Gibson, I.R. Silicon-containing apatites. In *Comprehensive Biomaterials II*; Chapter 1.20; Ducheyne, P., Ed.; Elsevier: Oxford, UK, 2017; pp. 428–459.

27. Jalota, S.; Bhaduri, S.B.; Tas, A.C. A new rhenanite (β-NaCaPO$_4$) and hydroxyapatite biphasic biomaterial for skeletal repair. *J. Biomed. Mater. Res. B Appl. Biomater.* **2007**, *80B*, 304–316. [CrossRef]

28. Kannan, S.; Ventura, J.M.G.; Ferreira, J.M.F. Synthesis and thermal stability of potassium substituted hydroxyapatites and hydroxyapatite/β-tricalcium phosphate mixtures. *Ceram. Int.* **2007**, *33*, 1489–1494. [CrossRef]

29. Kannan, S.; Rebelo, A.; Lemos, A.F.; Barba, A.; Ferreira, J.M.F. Synthesis and mechanical behaviour of chlorapatite and chlorapatite/β-TCP composites. *J. Eur. Ceram. Soc.* **2007**, *27*, 2287–2294. [CrossRef]

30. Kannan, S.; Goetz-Neunhoeffer, F.; Neubauer, J.; Ferreira, J.M.F. Ionic substitutions in biphasic hydroxyapatite and β-tricalcium phosphate mixtures: Structural analysis by Rietveld refinement. *J. Am. Ceram. Soc.* **2008**, *91*, 1–12. [CrossRef]

31. Meejoo, S.; Pon-On, W.; Charnchai, S.; Amornsakchai, T. Substitution of iron in preparation of enhanced thermal property and bioactivity of hydroxyapatite. *Adv. Mater. Res.* **2008**, *55–57*, 689–692. [CrossRef]

32. Kannan, S.; Goetz-Neunhoeffer, F.; Neubauer, J.; Ferreira, J.M.F. Synthesis and structure refinement of zinc-doped β-tricalcium phosphate powders. *J. Am. Ceram. Soc.* **2009**, *92*, 1592–1595. [CrossRef]

33. Matsumoto, N.; Yoshida, K.; Hashimoto, K.; Toda, Y. Thermal stability of β-tricalcium phosphate doped with monovalent metal ions. *Mater. Res. Bull.* **2009**, *44*, 1889–1894. [CrossRef]

34. Boanini, E.; Gazzano, M.; Bigi, A. Ionic substitutions in calcium phosphates synthesized at low temperature. *Acta Biomater.* **2010**, *6*, 1882–1894. [CrossRef] [PubMed]

35. Habibovic, P.; Barralet, J.E. Bioinorganics and biomaterials: Bone repair. *Acta Biomater.* **2011**, *7*, 3013–3026. [CrossRef] [PubMed]

36. Mellier, C.; Fayon, F.; Schnitzler, V.; Deniard, P.; Allix, M.; Quillard, S.; Massiot, D.; Bouler, J.M.; Bujoli, B.; Janvier, P. Characterization and properties of novel gallium-doped calcium phosphate ceramics. *Inorg. Chem.* **2011**, *50*, 8252–8260. [CrossRef] [PubMed]

37. Ansar, E.B.; Ajeesh, M.; Yokogawa, Y.; Wunderlich, W.; Varma, H. Synthesis and characterization of iron oxide embedded hydroxyapatite bioceramics. *J. Am. Ceram. Soc.* **2012**, *95*, 2695–2699. [CrossRef]

38. Shepherd, J.H.; Shepherd, D.V.; Best, S.M. Substituted hydroxyapatites for bone repair. *J. Mater. Sci. Mater. Med.* **2012**, *23*, 2335–2347. [CrossRef]

39. Šupová, M. Substituted hydroxyapatites for biomedical applications: A review. *Ceram. Int.* **2015**, *41*, 9203–9231. [CrossRef]

40. Ratnayake, J.T.B.; Mucalo, M.; Dias, G.J. Substituted hydroxyapatites for bone regeneration: A review of current trends. *J. Biomed. Mater. Res. B Appl. Biomater.* **2017**, *105B*, 1285–1299. [CrossRef]

41. Cacciotti, I. Multisubstituted hydroxyapatite powders and coatings: The influence of the codoping on the hydroxyapatite performances. *Int. J. Appl. Ceram. Technol.* **2019**, *16*, 1864–1884. [CrossRef]

42. Williams, D.F. *The Williams Dictionary of Biomaterials*; Liverpool University Press: Liverpool, UK, 1999; p. 368.

43. Williams, D.F. On the nature of biomaterials. *Biomaterials* **2009**, *30*, 5897–5909. [CrossRef]

44. Zhang, X.D.; Williams, D.F. (Eds.) Definitions of biomaterials for the twenty-first century. In Proceedings of the A Consensus Conference Held in Chengdu, Chengdu, China, 11–12 June 2018; Materials Today. Elsevier: Amsterdam, The Netherlands, 2019; p. 290.

45. Williams, D.F. Specifications for innovative, enabling biomaterials based on the principles of biocompatibility mechanisms. *Front. Bioeng. Biotechnol.* **2019**, *7*, 255. [CrossRef]

46. Mann, S. (Ed.) *Biomimetic Materials Chemistry*; Wiley-VCH: Weinheim, Germany, 1996; p. 400.

47. Vallet-Regí, M. Bioceramics: Where do we come from and which are the future expectations. *Key Eng. Mater.* **2008**, *377*, 1–18. [CrossRef]

48. Meyers, M.A.; Chen, P.Y.; Lin, A.Y.M.; Seki, Y. Biological materials: Structure and mechanical properties. *Prog. Mater. Sci.* **2008**, *53*, 1–206. [CrossRef]

49. Available online: https://en.wikipedia.org/wiki/Ceramic (accessed on 24 June 2022).

50. Hench, L.L. Bioceramics: From concept to clinic. *J. Am. Ceram. Soc.* **1991**, *74*, 1487–1510. [CrossRef]

51. Hench, L.L.; Day, D.E.; Höland, W.; Rheinberger, V.M. Glass and medicine. *Int. J. Appl. Glass Sci.* **2010**, *1*, 104–117. [CrossRef]

52. Pinchuk, N.D.; Ivanchenko, L.A. Making calcium phosphate biomaterials. *Powder Metall. Metal Ceram.* **2003**, *42*, 357–371.

53. Heimann, R.B. Materials science of crystalline bioceramics: A review of basic properties and applications. *CMU J.* **2002**, *1*, 23–46.

54. Tomoda, K.; Ariizumi, H.; Nakaji, T.; Makino, K. Hydroxyapatite particles as drug carriers for proteins. *Colloid Surf. B* **2010**, *76*, 226–235. [CrossRef]

55. Zamoume, O.; Thibault, S.; Regnié, G.; Mecherri, M.O.; Fiallo, M.; Sharrock, P. Macroporous calcium phosphate ceramic implants for sustained drug delivery. *Mater. Sci. Eng. C* **2011**, *31*, 1352–1356. [CrossRef]

56. Bose, S.; Tarafder, S. Calcium phosphate ceramic systems in growth factor and drug delivery for bone tissue engineering: A review. *Acta Biomater.* **2012**, *8*, 1401–1421. [CrossRef]

57. Arcos, D.; Vallet-Regí, M. Bioceramics for drug delivery. *Acta Mater.* **2013**, *61*, 890–911. [CrossRef]

58. Ducheyne, P.; Qiu, Q. Bioactive ceramics: The effect of surface reactivity on bone formation and bone cell function. *Biomaterials* **1999**, *20*, 2287–2303. [CrossRef]

59. Dorozhkin, S.V. Calcium orthophosphates and human beings. A historical perspective from the 1770s until 1940. *Biomatter* **2012**, *2*, 53–70. [CrossRef]

60. Dorozhkin, S.V. A detailed history of calcium orthophosphates from 1770-s till 1950. *Mater. Sci. Eng. C* **2013**, *33*, 3085–3110. [CrossRef] [PubMed]

61. De Groot, K. (Ed.) *Bioceramics of Calcium Phosphate*; CRC Press; Taylor & Francis: Boca Raton, FL, USA, 1983; p. 146.

62. Vallet-Regí, M.; González-Calbet, J.M. Calcium phosphates as substitution of bone tissues. *Progr. Solid State Chem.* **2004**, *32*, 1–31. [CrossRef]

63. Layrolle, P.; Ito, A.; Tateishi, T. Sol-gel synthesis of amorphous calcium phosphate and sintering into microporous hydroxyapatite bioceramics. *J. Am. Ceram. Soc.* **1998**, *81*, 1421–1428. [CrossRef]

64. Engin, N.O.; Tas, A.C. Manufacture of macroporous calcium hydroxyapatite bioceramics. *J. Eur. Ceram. Soc.* **1999**, *19*, 2569–2572. [CrossRef]

65. Ahn, E.S.; Gleason, N.J.; Nakahira, A.; Ying, J.Y. Nanostructure processing of hydroxyapatite-based bioceramics. *Nano Lett.* **2001**, *1*, 149–153. [CrossRef]

66. Khalil, K.A.; Kim, S.W.; Dharmaraj, N.; Kim, K.W.; Kim, H.Y. Novel mechanism to improve toughness of the hydroxyapatite bioceramics using high-frequency induction heat sintering. *J. Mater. Process. Technol.* **2007**, *187–188*, 417–420. [CrossRef]

67. Laasri, S.; Taha, M.; Laghzizil, A.; Hlil, E.K.; Chevalier, J. The affect of densification and dehydroxylation on the mechanical properties of stoichiometric hydroxyapatite bioceramics. *Mater. Res. Bull.* **2010**, *45*, 1433–1437. [CrossRef]

68. Kitamura, M.; Ohtsuki, C.; Ogata, S.; Kamitakahara, M.; Tanihara, M. Microstructure and bioresorbable properties of α-TCP ceramic porous body fabricated by direct casting method. *Mater. Trans.* **2004**, *45*, 983–988. [CrossRef]

69. Kawagoe, D.; Ioku, K.; Fujimori, H.; Goto, S. Transparent β-tricalcium phosphate ceramics prepared by spark plasma sintering. *J. Ceram. Soc. Jpn.* **2004**, *112*, 462–463. [CrossRef]

70. Wang, C.X.; Zhou, X.; Wang, M. Influence of sintering temperatures on hardness and Young's modulus of tricalcium phosphate bioceramic by nanoindentation technique. *Mater. Character.* **2004**, *52*, 301–307. [CrossRef]

71. Ioku, K.; Kawachi, G.; Nakahara, K.; Ishida, E.H.; Minagi, H.; Okuda, T.; Yonezawa, I.; Kurosawa, H.; Ikeda, T. Porous granules of β-tricalcium phosphate composed of rod-shaped particles. *Key Eng. Mater.* **2006**, *309–311*, 1059–1062. [CrossRef]

72. Kamitakahara, M.; Ohtsuki, C.; Miyazaki, T. Review paper: Behavior of ceramic biomaterials derived from tricalcium phosphate in physiological condition. *J. Biomater. Appl.* **2008**, *23*, 197–212. [CrossRef]

73. Vorndran, E.; Klarner, M.; Klammert, U.; Grover, L.M.; Patel, S.; Barralet, J.E.; Gbureck, U. 3D powder printing of β-tricalcium phosphate ceramics using different strategies. *Adv. Eng. Mater.* **2008**, *10*, B67–B71. [CrossRef]

74. Descamps, M.; Duhoo, T.; Monchau, F.; Lu, J.; Hardouin, P.; Hornez, J.C.; Leriche, A. Manufacture of macroporous β-tricalcium phosphate bioceramics. *J. Eur. Ceram. Soc.* **2008**, *28*, 149–157. [CrossRef]

75. Liu, Y.; Kim, J.H.; Young, D.; Kim, S.; Nishimoto, S.K.; Yang, Y. Novel template-casting technique for fabricating β-tricalcium phosphate scaffolds with high interconnectivity and mechanical strength and in vitro cell responses. *J. Biomed. Mater. Res. A* **2010**, *92A*, 997–1006.

76. Carrodeguas, R.G.; de Aza, S. α-tricalcium phosphate: Synthesis, properties and biomedical applications. *Acta Biomater.* **2011**, *7*, 3536–3546. [CrossRef]

77. Kim, I.Y.; Wen, J.; Ohtsuki, C. Fabrication of α-tricalcium phosphate ceramics through two-step sintering. *Key Eng. Mater.* **2015**, *631*, 78–82. [CrossRef]

78. Bohner, M.; Santoni, B.L.G.; Döbelin, N. β-tricalcium phosphate for bone substitution: Synthesis and properties. *Acta Biomater.* **2020**, *113*, 23–41. [CrossRef]

79. Dorozhkin, S.V. Multiphasic calcium orthophosphate ($CaPO_4$) bioceramics and their biomedical applications. *Ceram. Int.* **2016**, *42*, 6529–6554. [CrossRef]

80. LeGeros, R.Z.; Lin, S.; Rohanizadeh, R.; Mijares, D.; LeGeros, J.P. Biphasic calcium phosphate bioceramics: Preparation, properties and applications. *J. Mater. Sci. Mater. Med.* **2003**, *14*, 201–209. [CrossRef]

81. Daculsi, G.; Laboux, O.; Malard, O.; Weiss, P. Current state of the art of biphasic calcium phosphate bioceramics. *J. Mater. Sci. Mater. Med.* **2003**, *14*, 195–200. [CrossRef]

82. Dorozhkina, E.I.; Dorozhkin, S.V. Mechanism of the solid-state transformation of a calcium-deficient hydroxyapatite (CDHA) into biphasic calcium phosphate (BCP) at elevated temperatures. *Chem. Mater.* **2002**, *14*, 4267–4272. [CrossRef]

83. Daculsi, G. Biphasic calcium phosphate granules concept for injectable and mouldable bone substitute. *Adv. Sci. Technol.* **2006**, *49*, 9–13.

84. Lecomte, A.; Gautier, H.; Bouler, J.M.; Gouyette, A.; Pegon, Y.; Daculsi, G.; Merle, C. Biphasic calcium phosphate: A comparative study of interconnected porosity in two ceramics. *J. Biomed. Mater. Res. B Appl. Biomater.* **2008**, *84B*, 1–6. [CrossRef]

85. Daculsi, G.; Baroth, S.; LeGeros, R.Z. 20 years of biphasic calcium phosphate bioceramics development and applications. *Ceram. Eng. Sci. Proc.* **2010**, *30*, 45–58.

86. Lukić, M.; Stojanović, Z.; Škapin, S.D.; Maček-Kržmanc, M.; Mitrić, M.; Marković, S.; Uskoković, D. Dense fine-grained biphasic calcium phosphate (BCP) bioceramics designed by two-step sintering. *J. Eur. Ceram. Soc.* **2011**, *31*, 19–27. [CrossRef]

87. Descamps, M.; Boilet, L.; Moreau, G.; Tricoteaux, A.; Lu, J.; Leriche, A.; Lardot, V.; Cambier, F. Processing and properties of biphasic calcium phosphates bioceramics obtained by pressureless sintering and hot isostatic pressing. *J. Eur. Ceram. Soc.* **2013**, *33*, 1263–1270. [CrossRef]

88. Owen, R.G.; Dard, M.; Larjava, H. Hydoxyapatite/beta-tricalcium phosphate biphasic ceramics as regenerative material for the repair of complex bone defects. *J. Biomed. Mater. Res. B Appl. Biomater.* **2018**, *106B*, 2493–2512.

89. Li, Y.; Kong, F.; Weng, W. Preparation and characterization of novel biphasic calcium phosphate powders (α-TCP/HA) derived from carbonated amorphous calcium phosphates. *J. Biomed. Mater. Res. B Appl. Biomater.* **2009**, *89B*, 508–517. [CrossRef]

90. Sureshbabu, S.; Komath, M.; Varma, H.K. In situ formation of hydroxyapatite –alpha tricalcium phosphate biphasic ceramics with higher strength and bioactivity. *J. Am. Ceram. Soc.* **2012**, *95*, 915–924.

91. Radovanović, Ž.; Jokić, B.; Veljović, D.; Dimitrijević, S.; Kojić, V.; Petrović, R.; Janaćković, D. Antimicrobial activity and biocompatibility of Ag$^+$- and Cu^{2+}-doped biphasic hydroxyapatite/α-tricalcium phosphate obtained from hydrothermally synthesized Ag$^+$- and Cu^{2+}-doped hydroxyapatite. *Appl. Surf. Sci.* **2014**, *307*, 513–519.

92. Oishi, M.; Ohtsuki, C.; Kitamura, M.; Kamitakahara, M.; Ogata, S.; Miyazaki, T.; Tanihara, M. Fabrication and chemical durability of porous bodies consisting of biphasic tricalcium phosphates. *Phosphorus Res. Bull.* **2004**, *17*, 95–100. [CrossRef]

93. Kamitakahara, M.; Ohtsuki, C.; Oishi, M.; Ogata, S.; Miyazaki, T.; Tanihara, M. Preparation of porous biphasic tricalcium phosphate and its in vivo behavior. *Key Eng. Mater.* **2005**, *59*, 281–284. [CrossRef]

94. Wang, R.; Weng, W.; Deng, X.; Cheng, K.; Liu, X.; Du, P.; Shen, G.; Han, G. Dissolution behavior of submicron biphasic tricalcium phosphate powders. *Key Eng. Mater.* **2006**, *309–311*, 223–226. [CrossRef]

95. Li, Y.; Weng, W.; Tam, K.C. Novel highly biodegradable biphasic tricalcium phosphates composed of α-tricalcium phosphate and β-tricalcium phosphate. *Acta Biomater.* **2007**, *3*, 251–254. [CrossRef]

96. Zou, C.; Cheng, K.; Weng, W.; Song, C.; Du, P.; Shen, G.; Han, G. Characterization and dissolution–reprecipitation behavior of biphasic tricalcium phosphate powders. *J. Alloys Compd.* **2011**, *509*, 6852–6858. [CrossRef]

97. Xie, L.; Yu, H.; Deng, Y.; Yang, W.; Liao, L.; Long, Q. Preparation and in vitro degradation study of the porous dual alpha/beta-tricalcium phosphate bioceramics. *Mater. Res. Inn.* **2016**, *20*, 530–537. [CrossRef]

98. Albuquerque, J.S.V.; Nogueira, R.E.F.Q.; da Silva, T.D.P.; Lima, D.O.; da Silva, M.H.P. Porous triphasic calcium phosphate bioceramics. *Key Eng. Mater.* **2004**, *254–256*, 1021–1024. [CrossRef]

99. Mendonça, F.; Lourom, L.H.L.; de Campos, J.B.; da Silva, M.H.P. Porous biphasic and triphasic bioceramics scaffolds produced by gelcasting. *Key Eng. Mater.* **2008**, *361–363*, 27–30.

100. Vani, R.; Girija, E.K.; Elayaraja, K.; Parthiban, P.S.; Kesavamoorthy, R.; Narayana Kalkura, S. Hydrothermal synthesis of porous triphasic hydroxyapatite/(α and β) tricalcium phosphate. *J. Mater. Sci. Mater. Med.* **2009**, *20* (Suppl. 1), S43–S48. [CrossRef]

101. Ahn, M.K.; Moon, Y.W.; Koh, Y.H.; Kim, H.E. Production of highly porous triphasic calcium phosphate scaffolds with excellent in vitro bioactivity using vacuum-assisted foaming of ceramic suspension (VFC) technique. *Ceram. Int.* **2013**, *39*, 5879–5885. [CrossRef]

102. Dorozhkin, S.V. Self-setting calcium orthophosphate (CaPO$_4$) formulations and their biomedical applications. *Adv. Nano-Bio. Mater. Dev.* **2019**, *3*, 321–421.

103. Tamimi, F.; Sheikh, Z.; Barralet, J. Dicalcium phosphate cements: Brushite and monetite. *Acta Biomater.* **2012**, *8*, 474–487. [CrossRef]

104. Drouet, C.; Largeot, C.; Raimbeaux, G.; Estournès, C.; Dechambre, G.; Combes, C.; Rey, C. Bioceramics: Spark plasma sintering (SPS) of calcium phosphates. *Adv. Sci. Technol.* **2006**, *49*, 45–50.

105. Ishihara, S.; Matsumoto, T.; Onoki, T.; Sohmura, T.; Nakahira, A. New concept bioceramics composed of octacalcium phosphate (OCP) and dicarboxylic acid-intercalated OCP via hydrothermal hot-pressing. *Mater. Sci. Eng. C* **2009**, *29*, 1885–1888. [CrossRef]

106. Barinov, S.M.; Komlev, V.S. Osteoinductive ceramic materials for bone tissue restoration: Octacalcium phosphate (review). *Inorg. Mater. Appl. Res.* **2010**, *1*, 175–181. [CrossRef]

107. Moseke, C.; Gbureck, U. Tetracalcium phosphate: Synthesis, properties and biomedical applications. *Acta Biomater.* **2010**, *6*, 3815–3823. [CrossRef]

108. Morimoto, S.; Anada, T.; Honda, Y.; Suzuki, O. Comparative study on in vitro biocompatibility of synthetic octacalcium phosphate and calcium phosphate ceramics used clinically. *Biomed. Mater.* **2012**, *7*, 045020. [CrossRef]

109. Tamimi, F.; Nihouannen, D.L.; Eimar, H.; Sheikh, Z.; Komarova, S.; Barralet, J. The effect of autoclaving on the physical and biological properties of dicalcium phosphate dihydrate bioceramics: Brushite vs. monetite. *Acta Biomater.* **2012**, *8*, 3161–3169. [CrossRef]

110. Suzuki, O. Octacalcium phosphate (OCP)-based bone substitute materials. *Jpn. Dent. Sci. Rev.* **2013**, *49*, 58–71. [CrossRef]

111. Komlev, V.S.; Barinov, S.M.; Bozo, I.I.; Deev, R.V.; Eremin, I.I.; Fedotov, A.Y.; Gurin, A.N.; Khromova, N.V.; Kopnin, P.B.; Kuvshinova, E.A.; et al. Bioceramics composed of octacalcium phosphate demonstrate enhanced biological behavior. *ACS Appl. Mater. Interf.* **2014**, *6*, 16610–16620. [CrossRef]

112. Zhou, H.; Yang, L.; Gbureck, U.; Bhaduri, S.B.; Sikder, P. Monetite, an important calcium phosphate compound–its synthesis, properties and applications in orthopedics. *Acta Biomater.* **2021**, *127*, 41–55. [CrossRef] [PubMed]
113. LeGeros, R.Z. Calcium phosphates in oral biology and medicine. In *Monographs in Oral Science*; Karger: Basel, Switzerland, 1991; Volume 15, p. 201.
114. Narasaraju, T.S.B.; Phebe, D.E. Some physico-chemical aspects of hydroxylapatite. *J. Mater. Sci.* **1996**, *31*, 1–21. [CrossRef]
115. Elliott, J.C. Structure and chemistry of the apatites and other calcium orthophosphates. In *Studies in Inorganic Chemistry*; Elsevier: Amsterdam, The Netherlands, 1994; Volume 18, p. 389.
116. Brown, P.W.; Constantz, B. (Eds.) *Hydroxyapatite and Related Materials*; CRC Press: Boca Raton, FL, USA, 1994; p. 343.
117. Amjad, Z. (Ed.) *Calcium Phosphates in Biological and Industrial Systems*; Kluwer Academic Publishers: Boston, MA, USA, 1997; p. 529.
118. Da Silva, R.V.; Bertran, C.A.; Kawachi, E.Y.; Camilli, J.A. Repair of cranial bone defects with calcium phosphate ceramic implant or autogenous bone graft. *J. Craniofac. Surg.* **2007**, *18*, 281–286. [CrossRef] [PubMed]
119. Okanoue, Y.; Ikeuchi, M.; Takemasa, R.; Tani, T.; Matsumoto, T.; Sakamoto, M.; Nakasu, M. Comparison of in vivo bioactivity and compressive strength of a novel superporous hydroxyapatite with beta-tricalcium phosphates. *Arch. Orthop. Trauma Surg.* **2012**, *132*, 1603–1610. [CrossRef]
120. Draenert, M.; Draenert, A.; Draenert, K. Osseointegration of hydroxyapatite and remodeling-resorption of tricalciumphosphate ceramics. *Microsc. Res. Tech.* **2013**, *76*, 370–380. [CrossRef]
121. Okuda, T.; Ioku, K.; Yonezawa, I.; Minagi, H.; Gonda, Y.; Kawachi, G.; Kamitakahara, M.; Shibata, Y.; Murayama, H.; Kurosawa, H.; et al. The slow resorption with replacement by bone of a hydrothermally synthesized pure calcium-deficient hydroxyapatite. *Biomaterials* **2008**, *29*, 2719–2728. [CrossRef]
122. Daculsi, G.; Bouler, J.M.; LeGeros, R.Z. Adaptive crystal formation in normal and pathological calcifications in synthetic calcium phosphate and related biomaterials. *Int. Rev. Cytol.* **1997**, *172*, 129–191.
123. Zhu, X.D.; Zhang, H.J.; Fan, H.S.; Li, W.; Zhang, X.D. Effect of phase composition and microstructure of calcium phosphate ceramic particles on protein adsorption. *Acta Biomater.* **2010**, *6*, 1536–1541. [CrossRef]
124. Bohner, M. Calcium orthophosphates in medicine: From ceramics to calcium phosphate cements. *Injury* **2000**, *31* (Suppl. 4), D37–D47. [CrossRef]
125. Ahato, I. Reverse engineering the ceramic art of algae. *Science* **1999**, *286*, 1059–1061.
126. Popişter, F.; Popescu, D.; Hurgoiu, D. A new method for using reverse engineering in case of ceramic tiles. Qual. *Access Success* **2012**, *13* (Suppl. 5), 409–412.
127. Yang, S.; Leong, K.F.; Du, Z.; Chua, C.K. The design of scaffolds for use in tissue engineering. Part II. Rapid prototyping techniques. *Tissue Eng.* **2002**, *8*, 1–11. [CrossRef]
128. Yeong, W.Y.; Chua, C.K.; Leong, K.F.; Chandrasekaran, M. Rapid prototyping in tissue engineering: Challenges and potential. *Trends Biotechnol.* **2004**, *22*, 643–652. [CrossRef]
129. Ortona, A.; D'Angelo, C.; Gianella, S.; Gaia, D. Cellular ceramics produced by rapid prototyping and replication. *Mater. Lett.* **2012**, *80*, 95–98. [CrossRef]
130. Eufinger, H.; Wehmöller, M.; Machtens, E.; Heuser, L.; Harders, A.; Kruse, D. Reconstruction of craniofacial bone defects with individual alloplastic implants based on CAD/CAM-manipulated CT-data. *J. Cranio Maxillofac. Surg.* **1995**, *23*, 175–181. [CrossRef]
131. Klein, M.; Glatzer, C. Individual CAD/CAM fabricated glass-bioceramic implants in reconstructive surgery of the bony orbital floor. *Plastic Reconstruct. Surg.* **2006**, *117*, 565–570. [CrossRef]
132. Yin, L.; Song, X.F.; Song, Y.L.; Huang, T.; Li, J. An overview of in vitro abrasive finishing & CAD/CAM of bioceramics in restorative dentistry. *Int. J. Machine Tools Manufact.* **2006**, *46*, 1013–1026.
133. Li, J.; Hsu, Y.; Luo, E.; Khadka, A.; Hu, J. Computer-aided design and manufacturing and rapid prototyped nanoscale hydroxyapatite/polyamide (n-HA/PA) construction for condylar defect caused by mandibular angle ostectomy. *Aesthetic Plast. Surg.* **2011**, *35*, 636–640. [CrossRef]
134. Ciocca, L.; Donati, D.; Fantini, M.; Landi, E.; Piattelli, A.; Iezzi, G.; Tampieri, A.; Spadari, A.; Romagnoli, N.; Scotti, R. CAD-CAM-generated hydroxyapatite scaffold to replace the mandibular condyle in sheep: Preliminary results. *J. Biomater. Appl.* **2013**, *28*, 207–218. [CrossRef]
135. Janek, M.; Žilinská, V.; Kovár, V.; Hajdúchová, Z.; Tomanová, K.; Peciar, P.; Veteška, P.; Gabošová, T.; Fialka, R.; Feranc, J.; et al. Mechanical testing of hydroxyapatite filaments for tissue scaffolds preparation by fused deposition of ceramics. *J. Eur. Ceram. Soc.* **2020**, *40*, 4932–4938. [CrossRef]
136. Esslinger, S.; Grebhardt, A.; Jaeger, J.; Kern, F.; Killinger, A.; Bonten, C.; Gadow, R. Additive manufacturing of β-tricalcium phosphate components via fused deposition of ceramics (FDC). *Materials* **2021**, *14*, 156. [CrossRef] [PubMed]
137. Tan, K.H.; Chua, C.K.; Leong, K.F.; Cheah, C.M.; Cheang, P.; Abu Bakar, M.S.; Cha, S.W. Scaffold development using selective laser sintering of polyetheretherketone-hydroxyapatite biocomposite blends. *Biomaterials* **2003**, *24*, 3115–3123. [CrossRef]
138. Wiria, F.E.; Leong, K.F.; Chua, C.K.; Liu, Y. Poly-ε-caprolactone/hydroxyapatite for tissue engineering scaffold fabrication via selective laser sintering. *Acta Biomater.* **2007**, *3*, 1–12. [CrossRef]
139. Shuai, C.J.; Li, P.J.; Feng, P.; Lu, H.B.; Peng, S.P.; Liu, J.L. Analysis of transient temperature distribution during the selective laser sintering of β-tricalcium phosphate. *Laser Eng.* **2013**, *26*, 71–80.

140. Shuai, C.; Zhuang, J.; Hu, H.; Peng, S.; Liu, D.; Liu, J. In vitro bioactivity and degradability of β-tricalcium phosphate porous scaffold fabricated via selective laser sintering. *Biotechnol. Appl. Biochem.* **2013**, *60*, 266–273. [CrossRef] [PubMed]
141. Qin, T.; Li, X.; Long, H.; Bin, S.; Xu, Y. Bioactive tetracalcium phosphate scaffolds fabricated by selective laser sintering for bone regeneration applications. *Materials* **2020**, *13*, 2268. [CrossRef]
142. Bulina, N.V.; Titkov, A.I.; Baev, S.G.; Makarova, S.V.; Khusnutdinov, V.R.; Bessmeltsev, V.P.; Lyakhov, N.Z. Laser sintering of hydroxyapatite for potential fabrication of bioceramic scaffolds. *Mater. Today Proc.* **2021**, *37*, 4022–4026. [CrossRef]
143. Lusquiños, F.; Pou, J.; Boutinguiza, M.; Quintero, F.; Soto, R.; León, B.; Pérez-Amor, M. Main characteristics of calcium phosphate coatings obtained by laser cladding. *Appl. Surf. Sci.* **2005**, *247*, 486–492. [CrossRef]
144. Wang, D.G.; Chen, C.Z.; Ma, J.; Zhang, G. In situ synthesis of hydroxyapatite coating by laser cladding. *Colloid Surf. B* **2008**, *66*, 155–162. [CrossRef]
145. Comesaña, R.; Lusquiños, F.; del Val, J.; Malot, T.; López-Álvarez, M.; Riveiro, A.; Quintero, F.; Boutinguiza, M.; Aubry, P.; de Carlos, A.; et al. Calcium phosphate grafts produced by rapid prototyping based on laser cladding. *J. Eur. Ceram. Soc.* **2011**, *31*, 29–41. [CrossRef]
146. Jing, Z.; Cao, Q.; Jun, H. Corrosion, wear and biocompatibility of hydroxyapatite bio-functionally graded coating on titanium alloy surface prepared by laser cladding. *Ceram. Int.* **2021**, *47*, 24641–24651. [CrossRef]
147. Leukers, B.; Gülkan, H.; Irsen, S.H.; Milz, S.; Tille, C.; Schieker, M.; Seitz, H. Hydroxyapatite scaffolds for bone tissue engineering made by 3D printing. *J. Mater. Sci. Mater. Med.* **2005**, *16*, 1121–1124. [CrossRef]
148. Gbureck, U.; Hölzel, T.; Klammert, U.; Würzler, K.; Müller, F.A.; Barralet, J.E. Resorbable dicalcium phosphate bone substites prepared by 3D powder printing. *Adv. Funct. Mater.* **2007**, *17*, 3940–3945. [CrossRef]
149. Seitz, H.; Deisinger, U.; Leukers, B.; Detsch, R.; Ziegler, G. Different calcium phosphate granules for 3-D printing of bone tissue engineering scaffolds. *Adv. Eng. Mater.* **2009**, *11*, B41–B46. [CrossRef]
150. Butscher, A.; Bohner, M.; Roth, C.; Ernstberger, A.; Heuberger, R.; Doebelin, N.; von Rohr, R.P.; Müller, R. Printability of calcium phosphate powders for three-dimensional printing of tissue engineering scaffolds. *Acta Biomater.* **2012**, *8*, 373–385. [CrossRef]
151. Akkineni, A.R.; Luo, Y.; Schumacher, M.; Nies, B.; Lode, A.; Gelinsky, M. 3D plotting of growth factor loaded calcium phosphate cement scaffolds. *Acta Biomater.* **2015**, *27*, 264–274. [CrossRef]
152. Trombetta, R.; Inzana, J.A.; Schwarz, E.M.; Kates, S.L.; Awad, H.A. 3D printing of calcium phosphate ceramics for bone tissue engineering and drug delivery. *Ann. Biomed. Eng.* **2017**, *45*, 23–44. [CrossRef]
153. Ma, H.; Feng, C.; Chang, J.; Wu, C. 3D-printed bioceramic scaffolds: From bone tissue engineering to tumor therapy. *Acta Biomater.* **2018**, *79*, 37–59. [CrossRef]
154. Miranda, P.; Pajares, A.; Saiz, E.; Tomsia, A.P.; Guiberteau, F. Mechanical behaviour under uniaxial compression of robocast calcium phosphate scaffolds. *Eur. Cells Mater.* **2007**, *14* (Suppl. 1), 79.
155. Maazouz, Y.; Montufar, E.B.; Guillem-Marti, J.; Fleps, I.; Öhman, C.; Persson, C.; Ginebra, M.P. Robocasting of biomimetic hydroxyapatite scaffolds using self-setting inks. *J. Mater. Chem. B* **2014**, *2*, 5378–5386. [CrossRef]
156. Liu, Q.; Lu, W.F.; Zhai, W. Toward stronger robocast calcium phosphate scaffolds for bone tissue engineering: A mini-review and meta-analysis. *Biomater. Adv.* **2022**, *134*, 112578. [CrossRef]
157. Porter, N.L.; Pilliar, R.M.; Grynpas, M.D. Fabrication of porous calcium polyphosphate implants by solid freeform fabrication: A study of processing parameters and in vitro degradation characteristics. *J. Biomed. Mater. Res.* **2001**, *56*, 504–515. [CrossRef]
158. Leong, K.F.; Cheah, C.M.; Chua, C.K. Solid freeform fabrication of three-dimensional scaffolds for engineering replacement tissues and organs. *Biomaterials* **2003**, *24*, 2363–2378. [CrossRef]
159. Shanjani, Y.; de Croos, J.N.A.; Pilliar, R.M.; Kandel, R.A.; Toyserkani, E. Solid freeform fabrication and characterization of porous calcium polyphosphate structures for tissue engineering purposes. *J. Biomed. Mater. Res. B Appl. Biomater.* **2010**, *93B*, 510–519. [CrossRef] [PubMed]
160. Kim, J.; Lim, D.; Kim, Y.H.; Koh, Y.H.; Lee, M.H.; Han, I.; Lee, S.J.; Yoo, O.S.; Kim, H.S.; Park, J.C. A comparative study of the physical and mechanical properties of porous hydroxyapatite scaffolds fabricated by solid freeform fabrication and polymer replication method. *Int. J. Precision Eng. Manuf.* **2011**, *12*, 695–701. [CrossRef]
161. Shanjani, Y.; Hu, Y.; Toyserkani, E.; Grynpas, M.; Kandel, R.A.; Pilliar, R.M. Solid freeform fabrication of porous calcium polyphosphate structures for bone substitute applications: In vivo studies. *J. Biomed. Mater. Res. B Appl. Biomater.* **2013**, *101B*, 972–980. [CrossRef] [PubMed]
162. Kwon, B.J.; Kim, J.; Kim, Y.H.; Lee, M.H.; Baek, H.S.; Lee, D.H.; Kim, H.L.; Seo, H.J.; Lee, M.H.; Kwon, S.Y.; et al. Biological advantages of porous hydroxyapatite scaffold made by solid freeform fabrication for bone tissue regeneration. *Artif. Organs* **2013**, *37*, 663–670. [CrossRef] [PubMed]
163. Li, X.; Li, D.; Lu, B.; Wang, C. Fabrication of bioceramic scaffolds with pre-designed internal architecture by gel casting and indirect stereolithography techniques. *J. Porous Mater.* **2008**, *15*, 667–671. [CrossRef]
164. Ronca, A.; Ambrosio, L.; Grijpma, D.W. Preparation of designed poly(D,L-lactide)/nanosized hydroxyapatite composite structures by stereolithography. *Acta Biomater.* **2013**, *9*, 5989–5996. [CrossRef]
165. Wei, Y.; Zhao, D.; Cao, Q.; Wang, J.; Wu, Y.; Yuan, B.; Li, X.; Chen, X.; Zhou, Y.; Yang, X.; et al. Stereolithography-based additive manufacturing of high-performance osteoinductive calcium phosphate ceramics by a digital light-processing system. *ACS Biomater. Sci. Eng.* **2020**, *6*, 1787–1797. [CrossRef]

166. Ullah, I.; Cao, L.; Cui, W.; Xu, Q.; Yang, R.; Tang, K.L.; Zhang, X. Stereolithography printing of bone scaffolds using biofunctional calcium phosphate nanoparticles. *J. Mater. Sci. Technol.* **2021**, *88*, 99–108.
167. Paredes, C.; Martínez-Vázquez, F.J.; Elsayed, H.; Colombo, P.; Pajares, A.; Miranda, P. Evaluation of direct light processing for the fabrication of bioactive ceramic scaffolds: Effect of pore/strut size on manufacturability and mechanical performance. *J. Eur. Ceram. Soc.* **2021**, *41*, 892–900. [CrossRef]
168. Gladman, A.S.; Matsumoto, E.A.; Nuzzo, R.G.; Mahadevan, L.; Lewis, J.A. Biomimetic 4D printing. *Nat. Mater.* **2016**, *15*, 413–418. [CrossRef]
169. Hwangbo, H.; Lee, H.; Roh, E.J.; Kim, W.; Joshi, H.P.; Kwon, S.Y.; Choi, U.Y.; Han, I.B.; Kim, G.H. Bone tissue engineering via application of a collagen/hydroxyapatite 4D-printed biomimetic scaffold for spinal fusion. *Appl. Phys. Rev.* **2021**, *8*, 021403. [CrossRef]
170. Haleem, A.; Javaid, M.; Vaishya, R. 5D printing and its expected applications in orthopaedics. *J. Clin. Orthop. Trauma.* **2019**, *10*, 809–810. [CrossRef]
171. Du, X.; Fu, S.; Zhu, Y. 3D printing of ceramic-based scaffolds for bone tissue engineering: An overview. *J. Mater. Chem. B* **2018**, *6*, 4397–4412. [CrossRef]
172. Kumar, A.; Kargozar, S.; Baino, F.; Han, S.S. Additive manufacturing methods for producing hydroxyapatite and hydroxyapatite-based composite scaffolds: A review. *Front. Mater.* **2019**, *6*, 313. [CrossRef]
173. Chen, Z.; Li, Z.; Li, J.; Liu, C.; Lao, C.; Lao, C.; Fu, Y.; Liu, C.; Li, Y.; Wang, P.; et al. 3D printing of ceramics: A review. *J. Eur. Ceram. Soc.* **2019**, *39*, 661–687. [CrossRef]
174. Qu, H. Additive manufacturing for bone tissue engineering scaffolds. *Mater. Today Commun.* **2020**, *24*, 101024. [CrossRef]
175. Saber-Samandari, S.; Gross, K.A. The use of thermal printing to control the properties of calcium phosphate deposits. *Biomaterials* **2010**, *31*, 6386–6393. [CrossRef]
176. De Meira, C.R.; Gomes, D.T.; Braga, F.J.C.; de Moraes Purquerio, B.; Fortulan, C.A. Direct manufacture of hydroxyapatite scaffolds using blue laser. *Mater. Sci. Forum* **2015**, *805*, 128–133. [CrossRef]
177. Narayan, R.J.; Jin, C.; Doraiswamy, A.; Mihailescu, I.N.; Jelinek, M.; Ovsianikov, A.; Chichkov, B.; Chrisey, D.B. Laser processing of advanced bioceramics. *Adv. Eng. Mater.* **2005**, *7*, 1083–1098. [CrossRef]
178. Nather, A. (Ed.) *Bone Grafts and Bone Substitutes: Basic Science and Clinical Applications*; World Scientific: Singapore, 2005; p. 592.
179. Bártolo, P.; Bidanda, B. (Eds.) *Bio-Materials and Prototyping Applications in Medicine*; Springer: New York, NY, USA, 2008; p. 216.
180. Kokubo, T. (Ed.) *Bioceramics and Their Clinical Applications*; Woodhead Publishing: Cambridge, UK, 2008; p. 784.
181. Narayan, R. (Ed.) *Biomedical Materials*; Springer: New York, NY, USA, 2009; p. 566.
182. Park, J. *Bioceramics: Properties, Characterizations, and Applications*; Springer: New York, NY, USA, 2008; p. 364.
183. Rodríguez-Lorenzo, L.M.; Vallet-Regí, M.; Ferreira, J.M.F. Fabrication of hydroxyapatite bodies by uniaxial pressing from a precipitated powder. *Biomaterials* **2001**, *22*, 583–588. [CrossRef]
184. Indra, A.; Putra, A.B.; Handra, N.; Fahmi, H.; Nurzal; Asfarizal; Perdana, M.; Anrinal; Subardi, A.; Affi, J.; et al. Behavior of sintered body properties of hydroxyapatite ceramics: Effect of uniaxial pressure on green body fabrication. *Mater. Today Sustain.* **2022**, *17*, 100100. [CrossRef]
185. Uematsu, K.; Takagi, M.; Honda, T.; Uchida, N.; Saito, K. Transparent hydroxyapatite prepared by hot isostatic pressing of filter cake. *J. Am. Ceram. Soc.* **1989**, *72*, 1476–1478. [CrossRef]
186. Itoh, H.; Wakisaka, Y.; Ohnuma, Y.; Kuboki, Y. A new porous hydroxyapatite ceramic prepared by cold isostatic pressing and sintering synthesized flaky powder. *Dent. Mater.* **1994**, *13*, 25–35. [CrossRef]
187. Takikawa, K.; Akao, M. Fabrication of transparent hydroxyapatite and application to bone marrow derived cell/hydroxyapatite interaction observation in-vivo. *J. Mater. Sci. Mater. Med.* **1996**, *7*, 439–445. [CrossRef]
188. Gautier, H.; Merle, C.; Auget, J.L.; Daculsi, G. Isostatic compression, a new process for incorporating vancomycin into biphasic calcium phosphate: Comparison with a classical method. *Biomaterials* **2000**, *21*, 243–249. [CrossRef]
189. Tadic, D.; Epple, M. Mechanically stable implants of synthetic bone mineral by cold isostatic pressing. *Biomaterials* **2003**, *24*, 4565–4571. [CrossRef]
190. Onoki, T.; Hashida, T. New method for hydroxyapatite coating of titanium by the hydrothermal hot isostatic pressing technique. *Surf. Coat. Technol.* **2006**, *200*, 6801–6807. [CrossRef]
191. Ehsani, N.; Ruys, A.J.; Sorrell, C.C. Hot isostatic pressing (HIPing) of fecralloy-reinforced hydroxyapatite. *J. Biomimet. Biomater. Tissue Eng.* **2013**, *17*, 87–102. [CrossRef]
192. Irsen, S.H.; Leukers, B.; Höckling, C.; Tille, C.; Seitz, H. Bioceramic granulates for use in 3D printing: Process engineering aspects. *Materwiss. Werksttech.* **2006**, *37*, 533–537. [CrossRef]
193. Hsu, Y.H.; Turner, I.G.; Miles, A.W. Fabrication and mechanical testing of porous calcium phosphate bioceramic granules. *J. Mater. Sci. Mater. Med.* **2007**, *18*, 1931–1937. [CrossRef]
194. Zyman, Z.Z.; Glushko, V.; Dedukh, N.; Malyshkina, S.; Ashukina, N. Porous calcium phosphate ceramic granules and their behaviour in differently loaded areas of skeleton. *J. Mater. Sci. Mater. Med.* **2008**, *19*, 2197–2205. [CrossRef]
195. Viana, M.; Désiré, A.; Chevalier, E.; Champion, E.; Chotard, R.; Chulia, D. Interest of high shear wet granulation to produce drug loaded porous calcium phosphate pellets for bone filling. *Key Eng. Mater.* **2009**, *396–398*, 535–538. [CrossRef]
196. Chevalier, E.; Viana, M.; Cazalbou, S.; Chulia, D. Comparison of low-shear and high-shear granulation processes: Effect on implantable calcium phosphate granule properties. *Drug Dev. Ind. Pharm.* **2009**, *35*, 1255–1263. [CrossRef]

197. 197 Lakevics, V.; Locs, J.; Loca, D.; Stepanova, V.; Berzina-Cimdina, L.; Pelss, J. Bioceramic hydroxyapatite granules for purification of biotechnological products. *Adv. Mater. Res.* **2011**, *284–286*, 1764–1769. [CrossRef]

198. Camargo, N.H.A.; Franczak, P.F.; Gemelli, E.; da Costa, B.D.; de Moraes, A.N. Characterization of three calcium phosphate microporous granulated bioceramics. *Adv. Mater. Res.* **2014**, *936*, 687–694. [CrossRef]

199. Reikerås, O.; Johansson, C.B.; Sundfeldt, M. Bone ingrowths to press-fit and loose-fit implants: Comparisons between titanium and hydroxyapatite. *J. Long-Term Eff. Med. Implant.* **2006**, *16*, 157–164. [CrossRef]

200. Rao, R.R.; Kannan, T.S. Dispersion and slip casting of hydroxyapatite. *J. Am. Ceram. Soc.* **2001**, *84*, 1710–1716. [CrossRef]

201. Sakka, Y.; Takahashi, K.; Matsuda, N.; Suzuki, T.S. Effect of milling treatment on texture development of hydroxyapatite ceramics by slip casting in high magnetic field. *Mater. Trans.* **2007**, *48*, 2861–2866. [CrossRef]

202. Zhang, Y.; Yokogawa, Y.; Feng, X.; Tao, Y.; Li, Y. Preparation and properties of bimodal porous apatite ceramics through slip casting using different hydroxyapatite powders. *Ceram. Int.* **2010**, *36*, 107–113. [CrossRef]

203. Zhang, Y.; Kong, D.; Yokogawa, Y.; Feng, X.; Tao, Y.; Qiu, T. Fabrication of porous hydroxyapatite ceramic scaffolds with high flexural strength through the double slip-casting method using fine powders. *J. Am. Ceram. Soc.* **2012**, *95*, 147–152. [CrossRef]

204. Hagio, T.; Yamauchi, K.; Kohama, T.; Matsuzaki, T.; Iwai, K. Beta tricalcium phosphate ceramics with controlled crystal orientation fabricated by application of external magnetic field during the slip casting process. *Mater. Sci. Eng. C* **2013**, *33*, 2967–2970. [CrossRef] [PubMed]

205. Marçal, R.L.S.B.; da Rocha, D.N.; da Silva, M.H.P. Slip casting used as a forming technique for hydroxyapatite processing. *Key Eng. Mater.* **2017**, *720*, 219–222. [CrossRef]

206. Sepulveda, P.; Ortega, F.S.; Innocentini, M.D.M.; Pandolfelli, V.C. Properties of highly porous hydroxyapatite obtained by the gel casting of foams. *J. Am. Ceram. Soc.* **2000**, *83*, 3021–3024. [CrossRef]

207. Sánchez-Salcedo, S.; Werner, J.; Vallet-Regí, M. Hierarchical pore structure of calcium phosphate scaffolds by a combination of gel-casting and multiple tape-casting methods. *Acta Biomater.* **2008**, *4*, 913–922. [CrossRef]

208. Chen, B.; Zhang, T.; Zhang, J.; Lin, Q.; Jiang, D. Microstructure and mechanical properties of hydroxyapatite obtained by gel-casting process. *Ceram. Int.* **2008**, *34*, 359–364. [CrossRef]

209. Marcassoli, P.; Cabrini, M.; Tirillò, J.; Bartuli, C.; Palmero, P.; Montanaro, L. Mechanical characterization of hydroxiapatite micro/macro-porous ceramics obtained by means of innovative gel-casting process. *Key Eng. Mater.* **2010**, *417–418*, 565–568. [CrossRef]

210. Dash, S.R.; Sarkar, R.; Bhattacharyya, S. Gel casting of hydroxyapatite with naphthalene as pore former. *Ceram. Int.* **2015**, *41*, 3775–3790. [CrossRef]

211. Ramadas, M.; Ferreira, J.M.F.; Ballamurugan, A.M. Fabrication of three dimensional bioactive Sr^{2+} substituted apatite scaffolds by gel-casting technique for hard tissue regeneration. *J. Tissue Eng. Regen. Med.* **2021**, *15*, 577–585. [CrossRef]

212. Fomin, A.S.; Barinov, S.M.; Ievlev, V.M.; Smirnov, V.V.; Mikhailov, B.P.; Belonogov, E.K.; Drozdova, N.A. Nanocrystalline hydroxyapatite ceramics produced by low-temperature sintering after high-pressure treatment. *Dokl. Chem.* **2008**, *418*, 22–25. [CrossRef]

213. Zhang, J.; Yin, H.M.; Hsiao, B.S.; Zhong, G.J.; Li, Z.M. Biodegradable poly(lactic acid)/hydroxyl apatite 3D porous scaffolds using high-pressure molding and salt leaching. *J. Mater. Sci.* **2014**, *49*, 1648–1658. [CrossRef]

214. Zhang, J.; Liu, H.; Ding, J.X.; Wu, J.; Zhuang, X.; Chen, X.; Wang, J.; Yin, J.; Li, Z. High-pressure compression-molded porous resorbable polymer/hydroxyapatite composite scaffold for cranial bone regeneration. *ACS Biomater. Sci. Eng.* **2016**, *2*, 1471–1482. [CrossRef]

215. Kankawa, Y.; Kaneko, Y.; Saitou, K. Injection molding of highly-purified hydroxylapatite and TCP utilizing solid phase reaction method. *J. Ceram. Soc. Jpn.* **1991**, *99*, 438–442. [CrossRef]

216. Cihlář, J.; Trunec, M. Injection moulded hydroxyapatite ceramics. *Biomaterials* **1996**, *17*, 1905–1911. [CrossRef]

217. Jewad, R.; Bentham, C.; Hancock, B.; Bonfield, W.; Best, S.M. Dispersant selection for aqueous medium pressure injection moulding of anhydrous dicalcium phosphate. *J. Eur. Ceram. Soc.* **2008**, *28*, 547–553. [CrossRef]

218. McNamara, S.L.; McCarthy, E.M.; Schmidt, D.F.; Johnston, S.P.; Kaplan, D.L. Rheological characterization, compression, and injection molding of hydroxyapatite-silk fibroin composites. *Biomaterials* **2021**, *269*, 120643. [CrossRef]

219. Kwon, S.H.; Jun, Y.K.; Hong, S.H.; Lee, I.S.; Kim, H.E.; Won, Y.Y. Calcium phosphate bioceramics with various porosities and dissolution rates. *J. Am. Ceram. Soc.* **2002**, *85*, 3129–3131. [CrossRef]

220. Fooki, A.C.B.M.; Aparecida, A.H.; Fideles, T.B.; Costa, R.C.; Fook, M.V.L. Porous hydroxyapatite scaffolds by polymer sponge method. *Key Eng. Mater.* **2009**, *396–398*, 703–706. [CrossRef]

221. Sopyan, I.; Kaur, J. Preparation and characterization of porous hydroxyapatite through polymeric sponge method. *Ceram. Int.* **2009**, *35*, 3161–3168. [CrossRef]

222. Bellucci, D.; Cannillo, V.; Sola, A. Shell scaffolds: A new approach towards high strength bioceramic scaffolds for bone regeneration. *Mater. Lett.* **2010**, *64*, 203–206. [CrossRef]

223. Cunningham, E.; Dunne, N.; Walker, G.; Maggs, C.; Wilcox, R.; Buchanan, F. Hydroxyapatite bone substitutes developed via replication of natural marine sponges. *J. Mater. Sci. Mater. Med.* **2010**, *21*, 2255–2261. [CrossRef]

224. Sung, J.H.; Shin, K.H.; Koh, Y.H.; Choi, W.Y.; Jin, Y.; Kim, H.E. Preparation of the reticulated hydroxyapatite ceramics using carbon-coated polymeric sponge with elongated pores as a novel template. *Ceram. Int.* **2011**, *37*, 2591–2596. [CrossRef]

225. Mishima, F.D.; Louro, L.H.L.; Moura, F.N.; Gobbo, L.A.; da Silva, M.H.P. Hydroxyapatite scaffolds produced by hydrothermal deposition of monetite on polyurethane sponges substrates. *Key Eng. Mater.* **2012**, *493–494*, 820–825. [CrossRef]
226. Hannickel, A.; da Silva, M.H.P. Novel bioceramic scaffolds for regenerative medicine. *Bioceram. Dev. Appl.* **2015**, *5*, 1000082.
227. Das, S.; Kumar, S.; Doloi, B.; Bhattacharyya, B. Experimental studies of ultrasonic machining on hydroxyapatite bio-ceramics. *Int. J. Adv. Manuf. Tech.* **2016**, *86*, 829–839. [CrossRef]
228. Velayudhan, S.; Ramesh, P.; Sunny, M.C.; Varma, H.K. Extrusion of hydroxyapatite to clinically significant shapes. *Mater. Lett.* **2000**, *46*, 142–146. [CrossRef]
229. Yang, H.Y.; Thompson, I.; Yang, S.F.; Chi, X.P.; Evans, J.R.G.; Cook, R.J. Dissolution characteristics of extrusion freeformed hydroxyapatite–tricalcium phosphate scaffolds. *J. Mater. Sci. Mater. Med.* **2008**, *19*, 3345–3353. [CrossRef]
230. Yang, S.; Yang, H.; Chi, X.; Evans, J.R.G.; Thompson, I.; Cook, R.J.; Robinson, P. Rapid prototyping of ceramic lattices for hard tissue scaffolds. *Mater. Des.* **2008**, *29*, 1802–1809. [CrossRef]
231. Yang, H.Y.; Chi, X.P.; Yang, S.; Evans, J.R.G. Mechanical strength of extrusion freeformed calcium phosphate filaments. *J. Mater. Sci. Mater. Med.* **2010**, *21*, 1503–1510. [CrossRef]
232. Cortez, P.P.; Atayde, L.M.; Silva, M.A.; da Silva, P.A.; Fernandes, M.H.; Afonso, A.; Lopes, M.A.; Maurício, A.C.; Santos, J.D. Characterization and preliminary in vivo evaluation of a novel modified hydroxyapatite produced by extrusion and spheronization techniques. *J. Biomed. Mater. Res. B Appl. Biomater.* **2011**, *99B*, 170–179. [CrossRef]
233. Lim, S.; Chun, S.; Yang, D.; Kim, S. Comparison study of porous calcium phosphate blocks prepared by piston and screw type extruders for bone scaffold. *Tissue Eng. Regen. Med.* **2012**, *9*, 51–55. [CrossRef]
234. Blake, D.M.; Tomovic, S.; Jyung, R.W. Extrusion of hydroxyapatite ossicular prosthesis. *Ear Nose Throat J.* **2013**, *92*, 490–494. [CrossRef]
235. Muthutantri, A.I.; Huang, J.; Edirisinghe, M.J.; Bretcanu, O.; Boccaccini, A.R. Dipping and electrospraying for the preparation of hydroxyapatite foams for bone tissue engineering. *Biomed. Mater.* **2008**, *3*, 25009. [CrossRef]
236. Roncari, E.; Galassi, C.; Pinasco, P. Tape casting of porous hydroxyapatite ceramics. *J. Mater. Sci. Lett.* **2000**, *19*, 33–35. [CrossRef]
237. Tian, T.; Jiang, D.; Zhang, J.; Lin, Q. Aqueous tape casting process for hydroxyapatite. *J. Eur. Ceram. Soc.* **2007**, *27*, 2671–2677. [CrossRef]
238. Tanimoto, Y.; Shibata, Y.; Murakami, A.; Miyazaki, T.; Nishiyama, N. Effect of varying HAP/TCP ratios in tape-cast biphasic calcium phosphate ceramics on responcce in vitro. *J. Hard Tiss. Biol.* **2009**, *18*, 71–76. [CrossRef]
239. Tanimoto, Y.; Teshima, M.; Nishiyama, N.; Yamaguchi, M.; Hirayama, S.; Shibata, Y.; Miyazaki, T. Tape-cast and sintered β-tricalcium phosphate laminates for biomedical applications: Effect of milled Al_2O_3 fiber additives on microstructural and mechanical properties. *J. Biomed. Mater. Res. B Appl. Biomater.* **2012**, *100B*, 2261–2268. [CrossRef]
240. Khamkasem, C.; Chaijaruwanich, A. Effect of binder/plasticizer ratios in aqueous-based tape casting on mechanical properties of bovine hydroxyapatite tape. *Ferroelectrics* **2013**, *455*, 129–135. [CrossRef]
241. Suzuki, S.; Itoh, K.; Ohgaki, M.; Ohtani, M.; Ozawa, M. Preparation of sintered filter for ion exchange by a doctor blade method with aqueous slurries of needlelike hydroxyapatite. *Ceram. Int.* **1999**, *25*, 287–291. [CrossRef]
242. Nishikawa, H.; Hatanaka, R.; Kusunoki, M.; Hayami, T.; Hontsu, S. Preparation of freestanding hydroxyapatite membranes excellent biocompatibility and flexibility. *Appl. Phys. Express* **2008**, *1*, 088001. [CrossRef]
243. Padilla, S.; Roman, J.; Vallet-Regí, M. Synthesis of porous hydroxyapatites by combination of gel casting and foams burn out methods. *J. Mater. Sci. Mater. Med.* **2002**, *13*, 1193–1197. [CrossRef]
244. Yang, T.Y.; Lee, J.M.; Yoon, S.Y.; Park, H.C. Hydroxyapatite scaffolds processed using a TBA-based freeze-gel casting/polymer sponge technique. *J. Mater. Sci. Mater. Med.* **2010**, *21*, 1495–1502. [CrossRef] [PubMed]
245. Baradararan, S.; Hamdi, M.; Metselaar, I.H. Biphasic calcium phosphate (BCP) macroporous scaffold with different ratios of HA/β-TCP by combination of gel casting and polymer sponge methods. *Adv. Appl. Ceram.* **2012**, *111*, 367–373. [CrossRef]
246. Inoue, K.; Sassa, K.; Yokogawa, Y.; Sakka, Y.; Okido, M.; Asai, S. Control of crystal orientation of hydroxyapatite by imposition of a high magnetic field. *Mater. Trans.* **2003**, *44*, 1133–1137. [CrossRef]
247. Iwai, K.; Akiyama, J.; Tanase, T.; Asai, S. Alignment of HAp crystal using a sample rotation in a static magnetic field. *Mater. Sci. Forum.* **2007**, *539–543*, 716–719. [CrossRef]
248. Iwai, K.; Akiyama, J.; Asai, S. Structure control of hydroxyapatite using a magnetic field. *Mater. Sci. Forum.* **2007**, *561–565*, 1565–1568. [CrossRef]
249. Sakka, Y.; Takahashi, K.; Suzuki, T.S.; Ito, S.; Matsuda, N. Texture development of hydroxyapatite ceramics by colloidal processing in a high magnetic field followed by sintering. *Mater. Sci. Eng. A* **2008**, *475*, 27–33. [CrossRef]
250. Fleck, N.A. On the cold compaction of powders. *J. Mech. Phys. Solids* **1995**, *43*, 1409–1431. [CrossRef]
251. Kang, J.; Hadfield, M. Parameter optimization by Taguchi methods for finishing advanced ceramic balls using a novel eccentric lapping machine. *Proc. Inst. Mech. Eng. B* **2001**, *215*, 69–78. [CrossRef]
252. Kulkarni, S.S.; Yong, Y.; Rys, M.J.; Lei, S. Machining assessment of nano-crystalline hydroxyapatite bio-ceramic. *J. Manuf. Processes* **2013**, *15*, 666–672. [CrossRef]
253. Kurella, A.; Dahotre, N.B. Surface modification for bioimplants: The role of laser surface engineering. *J. Biomater. Appl.* **2005**, *20*, 5–50. [CrossRef]

254. Moriguchi, Y.; Lee, D.S.; Chijimatsu, R.; Thamina, K.; Masuda, K.; Itsuki, D.; Yoshikawa, H.; Hamaguchi, S.; Myoui, A. Impact of non-thermal plasma surface modification on porous calcium hydroxyapatite ceramics for bone regeneration. *PLoS ONE* **2018**, *13*, e0194303.

255. Bertol, L.S.; Schabbach, R.; dos Santos, L.A.L. Different post-processing conditions for 3D bioprinted α-tricalcium phosphate scaffolds. *J. Mater. Sci. Mater. Med.* **2017**, *28*, 168. [CrossRef]

256. Oktar, F.N.; Genc, Y.; Goller, G.; Erkmen, E.Z.; Ozyegin, L.S.; Toykan, D.; Demirkiran, H.; Haybat, H. Sintering of synthetic hydroxyapatite compacts. *Key Eng. Mater.* **2004**, *264–268*, 2087–2090. [CrossRef]

257. Georgiou, G.; Knowles, J.C.; Barralet, J.E. Dynamic shrinkage behavior of hydroxyapatite and glass-reinforced hydroxyapatites. *J. Mater. Sci.* **2004**, *39*, 2205–2208. [CrossRef]

258. Fellah, B.H.; Layrolle, P. Sol-gel synthesis and characterization of macroporous calcium phosphate bioceramics containing microporosity. *Acta Biomater.* **2009**, *5*, 735–742. [CrossRef]

259. Kutty, M.G.; Bhaduri, S.B.; Zhou, H.; Yaghoubi, A. In situ measurement of shrinkage and temperature profile in microwave- and conventionally-sintered hydroxyapatite bioceramic. *Mater. Lett.* **2015**, *161*, 375–378. [CrossRef]

260. Ben Ayed, F.; Bouaziz, J.; Bouzouita, K. Pressureless sintering of fluorapatite under oxygen atmosphere. *J. Eur. Ceram. Soc.* **2000**, *20*, 1069–1076. [CrossRef]

261. He, Z.; Ma, J.; Wang, C. Constitutive modeling of the densification and the grain growth of hydroxyapatite ceramics. *Biomaterials* **2005**, *26*, 1613–1621. [CrossRef] [PubMed]

262. Rahaman, M.N. *Sintering of Ceramics*; CRC Press: Boca Raton, FL, USA, 2007; p. 388.

263. Monroe, E.A.; Votava, W.; Bass, D.B.; McMullen, J. New calcium phosphate ceramic material for bone and tooth implants. *J. Dent. Res.* **1971**, *50*, 860–861. [CrossRef] [PubMed]

264. Landi, E.; Tampieri, A.; Celotti, G.; Sprio, S. Densification behaviour and mechanisms of synthetic hydroxyapatites. *J. Eur. Ceram. Soc.* **2000**, *20*, 2377–2387. [CrossRef]

265. Döbelin, N.; Maazouz, Y.; Heuberger, R.; Bohner, M.; Armstrong, A.A.; Wagoner Johnson, A.J.; Wanner, C. A thermodynamic approach to surface modification of calcium phosphate implants by phosphate evaporation and condensation. *J. Eur. Ceram. Soc.* **2020**, *40*, 6095–6106. [CrossRef]

266. Chen, S.; Wang, W.; Kono, H.; Sassa, K.; Asai, S. Abnormal grain growth of hydroxyapatite ceramic sintered in a high magnetic field. *J. Cryst. Growth* **2010**, *312*, 323–326. [CrossRef]

267. Ruys, A.J.; Wei, M.; Sorrell, C.C.; Dickson, M.R.; Brandwood, A.; Milthorpe, B.K. Sintering effect on the strength of hydroxyapatite. *Biomaterials* **1995**, *16*, 409–415. [CrossRef]

268. Van Landuyt, P.; Li, F.; Keustermans, J.P.; Streydio, J.M.; Delannay, F.; Munting, E. The influence of high sintering temperatures on the mechanical properties of hydroxylapatite. *J. Mater. Sci. Mater. Med.* **1995**, *6*, 8–13. [CrossRef]

269. Pramanik, S.; Agarwal, A.K.; Rai, K.N.; Garg, A. Development of high strength hydroxyapatite by solid-state-sintering process. *Ceram. Int.* **2007**, *33*, 419–426. [CrossRef]

270. Haberko, K.; Bućko, M.M.; Brzezińska-Miecznik, J.; Haberko, M.; Mozgawa, W.; Panz, T.; Pyda, A.; Zarebski, J. Natural hydroxyapatite–its behaviour during heat treatment. *J. Eur. Ceram. Soc.* **2006**, *26*, 537–542. [CrossRef]

271. Haberko, K.; Bućko, M.M.; Mozgawa, W.; Pyda, A.; Brzezińska-Miecznik, J.; Carpentier, J. Behaviour of bone origin hydroxyapatite at elevated temperatures and in O_2 and CO_2 atmospheres. *Ceram. Int.* **2009**, *35*, 2537–2540. [CrossRef]

272. Janus, A.M.; Faryna, M.; Haberko, K.; Rakowska, A.; Panz, T. Chemical and microstructural characterization of natural hydroxyapatite derived from pig bones. *Mikrochim. Acta* **2008**, *161*, 349–353. [CrossRef]

273. Bahrololoom, M.E.; Javidi, M.; Javadpour, S.; Ma, J. Characterisation of natural hydroxyapatite extracted from bovine cortical bone ash. *J. Ceram. Process. Res.* **2009**, *10*, 129–138.

274. Mostafa, N.Y. Characterization, thermal stability and sintering of hydroxyapatite powders prepared by different routes. *Mater. Chem. Phys.* **2005**, *94*, 333–341. [CrossRef]

275. Suchanek, W.; Yashima, M.; Kakihana, M.; Yoshimura, M. Hydroxyapatite ceramics with selected sintering additives. *Biomaterials* **1997**, *18*, 923–933. [CrossRef]

276. Kalita, S.J.; Bose, S.; Bandyopadhyay, A.; Hosick, H.L. Oxide based sintering additives for HAp ceramics. *Ceram. Trans.* **2003**, *147*, 63–72.

277. Kalita, S.J.; Bose, S.; Hosick, H.L.; Bandyopadhyay, A. $CaO–P_2O_5–Na_2O$-based sintering additives for hydroxyapatite (HAp) ceramics. *Biomaterials* **2004**, *25*, 2331–2339. [CrossRef]

278. Safronova, T.V.; Putlyaev, V.I.; Shekhirev, M.A.; Tretyakov, Y.D.; Kuznetsov, A.V.; Belyakov, A.V. Densification additives for hydroxyapatite ceramics. *J. Eur. Ceram. Soc.* **2009**, *29*, 1925–1932. [CrossRef]

279. Eskandari, A.; Aminzare, M.; Hassani, H.; Barounian, H.; Hesaraki, S.; Sadrnezhaad, S.K. Densification behavior and mechanical properties of biomimetic apatite nanocrystals. *Curr. Nanosci.* **2011**, *7*, 776–780. [CrossRef]

280. Ramesh, S.; Tolouei, R.; Tan, C.Y.; Aw, K.L.; Yeo, W.H.; Sopyan, I.; Teng, W.D. Sintering of hydroxyapatite ceramic produced by wet chemical method. *Adv. Mater. Res.* **2011**, *264–265*, 1856–1861.

281. Ou, S.F.; Chiou, S.Y.; Ou, K.L. Phase transformation on hydroxyapatite decomposition. *Ceram. Int.* **2013**, *39*, 3809–3816. [CrossRef]

282. Bernache-Assollant, D.; Ababou, A.; Champion, E.; Heughebaert, M. Sintering of calcium phosphate hydroxyapatite $Ca_{10}(PO_4)_6(OH)_2$ I. Calcination and particle growth. *J. Eur. Ceram. Soc.* **2003**, *23*, 229–241. [CrossRef]

283. Ramesh, S.; Tan, C.Y.; Bhaduri, S.B.; Teng, W.D.; Sopyan, I. Densification behaviour of nanocrystalline hydroxyapatite bioceramics. *J. Mater. Process. Technol.* **2008**, *206*, 221–230. [CrossRef]

284. Wang, J.; Shaw, L.L. Grain-size dependence of the hardness of submicrometer and nanometer hydroxyapatite. *J. Am. Ceram. Soc.* **2010**, *93*, 601–604. [CrossRef]

285. Kobayashi, S.; Kawai, W.; Wakayama, S. The effect of pressure during sintering on the strength and the fracture toughness of hydroxyapatite ceramics. *J. Mater. Sci. Mater. Med.* **2006**, *17*, 1089–1093. [CrossRef]

286. Chen, I.W.; Wang, X.H. Sintering dense nanocrystalline ceramics without final-stage grain growth. *Nature* **2000**, *404*, 168–170. [CrossRef]

287. Mazaheri, M.; Haghighatzadeh, M.; Zahedi, A.M.; Sadrnezhaad, S.K. Effect of a novel sintering process on mechanical properties of hydroxyapatite ceramics. *J. Alloys Compd.* **2009**, *471*, 180–184. [CrossRef]

288. Panyata, S.; Eitssayeam, S.; Rujijanagul, G.; Tunkasiri, T.; Pengpat, K. Property development of hydroxyapatite ceramics by two-step sintering. *Adv. Mater. Res.* **2012**, *506*, 190–193. [CrossRef]

289. Esnaashary, M.; Fathi, M.; Ahmadian, M. The effect of the two-step sintering process on consolidation of fluoridated hydroxyapatite and its mechanical properties and bioactivity. *Int. J. Appl. Ceram. Technol.* **2014**, *11*, 47–56. [CrossRef]

290. Feng, P.; Niu, M.; Gao, C.; Peng, S.; Shuai, C. A novel two-step sintering for nano-hydroxyapatite scaffolds for bone tissue engineering. *Sci. Rep.* **2014**, *4*, 5599. [CrossRef]

291. Halouani, R.; Bernache-Assolant, D.; Champion, E.; Ababou, A. Microstructure and related mechanical properties of hot pressed hydroxyapatite ceramics. *J. Mater. Sci. Mater. Med.* **1994**, *5*, 563–568. [CrossRef]

292. Kasuga, T.; Ota, Y.; Tsuji, K.; Abe, Y. Preparation of high-strength calcium phosphate ceramics with low modulus of elasticity containing β-Ca(PO$_3$)$_2$ fibers. *J. Am. Ceram. Soc.* **1996**, *79*, 1821–1824. [CrossRef]

293. Suchanek, W.L.; Yoshimura, M. Preparation of fibrous, porous hydroxyapatite ceramics from hydroxyapatite whiskers. *J. Am. Ceram. Soc.* **1998**, *81*, 765–767. [CrossRef]

294. Kim, Y.; Kim, S.R.; Song, H.; Yoon, H. Preparation of porous hydroxyapatite/TCP composite block using a hydrothermal hot pressing method. *Mater. Sci. Forum* **2005**, *486–487*, 117–120. [CrossRef]

295. Li, J.G.; Hashida, T. In situ formation of hydroxyapatite-whisker ceramics by hydrothermal hot-pressing method. *J. Am. Ceram. Soc.* **2006**, *89*, 3544–3546. [CrossRef]

296. Li, J.G.; Hashida, T. Preparation of hydroxyapatite ceramics by hydrothermal hot-pressing method at 300 °C. *J. Mater. Sci.* **2007**, *42*, 5013–5019. [CrossRef]

297. Petrakova, N.V.; Lysenkov, A.S.; Ashmarin, A.A.; Egorov, A.A.; Fedotov, A.Y.; Shvorneva, L.I.; Komlev, V.S.; Barinov, S.M. Effect of hot pressing temperature on the microstructure and strength of hydroxyapatite ceramic. *Inorg. Mater. Appl. Res.* **2013**, *4*, 362–367. [CrossRef]

298. Nakahira, A.; Murakami, T.; Onoki, T.; Hashida, T.; Hosoi, K. Fabrication of porous hydroxyapatite using hydrothermal hot pressing and post-sintering. *J. Am. Ceram. Soc.* **2005**, *88*, 1334–1336. [CrossRef]

299. Auger, M.A.; Savoini, B.; Muñoz, A.; Leguey, T.; Monge, M.A.; Pareja, R.; Victoria, J. Mechanical characteristics of porous hydroxyapatite/oxide composites produced by post-sintering hot isostatic pressing. *Ceram. Int.* **2009**, *35*, 2373–2380. [CrossRef]

300. Guo, N.; Shen, H.Z.; Jin, Q.; Shen, P. Hydrated precursor-assisted densification of hydroxyapatite and its composites by cold sintering. *Ceram. Int.* **2020**, *47*, 14348–14353. [CrossRef]

301. Silva, C.C.; Graça, M.P.F.; Sombra, A.S.B.; Valente, M.A. Structural and electrical study of calcium phosphate obtained by a microwave radiation assisted procedure. *Phys. Rev. B Condens. Matter* **2009**, *404*, 1503–1508. [CrossRef]

302. Veljović, D.; Zalite, I.; Palcevskis, E.; Smiciklas, I.; Petrović, R.; Janaćković, D. Microwave sintering of fine grained HAP and HAP/TCP bioceramics. *Ceram. Int.* **2010**, *36*, 595–603. [CrossRef]

303. Veljović, D.; Palcevskis, E.; Dindune, A.; Putić, S.; Balać, I.; Petrović, R.; Janaćković, D. Microwave sintering improves the mechanical properties of biphasic calcium phosphates from hydroxyapatite microspheres produced from hydrothermal processing. *J. Mater. Sci.* **2010**, *45*, 3175–3183. [CrossRef]

304. Tarafder, S.; Balla, V.K.; Davies, N.M.; Bandyopadhyay, A.; Bose, S. Microwave-sintered 3D printed tricalcium phosphate scaffolds for bone tissue engineering. *J. Tissue Eng. Regen. Med.* **2013**, *7*, 631–641. [CrossRef]

305. Thuault, A.; Savary, E.; Hornez, J.C.; Moreau, G.; Descamps, M.; Marinel, S.; Leriche, A. Improvement of the hydroxyapatite mechanical properties by direct microwave sintering in single mode cavity. *J. Eur. Ceram. Soc.* **2014**, *34*, 1865–1871. [CrossRef]

306. Sikder, P.; Ren, Y.; Bhaduri, S.B. Microwave processing of calcium phosphate and magnesium phosphate based orthopedic bioceramics: A state-of-the-art review. *Acta Biomater.* **2020**, *111*, 29–53. [CrossRef]

307. Nakamura, T.; Fukuhara, T.; Izui, H. Mechanical properties of hydroxyapatites sintered by spark plasma sintering. *Ceram. Trans.* **2006**, *194*, 265–272.

308. Grossin, D.; Rollin-Martinet, S.; Estournès, C.; Rossignol, F.; Champion, E.; Combes, C.; Rey, C.; Geoffroy, C.; Drouet, C. Biomimetic apatite sintered at very low temperature by spark plasma sintering: Physico-chemistry and microstructure aspects. *Acta Biomater.* **2010**, *6*, 577–585. [CrossRef]

309. Ortali, C.; Julien, I.; Vandenhende, M.; Drouet, C.; Champion, E. Consolidation of bone-like apatite bioceramics by spark plasma sintering of amorphous carbonated calcium phosphate at very low temperature. *J. Eur. Ceram. Soc.* **2018**, *38*, 2098–2109. [CrossRef]

310. Chesnaud, A.; Bogicevic, C.; Karolak, F.; Estournès, C.; Dezanneau, G. Preparation of transparent oxyapatite ceramics by combined use of freeze-drying and spark-plasma sintering. *Chem. Comm.* **2007**, 1550–1552. [CrossRef]

311. Eriksson, M.; Liu, Y.; Hu, J.; Gao, L.; Nygren, M.; Shen, Z. Transparent hydroxyapatite ceramics with nanograin structure prepared by high pressure spark plasma sintering at the minimized sintering temperature. *J. Eur. Ceram. Soc.* **2011**, *31*, 1533–1540. [CrossRef]
312. Yoshida, H.; Kim, B.N.; Son, H.W.; Han, Y.H.; Kim, S. Superplastic deformation of transparent hydroxyapatite. *Scripta Mater.* **2013**, *69*, 155–158. [CrossRef]
313. Kim, B.N.; Prajatelistia, E.; Han, Y.H.; Son, H.W.; Sakka, Y.; Kim, S. Transparent hydroxyapatite ceramics consolidated by spark plasma sintering. *Scripta Mater.* **2013**, *69*, 366–369. [CrossRef]
314. Yun, J.; Son, H.; Prajatelistia, E.; Han, Y.H.; Kim, S.; Kim, B.N. Characterisation of transparent hydroxyapatite nanoceramics prepared by spark plasma sintering. *Adv. Appl. Ceram.* **2014**, *113*, 67–72. [CrossRef]
315. Li, Z.; Khor, K.A. Transparent hydroxyapatite obtained through spark plasma sintering: Optical and mechanical properties. *Key Eng. Mater.* **2015**, *631*, 51–56. [CrossRef]
316. Frasnelli, M.; Sglavo, V.M. Flash sintering of tricalcium phosphate (TCP) bioceramics. *J. Eur. Ceram. Soc.* **2018**, *38*, 279–285. [CrossRef]
317. Hwang, C.; Yun, J. Flash sintering of hydroxyapatite ceramics. *J. Asian Ceram. Soc.* **2021**, *9*, 281–288. [CrossRef]
318. Biesuz, M.; Galotta, A.; Motta, A.; Kermani, M.; Grasso, S.; Vontorová, J.; Tyrpekl, V.; Vilémová, M.; Sglavo, V.M. Speedy bioceramics: Rapid densification of tricalcium phosphate by ultrafast high-temperature sintering. *Mater. Sci. Eng. C* **2021**, *127*, 112246. [CrossRef]
319. Yanagisawa, K.; Kim, J.H.; Sakata, C.; Onda, A.; Sasabe, E.; Yamamoto, T.; Matamoros-Veloza, Z.; Rendón-Angeles, J.C. Hydrothermal sintering under mild temperature conditions: Preparation of calcium-deficient hydroxyapatite compacts. *Z. Naturforsch. B* **2010**, *65*, 1038–1044. [CrossRef]
320. Hosoi, K.; Hashida, T.; Takahashi, H.; Yamasaki, N.; Korenaga, T. New processing technique for hydroxyapatite ceramics by the hydrothermal hot-pressing method. *J. Am. Ceram. Soc.* **1996**, *79*, 2771–2774. [CrossRef]
321. Champion, E. Sintering of calcium phosphate bioceramics. *Acta Biomater.* **2013**, *9*, 5855–5875. [CrossRef]
322. Indurkar, A.; Choudhary, R.; Rubenis, K.; Locs, J. Advances in sintering techniques for calcium phosphates ceramics. *Materials* **2021**, *14*, 6133.
323. Evans, J.R.G. Seventy ways to make ceramics. *J. Eur. Ceram. Soc.* **2008**, *28*, 1421–1432. [CrossRef]
324. Hench, L.L.; Polak, J.M. Third-generation biomedical materials. *Science* **2002**, *295*, 1014–1017. [CrossRef]
325. Black, J. *Biological Performance of Materials: Fundamentals of Biocompatibility*, 4th ed.; CRC Press: Boca Raton, FL, USA, 2005; p. 520.
326. Carter, C.B.; Norton, M.G. *Ceramic Materials: Science and Engineering*, 2nd ed.; Springer: New York, NY, USA, 2013; p. 766.
327. Benaqqa, C.; Chevalier, J.; Saâdaoui, M.; Fantozzi, G. Slow crack growth behaviour of hydroxyapatite ceramics. *Biomaterials* **2005**, *26*, 6106–6112. [CrossRef]
328. Pecqueux, F.; Tancret, F.; Payraudeau, N.; Bouler, J.M. Influence of microporosity and macroporosity on the mechanical properties of biphasic calcium phosphate bioceramics: Modelling and experiment. *J. Eur. Ceram. Soc.* **2010**, *30*, 819–829. [CrossRef]
329. Ramesh, S.; Tan, C.Y.; Sopyan, I.; Hamdi, M.; Teng, W.D. Consolidation of nanocrystalline hydroxyapatite powder. *Sci. Technol. Adv. Mater.* **2007**, *8*, 124–130. [CrossRef]
330. Wagoner Johnson, A.J.; Herschler, B.A. A review of the mechanical behavior of CaP and CaP/polymer composites for applications in bone replacement and repair. *Acta Biomater.* **2011**, *7*, 16–30. [CrossRef]
331. Suchanek, W.L.; Yoshimura, M. Processing and properties of hydroxyapatite-based biomaterials for use as hard tissue replacement implants. *J. Mater. Res.* **1998**, *13*, 94–117. [CrossRef]
332. Fan, X.; Case, E.D.; Ren, F.; Shu, Y.; Baumann, M.J. Part I: Porosity dependence of the Weibull modulus for hydroxyapatite and other brittle materials. *J. Mech. Behav. Biomed. Mater.* **2012**, *8*, 21–36. [CrossRef]
333. Fan, X.; Case, E.D.; Gheorghita, I.; Baumann, M.J. Weibull modulus and fracture strength of highly porous hydroxyapatite. *J. Mech. Behav. Biomed. Mater.* **2013**, *20*, 283–295. [CrossRef]
334. Cordell, J.; Vogl, M.; Johnson, A. The influence of micropore size on the mechanical properties of bulk hydroxyapatite and hydroxyapatite scaffolds. *J. Mech. Behav. Biomed. Mater.* **2009**, *2*, 560–570. [CrossRef]
335. Suzuki, S.; Sakamura, M.; Ichiyanagi, M.; Ozawa, M. Internal friction of hydroxyapatite and fluorapatite. *Ceram. Int.* **2004**, *30*, 625–627. [CrossRef]
336. Suzuki, S.; Takahiro, K.; Ozawa, M. Internal friction and dynamic modulus of polycrystalline ceramics prepared from stoichiometric and Ca-deficient hydroxyapatites. *Mater. Sci. Eng. B* **1998**, *55*, 68–70. [CrossRef]
337. Bouler, J.M.; Trecant, M.; Delecrin, J.; Royer, J.; Passuti, N.; Daculsi, G. Macroporous biphasic calcium phosphate ceramics: Influence of five synthesis parameters on compressive strength. *J. Biomed. Mater. Res.* **1996**, *32*, 603–609. [CrossRef]
338. Tancret, F.; Bouler, J.M.; Chamousset, J.; Minois, L.M. Modelling the mechanical properties of microporous and macroporous biphasic calcium phosphate bioceramics. *J. Eur. Ceram. Soc.* **2006**, *26*, 3647–3656. [CrossRef]
339. Le Huec, J.C.; Schaeverbeke, T.; Clement, D.; Faber, J.; le Rebeller, A. Influence of porosity on the mechanical resistance of hydroxyapatite ceramics under compressive stress. *Biomaterials* **1995**, *16*, 113–118. [CrossRef]
340. Hsu, Y.H.; Turner, I.G.; Miles, A.W. Mechanical properties of three different compositions of calcium phosphate bioceramic following immersion in Ringer's solution and distilled water. *J. Mater. Sci. Mater. Med.* **2009**, *20*, 2367–2374. [CrossRef]
341. Torgalkar, A.M. Resonance frequency technique to determine elastic modulus of hydroxyapatite. *J. Biomed. Mater. Res.* **1979**, *13*, 907–920. [CrossRef]

342. Gilmore, R.S.; Katz, J.L. Elastic properties of apatites. *J. Mater. Sci.* **1982**, *17*, 1131–1141. [CrossRef]

343. Fan, X.; Case, E.D.; Ren, F.; Shu, Y.; Baumann, M.J. Part II: Fracture strength and elastic modulus as a function of porosity for hydroxyapatite and other brittle materials. *J. Mech. Behav. Biomed. Mater.* **2012**, *8*, 99–110. [CrossRef]

344. Garcia-Prieto, A.; Hornez, J.C.; Leriche, A.; Pena, P.; Baudín, C. Influence of porosity on the mechanical behaviour of single phase β-TCP ceramics. *Ceram. Int.* **2017**, *43*, 6048–6053. [CrossRef]

345. De Aza, P.N.; de Aza, A.H.; de Aza, S. Crystalline bioceramic materials. *Bol. Soc. Esp. Ceram. V* **2005**, *44*, 135–145. [CrossRef]

346. Fritsch, A.; Dormieux, L.; Hellmich, C.; Sanahuja, J. Mechanical behavior of hydroxyapatite biomaterials: An experimentally validated micromechanical model for elasticity and strength. *J. Biomed. Mater. Res. A* **2009**, *88A*, 149–161. [CrossRef] [PubMed]

347. Ching, W.Y.; Rulis, P.; Misra, A. *Ab initio* elastic properties and tensile strength of crystalline hydroxyapatite. *Acta Biomater.* **2009**, *5*, 3067–3075. [CrossRef]

348. Fritsch, A.; Hellmich, C.; Dormieux, L. The role of disc-type crystal shape for micromechanical predictions of elasticity and strength of hydroxyapatite biomaterials. *Philos. Trans. R. Soc. Lond. A* **2010**, *368*, 1913–1935.

349. Menéndez-Proupin, E.; Cervantes-Rodríguez, S.; Osorio-Pulgar, R.; Franco-Cisterna, M.; Camacho-Montes, H.; Fuentes, M.E. Computer simulation of elastic constants of hydroxyapatite and fluorapatite. *J. Mech. Behav. Biomed. Mater.* **2011**, *4*, 1011–1120. [CrossRef]

350. Sun, J.P.; Song, Y.; Wen, G.W.; Wang, Y.; Yang, R. Softening of hydroxyapatite by vacancies: A first principles investigation. *Mater. Sci. Eng. C* **2013**, *33*, 1109–1115. [CrossRef]

351. Sha, M.C.; Li, Z.; Bradt, R.C. Single-crystal elastic constants of fluorapatite, $Ca_5F(PO_4)_3$. *J. Appl. Phys.* **1994**, *75*, 7784–7787. [CrossRef]

352. Wakai, F.; Kodama, Y.; Sakaguchi, S.; Nonami, T. Superplasticity of hot isostatically pressed hydroxyapatite. *J. Am. Ceram. Soc.* **1990**, *73*, 457–460. [CrossRef]

353. Tago, K.; Itatani, K.; Suzuki, T.S.; Sakka, Y.; Koda, S. Densification and superplasticity of hydroxyapatite ceramics. *J. Ceram. Soc. Jpn.* **2005**, *113*, 669–673. [CrossRef]

354. Burger, E.L.; Patel, V. Calcium phosphates as bone graft extenders. *Orthopedics* **2007**, *30*, 939–942.

355. Guillaume, B. Filling bone defects with β-TCP in maxillofacial surgery: A review. | Comblement osseux par β-TCP en chirurgie maxillofaciale: Revue des indications. *Morphologie* **2017**, *101*, 113–119. [CrossRef]

356. Song, J.; Liu, Y.; Zhang, Y.; Jiao, L. Mechanical properties of hydroxyapatite ceramics sintered from powders with different morphologies. *Mater. Sci. Eng. A* **2011**, *528*, 5421–5427. [CrossRef]

357. Dorozhkin, S.V. Calcium orthophosphate-containing biocomposites and hybrid biomaterials for biomedical applications. *J. Funct. Biomater.* **2015**, *6*, 708–832. [CrossRef]

358. Bouslama, N.; Ben Ayed, F.; Bouaziz, J. Sintering and mechanical properties of tricalcium phosphate–fluorapatite composites. *Ceram. Int.* **2009**, *35*, 1909–1917. [CrossRef]

359. Suchanek, W.; Yashima, M.; Kakihana, M.; Yoshimura, M. Processing and mechanical properties of hydroxyapatite reinforced with hydroxyapatite whiskers. *Biomaterials* **1996**, *17*, 1715–1723. [CrossRef]

360. Suchanek, W.; Yashima, M.; Kakihana, M.; Yoshimura, M. Hydroxyapatite/hydroxyapatite-whisker composites without sintering additives: Mechanical properties and microstructural evolution. *J. Am. Ceram. Soc.* **1997**, *80*, 2805–2813. [CrossRef]

361. Simsek, D.; Ciftcioglu, R.; Guden, M.; Ciftcioglu, M.; Harsa, S. Mechanical properties of hydroxyapatite composites reinforced with hydroxyapatite whiskers. *Key Eng. Mater.* **2004**, *264–268*, 1985–1988. [CrossRef]

362. Bose, S.; Banerjee, A.; Dasgupta, S.; Bandyopadhyay, A. Synthesis, processing, mechanical, and biological property characterization of hydroxyapatite whisker-reinforced hydroxyapatite composites. *J. Am. Ceram. Soc.* **2009**, *92*, 323–330. [CrossRef]

363. Lie-Feng, L.; Xiao-Yi, H.; Cai, Y.X.; Weng, J. Reinforcing of porous hydroxyapatite ceramics with hydroxyapatite fibres for enhanced bone tissue engineering. *J. Biomim. Biomater. Tissue Eng.* **2011**, *1314*, 67–73.

364. Shiota, T.; Shibata, M.; Yasuda, K.; Matsuo, Y. Influence of β-tricalcium phosphate dispersion on mechanical properties of hydroxyapatite ceramics. *J. Ceram. Soc. Jpn.* **2009**, *116*, 1002–1005. [CrossRef]

365. Shuai, C.; Feng, P.; Nie, Y.; Hu, H.; Liu, J.; Peng, S. Nano-hydroxyapatite improves the properties of β-tricalcium phosphate bone scaffolds. *Int. J. Appl. Ceram. Technol.* **2013**, *10*, 1003–1013. [CrossRef]

366. Dorozhkin, S.V.; Ajaal, T. Toughening of porous bioceramic scaffolds by bioresorbable polymeric coatings. *Proc. Inst. Mech. Eng. H* **2009**, *223*, 459–470. [CrossRef]

367. Dorozhkin, S.V.; Ajaal, T. Strengthening of dense bioceramic samples using bioresorbable polymers–a statistical approach. *J. Biomim. Biomater. Tissue Eng.* **2009**, *4*, 27–39. [CrossRef]

368. Dressler, M.; Dombrowski, F.; Simon, U.; Börnstein, J.; Hodoroaba, V.D.; Feigl, M.; Grunow, S.; Gildenhaar, R.; Neumann, M. Influence of gelatin coatings on compressive strength of porous hydroxyapatite ceramics. *J. Eur. Ceram. Soc.* **2011**, *31*, 523–529. [CrossRef]

369. Martinez-Vazquez, F.J.; Perera, F.H.; Miranda, P.; Pajares, A.; Guiberteau, F. Improving the compressive strength of bioceramic robocast scaffolds by polymer infiltration. *Acta Biomater.* **2010**, *6*, 4361–4368. [CrossRef]

370. Fedotov, A.Y.; Bakunova, N.V.; Komlev, V.S.; Barinov, S.M. High-porous calcium phosphate bioceramics reinforced by chitosan infiltration. *Dokl. Chem.* **2011**, *439*, 233–236. [CrossRef]

371. Martínez-Vázquez, F.J.; Pajares, A.; Guiberteau, F.; Miranda, P. Effect of polymer infiltration on the flexural behavior of β-tricalcium phosphate robocast scaffolds. *Materials* **2014**, *7*, 4001–4018. [CrossRef] [PubMed]

372. He, L.H.; Standard, O.C.; Huang, T.T.; Latella, B.A.; Swain, M.V. Mechanical behaviour of porous hydroxyapatite. *Acta Biomater.* **2008**, *4*, 577–586. [CrossRef] [PubMed]

373. Yamashita, K.; Owada, H.; Umegaki, T.; Kanazawa, T.; Futagami, T. Ionic conduction in apatite solid solutions. *Solid State Ionics* **1988**, *28–30*, 660–663. [CrossRef]

374. Nagai, M.; Nishino, T. Surface conduction of porous hydroxyapatite ceramics at elevated temperatures. *Solid State Ionics* **1988**, *28–30*, 1456–1461. [CrossRef]

375. Valdes, J.J.P.; Rodriguez, A.V.; Carrio, J.G. Dielectric properties and structure of hydroxyapatite ceramics sintered by different conditions. *J. Mater. Res.* **1995**, *10*, 2174–2177. [CrossRef]

376. Fanovich, M.A.; Castro, M.S.; Lopez, J.M.P. Analysis of the microstructural evolution in hydroxyapatite ceramics by electrical characterisation. *Ceram. Int.* **1999**, *25*, 517–522. [CrossRef]

377. Bensaoud, A.; Bouhaouss, A.; Ferhat, M. Electrical properties in compressed poorly crystalline apatite. *J. Solid State Electrochem.* **2001**, *5*, 362–365.

378. Mahabole, M.P.; Aiyer, R.C.; Ramakrishna, C.V.; Sreedhar, B.; Khairnar, R.S. Synthesis, characterization and gas sensing property of hydroxyapatite ceramic. *Bull. Mater. Sci.* **2005**, *28*, 535–545. [CrossRef]

379. Tanaka, Y.; Takata, S.; Shimoe, K.; Nakamura, M.; Nagai, A.; Toyama, T.; Yamashita, K. Conduction properties of non-stoichiometric hydroxyapatite whiskers for biomedical use. *J. Ceram. Soc. Jpn.* **2008**, *116*, 815–821. [CrossRef]

380. Tanaka, Y.; Nakamura, M.; Nagai, A.; Toyama, T.; Yamashita, K. Ionic conduction mechanism in Ca-deficient hydroxyapatite whiskers. *Mater. Sci. Eng. B* **2009**, *161*, 115–119. [CrossRef]

381. Wang, W.; Itoh, S.; Yamamoto, N.; Okawa, A.; Nagai, A.; Yamashita, K. Electrical polarization of β-tricalcium phosphate ceramics. *J. Am. Ceram. Soc.* **2010**, *93*, 2175–2177. [CrossRef]

382. Mahabole, M.P.; Mene, R.U.; Khairnar, R.S. Gas sensing and dielectric studies on cobalt doped hydroxyapatite thick films. *Adv. Mater. Lett.* **2013**, *4*, 46–52. [CrossRef]

383. Suresh, M.B.; Biswas, P.; Mahender, V.; Johnson, R. Comparative evaluation of electrical conductivity of hydroxyapatite ceramics densified through ramp and hold, spark plasma and post sinter hot isostatic pressing routes. *Mater. Sci. Eng. C* **2017**, *70*, 364–370. [CrossRef]

384. Das, A.; Pamu, D. A comprehensive review on electrical properties of hydroxyapatite based ceramic composites. *Mater. Sci. Eng. C* **2019**, *101*, 539–563. [CrossRef] [PubMed]

385. Prezas, P.R.; Dekhtyar, Y.; Sorokins, H.; Costa, M.M.; Soares, M.J.; Graça, M.P.F. Electrical charging of bioceramics by corona discharge. *J. Electrost.* **2022**, *115*, 103664. [CrossRef]

386. Gandhi, A.A.; Wojtas, M.; Lang, S.B.; Kholkin, A.L.; Tofail, S.A.M. Piezoelectricity in poled hydroxyapatite ceramics. *J. Am. Ceram. Soc.* **2014**, *97*, 2867–2872. [CrossRef]

387. Bystrov, V.S. Piezoelectricity in the ordered monoclinic hydroxyapatite. *Ferroelectrics* **2015**, *475*, 148–153. [CrossRef]

388. Horiuchi, N.; Madokoro, K.; Nozaki, K.; Nakamura, M.; Katayama, K.; Nagai, A.; Yamashita, K. Electrical conductivity of polycrystalline hydroxyapatite and its application to electret formation. *Solid State Ion.* **2018**, *315*, 19–25. [CrossRef]

389. Saxena, A.; Pandey, M.; Dubey, A.K. Induced electroactive response of hydroxyapatite: A review. *J. Indian I. Sci.* **2019**, *99*, 339–359. [CrossRef]

390. Available online: https://en.wikipedia.org/wiki/Electret (accessed on 24 June 2022).

391. Nakamura, S.; Takeda, H.; Yamashita, K. Proton transport polarization and depolarization of hydroxyapatite ceramics. *J. Appl. Phys.* **2001**, *89*, 5386–5392. [CrossRef]

392. Gittings, J.P.; Bowen, C.R.; Turner, I.G.; Baxter, F.R.; Chaudhuri, J.B. Polarisation behaviour of calcium phosphate based ceramics. *Mater. Sci. Forum.* **2008**, *587–588*, 91–95. [CrossRef]

393. Rivas, M.; del Valle, L.J.; Armelin, E.; Bertran, O.; Turon, P.; Puiggalí, J.; Alemán, C. Hydroxyapatite with permanent electrical polarization: Preparation, characterization, and response against inorganic adsorbates. *ChemPhysChem* **2018**, *19*, 1746–1755. [CrossRef]

394. Itoh, S.; Nakamura, S.; Kobayashi, T.; Shinomiya, K.; Yamashita, K.; Itoh, S. Effect of electrical polarization of hydroxyapatite ceramics on new bone formation. *Calcif. Tissue Int.* **2006**, *78*, 133–142. [CrossRef]

395. Iwasaki, T.; Tanaka, Y.; Nakamura, M.; Nagai, A.; Hashimoto, K.; Toda, Y.; Katayama, K.; Yamashita, K. Rate of bonelike apatite formation accelerated on polarized porous hydroxyapatite. *J. Am. Ceram. Soc.* **2008**, *91*, 3943–3949. [CrossRef]

396. Itoh, S.; Nakamura, S.; Kobayashi, T.; Shinomiya, K.; Yamashita, K. Enhanced bone ingrowth into hydroxyapatite with interconnected pores by electrical polarization. *Biomaterials* **2006**, *27*, 5572–5579. [CrossRef]

397. Kobayashi, T.; Itoh, S.; Nakamura, S.; Nakamura, M.; Shinomiya, K.; Yamashita, K. Enhanced bone bonding of hydroxyapatite-coated titanium implants by electrical polarization. *J. Biomed. Mater. Res. A* **2007**, *82A*, 145–151. [CrossRef]

398. Bodhak, S.; Bose, S.; Bandyopadhyay, A. Role of surface charge and wettability on early stage mineralization and bone cell-materials interactions of polarized hydroxyapatite. *Acta Biomater.* **2009**, *5*, 2178–2188. [CrossRef]

399. Sagawa, H.; Itoh, S.; Wang, W.; Yamashita, K. Enhanced bone bonding of the hydroxyapatite/β-tricalcium phosphate composite by electrical polarization in rabbit long bone. *Artif. Organs* **2010**, *34*, 491–497. [CrossRef]

400. Ohba, S.; Wang, W.; Itoh, S.; Nagai, A.; Yamashita, K. Enhanced effects of new bone formation by an electrically polarized hydroxyapatite microgranule/platelet-rich plasma composite gel. *Key Eng. Mater.* **2013**, *529–530*, 82–87. [CrossRef]

401. Yamashita, K.; Oikawa, N.; Umegaki, T. Acceleration and deceleration of bone-like crystal growth on ceramic hydroxyapatite by electric poling. *Chem. Mater.* **1996**, *8*, 2697–2700. [CrossRef]

402. Teng, N.C.; Nakamura, S.; Takagi, Y.; Yamashita, Y.; Ohgaki, M.; Yamashita, K. A new approach to enhancement of bone formation by electrically polarized hydroxyapatite. *J. Dent. Res.* **2001**, *80*, 1925–1929. [CrossRef]

403. Kobayashi, T.; Nakamura, S.; Yamashita, K. Enhanced osteobonding by negative surface charges of electrically polarized hydroxyapatite. *J. Biomed. Mater. Res.* **2001**, *57*, 477–484. [CrossRef]

404. Kato, R.; Nakamura, S.; Katayama, K.; Yamashita, K. Electrical polarization of plasma-spray-hydroxyapatite coatings for improvement of osteoconduction of implants. *J. Biomed. Mater. Res. A* **2005**, *74A*, 652–658. [CrossRef]

405. Nakamura, S.; Kobayashi, T.; Nakamura, M.; Itoh, S.; Yamashita, K. Electrostatic surface charge acceleration of bone ingrowth of porous hydroxyapatite/β-tricalcium phosphate ceramics. *J. Biomed. Mater. Res. A* **2010**, *92A*, 267–275. [CrossRef] [PubMed]

406. Tarafder, S.; Bodhak, S.; Bandyopadhyay, A.; Bose, S. Effect of electrical polarization and composition of biphasic calcium phosphates on early stage osteoblast interactions. *J. Biomed. Mater. Res. B Appl. Biomater.* **2011**, *97B*, 306–314. [CrossRef] [PubMed]

407. Ohba, S.; Wang, W.; Itoh, S.; Takagi, Y.; Nagai, A.; Yamashita, K. Acceleration of new bone formation by an electrically polarized hydroxyapatite microgranule/platelet-rich plasma composite. *Acta Biomater.* **2012**, *8*, 2778–2787. [CrossRef] [PubMed]

408. Tarafder, S.; Banerjee, S.; Bandyopadhyay, A.; Bose, S. Electrically polarized biphasic calcium phosphates: Adsorption and release of bovine serum albumin. *Langmuir* **2010**, *26*, 16625–16629. [CrossRef]

409. Itoh, S.; Nakamura, S.; Nakamura, M.; Shinomiya, K.; Yamashita, K. Enhanced bone regeneration by electrical polarization of hydroxyapatite. *Artif. Organs* **2006**, *30*, 863–869. [CrossRef]

410. Nakamura, M.; Nagai, A.; Ohashi, N.; Tanaka, Y.; Sekilima, Y.; Nakamura, S. Regulation of osteoblast-like cell behaviors on hydroxyapatite by electrical polarization. *Key Eng. Mater.* **2008**, *361–363*, 1055–1058.

411. Nakamura, M.; Nagai, A.; Tanaka, Y.; Sekilima, Y.; Yamashita, K. Polarized hydroxyapatite promotes spread and motility of osteoblastic cells. *J. Biomed. Mater. Res. A* **2010**, *92A*, 783–790. [CrossRef]

412. Nakamura, M.; Nagai, A.; Yamashita, K. Surface electric fields of apatite electret promote osteoblastic responses. *Key Eng. Mater.* **2013**, *529–530*, 357–360. [CrossRef]

413. Nakamura, S.; Kobayashi, T.; Yamashita, K. Extended bioactivity in the proximity of hydroxyapatite ceramic surfaces induced by polarization charges. *J. Biomed. Mater. Res.* **2002**, *61*, 593–599. [CrossRef]

414. Wang, W.; Itoh, S.; Tanaka, Y.; Nagai, A.; Yamashita, K. Comparison of enhancement of bone ingrowth into hydroxyapatite ceramics with highly and poorly interconnected pores by electrical polarization. *Acta Biomater.* **2009**, *5*, 3132–3140. [CrossRef]

415. Cartmell, S.H.; Thurstan, S.; Gittings, J.P.; Griffiths, S.; Bowen, C.R.; Turner, I.G. Polarization of porous hydroxyapatite scaffolds: Influence on osteoblast cell proliferation and extracellular matrix production. *J. Biomed. Mater. Res. A* **2014**, *102A*, 1047–1052. [CrossRef]

416. Nakamura, M.; Kobayashi, A.; Nozaki, K.; Horiuchi, N.; Nagai, A.; Yamashita, K. Improvement of osteoblast adhesion through polarization of plasma-sprayed hydroxyapatite coatings on metal. *J. Med. Biol. Eng.* **2014**, *34*, 44–48. [CrossRef]

417. Nagai, A.; Tanaka, K.; Tanaka, Y.; Nakamura, M.; Hashimoto, K.; Yamashita, K. Electric polarization and mechanism of B-type carbonated apatite ceramics. *J. Biomed. Mater. Res. A* **2011**, *99A*, 116–124. [CrossRef]

418. Nakamura, M.; Niwa, K.; Nakamura, S.; Sekijima, Y.; Yamashita, K. Interaction of a blood coagulation factor on electrically polarized hydroxyapatite surfaces. *J. Biomed. Mater. Res. B Appl. Biomater.* **2007**, *82B*, 29–36. [CrossRef]

419. Ioku, K. Tailored bioceramics of calcium phosphates for regenerative medicine. *J. Ceram. Soc. Jpn.* **2010**, *118*, 775–783. [CrossRef]

420. Fang, Y.; Agrawal, D.K.; Roy, D.M.; Roy, R. Fabrication of transparent hydroxyapatite ceramics by ambient-pressure sintering. *Mater. Lett.* **1995**, *23*, 147–151. [CrossRef]

421. Varma, H.; Vijayan, S.P.; Babu, S.S. Transparent hydroxyapatite ceramics through gel-casting and low-temperature sintering. *J. Am. Ceram. Soc.* **2002**, *85*, 493–495. [CrossRef]

422. Watanabe, Y.; Ikoma, T.; Monkawa, A.; Suetsugu, Y.; Yamada, H.; Tanaka, J.; Moriyoshi, Y. Fabrication of transparent hydroxyapatite sintered body with high crystal orientation by pulse electric current sintering. *J. Am. Ceram. Soc.* **2005**, *88*, 243–245. [CrossRef]

423. Kotobuki, N.; Ioku, K.; Kawagoe, D.; Fujimori, H.; Goto, S.; Ohgushi, H. Observation of osteogenic differentiation cascade of living mesenchymal stem cells on transparent hydroxyapatite ceramics. *Biomaterials* **2005**, *26*, 779–785. [CrossRef]

424. John, A.; Varma, H.K.; Vijayan, S.; Bernhardt, A.; Lode, A.; Vogel, A.; Burmeister, B.; Hanke, T.; Domaschke, H.; Gelinsky, M. in vitro investigations of bone remodeling on a transparent hydroxyapatite ceramic. *Biomed. Mater.* **2009**, *4*, 015007. [CrossRef]

425. Wang, J.; Shaw, L.L. Transparent nanocrystalline hydroxyapatite by pressure-assisted sintering. *Scripta Mater.* **2010**, *63*, 593–596. [CrossRef]

426. Boilet, L.; Descamps, M.; Rguiti, E.; Tricoteaux, A.; Lu, J.; Petit, F.; Lardot, V.; Cambier, F.; Leriche, A. Processing and properties of transparent hydroxyapatite and β tricalcium phosphate obtained by HIP process. *Ceram. Int.* **2013**, *39*, 283–288. [CrossRef]

427. Han, Y.H.; Kim, B.N.; Yoshida, H.; Yun, J.; Son, H.W.; Lee, J.; Kim, S. Spark plasma sintered superplastic deformed transparent ultrafine hydroxyapatite nanoceramics. *Adv. Appl. Ceram.* **2016**, *115*, 174–184. [CrossRef]

428. Kobune, M.; Mineshige, A.; Fujii, S.; Iida, H. Preparation of translucent hydroxyapatite ceramics by HIP and their physical properties. *J. Ceram. Soc. Jpn.* **1997**, *105*, 210–213. [CrossRef]

429. Barralet, J.E.; Fleming, G.J.P.; Campion, C.; Harris, J.J.; Wright, A.J. Formation of translucent hydroxyapatite ceramics by sintering in carbon dioxide atmospheres. *J. Mater. Sci.* **2003**, *38*, 3979–3993. [CrossRef]

430. Chaudhry, A.A.; Yan, H.; Gong, K.; Inam, F.; Viola, G.; Reece, M.J.; Goodall, J.B.M.; ur Rehman, I.; McNeil-Watson, F.K.; Corbett, J.C.W.; et al. High-strength nanograined and translucent hydroxyapatite monoliths via continuous hydrothermal synthesis and optimized spark plasma sintering. *Acta Biomater.* **2011**, *7*, 791–799. [CrossRef]

431. Tancred, D.C.; McCormack, B.A.; Carr, A.J. A synthetic bone implant macroscopically identical to cancellous bone. *Biomaterials* **1998**, *19*, 2303–2311. [CrossRef]

432. Miao, X.; Sun, D. Graded/gradient porous biomaterials. *Materials* **2010**, *3*, 26–47. [CrossRef]

433. Schliephake, H.; Neukam, F.W.; Klosa, D. Influence of pore dimensions on bone ingrowth into porous hydroxylapatite blocks used as bone graft substitutes. A histometric study. *Int. J. Oral Maxillofac. Surg.* **1991**, *20*, 53–58. [CrossRef]

434. Otsuki, B.; Takemoto, M.; Fujibayashi, S.; Neo, M.; Kokubo, T.; Nakamura, T. Pore throat size and connectivity determine bone and tissue ingrowth into porous implants: Three-dimensional micro-CT based structural analyses of porous bioactive titanium implants. *Biomaterials* **2006**, *27*, 5892–5900. [CrossRef]

435. Hing, K.A.; Best, S.M.; Bonfield, W. Characterization of porous hydroxyapatite. *J. Mater. Sci. Mater. Med.* **1999**, *10*, 135–145. [CrossRef]

436. Lu, J.X.; Flautre, B.; Anselme, K.; Hardouin, P.; Gallur, A.; Descamps, M.; Thierry, B. Role of interconnections in porous bioceramics on bone recolonization in vitro and in vivo. *J. Mater. Sci. Mater. Med.* **1999**, *10*, 111–120. [CrossRef]

437. Karageorgiou, V.; Kaplan, D. Porosity of 3D biomaterial scaffolds and osteogenesis. *Biomaterials* **2005**, *26*, 5474–5491. [CrossRef]

438. Tamai, N.; Myoui, A.; Kudawara, I.; Ueda, T.; Yoshikawa, H. Novel fully interconnected porous hydroxyapatite ceramic in surgical treatment of benign bone tumor. *J. Orthop. Sci.* **2010**, *15*, 560–568. [CrossRef]

439. Panzavolta, S.; Torricelli, P.; Amadori, S.; Parrilli, A.; Rubini, K.; Della Bella, E.; Fini, M.; Bigi, A. 3D interconnected porous biomimetic scaffolds: In vitro cell response. *J. Biomed. Mater. Res. A* **2013**, *101A*, 3560–3570. [CrossRef]

440. Jin, L.; Feng, Z.Q.; Wang, T.; Ren, Z.; Ma, S.; Wu, J.; Sun, D. A novel fluffy hydroxylapatite fiber scaffold with deep interconnected pores designed for three-dimensional cell culture. *J. Mater. Chem. B* **2014**, *2*, 129–136. [CrossRef]

441. Flautre, B.; Descamps, M.; Delecourt, C.; Blary, M.C.; Hardouin, P. Porous HA ceramic for bone replacement: Role of the pores and interconnections–experimental study in the rabbits. *J. Mater. Sci. Mater. Med.* **2001**, *12*, 679–682. [CrossRef]

442. Mastrogiacomo, M.; Scaglione, S.; Martinetti, R.; Dolcini, L.; Beltrame, F.; Cancedda, R.; Quarto, R. Role of scaffold internal structure on in vivo bone formation in macroporous calcium phosphate bioceramics. *Biomaterials* **2006**, *27*, 3230–3237. [CrossRef]

443. Okamoto, M.; Dohi, Y.; Ohgushi, H.; Shimaoka, H.; Ikeuchi, M.; Matsushima, A.; Yonemasu, K.; Hosoi, H. Influence of the porosity of hydroxyapatite ceramics on in vitro and in vivo bone formation by cultured rat bone marrow stromal cells. *J. Mater. Sci. Mater. Med.* **2006**, *17*, 327–336. [CrossRef]

444. Li, X.; Liu, H.; Niu, X.; Fan, Y.; Feng, Q.; Cui, F.Z.; Watari, F. Osteogenic differentiation of human adipose-derived stem cells induced by osteoinductive calcium phosphate ceramics. *J. Biomed. Mater. Res. B Appl. Biomater.* **2011**, *97B*, 10–19. [CrossRef]

445. Hong, M.H.; Kim, S.M.; Han, M.H.; Kim, Y.H.; Lee, Y.K.; Oh, D.S. Evaluation of microstructure effect of the porous spherical β-tricalcium phosphate granules on cellular responses. *Ceram. Int.* **2014**, *40*, 6095–6102. [CrossRef]

446. De Godoy, R.F.; Hutchens, S.; Campion, C.; Blunn, G. Silicate-substituted calcium phosphate with enhanced strut porosity stimulates osteogenic differentiation of human mesenchymal stem cells. *J. Mater. Sci. Mater. Med.* **2015**, *26*, 54. [CrossRef]

447. Omae, H.; Mochizuki, Y.; Yokoya, S.; Adachi, N.; Ochi, M. Effects of interconnecting porous structure of hydroxyapatite ceramics on interface between grafted tendon and ceramics. *J. Biomed. Mater. Res. A* **2006**, *79A*, 329–337. [CrossRef]

448. Yoshikawa, H.; Tamai, N.; Murase, T.; Myoui, A. Interconnected porous hydroxyapatite ceramics for bone tissue engineering. *J. R. Soc. Interface* **2009**, *6*, S341–S348. [CrossRef]

449. Ribeiro, G.B.M.; Trommer, R.M.; dos Santos, L.A.; Bergmann, C.P. Novel method to produce β-TCP scaffolds. *Mater. Lett.* **2011**, *65*, 275–277. [CrossRef]

450. Silva, T.S.N.; Primo, B.T.; Silva, A.N., Jr.; Machado, D.C.; Viezzer, C.; Santos, L.A. Use of calcium phosphate cement scaffolds for bone tissue engineering: In vitro study. *Acta Cir. Bras.* **2011**, *26*, 7–11. [CrossRef]

451. De Moraes MacHado, J.L.; Giehl, I.C.; Nardi, N.B.; dos Santos, L.A. Evaluation of scaffolds based on α-tricalcium phosphate cements for tissue engineering applications. *IEEE Trans. Biomed. Eng.* **2011**, *58*, 1814–1819. [CrossRef] [PubMed]

452. Li, S.H.; de Wijn, J.R.; Layrolle, P.; de Groot, K. Novel method to manufacture porous hydroxyapatite by dual-phase mixing. *J. Am. Ceram. Soc.* **2003**, *86*, 65–72. [CrossRef]

453. De Oliveira, J.F.; de Aguiar, P.F.; Rossi, A.M.; Soares, G.D.A. Effect of process parameters on the characteristics of porous calcium phosphate ceramics for bone tissue scaffolds. *Artif. Organs* **2003**, *27*, 406–411. [CrossRef]

454. Swain, S.K.; Bhattacharyya, S. Preparation of high strength macroporous hydroxyapatite scaffold. *Mater. Sci. Eng. C* **2013**, *33*, 67–71. [CrossRef]

455. Maeda, H.; Kasuga, T.; Nogami, M.; Kagami, H.; Hata, K.; Ueda, M. Preparation of bonelike apatite composite sponge. *Key Eng. Mater.* **2004**, *254–256*, 497–500. [CrossRef]

456. Le Ray, A.M.; Gautier, H.; Bouler, J.M.; Weiss, P.; Merle, C. A new technological procedure using sucrose as porogen compound to manufacture porous biphasic calcium phosphate ceramics of appropriate micro- and macrostructure. *Ceram. Int.* **2010**, *36*, 93–101. [CrossRef]

457. Li, S.H.; de Wijn, J.R.; Layrolle, P.; de Groot, K. Synthesis of macroporous hydroxyapatite scaffolds for bone tissue engineering. *J. Biomed. Mater. Res.* **2002**, *61*, 109–120.

458. Hesaraki, S.; Sharifi, D. Investigation of an effervescent additive as porogenic agent for bone cement macroporosity. *Biomed. Mater. Eng.* **2007**, *17*, 29–38.

459. Hesaraki, S.; Moztarzadeh, F.; Sharifi, D. Formation of interconnected macropores in apatitic calcium phosphate bone cement with the use of an effervescent additive. *J. Biomed. Mater. Res. A* **2007**, *83A*, 80–87. [CrossRef]

460. Tas, A.C. Preparation of porous apatite granules from calcium phosphate cement. *J. Mater. Sci. Mater. Med.* **2008**, *19*, 2231–2239. [CrossRef]

461. Şahin, E.; Çiftçioğlu, M. Compositional, microstructural and mechanical effects of NaCl porogens in brushite cement scaffolds. *J. Mech. Behav. Biomed. Mater.* **2021**, *116*, 104363. [CrossRef]

462. Yao, X.; Tan, S.; Jiang, D. Improving the properties of porous hydroxyapatite ceramics by fabricating methods. *J. Mater. Sci.* **2005**, *40*, 4939–4942. [CrossRef]

463. Song, H.Y.; Youn, M.H.; Kim, Y.H.; Min, Y.K.; Yang, H.M.; Lee, B.T. Fabrication of porous β-TCP bone graft substitutes using PMMA powder and their biocompatibility study. *Korean J. Mater. Res.* **2007**, *17*, 318–322. [CrossRef]

464. Indra, A.; Hadi, F.; Mulyadi, I.H.; Affi, J.; Gunawarman. A novel fabrication procedure for producing high strength hydroxyapatite ceramic scaffolds with high porosity. *Ceram. Int.* **2021**, *47*, 26991–27001. [CrossRef]

465. Almirall, A.; Larrecq, G.; Delgado, J.A.; Martínez, S.; Planell, J.A.; Ginebra, M.P. Fabrication of low temperature macroporous hydroxyapatite scaffolds by foaming and hydrolysis of an α-TCP paste. *Biomaterials* **2004**, *25*, 3671–3680. [CrossRef]

466. Huang, X.; Miao, X. Novel porous hydroxyapatite prepared by combining H_2O_2 foaming with PU sponge and modified with PLGA and bioactive glass. *J. Biomater. Appl.* **2007**, *21*, 351–374. [CrossRef]

467. Li, B.; Chen, X.; Guo, B.; Wang, X.; Fan, H.; Zhang, X. Fabrication and cellular biocompatibility of porous carbonated biphasic calcium phosphate ceramics with a nanostructure. *Acta Biomater.* **2009**, *5*, 134–143. [CrossRef] [PubMed]

468. Cheng, Z.; Zhao, K.; Wu, Z.P. Structure control of hydroxyapatite ceramics via an electric field assisted freeze casting method. *Ceram. Int.* **2015**, *41*, 8599–8604. [CrossRef]

469. Lyu, Y.; Asoh, T.A.; Uyama, H. Facile synthesis of a three-dimensional hydroxyapatite monolith for protein adsorption. *J. Mater. Chem. B* **2021**, *9*, 9711–9719. [CrossRef] [PubMed]

470. Tadic, D.; Beckmann, F.; Schwarz, K.; Epple, M. A novel method to produce hydroxylapatite objects with interconnecting porosity that avoids sintering. *Biomaterials* **2004**, *25*, 3335–3340. [CrossRef]

471. Sepulveda, P.; Binner, J.G.; Rogero, S.O.; Higa, O.Z.; Bressiani, J.C. Production of porous hydroxyapatite by the gel-casting of foams and cytotoxic evaluation. *J. Biomed. Mater. Res.* **2000**, *50*, 27–34. [CrossRef]

472. Chevalier, E.; Chulia, D.; Pouget, C.; Viana, M. Fabrication of porous substrates: A review of processes using pore forming agents in the biomaterial field. *J. Pharm. Sci.* **2008**, *97*, 1135–1154. [CrossRef]

473. Tang, Y.J.; Tang, Y.F.; Lv, C.T.; Zhou, Z.H. Preparation of uniform porous hydroxyapatite biomaterials by a new method. *Appl. Surf. Sci.* **2008**, *254*, 5359–5362. [CrossRef]

474. Abdulqader, S.T.; Rahman, I.A.; Ismail, H.; Ponnuraj Kannan, T.; Mahmood, Z. A simple pathway in preparation of controlled porosity of biphasic calcium phosphate scaffold for dentin regeneration. *Ceram. Int.* **2013**, *39*, 2375–2381. [CrossRef]

475. Wen, F.H.; Wang, F.; Gai, Y.; Wang, M.T.; Lai, Q.H. Preparation of mesoporous hydroxylapatite ceramics using polystyrene microspheres as template. *Appl. Mech. Mater.* **2013**, *389*, 194–198. [CrossRef]

476. Sari, M.; Hening, P.; Chotimah; Ana, I.D.; Yusuf, Y. Bioceramic hydroxyapatite-based scaffold with a porous structure using honeycomb as a natural polymeric porogen for bone tissue engineering. *Biomater. Res.* **2021**, *25*, 2. [CrossRef]

477. Castillo-Paz, A.M.; Cañon-Davila, D.F.; Londoño-Restrepo, S.M.; Jimenez-Mendoza, D.; Pfeiffer, H.; Ramírez-Bon, R.; Rodriguez-Garcia, M.E. Fabrication and characterization of bioinspired nanohydroxyapatite scaffolds with different porosities. *Ceram. Int.* **2022**, *in press*. [CrossRef]

478. Cho, S.; Kim, J.; Lee, S.B.; Choi, M.; Kim, D.H.; Jo, I.; Kwon, H.; Kim, Y. Fabrication of functionally graded hydroxyapatite and structurally graded porous hydroxyapatite by using multi-walled carbon nanotubes. *Compos. A* **2020**, *139*, 106138. [CrossRef]

479. Guda, T.; Appleford, M.; Oh, S.; Ong, J.L. A cellular perspective to bioceramic scaffolds for bone tissue engineering: The state of the art. *Curr. Top. Med. Chem.* **2008**, *8*, 290–299. [CrossRef]

480. Habraken, W.J.E.M.; Wolke, J.G.C.; Jansen, J.A. Ceramic composites as matrices and scaffolds for drug delivery in tissue engineering. *Adv. Drug Deliv. Rev.* **2007**, *59*, 234–248. [CrossRef]

481. Lee, J.B.; Maeng, W.Y.; Koh, Y.H.; Kim, H.E. Porous calcium phosphate ceramic scaffolds with tailored pore orientations and mechanical properties using lithography-based ceramic 3D printing technique. *Materials* **2018**, *11*, 1711. [CrossRef]

482. Tian, J.; Tian, J. Preparation of porous hydroxyapatite. *J. Mater. Sci.* **2001**, *36*, 3061–3066. [CrossRef]

483. Swain, S.K.; Bhattacharyya, S.; Sarkar, D. Preparation of porous scaffold from hydroxyapatite powders. *Mater. Sci. Eng. C* **2011**, *31*, 1240–1244. [CrossRef]

484. Zhao, K.; Tang, Y.F.; Qin, Y.S.; Luo, D.F. Polymer template fabrication of porous hydroxyapatite scaffolds with interconnected spherical pores. *J. Eur. Ceram. Soc.* **2011**, *31*, 225–229. [CrossRef]

485. Sung, J.H.; Shin, K.H.; Moon, Y.W.; Koh, Y.H.; Choi, W.Y.; Kim, H.E. Production of porous calcium phosphate (CaP) ceramics with highly elongated pores using carbon-coated polymeric templates. *Ceram. Int.* **2012**, *38*, 93–97. [CrossRef]

486. Morejón, L.; Delgado, J.A.; Ribeiro, A.A.; de Oliveira, M.V.; Mendizábal, E.; García, I.; Alfonso, A.; Poh, P.; van Griensven, M.; Balmayor, E.R. Development, characterization and in vitro biological properties of scaffolds fabricated from calcium phosphate nanoparticles. *Int. J. Mol. Sci.* **2019**, *20*, 1790. [CrossRef]

487. Deville, S.; Saiz, E.; Tomsia, A.P. Freeze casting of hydroxyapatite scaffolds for bone tissue engineering. *Biomaterials* **2006**, *27*, 5480–5489. [CrossRef]
488. Lee, E.J.; Koh, Y.H.; Yoon, B.H.; Kim, H.E.; Kim, H.W. Highly porous hydroxyapatite bioceramics with interconnected pore channels using camphene-based freeze casting. *Mater. Lett.* **2007**, *61*, 2270–2273. [CrossRef]
489. Fu, Q.; Rahaman, M.N.; Dogan, F.; Bal, B.S. Freeze casting of porous hydroxyapatite scaffolds. I. Processing and general microstructure. *J. Biomed. Mater. Res. B Appl. Biomater.* **2008**, *86B*, 125–135. [CrossRef]
490. Impens, S.; Schelstraete, R.; Luyten, J.; Mullens, S.; Thijs, I.; van Humbeeck, J.; Schrooten, J. Production and characterisation of porous calcium phosphate structures with controllable hydroxyapatite/β-tricalcium phosphate ratios. *Adv. Appl. Ceram.* **2009**, *108*, 494–500. [CrossRef]
491. Macchetta, A.; Turner, I.G.; Bowen, C.R. Fabrication of HA/TCP scaffolds with a graded and porous structure using a camphene-based freeze-casting method. *Acta Biomater.* **2009**, *5*, 1319–1327. [CrossRef]
492. Potoczek, M.; Zima, A.; Paszkiewicz, Z.; Ślósarczyk, A. Manufacturing of highly porous calcium phosphate bioceramics via gel-casting using agarose. *Ceram. Int.* **2009**, *35*, 2249–2254. [CrossRef]
493. Zuo, K.H.; Zeng, Y.P.; Jiang, D. Effect of polyvinyl alcohol additive on the pore structure and morphology of the freeze-cast hydroxyapatite ceramics. *Mater. Sci. Eng. C* **2010**, *30*, 283–287. [CrossRef]
494. Soon, Y.M.; Shin, K.H.; Koh, Y.H.; Lee, J.H.; Choi, W.Y.; Kim, H.E. Fabrication and compressive strength of porous hydroxyapatite scaffolds with a functionally graded core/shell structure. *J. Eur. Ceram. Soc.* **2011**, *31*, 13–18. [CrossRef]
495. Hesaraki, S. Freeze-casted nanostructured apatite scaffold obtained from low temperature biomineralization of reactive calcium phosphates. *Key Eng. Mater.* **2014**, *587*, 21–26. [CrossRef]
496. Ng, S.; Guo, J.; Ma, J.; Loo, S.C.J. Synthesis of high surface area mesostructured calcium phosphate particles. *Acta Biomater.* **2010**, *6*, 3772–3781. [CrossRef]
497. Walsh, D.; Hopwood, J.D.; Mann, S. Crystal tectonics: Construction of reticulated calcium phosphate frameworks in bicontinuous reverse microemulsions. *Science* **1994**, *264*, 1576–1578. [CrossRef] [PubMed]
498. Walsh, D.; Mann, S. Chemical synthesis of microskeletal calcium phosphate in bicontinuous microemulsions. *Chem. Mater.* **1996**, *8*, 1944–1953. [CrossRef]
499. Zhao, K.; Tang, Y.F.; Qin, Y.S.; Wei, J.Q. Porous hydroxyapatite ceramics by ice templating: Freezing characteristics and mechanical properties. *Ceram. Int.* **2011**, *37*, 635–639. [CrossRef]
500. Zhou, K.; Zhang, Y.; Zhang, D.; Zhang, X.; Li, Z.; Liu, G.; Button, T.W. Porous hydroxyapatite ceramics fabricated by an ice-templating method. *Scripta Mater.* **2011**, *64*, 426–429. [CrossRef]
501. Flauder, S.; Gbureck, U.; Muller, F.A. TCP scaffolds with an interconnected and aligned porosity fabricated via ice-templating. *Key Eng. Mater.* **2013**, *529–530*, 129–132. [CrossRef]
502. Zhang, Y.; Zhou, K.; Bao, Y.; Zhang, D. Effects of rheological properties on ice-templated porous hydroxyapatite ceramics. *Mater. Sci. Eng. C* **2013**, *33*, 340–346. [CrossRef]
503. White, E.; Shors, E.C. Biomaterial aspects of Interpore-200 porous hydroxyapatite. *Dent. Clin. North Am.* **1986**, *30*, 49–67. [CrossRef]
504. Aizawa, M.; Howell, S.F.; Itatani, K.; Yokogawa, Y.; Nishizawa, K.; Toriyama, M.; Kameyama, T. Fabrication of porous ceramics with well-controlled open pores by sintering of fibrous hydroxyapatite particles. *J. Ceram. Soc. Jpn.* **2000**, *108*, 249–253. [CrossRef]
505. Nakahira, A.; Tamai, M.; Sakamoto, K.; Yamaguchi, S. Sintering and microstructure of porous hydroxyapatite. *J. Ceram. Soc. Jpn.* **2000**, *108*, 99–104. [CrossRef]
506. Koh, Y.H.; Kim, H.W.; Kim, H.E.; Halloran, J.W. Fabrication of macrochannelled-hydroxyapatite bioceramic by a coextrusion process. *J. Am. Ceram. Soc.* **2002**, *85*, 2578–2580. [CrossRef]
507. Charriere, E.; Lemaitre, J.; Zysset, P. Hydroxyapatite cement scaffolds with controlled macroporosity: Fabrication protocol and mechanical properties. *Biomaterials* **2003**, *24*, 809–817. [CrossRef]
508. Gonzalez-McQuire, R.; Green, D.; Walsh, D.; Hall, S.; Chane-Ching, J.Y.; Oreffo, R.O.C.; Mann, S. Fabrication of hydroxyapatite sponges by dextran sulphate/amino acid templating. *Biomaterials* **2005**, *26*, 6652–6656. [CrossRef]
509. Eichenseer, C.; Will, J.; Rampf, M.; Wend, S.; Greil, P. Biomorphous porous hydroxyapatite-ceramics from rattan (*Calamus Rotang*). *J. Mater. Sci. Mater. Med.* **2010**, *21*, 131–137. [CrossRef]
510. Walsh, D.; Boanini, E.; Tanaka, J.; Mann, S. Synthesis of tri-calcium phosphate sponges by interfacial deposition and thermal transformation of self-supporting calcium phosphate films. *J. Mater. Chem.* **2005**, *15*, 1043–1048. [CrossRef]
511. Song, H.Y.; Islam, S.; Lee, B.T. A novel method to fabricate unidirectional porous hydroxyapatite body using ethanol bubbles in a viscous slurry. *J. Am. Ceram. Soc.* **2008**, *91*, 3125–3127. [CrossRef]
512. Zhou, L.; Wang, D.; Huang, W.; Yao, A.; Kamitakahara, M.; Ioku, K. Preparation and characterization of periodic porous frame of hydroxyapatite. *J. Ceram. Soc. Jpn.* **2009**, *117*, 521–524. [CrossRef]
513. Xu, S.; Li, D.; Lu, B.; Tang, Y.; Wang, C.; Wang, Z. Fabrication of a calcium phosphate scaffold with a three dimensional channel network and its application to perfusion culture of stem cells. *Rapid Prototyp. J.* **2007**, *13*, 99–106. [CrossRef]
514. Saiz, E.; Gremillard, L.; Menendez, G.; Miranda, P.; Gryn, K.; Tomsia, A.P. Preparation of porous hydroxyapatite scaffolds. *Mater. Sci. Eng. C* **2007**, *27*, 546–550. [CrossRef]
515. Sakamoto, M.; Nakasu, M.; Matsumoto, T.; Okihana, H. Development of superporous hydroxyapatites and their examination with a culture of primary rat osteoblasts. *J. Biomed. Mater. Res. A* **2007**, *82A*, 238–242. [CrossRef] [PubMed]

516. Wang, H.; Zhai, L.; Li, Y.; Shi, T. Preparation of irregular mesoporous hydroxyapatite. *Mater. Res. Bull.* **2008**, *43*, 1607–1614. [CrossRef]

517. Sakamoto, M. Development and evaluation of superporous hydroxyapatite ceramics with triple pore structure as bone tissue scaffold. *J. Ceram. Soc. Jpn.* **2010**, *118*, 753–757. [CrossRef]

518. Sakamoto, M.; Matsumoto, T. Development and evaluation of superporous ceramics bone tissue scaffold materials with triple pore structure a) hydroxyapatite, b) beta-tricalcium phosphate. In *Bone Regeneration*; Tal, H., Ed.; InTech Open: Rijeka, Croatia, 2012; pp. 301–320.

519. Deisinger, U. Generating porous ceramic scaffolds: Processing and properties. *Key Eng. Mater.* **2010**, *441*, 155–179. [CrossRef]

520. Ishikawa, K.; Tsuru, K.; Pham, T.K.; Maruta, M.; Matsuya, S. Fully-interconnected pore forming calcium phosphate cement. *Key Eng. Mater.* **2012**, *493–494*, 832–835.

521. Yoon, H.J.; Kim, U.C.; Kim, J.H.; Koh, Y.H.; Choi, W.Y.; Kim, H.E. Fabrication and characterization of highly porous calcium phosphate (CaP) ceramics by freezing foamed aqueous CaP suspensions. *J. Ceram. Soc. Jpn.* **2011**, *119*, 573–576. [CrossRef]

522. Ahn, M.K.; Shin, K.H.; Moon, Y.W.; Koh, Y.H.; Choi, W.Y.; Kim, H.E. Highly porous biphasic calcium phosphate (BCP) ceramics with large interconnected pores by freezing vigorously foamed BCP suspensions under reduced pressure. *J. Am. Ceram. Soc.* **2011**, *94*, 4154–4156. [CrossRef]

523. Schlosser, M.; Kleebe, H.J. Vapor transport sintering of porous calcium phosphate ceramics. *J. Am. Ceram. Soc.* **2012**, *95*, 1581–1587. [CrossRef]

524. Zheng, W.; Liu, G.; Yan, C.; Xiao, Y.; Miao, X.G. Strong and bioactive tri-calcium phosphate scaffolds with tube-like macropores. *J. Biomim. Biomater. Tissue Eng.* **2014**, *19*, 65–75. [CrossRef]

525. Tsuru, K.; Nikaido, T.; Munar, M.L.; Maruta, M.; Matsuya, S.; Nakamura, S.; Ishikawa, K. Synthesis of carbonate apatite foam using β-TCP foams as precursors. *Key Eng. Mater.* **2014**, *587*, 52–55. [CrossRef]

526. Chen, Z.C.; Zhang, X.L.; Zhou, K.; Cai, H.; Liu, C.Q. Novel fabrication of hierarchically porous hydroxyapatite scaffolds with refined porosity and suitable strength. *Adv. Appl. Ceram.* **2015**, *114*, 183–187. [CrossRef]

527. Swain, S.K.; Bhattacharyya, S.; Sarkar, D. Fabrication of porous hydroxyapatite scaffold via polyethylene glycol-polyvinyl alcohol hydrogel state. *Mater. Res. Bull.* **2015**, *64*, 257–261. [CrossRef]

528. Charbonnier, B.; Laurent, C.; Marchat, D. Porous hydroxyapatite bioceramics produced by impregnation of 3D-printed wax mold: Slurry feature optimization. *J. Eur. Ceram. Soc.* **2016**, *36*, 4269–4279. [CrossRef]

529. Roy, D.M.; Linnehan, S.K. Hydroxyapatite formed from coral skeletal carbonate by hydrothermal exchange. *Nature* **1974**, *247*, 220–222. [CrossRef]

530. Zhang, X.; Vecchio, K.S. Conversion of natural marine skeletons as scaffolds for bone tissue engineering. *Front. Mater. Sci.* **2013**, *7*, 103–117. [CrossRef]

531. Yang, Y.; Yao, Q.; Pu, X.; Hou, Z.; Zhang, Q. Biphasic calcium phosphate macroporous scaffolds derived from oyster shells for bone tissue engineering. *Chem. Eng. J.* **2011**, *173*, 837–845. [CrossRef]

532. Tampieri, A.; Sprio, S.; Ruffini, A.; Celotti, G.; Lesci, I.G.; Roveri, N. From wood to bone: Multi-step process to convert wood hierarchical structures into biomimetic hydroxyapatite scaffolds for bone tissue engineering. *J. Mater. Chem.* **2009**, *19*, 4973–4980. [CrossRef]

533. Thanh, T.N.X.; Maruta, M.; Tsuru, K.; Matsuya, S.; Ishikawa, K. Three–dimensional porous carbonate apatite with sufficient mechanical strength as a bone substitute material. *Adv. Mater. Res.* **2014**, *891–892*, 1559–1564. [CrossRef]

534. Moroni, L.; de Wijn, J.R.; van Blitterswijk, C.A. Integrating novel technologies to fabricate smart scaffolds. *J. Biomater. Sci. Polymer Edn.* **2008**, *19*, 543–572. [CrossRef]

535. Studart, A.R.; Gonzenbach, U.T.; Tervoort, E.; Gauckler, L.J. Processing routes to macroporous ceramics: A review. *J. Am. Ceram. Soc.* **2006**, *89*, 1771–1789. [CrossRef]

536. Hing, K.; Annaz, B.; Saeed, S.; Revell, P.; Buckland, T. Microporosity enhances bioactivity of synthetic bone graft substitutes. *J. Mater. Sci. Mater. Med.* **2005**, *16*, 467–475. [CrossRef]

537. Wang, Z.; Sakakibara, T.; Sudo, A.; Kasai, Y. Porosity of β-tricalcium phosphate affects the results of lumbar posterolateral fusion. *J. Spinal Disord. Tech.* **2013**, *26*, E40–E45. [CrossRef]

538. Lan Levengood, S.K.; Polak, S.J.; Wheeler, M.B.; Maki, A.J.; Clark, S.G.; Jamison, R.D.; Wagoner Johnson, A.J. Multiscale osteointegration as a new paradigm for the design of calcium phosphate scaffolds for bone regeneration. *Biomaterials* **2010**, *31*, 3552–3563. [CrossRef]

539. Ruksudjarit, A.; Pengpat, K.; Rujijanagul, G.; Tunkasiri, T. The fabrication of nanoporous hydroxyapatite ceramics. *Adv. Mater. Res.* 2008; *47–50*, 797–800.

540. Li, Y.; Tjandra, W.; Tam, K.C. Synthesis and characterization of nanoporous hydroxyapatite using cationic surfactants as templates. *Mater. Res. Bull.* **2008**, *43*, 2318–2326. [CrossRef]

541. El Asri, S.; Laghzizil, A.; Saoiabi, A.; Alaoui, A.; El Abassi, K.; M'hamdi, R.; Coradin, T. A novel process for the fabrication of nanoporous apatites from Moroccan phosphate rock. *Colloid Surf. A* **2009**, *350*, 73–78. [CrossRef]

542. Raksujarit, A.; Pengpat, K.; Rujijanagul, G.; Tunkasiri, T. Processing and properties of nanoporous hydroxyapatite ceramics. *Mater. Des.* **2010**, *31*, 1658–1660. [CrossRef]

543. Ramli, R.A.; Adnan, R.; Bakar, M.A.; Masudi, S.M. Synthesis and characterisation of pure nanoporous hydroxyapatite. *J. Phys. Sci.* **2011**, *22*, 25–37.

544. LeGeros, R.Z. Calcium phosphate-based osteoinductive materials. *Chem. Rev.* **2008**, *108*, 4742–4753. [CrossRef] [PubMed]
545. Prokopiev, O.; Sevostianov, I. Dependence of the mechanical properties of sintered hydroxyapatite on the sintering temperature. *Mater. Sci. Eng. A* **2006**, *431*, 218–227. [CrossRef]
546. Daculsi, G.; Jegoux, F.; Layrolle, P. The micro macroporous biphasic calcium phosphate concept for bone reconstruction and tissue engineering. In *Advanced Biomaterials: Fundamentals, Processing and Applications*; Basu, B., Katti, D.S., Kumar, A., Eds.; American Ceramic Society: Columbus, OH, USA; Wiley: Hoboken, NJ, USA, 2009; p. 768.
547. Shipman, P.; Foster, G.; Schoeninger, M. Burnt bones and teeth: An experimental study of color, morphology, crystal structure and shrinkage. *J. Archaeol. Sci.* **1984**, *11*, 307–325. [CrossRef]
548. Rice, R.W. *Porosity of Ceramics*; Marcel Dekker: New York, NY, USA, 1998; p. 560.
549. Fan, J.; Lei, J.; Yu, C.; Tu, B.; Zhao, D. Hard-templating synthesis of a novel rod-like nanoporous calcium phosphate bioceramics and their capacity as antibiotic carriers. *Mater. Chem. Phys.* **2007**, *103*, 489–493. [CrossRef]
550. Hsu, Y.H.; Turner, I.G.; Miles, A.W. Fabrication of porous bioceramics with porosity gradients similar to the bimodal structure of cortical and cancellous bone. *J. Mater. Sci. Mater. Med.* **2007**, *18*, 2251–2256. [CrossRef] [PubMed]
551. Munch, E.; Franco, J.; Deville, S.; Hunger, P.; Saiz, E.; Tomsia, A.P. Porous ceramic scaffolds with complex architectures. *JOM* **2008**, *60*, 54–59. [CrossRef]
552. Naqshbandi, A.R.; Sopyan, I.; Gunawan. Development of porous calcium phosphate bioceramics for bone implant applications: A review. *Rec. Pat. Mater. Sci.* **2013**, *6*, 238–252.
553. Jodati, H.; Yılmaz, B.; Evis, Z. A review of bioceramic porous scaffolds for hard tissue applications: Effects of structural features. *Ceram. Int.* **2020**, *46*, 15725–15739. [CrossRef]
554. Yan, X.; Yu, C.; Zhou, X.; Tang, J.; Zhao, D. Highly ordered mesoporous bioactive glasses with superior in vitro bone-forming bioactivities. *Angew. Chem. Int. Ed. Engl.* **2004**, *43*, 5980–5984. [CrossRef] [PubMed]
555. Barba, A.; Maazouz, Y.; Diez-Escudero, A.; Rappe, K.; Espanol, M.; Montufar, E.B.; Öhman-Mägi, C.; Persson, C.; Fontecha, P.; Manzanares, M.C.; et al. Osteogenesis by foamed and 3D-printed nanostructured calcium phosphate scaffolds: Effect of pore architecture. *Acta Biomater.* **2018**, *79*, 135–147. [CrossRef] [PubMed]
556. Cosijns, A.; Vervaet, C.; Luyten, J.; Mullens, S.; Siepmann, F.; van Hoorebeke, L.; Masschaele, B.; Cnudde, V.; Remon, J.P. Porous hydroxyapatite tablets as carriers for low-dosed drugs. *Eur. J. Pharm. Biopharm.* **2007**, *67*, 498–506. [CrossRef]
557. Uchida, A.; Shinto, Y.; Araki, N.; Ono, K. Slow release of anticancer drugs from porous calcium hydroxyapatite ceramic. *J. Orthop. Res.* **1992**, *10*, 440–445. [CrossRef]
558. Shinto, Y.; Uchida, A.; Korkusuz, F.; Araki, N.; Ono, K. Calcium hydroxyapatite ceramic used as a delivery system for antibiotics. *J. Bone. Joint. Surg. Br.* **1992**, *74*, 600–604. [CrossRef]
559. Martin, R.B.; Chapman, M.W.; Sharkey, N.A.; Zissimos, S.L.; Bay, B.; Shors, E.C. Bone ingrowth and mechanical properties of coralline hydroxyapatite 1 yr after implantation. *Biomaterials* **1993**, *14*, 341–348. [CrossRef]
560. Kazakia, G.J.; Nauman, E.A.; Ebenstein, D.M.; Halloran, B.P.; Keaveny, T.M. Effects of in vitro bone formation on the mechanical properties of a trabeculated hydroxyapatite bone substitute. *J. Biomed. Mater. Res. A* **2006**, *77A*, 688–699. [CrossRef]
561. Hing, K.A.; Best, S.M.; Tanner, K.E.; Bonfield, W.; Revell, P.A. Mediation of bone ingrowth in porous hydroxyapatite bone graft substitutes. *J. Biomed. Mater. Res. A* **2004**, *68A*, 187–200. [CrossRef]
562. Vuola, J.; Taurio, R.; Goransson, H.; Asko-Seljavaara, S. Compressive strength of calcium carbonate and hydroxyapatite implants after bone marrow induced osteogenesis. *Biomaterials* **1998**, *19*, 223–227. [CrossRef]
563. Von Doernberg, M.C.; von Rechenberg, B.; Bohner, M.; Grünenfelder, S.; van Lenthe, G.H.; Müller, R.; Gasser, B.; Mathys, R.; Baroud, G.; Auer, J. In vivo behavior of calcium phosphate scaffolds with four different pore sizes. *Biomaterials* **2006**, *27*, 5186–5198. [CrossRef]
564. Mygind, T.; Stiehler, M.; Baatrup, A.; Li, H.; Zou, X.; Flyvbjerg, A.; Kassem, M.; Bunger, C. Mesenchymal stem cell ingrowth and differentiation on coralline hydroxyapatite scaffolds. *Biomaterials* **2007**, *28*, 1036–1047. [CrossRef]
565. Mankani, M.H.; Afghani, S.; Franco, J.; Launey, M.; Marshall, S.; Marshall, G.W.; Nissenson, R.; Lee, J.; Tomsia, A.P.; Saiz, E. Lamellar spacing in cuboid hydroxyapatite scaffolds regulates bone formation by human bone marrow stromal cells. *Tissue Eng. A* **2011**, *17*, 1615–1623. [CrossRef]
566. Chan, O.; Coathup, M.J.; Nesbitt, A.; Ho, C.Y.; Hing, K.A.; Buckland, T.; Campion, C.; Blunn, G.W. The effects of microporosity on osteoinduction of calcium phosphate bone graft substitute biomaterials. *Acta Biomater.* **2012**, *8*, 2788–2794. [CrossRef]
567. Holmes, R.E. Bone regeneration within a coralline hydroxyapatite implant. *Plast. Reconstr. Surg.* **1979**, *63*, 626–633. [CrossRef]
568. Tsuruga, E.; Takita, H.; Wakisaka, Y.; Kuboki, Y. Pore size of porous hydoxyapatite as the cell-substratum controls BMP-induced osteogenesis. *J. Biochem.* **1997**, *121*, 317–324. [CrossRef]
569. LeGeros, R.Z.; LeGeros, J.P. Calcium phosphate bioceramics: Past, present, future. *Key Eng. Mater.* **2003**, *240–242*, 3–10. [CrossRef]
570. Woodard, J.R.; Hilldore, A.J.; Lan, S.K.; Park, C.J.; Morgan, A.W.; Eurell, J.A.C.; Clark, S.G.; Wheeler, M.B.; Jamison, R.D.; Wagoner Johnson, A.J. The mechanical properties and osteoconductivity of hydroxyapatite bone scaffolds with multi-scale porosity. *Biomaterials* **2007**, *28*, 45–54. [CrossRef] [PubMed]
571. Jacovella, P.F.; Peiretti, C.B.; Cunille, D.; Salzamendi, M.; Schechtel, S.A. Long-lasting results with hydroxylapatite (Radiesse) facial filler. *Plast. Reconstr. Surg.* **2006**, *118*, 15S–21S. [CrossRef] [PubMed]
572. Rustom, L.E.; Poellmann, M.J.; Wagoner Johnson, A.J. Mineralization in micropores of calcium phosphate scaffolds. *Acta Biomater.* **2019**, *83*, 435–455. [CrossRef] [PubMed]

573. Dubok, V.A. Bioceramics–yesterday, today, tomorrow. *Powder Metall. Met. Ceram.* **2000**, *39*, 381–394. [CrossRef]
574. Heness, G.; Ben-Nissan, B. Innovative bioceramics. *Mater. Forum* **2004**, *27*, 104–114.
575. Williams, D.F. There is no such thing as a biocompatible material. *Biomaterials* **2014**, *35*, 10009–10014. [CrossRef]
576. Punj, S.; Singh, J.; Singh, K. Ceramic biomaterials: Properties, state of the art and future prospectives. *Ceram. Int.* **2021**, *47*, 28059–28074. [CrossRef]
577. Greenspan, D.C. Bioactive ceramic implant materials. *Curr. Opin. Solid State Mater. Sci.* **1999**, *4*, 389–393. [CrossRef]
578. Blokhuis, T.J.; Termaat, M.F.; den Boer, F.C.; Patka, P.; Bakker, F.C.; Haarman, H.J.T.M. Properties of calcium phosphate ceramics in relation to their in vivo behavior. *J. Trauma* **2000**, *48*, 179–189. [CrossRef]
579. Kim, H.M. Bioactive ceramics: Challenges and perspectives. *J. Ceram. Soc. Jpn.* **2001**, *109*, S49–S57. [CrossRef]
580. Seeley, Z.; Bandyopadhyay, A.; Bose, S. Tricalcium phosphate based resorbable ceramics: Influence of NaF and CaO addition. *Mater. Sci. Eng. C* **2008**, *28*, 11–17. [CrossRef]
581. Descamps, M.; Richart, O.; Hardouin, P.; Hornez, J.C.; Leriche, A. Synthesis of macroporous β-tricalcium phosphate with controlled porous architectural. *Ceram. Int.* **2008**, *34*, 1131–1137. [CrossRef]
582. Cushnie, E.K.; Khan, Y.M.; Laurencin, C.T. Amorphous hydroxyapatite-sintered polymeric scaffolds for bone tissue regeneration: Physical characterization studies. *J. Biomed. Mater. Res. A* **2008**, *84A*, 54–62. [CrossRef]
583. Hench, L.L.; Thompson, I. Twenty-first century challenges for biomaterials. *J. R. Soc. Interface* **2010**, *7*, S379–S391. [CrossRef]
584. Nagase, M.; Baker, D.G.; Schumacher, H.R. Prolonged inflammatory reactions induced by artificial ceramics in the rat pouch model. *J. Rheumatol.* **1988**, *15*, 1334–1338.
585. Rooney, T.; Berman, S.; Indersano, A.T. Evaluation of porous block hydroxylapatite for augmentation of alveolar ridges. *J. Oral Maxillofac. Surg.* **1988**, *46*, 15–18. [CrossRef]
586. Prudhommeaux, F.; Schiltz, C.; Lioté, F.; Hina, A.; Champy, R.; Bucki, B.; Ortiz-Bravo, E.; Meunier, A.; Rey, C.; Bardin, T. Variation in the inflammatory properties of basic calcium phosphate crystals according to crystal type. *Arthritis Rheum.* **1996**, *39*, 1319–1326. [CrossRef]
587. Ghanaati, S.; Barbeck, M.; Orth, C.; Willershausen, I.; Thimm, B.W.; Hoffmann, C.; Rasic, A.; Sader, R.A.; Unger, R.E.; Peters, F.; et al. Influence of β-tricalcium phosphate granule size and morphology on tissue reaction in vivo. *Acta Biomater.* **2010**, *6*, 4476–4487. [CrossRef]
588. Lin, K.; Yuan, W.; Wang, L.; Lu, J.; Chen, L.; Wang, Z.; Chang, J. Evaluation of host inflammatory responses of β-tricalcium phosphate bioceramics caused by calcium pyrophosphate impurity using a subcutaneous model. *J. Biomed. Mater. Res. B Appl. Biomater.* **2011**, *99B*, 350–358. [CrossRef]
589. Velard, F.; Braux, J.; Amedee, J.; Laquerriere, P. Inflammatory cell response to calcium phosphate biomaterial particles: An overview. *Acta Biomater.* **2013**, *9*, 4956–4963. [CrossRef] [PubMed]
590. Rydén, L.; Molnar, D.; Esposito, M.; Johansson, A.; Suska, F.; Palmquist, A.; Thomsen, P. Early inflammatory response in soft tissues induced by thin calcium phosphates. *J. Biomed. Mater. Res. A* **2013**, *101A*, 2712–2717. [CrossRef] [PubMed]
591. Chatterjea, A.; van der Stok, J.; Danoux, C.B.; Yuan, H.; Habibovic, P.; van Blitterswijk, C.A.; Weinans, H.; de Boer, J. Inflammatory response and bone healing capacity of two porous calcium phosphate ceramics in a critical size cortical bone defects. *J. Biomed. Mater. Res. A* **2014**, *102A*, 1399–1407. [CrossRef]
592. Friesenbichler, J.; Maurer-Ertl, W.; Sadoghi, P.; Pirker-Fruehauf, U.; Bodo, K.; Leithner, A. Adverse reactions of artificial bone graft substitutes: Lessons learned from using tricalcium phosphate geneX®. *Clin. Orthop. Relat. Res.* **2014**, *472*, 976–982. [CrossRef] [PubMed]
593. Chang, T.Y.; Pan, S.C.; Huang, Y.H.; Hsueh, Y.Y. Blindness after calcium hydroxylapatite injection at nose. *J. Plast. Reconstr. Aesthet. Surg.* **2014**, *67*, 1755–1757. [CrossRef]
594. Ghanaati, S.; Barbeck, M.; Detsch, R.; Deisinger, U.; Hilbig, U.; Rausch, V.; Sader, R.; Unger, R.E.; Ziegler, G.; Kirkpatrick, C.J. The chemical composition of synthetic bone substitutes influences tissue reactions in vivo: Histological and histomorphometrical analysis of the cellular inflammatory response to hydroxyapatite, beta-tricalcium phosphate and biphasic calcium phosphate ceramics. *Biomed. Mater.* **2012**, *7*, 015005.
595. Draenert, K.; Draenert, M.; Erler, M.; Draenert, A.; Draenert, Y. How bone forms in large cancellous defects: Critical analysis based on experimental work and literature. *Injury* **2011**, *42* (Suppl. 2), S47–S55. [CrossRef]
596. Albrektsson, T.; Johansson, C. Osteoinduction, osteoconduction and osseointegration. *Eur. Spine. J.* **2001**, *10*, S96–S101.
597. Yuan, H.; Kurashina, K.; de Bruijn, D.J.; Li, Y.; de Groot, K.; Zhang, X. A preliminary study of osteoinduction of two kinds of calcium phosphate bioceramics. *Biomaterials* **1999**, *20*, 1799–1806. [CrossRef]
598. Yuan, H.P.; de Bruijn, J.D.; Li, Y.B.; Feng, J.Q.; Yang, Z.J.; de Groot, K.; Zhang, X.D. Bone formation induced by calcium phosphate ceramics in soft tissue of dogs: A comparative study between porous α-TCP and β-TCP. *J. Mater. Sci. Mater. Med.* **2001**, *12*, 7–13. [CrossRef]
599. Barrere, F.; van der Valk, C.M.; Dalmeijer, R.A.; Meijer, G.; van Blitterswijk, C.A.; de Groot, K.; Layrolle, P. Osteogenecity of octacalcium phosphate coatings applied on porous metal implants. *J. Biomed. Mater. Res. A* **2003**, *66A*, 779–788. [CrossRef]
600. Habibovic, P.; van der Valk, C.M.; van Blitterswijk, C.A.; de Groot, K.; Meijer, G. Influence of octacalcium phosphate coating on osteoinductive properties of biomaterials. *J. Mater. Sci. Mater. Med.* **2004**, *15*, 373–380. [CrossRef]
601. Ripamonti, U.; Richter, P.W.; Nilen, R.W.; Renton, L. The induction of bone formation by smart biphasic hydroxyapatite tricalcium phosphate biomimetic matrices in the non human primate *Papio ursinus*. *J. Cell. Mol. Med.* **2008**, *12*, 2609–2621. [CrossRef]

602. Cheng, L.; Ye, F.; Yang, R.; Lu, X.; Shi, Y.; Li, L.; Fan, H.; Bu, H. Osteoinduction of hydroxyapatite/β-tricalcium phosphate bioceramics in mice with a fractured fibula. *Acta Biomater.* **2010**, *6*, 1569–1574. [CrossRef]
603. Yuan, H.; Fernandes, H.; Habibovic, P.; de Boer, J.; Barradas, A.M.C.; de Ruiter, A.; Walsh, W.R.; van Blitterswijk, C.A.; de Bruijn, J.D. Osteoinductive ceramics as a synthetic alternative to autologous bone grafting. *Proc. Natl. Acad. Sci. USA* **2010**, *107*, 13614–13619. [CrossRef]
604. Barradas, A.M.; Yuan, H.; van der Stok, J.; le Quang, B.; Fernandes, H.; Chaterjea, A.; Hogenes, M.C.; Shultz, K.; Donahue, L.R.; van Blitterswijk, C.; et al. The influence of genetic factors on the osteoinductive potential of calcium phosphate ceramics in mice. *Biomaterials* **2012**, *33*, 5696–5705. [CrossRef]
605. Li, B.; Liao, X.; Zheng, L.; Zhu, X.; Wang, Z.; Fan, H.; Zhang, X. Effect of nanostructure on osteoinduction of porous biphasic calcium phosphate ceramics. *Acta Biomater.* **2012**, *8*, 3794–3804. [CrossRef]
606. Cheng, L.; Shi, Y.; Ye, F.; Bu, H. Osteoinduction of calcium phosphate biomaterials in small animals. *Mater. Sci. Eng. C* **2013**, *33*, 1254–1260. [CrossRef]
607. Davison, N.L.; Gamblin, A.L.; Layrolle, P.; Yuan, H.; de Bruijn, J.D.; Barrère-de Groot, F. Liposomal clodronate inhibition of osteoclastogenesis and osteoinduction by submicrostructured beta-tricalcium phosphate. *Biomaterials* **2014**, *35*, 5088–5097. [CrossRef]
608. Wang, L.; Barbieri, D.; Zhou, H.; de Bruijn, J.D.; Bao, C.; Yuan, H. Effect of particle size on osteoinductive potential of microstructured biphasic calcium phosphate ceramic. *J. Biomed. Mater. Res. A* **2015**, *103A*, 1919–1929. [CrossRef]
609. He, Y.; Peng, Y.; Liu, L.; Hou, S.; Mu, J.; Lan, L.; Cheng, L.; Shi, Z. The relationship between osteoinduction and vascularization: Comparing the ectopic bone formation of five different calcium phosphate biomaterials. *Materials* **2022**, *15*, 3440. [CrossRef]
610. Yuan, H.; Barbieri, D.; Luo, X.; van Blitterswijk, C.A.; de Bruijn, J.D. Calcium phosphates and bone induction. In *Comprehensive Biomaterials II*; Chapter 1.14; Ducheyne, P., Ed.; Elsevier: Oxford, UK, 2017; pp. 339–349.
611. Murata, M.; Hino, J.; Kabir, M.A.; Yokozeki, K.; Sakamoto, M.; Nakajima, T.; Akazawa, T. Osteoinduction in novel micropores of partially dissolved and precipitated hydroxyapatite block in scalp of young rats. *Materials* **2021**, *14*, 196. [CrossRef]
612. Iaquinta, M.R.; Torreggiani, E.; Mazziotta, C.; Ruffini, A.; Sprio, S.; Tampieri, A.; Tognon, M.; Martini, F.; Mazzoni, E. In vitro osteoinductivity assay of hydroxylapatite scaffolds, obtained with biomorphic transformation processes, assessed using human adipose stem cell cultures. *Int. J. Mol. Sci.* **2021**, *22*, 7092. [CrossRef]
613. Habibovic, P.; Yuan, H.; van der Valk, C.M.; Meijer, G.; van Blitterswijk, C.A.; de Groot, K. 3D microenvironment as essential element for osteoinduction by biomaterials. *Biomaterials* **2005**, *26*, 3565–3575. [CrossRef]
614. Habibovic, P.; Sees, T.M.; van den Doel, M.A.; van Blitterswijk, C.A.; de Groot, K. Osteoinduction by biomaterials—physicochemical and structural influences. *J. Biomed. Mater. Res. A* **2006**, *77A*, 747–762. [CrossRef] [PubMed]
615. Reddi, A.H. Morphogenesis and tissue engineering of bone and cartilage: Inductive signals, stem cells and biomimetic biomaterials. *Tissue Eng.* **2000**, *6*, 351–359. [CrossRef] [PubMed]
616. Ripamonti, U. The morphogenesis of bone in replicas of porous hydroxyapatite obtained by conversion of calcium carbonate exoskeletons of coral. *J. Bone Joint Surg. A* **1991**, *73*, 692–703. [CrossRef]
617. Kuboki, Y.; Takita, H.; Kobayashi, D. BMP-induced osteogenesis on the surface of hydroxyapatite with geometrically feasible and nonfeasible structures: Topology of osteogenesis. *J. Biomed. Mater. Res.* **1998**, *39*, 190–199. [CrossRef]
618. Zhang, J.; Luo, X.; Barbieri, D.; Barradas, A.M.C.; de Bruijn, J.D.; van Blitterswijk, C.A.; Yuan, H. The size of surface microstructures as an osteogenic factor in calcium phosphate ceramics. *Acta Biomater.* **2014**, *10*, 3254–3263. [CrossRef]
619. Zhang, J.; Barbieri, D.; Ten Hoopen, H.; de Bruijn, J.D.; van Blitterswijk, C.A.; Yuan, H. Microporous calcium phosphate ceramics driving osteogenesis through surface architecture. *J. Biomed. Mater. Res. A* **2015**, *103A*, 1188–1199. [CrossRef]
620. Xiao, D.; Zhang, J.; Zhang, C.; Barbieri, D.; Yuan, H.; Moroni, L.; Feng, G. The role of calcium phosphate surface structure in osteogenesis and the mechanisms involved. *Acta Biomater.* **2020**, *106*, 22–33. [CrossRef]
621. Daskalova, A.; Angelova, L.; Trifonov, A.; Lasgorceix, M.; Hocquet, S.; Minne, M.; Declercq, H.; Leriche, A.; Aceti, D.; Buchvarov, I. Development of femtosecond laser-engineered β-tricalcium phosphate (β-TCP) biomimetic templates for orthopaedic tissue engineering. *Appl. Sci.* **2021**, *11*, 2565. [CrossRef]
622. Diaz-Flores, L.; Gutierrez, R.; Lopez-Alonso, A.; Gonzalez, R.; Varela, H. Pericytes as a supplementary source of osteoblasts in periosteal osteogenesis. *Clin. Orthop. Relat. Res.* **1992**, *275*, 280–286. [CrossRef]
623. Bohner, M.; Miron, R.J. A proposed mechanism for material-induced heterotopic ossification. *Mater. Today* **2019**, *22*, 132–141. [CrossRef]
624. Boyan, B.D.; Schwartz, Z. Are calcium phosphate ceramics 'smart' biomaterials? *Nat. Rev. Rheumatol.* **2011**, *7*, 8–9. [CrossRef]
625. Li, X.; Ma, B.; Li, J.; Shang, L.; Liu, H.; Ge, S. A method to visually observe the degradation-diffusion-reconstruction behavior of hydroxyapatite in the bone repair process. *Acta Biomater.* **2020**, *101*, 554–564. [CrossRef]
626. Lu, J.; Descamps, M.; Dejou, J.; Koubi, G.; Hardouin, P.; Lemaitre, J.; Proust, J.P. The biodegradation mechanism of calcium phosphate biomaterials in bone. *J. Biomed. Mater. Res. Appl. Biomater.* **2002**, *63*, 408–412. [CrossRef]
627. Wang, H.; Lee, J.K.; Moursi, A.; Lannutti, J.J. Ca/P ratio effects on the degradation of hydroxyapatite in vitro. *J. Biomed. Mater. Res. A* **2003**, *67A*, 599–608. [CrossRef]
628. Dorozhkin, S.V. Inorganic chemistry of the dissolution phenomenon, the dissolution mechanism of calcium apatites at the atomic (ionic) level. *Comments Inorg. Chem.* **1999**, *20*, 285–299. [CrossRef]

629. Dorozhkin, S.V. Dissolution mechanism of calcium apatites in acids: A review of literature. *World J. Methodol.* **2012**, *2*, 1–17. [CrossRef]

630. Sakai, S.; Anada, T.; Tsuchiya, K.; Yamazaki, H.; Margolis, H.C.; Suzuki, O. Comparative study on the resorbability and dissolution behavior of octacalcium phosphate, β-tricalcium phosphate, and hydroxyapatite under physiological conditions. *Dent. Mater. J.* **2016**, *35*, 216–224. [CrossRef] [PubMed]

631. Wenisch, S.; Stahl, J.P.; Horas, U.; Heiss, C.; Kilian, O.; Trinkaus, K.; Hild, A.; Schnettler, R. In vivo mechanisms of hydroxyapatite ceramic degradation by osteoclasts: Fine structural microscopy. *J. Biomed. Mater. Res. A* **2003**, *67A*, 713–718. [CrossRef] [PubMed]

632. Riihonen, R.; Nielsen, S.; Väänänen, H.K.; Laitala-Leinonen, T.; Kwon, T.H. Degradation of hydroxyapatite in vivo and in vitro requires osteoclastic sodium-bicarbonate co-transporter NBCn1. *Matrix Biol.* **2010**, *29*, 287–294. [CrossRef] [PubMed]

633. Meille, S.; Gallo, M.; Clément, P.; Tadier, S.; Chevalier, J. Spherical instrumented indentation as a tool to characterize porous bioceramics and their resorption. *J. Eur. Ceram. Soc.* **2019**, *39*, 4459–4472. [CrossRef]

634. Teitelbaum, S.L. Bone resorption by osteoclasts. *Science* **2000**, *289*, 1504–1508. [CrossRef]

635. Matsunaga, A.; Takami, M.; Irié, T.; Mishima, K.; Inagaki, K.; Kamijo, R. Microscopic study on resorption of β-tricalcium phosphate materials by osteoclasts. *Cytotechnology* **2015**, *67*, 727–732. [CrossRef]

636. Narducci, P.; Nicolin, V. Differentiation of activated monocytes into osteoclast-like cells on a hydroxyapatite substrate: An in vitro study. *Ann. Anat.* **2009**, *191*, 349–355. [CrossRef]

637. Wu, V.M.; Uskoković, V. Is there a relationship between solubility and resorbability of different calcium phosphate phases in vitro? *Biochim. Biophys. Acta* **2016**, *1860*, 2157–2168. [CrossRef]

638. Stastny, P.; Sedlacek, R.; Suchy, T.; Lukasova, V.; Rampichova, M.; Trunec, M. Structure degradation and strength changes of sintered calcium phosphate bone scaffolds with different phase structures during simulated biodegradation in vitro. *Mater. Sci. Eng. C* **2019**, *100*, 544–553. [CrossRef]

639. Tamimi, F.; Torres, J.; Bassett, D.; Barralet, J.; Cabarcos, E.L. Resorption of monetite granules in alveolar bone defects in human patients. *Biomaterials* **2010**, *31*, 2762–2769. [CrossRef]

640. Sheikh, Z.; Abdallah, M.N.; Hanafi, A.A.; Misbahuddin, S.; Rashid, H.; Glogauer, M. Mechanisms of in vivo degradation and resorption of calcium phosphate based biomaterials. *Materials* **2015**, *8*, 7913–7925. [CrossRef]

641. Raynaud, S.; Champion, E.; Lafon, J.P.; Bernache-Assollant, D. Calcium phosphate apatites with variable Ca/P atomic ratio. III. Mechanical properties and degradation in solution of hot pressed ceramics. *Biomaterials* **2002**, *23*, 1081–1089. [CrossRef]

642. Barrère, F.; van der Valk, C.M.; Dalmeijer, R.A.J.; van Blitterswijk, C.A.; de Groot, K.; Layrolle, P. In vitro and in vivo degradation of biomimetic octacalcium phosphate and carbonate apatite coatings on titanium implants. *J. Biomed. Mater. Res. A* **2003**, *64A*, 378–387. [CrossRef]

643. Souto, R.M.; Laz, M.M.; Reis, R.L. Degradation characteristics of hydroxyapatite coatings on orthopaedic TiAlV in simulated physiological media investigated by electrochemical impedance spectroscopy. *Biomaterials* **2003**, *24*, 4213–4221. [CrossRef]

644. Dellinger, J.G.; Wojtowicz, A.M.; Jamison, R.D. Effects of degradation and porosity on the load bearing properties of model hydroxyapatite bone scaffolds. *J. Biomed. Mater. Res. A* **2006**, *77A*, 563–571. [CrossRef]

645. Gallo, M.; Tadier, S.; Meille, S.; Chevalier, J. Resorption of calcium phosphate materials: Considerations on the in vitro evaluation. *J. Eur. Ceram. Soc.* **2018**, *38*, 899–914. [CrossRef]

646. Okuda, T.; Ioku, K.; Yonezawa, I.; Minagi, H.; Kawachi, G.; Gonda, Y.; Murayama, H.; Shibata, Y.; Minami, S.; Kamihara, S.; et al. The effect of the microstructure of β-tricalcium phosphate on the metabolism of subsequently formed bone tissue. *Biomaterials* **2007**, *28*, 2612–2621. [CrossRef]

647. Maazouz, Y.; Rentsch, I.; Lu, B.; Santoni, B.L.G.; Doebelin, N.; Bohner, M. In vitro measurement of the chemical changes occurring within β-tricalcium phosphate bone graft substitutes. *Acta Biomater.* **2020**, *102*, 440–457. [CrossRef]

648. Li, X.; Zou, Q.; Chen, H.; Li, W. In vivo changes of nanoapatite crystals during bone reconstruction and the differences with native bone apatite. *Sci. Adv.* **2019**, *5*, eaay6484. [CrossRef]

649. Orly, I.; Gregoire, M.; Menanteau, J.; Heughebaert, M.; Kerebel, B. Chemical changes in hydroxyapatite biomaterial under in vivo and in vitro biological conditions. *Calcif. Tissue Int.* **1989**, *45*, 20–26. [CrossRef]

650. Sun, L.; Berndt, C.C.; Gross, K.A.; Kucuk, A. Review: Material fundamentals and clinical performance of plasma sprayed hydroxyapatite coatings. *J. Biomed. Mater. Res. Appl. Biomater.* **2001**, *58*, 570–592. [CrossRef]

651. Bertazzo, S.; Zambuzzi, W.F.; Campos, D.D.P.; Ogeda, T.L.; Ferreira, C.V.; Bertran, C.A. Hydroxyapatite surface solubility and effect on cell adhesion. *Colloid Surf. B* **2010**, *78*, 177–184. [CrossRef]

652. Arbez, B.; Manero, F.; Mabilleau, G.; Libouban, H.; Chappard, D. Human macrophages and osteoclasts resorb β-tricalcium phosphate in vitro but not mouse macrophages. *Micron* **2019**, *125*, 102730. [CrossRef]

653. Schwartz, Z.; Boyan, B.D. Underlying mechanisms at the bone-biomaterial interface. *J. Cell. Biochem.* **1994**, *56*, 340–347. [CrossRef]

654. Puleo, D.A.; Nanci, A. Understanding and controlling the bone-implant interface. *Biomaterials* **1999**, *20*, 2311–2321. [CrossRef]

655. Xin, R.; Leng, Y.; Chen, J.; Zhang, Q. A comparative study of calcium phosphate formation on bioceramics in vitro and in vivo. *Biomaterials* **2005**, *26*, 6477–6486. [CrossRef]

656. Knabe, C.; Adel-Khattab, D.; Ducheyne, P. Bioactivity: Mechanisms. In *Comprehensive Biomaterials II*; Chapter 1.12; Ducheyne, P., Ed.; Elsevier: Oxford, UK, 2017; pp. 291–310.

657. Girija, E.K.; Parthiban, S.P.; Suganthi, R.V.; Elayaraja, K.; Joshy, M.I.A.; Vani, R.; Kularia, P.; Asokan, K.; Kanjilal, D.; Yokogawa, Y.; et al. High energy irradiation–a tool for enhancing the bioactivity of hydroxyapatite. *J. Ceram. Soc. Jpn.* **2008**, *116*, 320–324. [CrossRef]

658. Okada, M.; Furukawa, K.; Serizawa, T.; Yanagisawa, Y.; Tanaka, H.; Kawai, T.; Furuzono, T. Interfacial interactions between calcined hydroxyapatite nanocrystals and substrates. *Langmuir* **2009**, *25*, 6300–6306. [CrossRef]

659. Callis, P.D.; Donaldson, K.; McCord, J.F. Early cellular responses to calcium phosphate ceramics. *Clin. Mater.* **1988**, *3*, 183–190. [CrossRef]

660. Okumura, M.; Ohgushi, H.; Tamai, S. Bonding osteogenesis in coralline hydroxyapatite combined with bone marrow cells. *Biomaterials* **1990**, *12*, 28–37. [CrossRef]

661. Doi, Y.; Iwanaga, H.; Shibutani, T.; Moriwaki, Y.; Iwayama, Y. Osteoclastic responses to various calcium phosphates in cell cultures. *J. Biomed. Mater. Res.* **1999**, *47*, 424–433. [CrossRef]

662. Guo, X.; Gough, J.E.; Xiao, P.; Liu, J.; Shen, Z. Fabrication of nanostructured hydroxyapatite and analysis of human osteoblastic cellular response. *J. Biomed. Mater. Res. A* **2007**, *82A*, 1022–1032. [CrossRef] [PubMed]

663. Wang, Y.; Zhang, S.; Zeng, X.; Ma, L.L.; Weng, W.; Yan, W.; Qian, M. Osteoblastic cell response on fluoridated hydroxyapatite coatings. *Acta Biomater.* **2007**, *3*, 191–197. [CrossRef] [PubMed]

664. Bae, W.J.; Chang, S.W.; Lee, S.I.; Kum, K.Y.; Bae, K.S.; Kim, E.C. Human periodontal ligament cell response to a newly developed calcium phosphate-based root canal sealer. *J. Endod.* **2010**, *36*, 1658–1663. [CrossRef] [PubMed]

665. Zhao, X.; Heng, B.C.; Xiong, S.; Guo, J.; Tan, T.T.Y.; Boey, F.Y.C.; Ng, K.W.; Loo, J.S.C. In vitro assessment of cellular responses to rod-shaped hydroxyapatite nanoparticles of varying lengths and surface areas. *Nanotoxicology* **2011**, *5*, 182–194. [CrossRef] [PubMed]

666. Liu, X.; Zhao, M.; Lu, J.; Ma, J.; Wei, J.; Wei, S. Cell responses to two kinds of nanohydroxyapatite with different sizes and crystallinities. *Int. J. Nanomed.* **2012**, *7*, 1239–1250. [CrossRef] [PubMed]

667. Lobo, S.E.; Glickman, R.; da Silva, W.N.; Arinzeh, T.L.; Kerkis, I. Response of stem cells from different origins to biphasic calcium phosphate bioceramics. *Cell Tiss. Res.* **2015**, *361*, 477–495. [CrossRef] [PubMed]

668. Oliveira, H.L.; da Rosa, W.L.O.; Cuevas-Suárez, C.E.; Carreño, N.L.V.; da Silva, A.F.; Guim, T.N.; Dellagostin, O.A.; Piva, E. Histological evaluation of bone repair with hydroxyapatite: A systematic review. *Calcif. Tissue Int.* **2017**, *101*, 341–354. [CrossRef]

669. Sadowska, J.M.; Wei, F.; Guo, J.; Guillem-Marti, J.; Lin, Z.; Ginebra, M.P.; Xiao, Y. The effect of biomimetic calcium deficient hydroxyapatite and sintered β-tricalcium phosphate on osteoimmune reaction and osteogenesis. *Acta Biomater.* **2019**, *96*, 605–618. [CrossRef]

670. Šupová, M.; Suchý, T.; Sucharda, Z.; Filová, E.; der Kinderen, J.N.L.M.; Steinerová, M.; Bačáková, L.; Martynková, G.S. The comprehensive in vitro evaluation of eight different calcium phosphates: Significant parameters for cell behavior. *J. Amer. Ceram. Soc.* **2019**, *102*, 2882–2904. [CrossRef]

671. Wang, X.; Yu, Y.; Ji, L.; Geng, Z.; Wang, J.; Liu, C. Calcium phosphate-based materials regulate osteoclast-mediated osseointegration. *Bioact. Mater.* **2021**, *6*, 4517–4530. [CrossRef]

672. Suzuki, T.; Ohashi, R.; Yokogawa, Y.; Nishizawa, K.; Nagata, F.; Kawamoto, Y.; Kameyama, T.; Toriyama, M. Initial anchoring and proliferation of fibroblast L-929 cells on unstable surface of calcium phosphate ceramics. *J. Biosci. Bioeng.* **1999**, *87*, 320–327. [CrossRef]

673. Arinzeh, T.L.; Tran, T.; McAlary, J.; Daculsi, G. A comparative study of biphasic calcium phosphate ceramics for human mesenchymal stem-cell-induced bone formation. *Biomaterials* **2005**, *26*, 3631–3638. [CrossRef]

674. Oh, S.; Oh, N.; Appleford, M.; Ong, J.L. Bioceramics for tissue engineering applications–a review. *Am. J. Biochem. Biotechnol.* **2006**, *2*, 49–56. [CrossRef]

675. Appleford, M.; Oh, S.; Cole, J.A.; Carnes, D.L.; Lee, M.; Bumgardner, J.D.; Haggard, W.O.; Ong, J.L. Effects of trabecular calcium phosphate scaffolds on stress signaling in osteoblast precursor cells. *Biomaterials* **2007**, *28*, 2747–2753. [CrossRef]

676. Gamie, Z.; Tran, G.T.; Vyzas, G.; Korres, N.; Heliotis, M.; Mantalaris, A.; Tsiridis, E. Stem cells combined with bone graft substitutes in skeletal tissue engineering. *Expert Opin. Biol. Ther.* **2012**, *12*, 713–729. [CrossRef]

677. Manfrini, M.; di Bona, C.; Canella, A.; Lucarelli, E.; Pellati, A.; d'Agostino, A.; Barbanti-Bròdano, G.; Tognon, M. Mesenchymal stem cells from patients to assay bone graft substitutes. *J. Cell. Physiol.* **2013**, *228*, 1229–1237. [CrossRef]

678. Unger, R.E.; Sartoris, A.; Peters, K.; Motta, A.; Migliaresi, C.; Kunkel, M.; Bulnheim, U.; Rychly, J.; Kirkpatrick, C.J. Tissue-like self-assembly in cocultures of endothelial cells and osteoblasts and the formation of microcapillary like structures on three-dimensional porous biomaterials. *Biomaterials* **2007**, *28*, 3965–3976. [CrossRef]

679. Nazir, N.M.; Dasmawati, M.; Azman, S.M.; Omar, N.S.; Othman, R. Biocompatibility of in house β-tricalcium phosphate ceramics with normal human osteoblast cell. *J. Eng. Sci. Technol.* **2012**, *7*, 169–176.

680. Tan, F.; O'Neill, F.; Naciri, M.; Dowling, D.; Al-Rubeai, M. Cellular and transcriptomic analysis of human mesenchymal stem cell response to plasma-activated hydroxyapatite coating. *Acta Biomater.* **2012**, *8*, 1627–1638. [CrossRef]

681. Arbez, B.; Libouban, H. Behavior of macrophage and osteoblast cell lines in contact with the β-TCP biomaterial (beta-tricalcium phosphate) | Comportement de lignées cellulaires de macrophages et d'ostéoblastes en contact avec le biomatériau β-TCP (phosphate bêta tricalcique).]. *Morphologie* **2017**, *101*, 154–163. [CrossRef] [PubMed]

682. Frayssinet, P.; Rouquet, N.; Vidalain, P.O. Calcium phosphates for cell transfection. In *Comprehensive Biomaterials II*; Chapter 1.15.; Ducheyne, P., Ed.; Elsevier: Oxford, UK, 2017; pp. 350–359.

683. Teixeira, S.; Fernandes, M.H.; Ferraz, M.P.; Monteiro, F.J. Proliferation and mineralization of bone marrow cells cultured on macroporous hydroxyapatite scaffolds functionalized with collagen type I for bone tissue regeneration. *J. Biomed. Mater. Res. A* **2010**, *95A*, 1–8. [CrossRef] [PubMed]

684. Treccani, L.; Klein, T.Y.; Meder, F.; Pardun, K.; Rezwan, K. Functionalized ceramics for biomedical, biotechnological and environmental applications. *Acta Biomater.* **2013**, *9*, 7115–7150. [CrossRef] [PubMed]

685. Bigi, A.; Boanini, E. Functionalized biomimetic calcium phosphates for bone tissue repair. *J. Appl. Biomater. Funct. Mater.* **2017**, *15*, e313–e325. [CrossRef] [PubMed]

686. Dorozhkin, S.V. Functionalized calcium orthophosphates ($CaPO_4$) and their biomedical applications. *J. Mater. Chem. B* **2019**, *7*, 7471–7489. [CrossRef]

687. Zhuang, Z.; Yoshimura, H.; Aizawa, M. Synthesis and ultrastructure of plate-like apatite single crystals as a model for tooth enamel. *Mater. Sci. Eng. C* **2013**, *33*, 2534–2540. [CrossRef] [PubMed]

688. Zhuang, Z.; Fujimi, T.J.; Nakamura, M.; Konishi, T.; Yoshimura, H.; Aizawa, M. Development of *a,b*-plane-oriented hydroxyapatite ceramics as models for living bones and their cell adhesion behavior. *Acta Biomater.* **2013**, *9*, 6732–6740. [CrossRef]

689. Aizawa, M.; Matsuura, T.; Zhuang, Z. Syntheses of single-crystal apatite particles with preferred orientation to the *a*- and *c*-axes as models of hard tissue and their applications. *Biol. Pharm. Bull.* **2013**, *36*, 1654–1661. [CrossRef]

690. Lin, K.; Wu, C.; Chang, J. Advances in synthesis of calcium phosphate crystals with controlled size and shape. *Acta Biomater.* **2014**, *10*, 4071–4102. [CrossRef]

691. Chen, W.; Long, T.; Guo, Y.J.; Zhu, Z.A.; Guo, Y.P. Hydrothermal synthesis of hydroxyapatite coatings with oriented nanorod arrays. *RSC Adv.* **2014**, *4*, 185–191. [CrossRef]

692. Guan, J.J.; Tian, B.; Tang, S.; Ke, Q.F.; Zhang, C.Q.; Zhu, Z.A.; Guo, Y.P. Hydroxyapatite coatings with oriented nanoplate arrays: Synthesis, formation mechanism and cytocompatibility. *J. Mater. Chem. B* **2015**, *3*, 1655–1666. [CrossRef]

693. Ribeiro, C.; Rigo, E.C.S.; Sepúlveda, P.; Bressiani, J.C.; Bressiani, A.H.A. Formation of calcium phosphate layer on ceramics with different reactivities. *Mater. Sci. Eng. C* **2004**, *24*, 631–636. [CrossRef]

694. Barba, A.; Diez-Escudero, A.; Maazouz, Y.; Rappe, K.; Espanol, M.; Montufar, E.B.; Bonany, M.; Sadowska, J.M.; Guillem-Marti, J.; Öhman-Mägi, C.; et al. Osteoinduction by foamed and 3D-printed calcium phosphate scaffolds: Effect of nanostructure and pore architecture. *ACS Appl. Mater. Interfaces* **2017**, *9*, 41722–41736. [CrossRef]

695. Tsukanaka, M.; Fujibayashi, S.; Otsuki, B.; Takemoto, M.; Matsuda, S. Osteoinductive potential of highly purified porous β-TCP in mice. *J. Mater. Sci. Mater. Med.* **2015**, *26*, 132. [CrossRef]

696. Duan, R.; Barbieri, D.; Luo, X.; Weng, J.; Bao, C.; de Bruijn, J.D.; Yuan, H. Variation of the bone forming ability with the physicochemical properties of calcium phosphate bone substitutes. *Biomater. Sci.* **2018**, *6*, 136–145. [CrossRef]

697. Davison, N.L.; ten Harkel, B.; Schoenmaker, T.; Luo, X.; Yuan, H.; Everts, V.; Barrère-de Groot, F.; de Bruijn, J.D. Osteoclast resorption of beta-tricalcium phosphate controlled by surface architecture. *Biomaterials* **2014**, *35*, 7441–7451. [CrossRef]

698. Al-Maawi, S.; Barbeck, M.; Vizcaíno, C.H.; Egli, R.; Sader, R.; Kirkpatrick, C.J.; Bohner, M.; Ghanaati, S. Thermal treatment at 500 °C significantly reduces the reaction to irregular tricalcium phosphate granules as foreign bodies: An in vivo study. *Acta Biomater.* **2021**, *121*, 621–636. [CrossRef]

699. Levitt, G.E.; Crayton, P.H.; Monroe, E.A.; Condrate, R.A. Forming methods for apatite prosthesis. *J. Biomed. Mater. Res.* **1969**, *3*, 683–685. [CrossRef]

700. Fukuba, S.; Okada, M.; Nohara, K.; Iwata, T. Alloplastic bone substitutes for periodontal and bone regeneration in dentistry: Current status and prospects. *Materials* **2021**, *14*, 1096. [CrossRef]

701. Roca-Millan, E.; Jané-Salas, E.; Marí-Roig, A.; Jiménez-Guerra, Á.; Ortiz-García, I.; Velasco-Ortega, E.; López-López, J.; Monsalve-Guil, L. The application of beta-tricalcium phosphate in implant dentistry: A systematic evaluation of clinical studies. *Materials* **2022**, *15*, 655. [CrossRef]

702. Easwer, H.V.; Rajeev, A.; Varma, H.K.; Vijayan, S.; Bhattacharya, R.N. Cosmetic and radiological outcome following the use of synthetic hydroxyapatite porous-dense bilayer burr-hole buttons. *Acta Neurochir.* **2007**, *149*, 481–485. [CrossRef] [PubMed]

703. Kashimura, H.; Ogasawara, K.; Kubo, Y.; Yoshida, K.; Sugawara, A.; Ogawa, A. A newly designed hydroxyapatite ceramic burr-hole button. *Vasc. Health Risk Manag.* **2010**, *6*, 105–108. [CrossRef]

704. Jordan, D.R.; Gilberg, S.; Bawazeer, A. Coralline hydroxyapatite orbital implant (Bio-Eye): Experience with 158 patients. *Ophthal. Plast. Reconstr. Surg.* **2004**, *20*, 69–74. [CrossRef]

705. Yoon, J.S.; Lew, H.; Kim, S.J.; Lee, S.Y. Exposure rate of hydroxyapatite orbital implants. A 15-year experience of 802 cases. *Ophthalmology* **2008**, *115*, 566–572. [CrossRef] [PubMed]

706. Tabatabaee, Z.; Mazloumi, M.; Rajabi, T.M.; Khalilzadeh, O.; Kassaee, A.; Moghimi, S.; Eftekhar, H.; Goldberg, R.A. Comparison of the exposure rate of wrapped hydroxyapatite (Bio-Eye) versus unwrapped porous polyethylene (Medpor) orbital implants in enucleated patients. *Ophthal. Plast. Reconstr. Surg.* **2011**, *27*, 114–118. [CrossRef] [PubMed]

707. Ma, X.Z.; Bi, H.S.; Zhang, X. Effect of hydroxyapatite orbital implant for plastic surgery of eye in 52 cases. *Int. Eye Sci.* **2012**, *12*, 988–990.

708. Baino, F.; Vitale-Brovarone, C. Bioceramics in ophthalmology. *Acta Biomater.* **2014**, *10*, 3372–3397. [CrossRef]

709. Thiesmann, R.; Anagnostopoulos, A.; Stemplewitz, B. Long-term results of the compatibility of a coralline hydroxyapatite implant as eye replacement | Langzeitergebnisse zur Verträglichkeit eines korallinen Hydroxylapatitimplantats als Bulbusersatz. *Ophthalmologe* **2018**, *115*, 131–136. [CrossRef]

710. Wehrs, R.E. Hearing results with incus and incus stapes prostheses of hydroxylapatite. *Laryngoscope* **1991**, *101*, 555–556. [CrossRef]
711. Smith, J.; Gardner, E.; Dornhoffer, J.L. Hearing results with a hydroxylapatite/titanium bell partial ossicular replacement prosthesis. *Laryngoscope* **2002**, *112*, 1796–1799. [CrossRef]
712. Doi, T.; Hosoda, Y.; Kaneko, T.; Munemoto, Y.; Kaneko, A.; Komeda, M.; Furukawa, M.; Kuriyama, H.; Kitajiri, M.; Tomoda, K.; et al. Hearing results for ossicular reconstruction using a cartilage-connecting hydroxyapatite prosthesis with a spearhead. *Otol. Neurotol.* **2007**, *28*, 1041–1044. [CrossRef]
713. Thalgott, J.S.; Fritts, K.; Giuffre, J.M.; Timlin, M. Anterior interbody fusion of the cervical spine with coralline hydroxyapatite. *Spine* **1999**, *24*, 1295–1299. [CrossRef]
714. Mashoof, A.A.; Siddiqui, S.A.; Otero, M.; Tucci, J.J. Supplementation of autogenous bone graft with coralline hydroxyapatite in posterior spine fusion for idiopathic adolescent scoliosis. *Orthopedics* **2002**, *25*, 1073–1076. [CrossRef]
715. Liu, W.Y.; Mo, J.W.; Gao, H.; Liu, H.L.; Wang, M.Y.; He, C.L.; Tang, W.; Ye, Y.J. Nano-hydroxyapatite artificial bone serves as a spacer for fusion with the cervical spine after bone grafting. *Chin. J. Tissue Eng. Res.* **2012**, *16*, 5327–5330.
716. Litak, J.; Czyzewski, W.; Szymoniuk, M.; Pastuszak, B.; Litak, J.; Litak, G.; Grochowski, C.; Rahnama-Hezavah, M.; Kamieniak, P. Hydroxyapatite use in spine surgery–molecular and clinical aspect. *Materials* **2022**, *15*, 2906. [CrossRef]
717. Silva, R.V.; Camilli, J.A.; Bertran, C.A.; Moreira, N.H. The use of hydroxyapatite and autogenous cancellous bone grafts to repair bone defects in rats. *Int. J. Oral Maxillofac. Surg.* **2005**, *34*, 178–184. [CrossRef]
718. Damron, T.A. Use of 3D β-tricalcium phosphate (Vitoss®) scaffolds in repairing bone defects. *Nanomedicine* **2007**, *2*, 763–775. [CrossRef]
719. Zaed, I.; Cardia, A.; Stefini, R. From reparative surgery to regenerative surgery: State of the art of porous hydroxyapatite in cranioplasty. *Int. J. Mol. Sci.* **2022**, *23*, 5434. [CrossRef]
720. Alsahafi, R.A.; Mitwalli, H.A.; Balhaddad, A.A.; Weir, M.D.; Xu, H.H.K.; Melo, M.A.S. Regenerating craniofacial dental defects with calcium phosphate cement scaffolds: Current status and innovative scope review. *Front. Dent. Med.* **2021**, *2*, 743065. [CrossRef]
721. Bass, L.S.; Smith, S.; Busso, M.; McClaren, M. Calcium hydroxylapatite (Radiesse) for treatment of nasolabial folds: Long-term safety and efficacy results. *Aesthetic Surg. J.* **2010**, *30*, 235–238. [CrossRef]
722. Low, K.L.; Tan, S.H.; Zein, S.H.S.; Roether, J.A.; Mouriño, V.; Boccaccini, A.R. Calcium phosphate-based composites as injectable bone substitute materials. *J. Biomed. Mater. Res. B Appl. Biomater.* **2010**, *94B*, 273–286. [CrossRef] [PubMed]
723. Daculsi, G.; Uzel, A.P.; Weiss, P.; Goyenvalle, E.; Aguado, E. Developments in injectable multiphasic biomaterials. The performance of microporous biphasic calcium phosphate granules and hydrogels. *J. Mater. Sci. Mater. Med.* **2010**, *21*, 855–861. [CrossRef] [PubMed]
724. Suzuki, K.; Anada, T.; Honda, Y.; Kishimoto, K.N.; Miyatake, N.; Hosaka, M.; Imaizumi, H.; Itoi, E.; Suzuki, O. Cortical bone tissue response of injectable octacalcium phosphate-hyaluronic acid complexes. *Key Eng. Mater.* 2013; *529–530*, 296–299.
725. Pastorino, D.; Canal, C.; Ginebra, M.P. Drug delivery from injectable calcium phosphate foams by tailoring the macroporosity-drug interaction. *Acta Biomater.* **2015**, *12*, 250–259. [CrossRef] [PubMed]
726. Moussi, H.; Weiss, P.; le Bideau, J.; Gautier, H.; Charbonnier, B. Injectable macromolecules-based calcium phosphate bone substitutes. *Mater. Adv.* **2022**, *3*, 6125–6141. [CrossRef]
727. Bohner, M.; Baroud, G. Injectability of calcium phosphate pastes. *Biomaterials* **2005**, *26*, 1553–1563. [CrossRef]
728. Laschke, M.W.; Witt, K.; Pohlemann, T.; Menger, M.D. Injectable nanocrystalline hydroxyapatite paste for bone substitution: In vivo analysis of biocompatibility and vascularization. *J. Biomed. Mater. Res. B Appl. Biomater.* **2007**, *82B*, 494–505. [CrossRef]
729. Lopez-Heredia, M.A.; Barnewitz, D.; Genzel, A.; Stiller, M.; Peters, F.; Huebner, W.D.; Stang, B.; Kuhr, A.; Knabe, C. In vivo osteogenesis assessment of a tricalcium phosphate paste and a tricalcium phosphate foam bone grafting materials. *Key Eng. Mater.* **2015**, *631*, 426–429. [CrossRef]
730. Torres, P.M.C.; Gouveia, S.; Olhero, S.; Kaushal, A.; Ferreira, J.M.F. Injectability of calcium phosphate pastes: Effects of particle size and state of aggregation of β-tricalcium phosphate powders. *Acta Biomater.* **2015**, *21*, 204–216. [CrossRef]
731. Salinas, A.J.; Esbrit, P.; Vallet-Regí, M. A tissue engineering approach based on the use of bioceramics for bone repair. *Biomater. Sci.* **2013**, *1*, 40–51. [CrossRef]
732. ISO 13175-3:2012 Implants for Surgery—Calcium Phosphates—Part 3: Hydroxyapatite and Beta-Tricalcium Phosphate Bone Substitutes. Available online: https://www.iso.org/obp/ui/#iso:std:iso:13175:-3:ed-1:v1:en (accessed on 24 June 2022).
733. Chow, L.C. Next generation calcium phosphate-based biomaterials. *Dent. Mater. J.* **2009**, *28*, 1–10. [CrossRef]
734. Victor, S.P.; Kumar, T.S.S. Processing and properties of injectable porous apatitic cements. *J. Ceram. Soc. Jpn.* **2008**, *116*, 105–107. [CrossRef]
735. Hesaraki, S.; Nemati, R.; Nosoudi, N. Preparation and characterisation of porous calcium phosphate bone cement as antibiotic carrier. *Adv. Appl. Ceram.* **2009**, *108*, 231–240. [CrossRef]
736. Stulajterova, R.; Medvecky, L.; Giretova, M.; Sopcak, T. Structural and phase characterization of bioceramics prepared from tetracalcium phosphate–monetite cement and in vitro osteoblast response. *J. Mater. Sci. Mater. Med.* **2015**, *26*, 183. [CrossRef]
737. Bohner, M. Resorbable biomaterials as bone graft substitutes. *Mater. Today* **2010**, *13*, 24–30. [CrossRef]
738. Moussa, H.; Jiang, W.; Alsheghri, A.; Mansour, A.; Hadad, A.E.; Pan, H.; Tang, R.; Song, J.; Vargas, J.; McKee, M.D.; et al. High strength brushite bioceramics obtained by selective regulation of crystal growth with chiral biomolecules. *Acta Biomater.* **2020**, *106*, 351–359. [CrossRef]

739. Schröter, L.; Kaiser, F.; Stein, S.; Gbureck, U.; Ignatius, A. Biological and mechanical performance and degradation characteristics of calcium phosphate cements in large animals and humans. *Acta Biomater.* **2020**, *117*, 1–20. [CrossRef]

740. Paital, S.R.; Dahotre, N.B. Calcium phosphate coatings for bio-implant applications: Materials, performance factors, and methodologies. *Mater. Sci. Eng. R* **2009**, *66*, 1–70. [CrossRef]

741. León, B.; Jansen, J.A. (Eds.) *Thin Calcium Phosphate Coatings for Medical Implants*; Springer: New York, NY, USA, 2009; p. 326.

742. Dorozhkin, S.V. Calcium orthophosphate deposits: Preparation, properties and biomedical applications. *Mater. Sci. Eng. C* **2015**, *55*, 272–326. [CrossRef]

743. Kon, M.; Ishikawa, K.; Miyamoto, Y.; Asaoka, K. Development of calcium phosphate based functional gradient bioceramics. *Biomaterials* **1995**, *16*, 709–714. [CrossRef]

744. Wong, L.H.; Tio, B.; Miao, X. Functionally graded tricalcium phosphate/fluoroapatite composites. *Mater. Sci. Eng. C* **2002**, *20*, 111–115. [CrossRef]

745. Tampieri, A.; Celotti, G.; Sprio, S.; Delcogliano, A.; Franzese, S. Porosity-graded hydroxyapatite ceramics to replace natural bone. *Biomaterials* **2001**, *22*, 1365–1370. [CrossRef]

746. Werner, J.; Linner-Krcmar, B.; Friess, W.; Greil, P. Mechanical properties and in vitro cell compatibility of hydroxyapatite ceramics with graded pore structure. *Biomaterials* **2002**, *23*, 4285–4294. [CrossRef]

747. Watanabe, T.; Fukuhara, T.; Izui, H.; Fukase, Y.; Okano, M. Properties of HAp/β-TCP functionally graded material by spark plasma sintering. *Trans. Jpn. Soc. Mech. Eng. A* **2009**, *75*, 612–618. [CrossRef]

748. Bai, X.; Sandukas, S.; Appleford, M.R.; Ong, J.L.; Rabiei, A. Deposition and investigation of functionally graded calcium phosphate coatings on titanium. *Acta Biomater.* **2009**, *5*, 3563–3572. [CrossRef]

749. Tamura, A.; Asaoka, T.; Furukawa, K.; Ushida, T.; Tateishi, T. Application of α-TCP/HAp functionally graded porous beads for bone regenerative scaffold. *Adv. Sci. Technol.* **2013**, *86*, 63–69.

750. Gasik, M.; Keski-Honkola, A.; Bilotsky, Y.; Friman, M. Development and optimisation of hydroxyapatite-β-TCP functionally gradated biomaterial. *J. Mech. Behav. Biomed. Mater.* **2014**, *30*, 266–273. [CrossRef]

751. Zhou, C.; Deng, C.; Chen, X.; Zhao, X.; Chen, Y.; Fan, Y.; Zhang, X. Mechanical and biological properties of the micro-/nano-grain functionally graded hydroxyapatite bioceramics for bone tissue engineering. *J. Mech. Behav. Biomed. Mater.* **2015**, *48*, 1–11. [CrossRef]

752. Marković, S.; Lukić, M.J.; Škapin, S.D.; Stojanović, B.; Uskoković, D. Designing, fabrication and characterization of nanostructured functionally graded HAp/BCP ceramics. *Ceram. Int.* **2015**, *41*, 2654–2667. [CrossRef]

753. Salimi, E. Functionally graded calcium phosphate bioceramics: An overview of preparation and properties. *Ceram. Int.* **2020**, *46*, 19664–19668. [CrossRef]

754. Hollister, S.J. Porous scaffold design for tissue engineering. *Nat. Mater.* **2005**, *4*, 518–524. [CrossRef]

755. Jones, J.R.; Hench, L.L. Regeneration of trabecular bone using porous ceramics. *Curr. Opin. Solid State Mater. Sci.* **2003**, *7*, 301–307. [CrossRef]

756. Williams, D.F. To engineer is to create; the link between engineering and regeneration. *Trends Biotech.* **2006**, *24*, 4–8. [CrossRef] [PubMed]

757. Griffith, L.G.; Naughton, G. Tissue engineering–current challenges and expanding opportunities. *Science* **2002**, *295*, 1009–1014. [CrossRef] [PubMed]

758. Goldberg, V.M.; Caplan, A.I. *Orthopedic Tissue Engineering Basic Science and Practice*; Marcel Dekker: New York, NY, USA, 2004; p. 338.

759. Bahraminasab, M.; Janmohammadi, M.; Arab, S.; Talebi, A.; Nooshabadi, V.T.; Koohsarian, P.; Nourbakhsh, M.S. Bone scaffolds: An incorporation of biomaterials, cells, and biofactors. *ACS Biomater. Sci. Eng.* **2021**, *7*, 5397–5431. [CrossRef]

760. Ikada, Y. Challenges in tissue engineering. *J. R. Soc. Interface* **2006**, *3*, 589–601. [CrossRef]

761. Cima, L.G.; Langer, R. Engineering human tissue. *Chem. Eng. Prog.* **1993**, *89*, 46–54.

762. Langer, R.; Vacanti, J.P. Tissue engineering. *Science* **1993**, *260*, 920–926. [CrossRef]

763. El-Ghannam, A. Bone reconstruction: From bioceramics to tissue engineering. *Expert Rev. Med. Dev.* **2005**, *2*, 87–101. [CrossRef]

764. Kneser, U.; Schaefer, D.J.; Polykandriotis, E.; Horch, R.E. Tissue engineering of bone: The reconstructive surgeon's point of view. *J. Cell. Mol. Med.* **2006**, *10*, 7–19. [CrossRef]

765. Boyan, B.D.; Cohen, D.J.; Schwartz, Z. Bone tissue grafting and tissue engineering concepts. In *Comprehensive Biomaterials II*; Chapter 7.17.; Ducheyne, P., Ed.; Elsevier: Oxford, UK, 2017; pp. 298–313.

766. Koons, G.L.; Diba, M.; Mikos, A.G. Materials design for bone-tissue engineering. *Nat. Rev. Mater.* **2020**, *5*, 584–603. [CrossRef]

767. Lutolf, M.P.; Hubbell, J.A. Synthetic biomaterials as instructive extracellular microenvironments for morphogenesis in tissue engineering. *Nat. Biotechnol.* **2005**, *23*, 47–55. [CrossRef]

768. Ma, P.X. Biomimetic materials for tissue engineering. *Adv. Drug. Deliv. Rev.* **2008**, *60*, 184–198. [CrossRef]

769. Yang, S.; Leong, K.F.; Du, Z.; Chua, C.K. The design of scaffolds for use in tissue engineering. Part, I. Traditional factors. *Tissue Eng.* **2001**, *7*, 679–689. [CrossRef]

770. Ma, P.X.; Elisseeff, J. (Eds.) *Scaffolding in Tissue Engineering*; CRC Press: Boca Raton, FL, USA, 2006; p. 638.

771. Alvarez-Urena, P.; Kim, J.; Bhattacharyya, S.; Ducheyne, P. Bioactive ceramics and bioactive ceramic composite-based scaffolds. In *Comprehensive Biomaterials II*; Chapter 6.1.; Ducheyne, P., Ed.; Elsevier: Oxford, UK, 2017; pp. 1–19.

772. Baldwin, J.; Henkel, J.; Hutmacher, D.W. Engineering the organ bone. In *Comprehensive Biomaterials II*; Chapter 6.3.; Ducheyne, P., Ed.; Elsevier: Oxford, UK, 2017; pp. 54–74.

773. Williams, D.F. The biomaterials conundrum in tissue engineering. *Tissue Eng. A* **2014**, *20*, 1129–1131. [CrossRef]

774. Freed, L.E.; Guilak, F.; Guo, X.E.; Gray, M.L.; Tranquillo, R.; Holmes, J.W.; Radisic, M.; Sefton, M.V.; Kaplan, D.; Vunjak-Novakovic, G. Advanced tools for tissue engineering: Scaffolds, bioreactors, and signaling. *Tissue Eng.* **2006**, *12*, 3285–3305. [CrossRef]

775. Gandaglia, A.; Bagno, A.; Naso, F.; Spina, M.; Gerosa, G. Cells, scaffolds and bioreactors for tissue-engineered heart valves: A journey from basic concepts to contemporary developmental innovations. *Eur. J. Cardiothorac. Surg.* **2011**, *39*, 523–531. [CrossRef]

776. Hui, J.H.P.; Buhary, K.S.; Chowdhary, A. Implantation of orthobiologic, biodegradable scaffolds in osteochondral repair. *Orthop. Clin. North Am.* **2012**, *43*, 255–261. [CrossRef]

777. Vanderleyden, E.; Mullens, S.; Luyten, J.; Dubruel, P. Implantable (bio)polymer coated titanium scaffolds: A review. *Curr. Pharm. Des.* **2012**, *18*, 2576–2590. [CrossRef]

778. Service, R.F. Tissue engineers build new bone. *Science* **2000**, *289*, 1498–1500. [CrossRef]

779. Deligianni, D.D.; Katsala, N.D.; Koutsoukos, P.G.; Missirlis, Y.F. Effect of surface roughness of hydroxyapatite on human bone marrow cell adhesion, proliferation, differentiation and detachment strength. *Biomaterials* **2001**, *22*, 87–96. [CrossRef]

780. Fini, M.; Giardino, R.; Borsari, V.; Torricelli, P.; Rimondini, L.; Giavaresi, G.; Aldini, N.N. In vitro behaviour of osteoblasts cultured on orthopaedic biomaterials with different surface roughness, uncoated and fluorohydroxyapatite-coated, relative to the in vivo osteointegration rate. *Int. J. Artif. Organs* **2003**, *26*, 520–528. [CrossRef] [PubMed]

781. Kumar, G.; Waters, M.S.; Farooque, T.M.; Young, M.F.; Simon, C.G. Freeform fabricated scaffolds with roughened struts that enhance both stem cell proliferation and differentiation by controlling cell shape. *Biomaterials* **2012**, *33*, 4022–4030. [CrossRef] [PubMed]

782. Holthaus, M.G.; Treccani, L.; Rezwan, K. Osteoblast viability on hydroxyapatite with well-adjusted submicron and micron surface roughness as monitored by the proliferation reagent WST2-1. *J. Biomater. Appl.* **2013**, *27*, 791–800. [CrossRef] [PubMed]

783. Bianchi, M.; Edreira, E.R.U.; Wolke, J.G.C.; Birgani, Z.T.; Habibovic, P.; Jansen, J.A.; Tampieri, A.; Marcacci, M.; Leeuwenburgh, S.C.G.; van den Beucken, J.J.J.P. Substrate geometry directs the in vitro mineralization of calcium phosphate ceramics. *Acta Biomater.* **2014**, *10*, 661–669. [CrossRef]

784. Sadowska, J.M.; Wei, F.; Guo, J.; Guillem-Marti, J.; Ginebra, M.P.; Xiao, Y. Effect of nano-structural properties of biomimetic hydroxyapatite on osteoimmunomodulation. *Biomaterials* **2018**, *181*, 318–332. [CrossRef]

785. Ebaretonbofa, E.; Evans, J.R. High porosity hydroxyapatite foam scaffolds for bone substitute. *J. Porous Mater.* **2002**, *9*, 257–263. [CrossRef]

786. Hing, K.A. Bioceramic bone graft substitutes: Influence of porosity and chemistry. *Int. J. Appl. Ceram. Technol.* **2005**, *2*, 184–199. [CrossRef]

787. Malmström, J.; Adolfsson, E.; Arvidsson, A.; Thomsen, P. Bone response inside free-form fabricated macroporous hydroxyapatite scaffolds with and without an open microporosity. *Clin. Implant. Dent. Rel. Res.* **2007**, *9*, 79–88.

788. Lew, K.S.; Othman, R.; Ishikawa, K.; Yeoh, F.Y. Macroporous bioceramics: A remarkable material for bone regeneration. *J. Biomater. Appl.* **2012**, *27*, 345–358. [CrossRef]

789. Ren, L.M.; Todo, M.; Arahira, T.; Yoshikawa, H.; Myoui, A. A comparative biomechanical study of bone ingrowth in two porous hydroxyapatite bioceramics. *Appl. Surf. Sci.* **2012**, *262*, 81–88. [CrossRef]

790. Guda, T.; Walker, J.A.; Singleton, B.; Hernandez, J.; Oh, D.S.; Appleford, M.R.; Ong, J.L.; Wenke, J.C. Hydroxyapatite scaffold pore architecture effects in large bone defects in vivo. *J. Biomater. Appl.* **2014**, *28*, 1016–1027. [CrossRef]

791. Shao, R.; Quan, R.; Zhang, L.; Wei, X.; Yang, D.; Xie, S. Porous hydroxyapatite bioceramics in bone tissue engineering: Current uses and perspectives. *J. Ceram. Soc. Jpn.* **2015**, *123*, 17–20. [CrossRef]

792. Zhou, Y.; Chen, F.; Ho, S.T.; Woodruff, M.A.; Lim, T.M.; Hutmacher, D.W. Combined marrow stromal cell-sheet techniques and high-strength biodegradable composite scaffolds for engineered functional bone grafts. *Biomaterials* **2007**, *28*, 814–824. [CrossRef]

793. Vitale-Brovarone, C.; Baino, F.; Verné, E. High strength bioactive glass-ceramic scaffolds for bone regeneration. *J. Mater. Sci. Mater. Med.* **2009**, *20*, 643–653. [CrossRef]

794. Stevens, M.M. Biomaterials for bone tissue engineering. *Mater. Today* **2008**, *11*, 18–25. [CrossRef]

795. Artzi, Z.; Weinreb, M.; Givol, N.; Rohrer, M.D.; Nemcovsky, C.E.; Prasad, H.S.; Tal, H. Biomaterial resorbability and healing site morphology of inorganic bovine bone and beta tricalcium phosphate in the canine: A 24-month longitudinal histologic study and morphometric analysis. *Int. J. Oral Max. Impl.* **2004**, *19*, 357–368.

796. Burg, K.J.L.; Porter, S.; Kellam, J.F. Biomaterial developments for bone tissue engineering. *Biomaterials* **2000**, *21*, 2347–2359. [CrossRef]

797. Huebsch, N.; Mooney, D.J. Inspiration and application in the evolution of biomaterials. *Nature* **2009**, *462*, 426–432. [CrossRef]

798. Ajaal, T.T.; Smith, R.W. Employing the Taguchi method in optimizing the scaffold production process for artificial bone grafts. *J. Mater. Process. Technol.* **2009**, *209*, 1521–1532. [CrossRef]

799. Daculsi, G. Smart scaffolds: The future of bioceramic. *J. Mater. Sci. Mater. Med.* **2015**, *26*, 154. [CrossRef]

800. Daculsi, G.; Miramond, T.; Borget, P.; Baroth, S. Smart calcium phosphate bioceramic scaffold for bone tissue engineering. *Key Eng. Mater.* **2013**, *529–530*, 19–23. [CrossRef]

801. Bohner, M.; Loosli, Y.; Baroud, G.; Lacroix, D. Commentary: Deciphering the link between architecture and biological response of a bone graft substitute. *Acta Biomater.* **2011**, *7*, 478–484. [CrossRef]
802. Peppas, N.A.; Langer, R. New challenges in biomaterials. *Science* **1994**, *263*, 1715–1720. [CrossRef] [PubMed]
803. Hench, L.L. Biomaterials: A forecast for the future. *Biomaterials* **1998**, *19*, 1419–1423. [CrossRef]
804. Barrère, F.; Mahmood, T.A.; de Groot, K.; van Blitterswijk, C.A. Advanced biomaterials for skeletal tissue regeneration: Instructive and smart functions. *Mater. Sci. Eng. R* **2008**, *59*, 38–71. [CrossRef]
805. Liu, H.; Webster, T.J. Nanomedicine for implants: A review of studies and necessary experimental tools. *Biomaterials* **2007**, *28*, 354–369. [CrossRef] [PubMed]
806. Wang, C.; Duan, Y.; Markovic, B.; Barbara, J.; Howlett, C.R.; Zhang, X.; Zreiqat, H. Proliferation and bone-related gene expression of osteoblasts grown on hydroxyapatite ceramics sintered at different temperature. *Biomaterials* **2004**, *25*, 2949–2956. [CrossRef] [PubMed]
807. Samavedi, S.; Whittington, A.R.; Goldstein, A.S. Calcium phosphate ceramics in bone tissue engineering: A review of properties and their influence on cell behavior. *Acta Biomater.* **2013**, *9*, 8037–8045. [CrossRef]
808. Matsumoto, T.; Okazaki, M.; Nakahira, A.; Sasaki, J.; Egusa, H.; Sohmura, T. Modification of apatite materials for bone tissue engineering and drug delivery carriers. *Curr. Med. Chem.* **2007**, *14*, 2726–2733. [CrossRef]
809. Chai, Y.C.; Carlier, A.; Bolander, J.; Roberts, S.J.; Geris, L.; Schrooten, J.; van Oosterwyck, H.; Luyten, F.P. Current views on calcium phosphate osteogenicity and the translation into effective bone regeneration strategies. *Acta Biomater.* **2012**, *8*, 3876–3887. [CrossRef]
810. Denry, I.; Kuhn, L.T. Design and characterization of calcium phosphate ceramic scaffolds for bone tissue engineering. *Dent. Mater.* **2016**, *32*, 43–53. [CrossRef]
811. Lodoso-Torrecilla, I.; Gunnewiek, R.K.; Grosfeld, E.C.; de Vries, R.B.M.; Habibović, P.; Jansen, J.A.; van den Beucken, J.J.J.P. Bioinorganic supplementation of calcium phosphate-based bone substitutes to improve in vivo performance: A systematic review and meta-analysis of animal studies. *Biomater. Sci.* **2020**, *8*, 4792–4809. [CrossRef]
812. Ishikawa, K.; Miyamoto, Y.; Tsuchiya, A.; Hayashi, K.; Tsuru, K.; Ohe, G. Physical and histological comparison of hydroxyapatite, carbonate apatite, and β-tricalcium phosphate bone substitutes. *Materials* **2018**, *11*, 1993. [CrossRef]
813. Traykova, T.; Aparicio, C.; Ginebra, M.P.; Planell, J.A. Bioceramics as nanomaterials. *Nanomedicine* **2006**, *1*, 91–106. [CrossRef]
814. Kalita, S.J.; Bhardwaj, A.; Bhatt, H.A. Nanocrystalline calcium phosphate ceramics in biomedical engineering. *Mater. Sci. Eng. C* **2007**, *27*, 441–449. [CrossRef]
815. Dorozhkin, S.V. Nanometric calcium orthophosphates ($CaPO_4$): Preparation, properties and biomedical applications. *Adv. Nano-Bio. Mater. Dev.* **2019**, *3*, 422–513.
816. Šupová, M. Isolation and preparation of nanoscale bioapatites from natural sources: A review. *J. Nanosci. Nanotechnol.* **2014**, *14*, 546–563. [CrossRef]
817. Zhao, J.; Liu, Y.; Sun, W.B.; Zhang, H. Amorphous calcium phosphate and its application in dentistry. *Chem. Cent. J.* **2011**, *5*, 40. [CrossRef]
818. Dorozhkin, S.V. Synthetic amorphous calcium phosphates (ACPs): Preparation, structure, properties, and biomedical applications. *Biomater. Sci.* **2021**, *9*, 7748–7798. [CrossRef]
819. Venkatesan, J.; Kim, S.K. Nano-hydroxyapatite composite biomaterials for bone tissue engineering–A review. *J. Biomed. Nanotechnol.* **2014**, *10*, 3124–3140. [CrossRef]
820. Holzmeister, I.; Schamel, M.; Groll, J.; Gbureck, U.; Vorndran, E. Artificial inorganic biohybrids: The functional combination of microorganisms and cells with inorganic materials. *Acta Biomater.* **2018**, *74*, 17–35. [CrossRef]
821. Wu, Y.; Hench, L.L.; Du, J.; Choy, K.L.; Guo, J. Preparation of hydroxyapatite fibers by electrospinning technique. *J. Am. Ceram. Soc.* **2004**, *87*, 1988–1991. [CrossRef]
822. Ramanan, S.R.; Venkatesh, R. A study of hydroxyapatite fibers prepared via sol-gel route. *Mater. Lett.* **2004**, *58*, 3320–3323. [CrossRef]
823. Aizawa, M.; Porter, A.E.; Best, S.M.; Bonfield, W. Ultrastructural observation of single-crystal apatite fibres. *Biomaterials* **2005**, *26*, 3427–3433. [CrossRef]
824. Park, Y.M.; Ryu, S.C.; Yoon, S.Y.; Stevens, R.; Park, H.C. Preparation of whisker-shaped hydroxyapatite/β-tricalcium phosphate composite. *Mater. Chem. Phys.* **2008**, *109*, 440–447. [CrossRef]
825. Aizawa, M.; Ueno, H.; Itatani, K.; Okada, I. Syntheses of calcium-deficient apatite fibres by a homogeneous precipitation method and their characterizations. *J. Eur. Ceram. Soc.* **2006**, *26*, 501–507. [CrossRef]
826. Seo, D.S.; Lee, J.K. Synthesis of hydroxyapatite whiskers through dissolution-reprecipitation process using EDTA. *J. Cryst. Growth* **2008**, *310*, 2162–2167. [CrossRef]
827. Tas, A.C. Formation of calcium phosphate whiskers in hydrogen peroxide (H_2O_2) solutions at 90 °C. *J. Am. Ceram. Soc.* **2007**, *90*, 2358–2362. [CrossRef]
828. Neira, I.S.; Guitián, F.; Taniguchi, T.; Watanabe, T.; Yoshimura, M. Hydrothermal synthesis of hydroxyapatite whiskers with sharp faceted hexagonal morphology. *J. Mater. Sci.* **2008**, *43*, 2171–2178. [CrossRef]
829. Yang, H.Y.; Yang, S.F.; Chi, X.P.; Evans, J.R.G.; Thompson, I.; Cook, R.J.; Robinson, P. Sintering behaviour of calcium phosphate filaments for use as hard tissue scaffolds. *J. Eur. Ceram. Soc.* **2008**, *28*, 159–167. [CrossRef]

830. Junginger, M.; Kübel, C.; Schacher, F.H.; Müller, A.H.E.; Taubert, A. Crystal structure and chemical composition of biomimetic calcium phosphate nanofibers. *RSC Adv.* **2013**, *3*, 11301–11308. [CrossRef]
831. Cui, Y.S.; Yan, T.T.; Wu, X.P.; Chen, Q.H. Preparation and characterization of hydroxyapatite whiskers. *Appl. Mech. Mater.* **2013**, *389*, 21–24. [CrossRef]
832. Lee, J.H.; Kim, Y.J. Hydroxyapatite nanofibers fabricated through electrospinning and sol-gel process. *Ceram. Int.* **2014**, *40*, 3361–3369. [CrossRef]
833. Zhang, H.; Zhu, Q. Synthesis of nanospherical and ultralong fibrous hydroxyapatite and reinforcement of biodegradable chitosan/hydroxyapatite composite. *Modern Phys. Lett. B* **2009**, *23*, 3967–3976. [CrossRef]
834. Wijesinghe, W.P.S.L.; Mantilaka, M.M.M.G.P.G.; Premalal, E.V.A.; Herath, H.M.T.U.; Mahalingam, S.; Edirisinghe, M.; Rajapakse, R.P.V.J.; Rajapakse, R.M.G. Facile synthesis of both needle-like and spherical hydroxyapatite nanoparticles: Effect of synthetic temperature and calcination on morphology, crystallite size and crystallinity. *Mater. Sci. Eng. C* **2014**, *42*, 83–90. [CrossRef] [PubMed]
835. Zhou, W.Y.; Wang, M.; Cheung, W.L.; Guo, B.C.; Jia, D.M. Synthesis of carbonated hydroxyapatite nanospheres through nanoemulsion. *J. Mater. Sci. Mater. Med.* **2008**, *19*, 103–110. [CrossRef]
836. Lim, J.H.; Park, J.H.; Park, E.K.; Kim, H.J.; Park, I.K.; Shin, H.Y.; Shin, H.I. Fully interconnected globular porous biphasic calcium phosphate ceramic scaffold facilitates osteogenic repair. *Key Eng. Mater*, 2008; *361–363*, 119–122.
837. Kawai, T.; Sekikawa, H.; Unuma, H. Preparation of hollow hydroxyapatite microspheres utilizing poly(divinylbenzene) as a template. *J. Ceram. Soc. Jpn.* **2009**, *117*, 340–343. [CrossRef]
838. Cho, J.S.; Ko, Y.N.; Koo, H.Y.; Kang, Y.C. Synthesis of nano-sized biphasic calcium phosphate ceramics with spherical shape by flame spray pyrolysis. *J. Mater. Sci. Mater. Med.* **2010**, *21*, 1143–1149. [CrossRef]
839. Ye, F.; Guo, H.; Zhang, H.; He, X. Polymeric micelle-templated synthesis of hydroxyapatite hollow nanoparticles for a drug delivery system. *Acta Biomater.* **2010**, *6*, 2212–2218. [CrossRef]
840. Itatani, K.; Tsugawa, T.; Umeda, T.; Musha, Y.; Davies, I.J.; Koda, S. Preparation of submicrometer-sized porous spherical hydroxyapatite agglomerates by ultrasonic spray pyrolysis technique. *J. Ceram. Soc. Jpn.* **2010**, *118*, 462–466. [CrossRef]
841. Xiao, W.; Fu, H.; Rahaman, M.N.; Liu, Y.; Bal, B.S. Hollow hydroxyapatite microspheres: A novel bioactive and osteoconductive carrier for controlled release of bone morphogenetic protein-2 in bone regeneration. *Acta Biomater.* **2013**, *9*, 8374–8383. [CrossRef]
842. Rahaman, M.N.; Fu, H.; Xiao, W.; Liu, Y. Bioactive ceramic implants composed of hollow hydroxyapatite microspheres for bone regeneration. *Ceram. Eng. Sci. Proc.* **2014**, *34*, 67–76.
843. Ito, N.; Kamitakahara, M.; Ioku, K. Preparation and evaluation of spherical porous granules of octacalcium phosphate/hydroxyapatite as drug carriers in bone cancer treatment. *Mater. Lett.* **2014**, *120*, 94–96. [CrossRef]
844. Li, Z.; Wen, T.; Su, Y.; Wei, X.; He, C.; Wang, D. Hollow hydroxyapatite spheres fabrication with three-dimensional hydrogel template. *CrystEngComm* **2014**, *16*, 4202–4209. [CrossRef]
845. Kovach, I.; Kosmella, S.; Prietzel, C.; Bagdahn, C.; Koetz, J. Nano-porous calcium phosphate balls. *Colloid Surf. B* **2015**, *132*, 246–252. [CrossRef]
846. He, F.; Tian, Y.; Fang, X.; Xu, Y.; Ye, J. Porous calcium phosphate composite bioceramic beads. *Ceram. Int.* **2018**, *44*, 13430–13433. [CrossRef]
847. Hettich, G.; Schierjott, R.A.; Epple, M.; Gbureck, U.; Heinemann, S.; Mozaffari-Jovein, H.; Grupp, T.M. Calcium phosphate bone graft substitutes with high mechanical load capacity and high degree of interconnecting porosity. *Materials* **2019**, *12*, 3471. [CrossRef]
848. Chandanshive, B.; Dyondi, D.; Ajgaonkar, V.R.; Banerjee, R.; Khushalani, D. Biocompatible calcium phosphate based tubes. *J. Mater. Chem.* **2010**, *20*, 6923–6928. [CrossRef]
849. Kamitakahara, M.; Takahashi, H.; Ioku, K. Tubular hydroxyapatite formation through a hydrothermal process from α-tricalcium phosphate with anatase. *J. Mater. Sci.* **2012**, *47*, 4194–4199. [CrossRef]
850. Ustundag, C.B.; Kaya, F.; Kamitakahara, M.; Kaya, C.; Ioku, K. Production of tubular porous hydroxyapatite using electrophoretic deposition. *J. Ceram. Soc. Jpn.* **2012**, *120*, 569–573. [CrossRef]
851. Li, C.; Ge, X.; Li, G.; Lu, H.; Ding, R. In situ hydrothermal crystallization of hexagonal hydroxyapatite tubes from yttrium ion-doped hydroxyapatite by the Kirkendall effect. *Mater. Sci. Eng. C* **2014**, *45*, 191–195. [CrossRef] [PubMed]
852. Zhang, Y.G.; Zhu, Y.J.; Chen, F.; Sun, T.W. Biocompatible, ultralight, strong hydroxyapatite networks based on hydroxyapatite microtubes with excellent permeability and ultralow thermal conductivity. *ACS Appl. Mater. Interfaces* **2017**, *9*, 7918–7928.
853. Zhang, Y.G.; Zhu, Y.J.; Xiong, Z.C.; Wu, J.; Chen, F. Bioinspired ultralight inorganic aerogel for highly efficient air filtration and oil-water separation. *ACS Appl. Mater. Interfaces* **2018**, *10*, 13019–13027. [CrossRef]
854. Dadhich, P.; Dhara, S. Calcium phosphate flowers: A bone filler substitute. *Mater. Today* **2017**, *20*, 657–658. [CrossRef]
855. Nonoyama, T.; Kinoshita, T.; Higuchi, M.; Nagata, K.; Tanaka, M.; Kamada, M.; Sato, K.; Kato, K. Arrangement techniques of proteins and cells using amorphous calcium phosphate nanofiber scaffolds. *Appl. Surf. Sci.* **2012**, *262*, 8–12. [CrossRef]
856. Galea, L.G.; Bohner, M.; Lemaître, J.; Kohler, T.; Müller, R. Bone substitute: Transforming β-tricalcium phosphate porous scaffolds into monetite. *Biomaterials* **2008**, *29*, 3400–3407. [CrossRef] [PubMed]
857. Tamimi, F.; Torres, J.; Gbureck, U.; Lopez-Cabarcos, E.; Bassett, D.C.; Alkhraisat, M.H.; Barralet, J.E. Craniofacial vertical bone augmentation: A comparison between 3D printed monolithic monetite blocks and autologous onlay grafts in the rabbit. *Biomaterials* **2009**, *30*, 6318–6326. [CrossRef] [PubMed]

858. Sheikh, Z.; Drager, J.; Zhang, Y.L.; Abdallah, M.N.; Tamimi, F.; Barralet, J. Controlling bone graft substitute microstructure to improve bone augmentation. *Adv. Healthc. Mater.* **2016**, *5*, 1646–1655. [CrossRef]

859. Oryan, A.; Alidadi, S.; Bigham-Sadegh, A. Dicalcium phosphate anhydrous: An appropriate bioceramic in regeneration of critical-sized radial bone defects in rats. *Calcif. Tissue Int.* **2017**, *101*, 530–544. [CrossRef]

860. Sugiura, Y.; Ishikawa, K. Fabrication of pure octacalcium phosphate blocks from dicalcium hydrogen phosphate dihydrate blocks via a dissolution–precipitation reaction in a basic solution. *Mater. Lett.* **2019**, *239*, 143–146. [CrossRef]

861. Kim, J.S.; Jang, T.S.; Kim, S.Y.; Lee, W.P. Octacalcium phosphate bone substitute (Bontree®): From basic research to clinical case study. *Appl. Sci.* **2021**, *11*, 7921. [CrossRef]

862. Sohier, J.; Daculsi, G.; Sourice, S.; de Groot, K.; Layrolle, P. Porous beta tricalcium phosphate scaffolds used as a BMP-2 delivery system for bone tissue engineering. *J. Biomed. Mater. Res. A* **2010**, *92A*, 1105–1114. [CrossRef]

863. Stähli, C.; Bohner, M.; Bashoor-Zadeh, M.; Doebelin, N.; Baroud, G. Aqueous impregnation of porous β-tricalcium phosphate scaffolds. *Acta Biomater.* **2010**, *6*, 2760–2772. [CrossRef]

864. Lin, K.; Chen, L.; Qu, H.; Lu, J.; Chang, J. Improvement of mechanical properties of macroporous β-tricalcium phosphate bioceramic scaffolds with uniform and interconnected pore structures. *Ceram. Int.* **2011**, *37*, 2397–2403. [CrossRef]

865. Wójtowicz, J.; Leszczyńska, J.; Chróścicka, A.; Ślósarczyk, A.; Paszkiewicz, Z.; Zima, A.; Rozniatowski, K.; Jeleń, P.; Lewandowska-Szumieł, M. Comparative in vitro study of calcium phosphate ceramics for their potency as scaffolds for tissue engineering. *Bio-Med. Mater. Eng.* **2014**, *24*, 1609–1623. [CrossRef]

866. Simon, J.L.; Michna, S.; Lewis, J.A.; Rekow, E.D.; Thompson, V.P.; Smay, J.E.; Yampolsky, A.; Parsons, J.R.; Ricci, J.L. In vivo bone response to 3D periodic hydroxyapatite scaffolds assembled by direct ink writing. *J. Biomed. Mater. Res. A* **2007**, *83A*, 747–758. [CrossRef] [PubMed]

867. Yoshikawa, H.; Myoui, A. Bone tissue engineering with porous hydroxyapatite ceramics. *J. Artif. Organs* **2005**, *8*, 131–136. [CrossRef]

868. Min, S.H.; Jin, H.H.; Park, H.Y.; Park, I.M.; Park, H.C.; Yoon, S.Y. Preparation of porous hydroxyapatite scaffolds for bone tissue engineering. *Mater. Sci. Forum*, 2006; *510–511*, 754–757.

869. Deville, S.; Saiz, E.; Nalla, R.K.; Tomsia, A.P. Strong biomimetic hydroxyapatite scaffolds. *Adv. Sci. Technol.* **2006**, *49*, 148–152.

870. Buckley, C.T.; O'Kelly, K.U. Fabrication and characterization of a porous multidomain hydroxyapatite scaffold for bone tissue engineering investigations. *J. Biomed. Mater. Res. B Appl. Biomater.* **2010**, *93B*, 459–467. [CrossRef]

871. Ramay, H.R.R.; Zhang, M. Biphasic calcium phosphate nanocomposite porous scaffolds for load-bearing bone tissue engineering. *Biomaterials* **2004**, *25*, 5171–5180. [CrossRef]

872. Chen, G.; Li, W.; Zhao, B.; Sun, K. A novel biphasic bone scaffold: β-calcium phosphate and amorphous calcium polyphosphate. *J. Am. Ceram. Soc.* **2009**, *92*, 945–948. [CrossRef]

873. Guo, D.; Xu, K.; Han, Y. The in situ synthesis of biphasic calcium phosphate scaffolds with controllable compositions, structures, and adjustable properties. *J. Biomed. Mater. Res. A* **2009**, *88A*, 43–52. [CrossRef] [PubMed]

874. Sarin, P.; Lee, S.J.; Apostolov, Z.D.; Kriven, W.M. Porous biphasic calcium phosphate scaffolds from cuttlefish bone. *J. Am. Ceram. Soc.* **2011**, *94*, 2362–2370. [CrossRef]

875. Kim, D.H.; Kim, K.L.; Chun, H.H.; Kim, T.W.; Park, H.C.; Yoon, S.Y. In vitro biodegradable and mechanical performance of biphasic calcium phosphate porous scaffolds with unidirectional macro-pore structure. *Ceram. Int.* **2014**, *40*, 8293–8300. [CrossRef]

876. Marques, C.F.; Perera, F.H.; Marote, A.; Ferreira, S.; Vieira, S.I.; Olhero, S.; Miranda, P.; Ferreira, J.M.F. Biphasic calcium phosphate scaffolds fabricated by direct write assembly: Mechanical, anti-microbial and osteoblastic properties. *J. Eur. Ceram. Soc.* **2017**, *37*, 359–368. [CrossRef]

877. Furuichi, K.; Oaki, Y.; Ichimiya, H.; Komotori, J.; Imai, H. Preparation of hierarchically organized calcium phosphate-organic polymer composites by calcification of hydrogel. *Sci. Technol. Adv. Mater.* **2006**, *7*, 219–225. [CrossRef]

878. Wei, J.; Jia, J.; Wu, F.; Wei, S.; Zhou, H.; Zhang, H.; Shin, J.W.; Liu, C. Hierarchically microporous/macroporous scaffold of magnesium-calcium phosphate for bone tissue regeneration. *Biomaterials* **2010**, *31*, 1260–1269. [CrossRef]

879. Ye, X.; Zhou, C.; Xiao, Z.; Fan, Y.; Zhu, X.; Sun, Y.; Zhang, X. Fabrication and characterization of porous 3D whisker-covered calcium phosphate scaffolds. *Mater. Lett.* **2014**, *128*, 179–182. [CrossRef]

880. Zhao, R.; Shang, T.; Yuan, B.; Yang, X.; Zhu, X.; Zhang, X. Osteoporotic bone recovery by a bamboo-structured bioceramic with controlled release of hydroxyapatite nanoparticles. *Bioact. Mater.* **2022**, *17*, 379–393. [CrossRef]

881. Zhu, Y.J.; Lu, B.Q. Deformable biomaterials based on ultralong hydroxyapatite nanowires. *ACS Biomater. Sci. Eng.* **2019**, *5*, 4951–4961. [CrossRef]

882. Huang, G.J.; Yu, H.P.; Wang, X.L.; Ning, B.B.; Gao, J.; Shi, Y.Q.; Zhu, Y.J.; Duan, J.L. Highly porous and elastic aerogel based on ultralong hydroxyapatite nanowires for high-performance bone regeneration and neovascularization. *J. Mater. Chem. B* **2021**, *9*, 1277–1287, Erratum *J. Mater. Chem. B* **2021**, *9*, 7566. [CrossRef]

883. Gbureck, U.; Grolms, O.; Barralet, J.E.; Grover, L.M.; Thull, R. Mechanical activation and cement formation of β-tricalcium phosphate. *Biomaterials* **2003**, *24*, 4123–4131. [CrossRef]

884. Gbureck, U.; Barralet, J.E.; Hofmann, M.; Thull, R. Mechanical activation of tetracalcium phosphate. *J. Am. Ceram. Soc.* **2004**, *87*, 311–313. [CrossRef]

885. Bohner, M.; Luginbühl, R.; Reber, C.; Doebelin, N.; Baroud, G.; Conforto, E. A physical approach to modify the hydraulic reactivity of α-tricalcium phosphate powder. *Acta Biomater.* **2009**, *5*, 3524–3535. [CrossRef]

886. Hagio, T.; Tanase, T.; Akiyama, J.; Iwai, K.; Asai, S. Formation and biological affinity evaluation of crystallographically aligned hydroxyapatite. *J. Ceram. Soc. Jpn.* **2008**, *116*, 79–82. [CrossRef]

887. Blawas, A.S.; Reichert, W.M. Protein patterning. *Biomaterials* **1998**, *19*, 595–609. [CrossRef]

888. Kasai, T.; Sato, K.; Kanematsu, Y.; Shikimori, M.; Kanematsu, N.; Doi, Y. Bone tissue engineering using porous carbonate apatite and bone marrow cells. *J. Craniofac. Surg.* **2010**, *21*, 473–478. [CrossRef]

889. Wang, L.; Fan, H.; Zhang, Z.Y.; Lou, A.J.; Pei, G.X.; Jiang, S.; Mu, T.W.; Qin, J.J.; Chen, S.Y.; Jin, D. Osteogenesis and angiogenesis of tissue-engineered bone constructed by prevascularized β-tricalcium phosphate scaffold and mesenchymal stem cells. *Biomaterials* **2010**, *31*, 9452–9461. [CrossRef]

890. Wijnhoven, I.B.; Vallejos, R.; Santibanez, J.F.; Millán, C.; Vivanco, J.F. Analysis of cell-biomaterial interaction through cellular bridge formation in the interface between hGMSCs and CaP bioceramics. *Sci. Rep.* **2020**, *10*, 16493. [CrossRef]

891. Sánchez-Salcedo, S.; Izquierdo-Barba, I.; Arcos, D.; Vallet-Regí, M. In vitro evaluation of potential calcium phosphate scaffolds for tissue engineering. *Tissue Eng.* **2006**, *12*, 279–290. [CrossRef]

892. Meganck, J.A.; Baumann, M.J.; Case, E.D.; McCabe, L.R.; Allar, J.N. Biaxial flexure testing of calcium phosphate bioceramics for use in tissue engineering. *J. Biomed. Mater. Res. A* **2005**, *72A*, 115–126. [CrossRef] [PubMed]

893. Case, E.D.; Smith, I.O.; Baumann, M.J. Microcracking and porosity in calcium phosphates and the implications for bone tissue engineering. *Mater. Sci. Eng. A* **2005**, *390*, 246–254. [CrossRef]

894. Tripathi, G.; Basu, B. A porous hydroxyapatite scaffold for bone tissue engineering: Physico-mechanical and biological evaluations. *Ceram. Int.* **2012**, *38*, 341–349. [CrossRef]

895. Sibilla, P.; Sereni, A.; Aguiari, G.; Banzi, M.; Manzati, E.; Mischiati, C.; Trombelli, L.; del Senno, L. Effects of a hydroxyapatite-based biomaterial on gene expression in osteoblast-like cells. *J. Dent. Res.* **2006**, *85*, 354–358. [CrossRef]

896. Verron, E.; Bouler, J.M. Calcium phosphate ceramics as bone drug-combined devices. *Key Eng. Mater.* **2010**, *441*, 181–201. [CrossRef]

897. Kolmas, J.; Krukowski, S.; Laskus, A.; Jurkitewicz, M. Synthetic hydroxyapatite in pharmaceutical applications. *Ceram. Int.* **2016**, *42*, 2472–2487. [CrossRef]

898. Parent, M.; Baradari, H.; Champion, E.; Damia, C.; Viana-Trecant, M. Design of calcium phosphate ceramics for drug delivery applications in bone diseases: A review of the parameters affecting the loading and release of the therapeutic substance. *J. Control. Release* **2017**, *252*, 1–17. [CrossRef]

899. Mondal, S.; Pal, U. 3D hydroxyapatite scaffold for bone regeneration and local drug delivery applications. *J. Drug Deliv. Sci. Tec.* **2019**, *53*, 101131. [CrossRef]

900. Zhao, Q.; Zhang, D.; Sun, R.; Shang, S.; Wang, H.; Yang, Y.; Wang, L.; Liu, X.; Sun, T.; Chen, K. Adsorption behavior of drugs on hydroxyapatite with different morphologies: A combined experimental and molecular dynamics simulation study. *Ceram. Int.* **2019**, *45*, 19522–19527, Correction *Ceram. Int.* **2020**, *46*, 27909. [CrossRef]

901. Rapoport, A.; Borovikova, D.; Kokina, A.; Patmalnieks, A.; Polyak, N.; Pavlovska, I.; Mezinskis, G.; Dekhtyar, Y. Immobilisation of yeast cells on the surface of hydroxyapatite ceramics. *Process. Biochem.* **2011**, *46*, 665–670. [CrossRef]

902. Ghiasi, B.; Sefidbakht, Y.; Mozaffari-Jovin, S.; Gharehcheloo, B.; Mehrarya, M.; Khodadadi, A.; Rezaei, M.; Siadat, S.O.R.; Uskoković, V. Hydroxyapatite as a biomaterial–a gift that keeps on giving. *Drug Dev. Ind. Pharm.* **2020**, *46*, 1035–1062. [CrossRef]

903. Mastrogiacomo, M.; Muraglia, A.; Komlev, V.; Peyrin, F.; Rustichelli, F.; Crovace, A.; Cancedda, R. Tissue engineering of bone: Search for a better scaffold. *Orthod. Craniofac. Res.* **2005**, *8*, 277–284. [CrossRef]

904. Quarto, R.; Mastrogiacomo, M.; Cancedda, R.; Kutepov, S.M.; Mukhachev, V.; Lavroukov, A.; Kon, E.; Marcacci, M. Repair of large bone defects with the use of autologous bone marrow stromal cells. *N. Engl. J. Med.* **2001**, *344*, 385–386. [CrossRef]

905. Vacanti, C.A.; Bonassar, L.J.; Vacanti, M.P.; Shufflebarger, J. Replacement of an avulsed phalanx with tissue-engineered bone. *N. Engl. J. Med.* **2001**, *344*, 1511–1514. [CrossRef]

906. Morishita, T.; Honoki, K.; Ohgushi, H.; Kotobuki, N.; Matsushima, A.; Takakura, Y. Tissue engineering approach to the treatment of bone tumors: Three cases of cultured bone grafts derived from patients' mesenchymal stem cells. *Artif. Organs* **2006**, *30*, 115–118. [CrossRef]

907. Eniwumide, J.O.; Yuan, H.; Cartmell, S.H.; Meijer, G.J.; de Bruijn, J.D. Ectopic bone formation in bone marrow stem cell seeded calcium phosphate scaffolds as compared to autograft and (cell seeded) allograft. *Eur. Cell Mater.* **2007**, *14*, 30–39. [CrossRef]

908. Zuolin, J.; Hong, Q.; Jiali, T. Dental follicle cells combined with beta-tricalcium phosphate ceramic: A novel available therapeutic strategy to restore periodontal defects. *Med. Hypotheses* **2010**, *75*, 669–670. [CrossRef]

909. Ge, S.; Zhao, N.; Wang, L.; Yu, M.; Liu, H.; Song, A.; Huang, J.; Wang, G.; Yang, P. Bone repair by periodontal ligament stem cell-seeded nanohydroxyapatite-chitosan scaffold. *Int. J. Nanomed.* **2012**, *7*, 5405–5414. [CrossRef]

910. Farré-Guasch, E.; Bravenboer, N.; Helder, M.N.; Schulten, E.A.J.M.; Ten Bruggenkate, C.M.; Klein-Nulend, J. Blood vessel formation and bone regeneration potential of the stromal vascular fraction seeded on a calcium phosphate scaffold in the human maxillary sinus floor elevation model. *Materials* **2018**, *11*, 161. [CrossRef]

911. Tanaka, T.; Komaki, H.; Chazono, M.; Kitasato, S.; Kakuta, A.; Akiyama, S.; Marumo, K. Basic research and clinical application of beta-tricalcium phosphate (β-TCP). | Recherche fondamentale et application clinique du bêta-tricalcium phosphate (β-TCP). *Morphologie* **2017**, *101*, 164–172. [CrossRef]

912. Franch, J.; Díaz-Bertrana, C.; Lafuente, P.; Fontecha, P.; Durall, I. Beta-tricalcium phosphate as a synthetic cancellous bone graft in veterinary orthopaedics: A retrospective study of 13 clinical cases. *Vet. Comp. Orthop. Traumatol.* **2006**, *19*, 196–204.

913. Vertenten, G.; Gasthuys, F.; Cornelissen, M.; Schacht, E.; Vlaminck, L. Enhancing bone healing and regeneration: Present and future perspectives in veterinary orthopaedics. *Vet. Comp. Orthop. Traumatol.* **2010**, *23*, 153–162.

914. Freidlin, L.K.; Sharf, V.Z. Two paths for the dehydration of 1,4-butandiol to divinyl with a tricalcium phosphate catalyst. *Bull. Acad. Sci. USSR Div. Chem. Sci.* **1960**, *9*, 1577–1579. [CrossRef]

915. Bett, J.A.S.; Christner, L.G.; Hall, W.K. Studies of the hydrogen held by solids. XII. Hydroxyapatite catalysts. *J. Am. Chem. Soc.* **1967**, *89*, 5535–5541. [CrossRef]

916. Monma, H. Catalytic behavior of calcium phosphates for decompositions of 2-propanol and ethanol. *J. Catal.* **1982**, *75*, 200–203. [CrossRef]

917. Tsuchida, T.; Yoshioka, T.; Sakuma, S.; Takeguchi, T.; Ueda, W. Synthesis of biogasoline from ethanol over hydroxyapatite catalyst. *Ind. Eng. Chem. Res.* **2008**, *47*, 1443–1452. [CrossRef]

918. Tsuchida, T.; Kubo, J.; Yoshioka, T.; Sakuma, S.; Takeguchi, T.; Ueda, W. Reaction of ethanol over hydroxyapatite affected by Ca/P ratio of catalyst. *J. Catal.* **2008**, *259*, 183–189. [CrossRef]

919. Xu, J.; White, T.; Li, P.; He, C.; Han, Y.F. Hydroxyapatite foam as a catalyst for formaldehyde combustion at room temperature. *J. Am. Chem. Soc.* **2010**, *132*, 13172–13173. [CrossRef] [PubMed]

920. Hatano, M.; Moriyama, K.; Maki, T.; Ishihara, K. Which is the actual catalyst: Chiral phosphoric acid or chiral calcium phosphate? *Angew. Chem. Int. Ed. Engl.* **2010**, *49*, 3823–3826. [CrossRef]

921. Zhang, D.; Zhao, H.; Zhao, X.; Liu, Y.; Chen, H.; Li, X. Application of hydroxyapatite as catalyst and catalyst carrier. *Prog. Chem.* **2011**, *23*, 687–694.

922. Gruselle, M.; Kanger, T.; Thouvenot, R.; Flambard, A.; Kriis, K.; Mikli, V.; Traksmaa, R.; Maaten, B.; Tõnsuaadu, K. Calcium hydroxyapatites as efficient catalysts for the Michael C–C bond formation. *ACS Catal.* **2011**, *1*, 1729–1733. [CrossRef]

923. Stošić, D.; Bennici, S.; Sirotin, S.; Calais, C.; Couturier, J.L.; Dubois, J.L.; Travert, A.; Auroux, A. Glycerol dehydration over calcium phosphate catalysts: Effect of acidic-basic features on catalytic performance. *Appl. Catal. A* **2012**, *447–448*, 124–134. [CrossRef]

924. Ghantani, V.C.; Lomate, S.T.; Dongare, M.K.; Umbarkar, S.B. Catalytic dehydration of lactic acid to acrylic acid using calcium hydroxyapatite catalysts. *Green Chem.* **2013**, *15*, 1211–1217. [CrossRef]

925. Chen, G.; Shan, R.; Shi, J.; Liu, C.; Yan, B. Biodiesel production from palm oil using active and stable K doped hydroxyapatite catalysts. *Energ. Convers. Manage.* **2015**, *98*, 463–469. [CrossRef]

926. Gruselle, M. Apatites: A new family of catalysts in organic synthesis. *J. Organomet. Chem.* **2015**, *793*, 93–101. [CrossRef]

927. Urist, M.R.; Huo, Y.K.; Brownell, A.G.; Hohl, W.M.; Buyske, J.; Lietze, A.; Tempst, P.; Hunkapiller, M.; de Lange, R.J. Purification of bovine bone morphogenetic protein by hydroxyapatite chromatography. *Proc. Natl. Acad. Sci. USA* **1984**, *81*, 371–375. [CrossRef]

928. Kawasaki, T. Hydroxyapatite as a liquid chromatographic packing. *J. Chromatogr.* **1991**, *544*, 147–184. [CrossRef]

929. Kuiper, M.; Sanches, R.M.; Walford, J.A.; Slater, N.K.H. Purification of a functional gene therapy vector derived from moloney murine leukaemia virus using membrane filtration and ceramic hydroxyapatite chromatography. *Biotechnol. Bioeng.* **2002**, *80*, 445–453. [CrossRef]

930. Jungbauer, A.; Hahn, R.; Deinhofer, K.; Luo, P. Performance and characterization of a nanophased porous hydroxyapatite for protein chromatography. *Biotechnol. Bioeng.* **2004**, *87*, 364–375. [CrossRef]

931. Wensel, D.L.; Kelley, B.D.; Coffman, J.L. High-throughput screening of chromatographic separations: III. Monoclonal antibodies on ceramic hydroxyapatite. *Biotechnol. Bioeng.* **2008**, *100*, 839–854. [CrossRef]

932. Hilbrig, F.; Freitag, R. Isolation and purification of recombinant proteins, antibodies and plasmid DNA with hydroxyapatite chromatography. *Biotechnol. J.* **2012**, *7*, 90–102. [CrossRef]

933. Cummings, L.J.; Frost, R.G.; Snyder, M.A. Monoclonal antibody purification by ceramic hydroxyapatite chromatography. *Method Mol. Biol.* **2014**, *1131*, 241–251.

934. Nagai, M.; Nishino, T.; Saeki, T. A new type of CO_2 gas sensor comprising porous hydroxyapatite ceramics. *Sens. Actuator* **1988**, *15*, 145–151. [CrossRef]

935. Petrucelli, G.C.; Kawachi, E.Y.; Kubota, L.T.; Bertran, C.A. Hydroxyapatite-based electrode: A new sensor for phosphate. *Anal. Commun.* **1996**, *33*, 227–229. [CrossRef]

936. Tagaya, M.; Ikoma, T.; Hanagata, N.; Chakarov, D.; Kasemo, B.; Tanaka, J. Reusable hydroxyapatite nanocrystal sensors for protein adsorption. *Sci. Technol. Adv. Mater.* **2010**, *11*, 045002. [CrossRef]

937. Khairnar, R.S.; Mene, R.U.; Munde, S.G.; Mahabole, M.P. Nano-hydroxyapatite thick film gas sensors. *AIP Conf. Proc.* **2011**, *1415*, 189–192.

938. López, M.S.P.; Redondo-Gómez, E.; López-Ruiz, B. Electrochemical enzyme biosensors based on calcium phosphate materials for tyramine detection in food samples. *Talanta* **2017**, *175*, 209–216. [CrossRef]

939. Nijhawan, A.; Butler, E.C.; Sabatini, D.A. Hydroxyapatite ceramic adsorbents: Effect of pore size, regeneration, and selectivity for fluoride. *J. Environ. Eng.* **2018**, *144*, 04018117. [CrossRef]

940. Ibrahim, M.; Labaki, M.; Giraudon, J.M.; Lamonier, J.F. Hydroxyapatite, a multifunctional material for air, water and soil pollution control: A review. *J. Hazard. Mater.* **2020**, *383*, 121139. [CrossRef]

941. Hench, L.L.; Wilson, J. Surface-active biomaterials. *Science* **1984**, *226*, 630–636. [CrossRef]

942. Bongio, M.; van den Beucken, J.J.J.P.; Leeuwenburgh, S.C.G.; Jansen, J.A. Development of bone substitute materials: From 'biocompatible' to 'instructive'. *J. Mater. Chem.* **2010**, *20*, 8747–8759. [CrossRef]

943. Navarro, M.; Michiardi, A.; Castano, O.; Planell, J.A. Biomaterials in orthopaedics. *J. R. Soc. Interface* **2008**, *5*, 1137–1158. [CrossRef] [PubMed]
944. Anderson, J.M. The future of biomedical materials. *J. Mater. Sci. Mater. Med.* **2006**, *17*, 1025–1028. [CrossRef] [PubMed]
945. Zadpoor, A.A. Meta-biomaterials. *Biomater. Sci.* **2020**, *8*, 18–38. [CrossRef]
946. Sanchez-Sálcedo, S.; Arcos, D.; Vallet-Regí, M. Upgrading calcium phosphate scaffolds for tissue engineering applications. *Key Eng. Mater.* **2008**, *377*, 19–42. [CrossRef]
947. Chevalier, J.; Gremillard, L. Ceramics for medical applications: A picture for the next 20 years. *J. Eur. Ceram. Soc.* **2009**, *29*, 1245–1255. [CrossRef]
948. Salgado, P.C.; Sathler, P.C.; Castro, H.C.; Alves, G.G.; de Oliveira, A.M.; de Oliveira, R.C.; Maia, M.D.C.; Rodrigues, C.R.; Coelh, P.G.; Fuly, A.; et al. Bone remodeling, biomaterials and technological applications: Revisiting basic concepts. *J. Biomater. Nanobiotechnol.* **2011**, *2*, 318–328. [CrossRef]
949. Vallet-Regí, M. Evolution of bioceramics within the field of biomaterials. *C. R. Chimie* **2010**, *13*, 174–185. [CrossRef]
950. Hartgerink, J.D.; Beniash, E.; Stupp, S.I. Self-assembly and mineralization of peptide-amphiphile nanofibers. *Science* **2001**, *294*, 1684–1688. [CrossRef]
951. Malhotra, A.; Habibovic, P. Calcium phosphates and angiogenesis: Implications and advances for bone regeneration. *Trends Biotechnol.* **2016**, *34*, 983–992. [CrossRef]
952. Kim, J. Multilayered injection of calcium hydroxylapatite filler on ischial soft tissue to rejuvenate the previous phase of chronic sitting pressure sore. *Clin. Cosmet. Investig. Dermatol.* **2019**, *12*, 771–784. [CrossRef]
953. Kargozar, S.; Singh, R.K.; Kim, H.W.; Baino, F. "Hard" ceramics for "soft" tissue engineering: Paradox or opportunity? *Acta Biomater.* **2020**, *115*, 1–28. [CrossRef]
954. Jansen, D.A.; Graivier, M.H. Evaluation of a calcium hydroxylapatite-based implant (Radiesse) for facial soft-tissue augmentation. *Plast. Reconstr. Surg.* **2006**, *118* (Suppl. 3), 22S–30S. [CrossRef]
955. Wasylkowski, V.C. Body vectoring technique with Radiesse® for tightening of the abdomen, thighs, and brachial zone. *Clin. Cosmet. Invest. Dermatol.* **2015**, *8*, 267–273. [CrossRef]
956. Shi, X.H.; Zhou, X.; Zhang, Y.M.; Lei, Z.Y.; Liu, T.; Fan, D.L. Complications from nasolabial fold injection of calcium hydroxylapatite for facial soft-tissue augmentation: A systematic review and meta-analysis. *Aesthet. Surg. J.* **2016**, *36*, 712–717. [CrossRef]
957. Qi, C.; Lin, J.; Fu, L.H.; Huang, P. Calcium-based biomaterials for diagnosis, treatment, and theranostics. *Chem. Soc. Rev.* **2018**, *47*, 357–403. [CrossRef]
958. Chyzhma, R.; Piddubnyi, A.; Danilchenko, S.; Kravtsova, O.; Moskalenko, R. Potential role of hydroxyapatite nanocrystalline for early diagnostics of ovarian cancer. *Diagnostics* **2021**, *11*, 1741. [CrossRef]
959. Du, M.; Chen, J.; Liu, K.; Xing, H.; Song, C. Recent advances in biomedical engineering of nano-hydroxyapatite including dentistry, cancer treatment and bone repair. *Composites B* **2021**, *215*, 108790. [CrossRef]
960. Yu, W.; Jiang, Y.Y.; Sun, T.W.; Qi, C.; Zhao, H.; Chen, F.; Shi, Z.; Zhu, Y.J.; Chen, D.; He, Y. Design of a novel wound dressing consisting of alginate hydrogel and simvastatin-incorporated mesoporous hydroxyapatite microspheres for cutaneous wound healing. *RSC Adv.* **2016**, *106*, 104375–104387. [CrossRef]
961. Carella, F.; Esposti, L.D.; Adamiano, A.; Iafisco, M. The use of calcium phosphates in cosmetics, state of the art and future perspectives. *Materials* **2021**, *14*, 6398. [CrossRef]
962. Masson, J.D.; Thibaudon, M.; Belec, L.; Crepeaux, G. Calcium phosphate: A substitute for aluminum adjuvants? *Expert Rev. Vaccines* **2017**, *16*, 289–299. [CrossRef]
963. Wang, X.; Xue, J.; Ma, B.; Wu, J.; Chang, J.; Gelinsky, M.; Wu, C. Black bioceramics: Combining regeneration with therapy. *Adv. Mater.* **2020**, *32*, 2005140. [CrossRef]
964. Safronova, T.; Kiselev, A.; Selezneva, I.; Shatalova, T.; Lukina, Y.; Filippov, Y.; Toshev, O.; Tikhonova, S.; Antonova, O.; Knotko, A. Bioceramics based on β-calcium pyrophosphate. *Materials* **2022**, *15*, 3105. [CrossRef]
965. Germaini, M.M.; Belhabib, S.; Guessasma, S.; Deterre, R.; Corre, P.; Weiss, P. Additive manufacturing of biomaterials for bone tissue engineering–A critical review of the state of the art and new concepts. *Prog. Mater. Sci.* **2022**, *130*, 100963. [CrossRef]
966. Galván-Chacón, V.P.; Vasilevich, A.; Yuan, H.; Groen, N.; de Boer, J.; Habibovic, P.; Vermeulen, S. A bioinformatics approach to study the role of calcium phosphate properties in bone regeneration. *Eur. J. Mater.* **2022**. [CrossRef]
967. Mohd, N.; Razali, M.; Ghazali, M.J.; Kasim, N.H.A. 3D-printed hydroxyapatite and tricalcium phosphates-based scaffolds for alveolar bone regeneration in animal models: A scoping review. *Materials* **2022**, *15*, 2621. [CrossRef]
968. Ginebra, M.P.; Espanol, M.; Maazouz, Y.; Bergez, V.; Pastorino, D. Bioceramics and bone healing. *EFORT Open Rev.* **2018**, *3*, 173–183. [CrossRef]
969. Albulescu, R.; Popa, A.C.; Enciu, A.M.; Albulescu, L.; Dudau, M.; Popescu, I.D.; Mihai, S.; Codrici, E.; Pop, S.; Lupu, A.R.; et al. Comprehensive in vitro testing of calcium phosphate-based bioceramics with orthopedic and dentistry applications. *Materials* **2019**, *12*, 3704. [CrossRef]

MDPI AG
Grosspeteranlage 5
4052 Basel
Switzerland
Tel.: +41 61 683 77 34

Coatings Editorial Office
E-mail: coatings@mdpi.com
www.mdpi.com/journal/coatings